Cell Biology of Addiction

ALSO FROM COLD SPRING HARBOR LABORATORY PRESS

RELATED LABORATORY MANUALS

Molecular Cloning: A Laboratory Manual

Live Cell Imaging: A Laboratory Manual

Imaging in Neuroscience and Development: A Laboratory Manual

Proteins and Proteomics: A Laboratory Manual

OTHER RELATED TITLES

Bioinformatics Sequence and Genome Analysis

Prion Biology and Diseases, Second Edition

*Lab Math: A Handbook of Measurements, Calculations, and Other Quantitative Skills
    for Use at the Bench*

*Lab Ref: A Handbook of Recipes, Reagents, and Other Reference Tools
    for Use at the Bench*

Lab Dynamics: Management Skills for Scientists

At the Bench: A Laboratory Navigator

At the Helm: A Laboratory Navigator

Cell Biology of Addiction

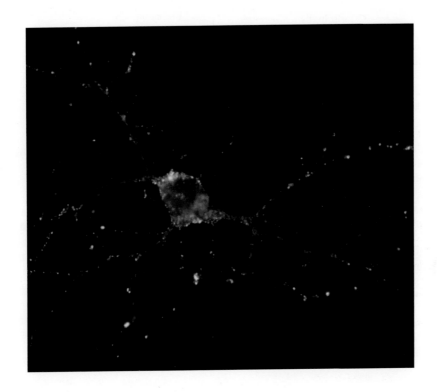

EDITED BY

Bertha K. Madras
Harvard Medical School

Christine M. Colvis
National Institute on Drug Abuse

Jonathan D. Pollock
National Institute on Drug Abuse

Joni L. Rutter
National Institute on Drug Abuse

David Shurtleff
National Institute on Drug Abuse

Mark von Zastrow
University of California at San Francisco

COLD SPRING HARBOR LABORATORY PRESS
Cold Spring Harbor, New York

Cell Biology of Addiction

Publisher	John Inglis
Acquisition Editors	John Inglis and David Crotty
Development Manager	Jan Argentine
Project Coordinator	Mary Cozza
Production Editor	Rena Steuer
Desktop Editor	Susan Schaefer
Production Manager	Denise Weiss
Cover Designer	Michael Albano

Front cover: Localization of μ-opioid peptide receptors in a cultured rat medium spiny neuron exposed acutely (for 30 minutes) to morphine. Receptors present in the plasma membrane are shown in *yellow-green;* those present in endosomes are shown in *red.* (Courtesy of Joy Yu, University of California at San Francisco.)

Back cover: Brain images obtained with positron emission topography and [^{11}C]cocaine in a cocaine abuser showing dopamine transporter availability at baseline (*left*) and after snorting cocaine (*right*). Note an almost complete decrease in binding of [^{11}C]cocaine in the stratium after a 96-mg snorted dose of cocaine, indicating almost complete occupancy of dopamine transporters in the striatum by the unlabeled snorted cocaine. (Reprinted, with permission, from Volkow N.D., Wang G.-J., Fischman M.W., Foltin R., Fowler J.S., Franceschi D., Franceschi M., Logan J., Gatley S.J., Wong C., Ding Y.S., Hitzemann R., and Pappas N. 2000. Effects of route of administration on cocaine induced dopamine transporter blockage in the human brain. *Life Sci.* **67:** 1507–1515.)

Library of Congress Cataloging-in-Publication Data

The cell biology of addiction / edited by Bertha Madras ... [et al.].
    p. cm.
  Includes index.
  ISBN 0-87969-753-9 (hardcover : alk. paper)
  1. Drug abuse--Molecular aspects. 2. Drug abuse--Physiological aspects. 3. Substance abuse--Physiological aspects. 4. Cytology.
I. Madras, Bertha.
  [DNLM: 1. Substance-Related Disorders--physiopathology.
2. Substance-Related Disorders--etiology. 3. Brain--drug effects.
4. Psychotropic Drugs--pharmacokinetics. 5. Psychotropic Drugs
--metabolism.   WM 270 C3925 2006]
  RC564.C46 2006
  616.86--dc22
2005019982

10 9 8 7 6 5 4 3 2 1

This book is dedicated to our families, who understand our commitment to neuroscience research and its magnetic enticements. We also fervently praise substance abuse investigators, treatment providers, and others who courageously work in a field shadowed by stigma.

Table of Contents

PART 4: Synaptic Plasticity and Addiction, 339

PART 5: Systems Analysis of Drug Abuse, 413

Preface

CELL BIOLOGY OF ADDICTION EMERGES FROM A COURSE presented at Cold Spring Harbor Laboratory in August of 2001, 2003, and 2005. The book is not a compilation of transcribed lectures based solely on the research of its contributors, but was written de novo to provide readers with an intensive overview of the fundamentals, state-of-the-art advances, and major gaps in the cell and molecular biology of drug addiction within the broader context of neuroscience. Targeted to both new and experienced investigators, the book conveys the merits and excitement of cellular and molecular approaches to drug addiction research. The research presented is not only applicable to the study of drug abuse and addiction, but also has clear implications for clarifying mechanisms of learning and memory, neuroadaptation, perception, volitional behavior, motivation, reward, and other disciplines of neuroscience.

The authors and editors are most grateful to Dr. Jonathan Pollock of the National Institute on Drug Abuse (NIDA), who initially saw the need for the course and enlisted the enthusiastic help of the course originator, Dr. Bertha K. Madras. Dr. Pollock's active contributions to the success of the course and to the production of this book are immeasurable. We also appreciate the efforts of the first co-course directors in 2001, Dr. Randy Blakely and Dr. Nora Volkow.

The course directors and the editors of the book are also very grateful to NIDA, its current director Dr. Nora Volkow, former director Dr. Alan I. Leshner, and deputy director Dr. Timothy Condon, whose visionary recognition of the value of molecular and cell biological approaches to substance abuse issues fostered a supportive environment for novel ideas, state-of-the-art technologies, and integrative, multidisciplinary research in a field dominated by behavioral approaches. Collectively, they encouraged translational research and objective solutions to the vast public health problem of substance abuse, which is amply reflected in this book. We also appreciate NIDA staff members who contributed to the book as editors and mentors, and members of the staff at Cold Spring Harbor Laboratory Press, including Mary Cozza, Rena Steuer, Susan Schaefer, and David A. Crotty, Ph.D., whose wise counsel facilitated this initiative.

Each chapter was written by a leading neuroscientist with particular expertise in the topic covered. All of these individuals have many demands on their time, and their informative contributions and cooperation in meeting deadlines are highly appreciated. We also collectively thank our family members who displayed patience and forbearance during this process. Finally, members of the neuroscience community not included on the list of authors should be recognized and acknowledged, as they are the repository of fundamental contributions to our conceptual framework, methodologies, and knowledge base.

BERTHA K. MADRAS AND MARK VON ZASTROW

1 | Introduction

Bertha K. Madras

Department of Psychiatry, Harvard Medical School, New England Primate Research Center, Southborough, Massachusetts 01722

Substance abuse and addiction disorders have the ostensible distinction of being the most costly of all neuropsychiatric disorders (Table 1). Cellular and molecular biological research is of fundamental relevance to reducing this public health burden, by developing evidence-based insights into the etiology and neurobiology of drug addiction and contributing to novel approaches to treatment strategies. This level of analysis can provide a comprehensive view of genetic influences, biological targets of drugs, neurotoxicity, and the signaling pathways that trigger neuroadaptive processes to drive or contribute to compulsive and uncontrollable drug use, withdrawal, craving, and relapse.

This compendium conveys the merits and excitement of cellular and molecular approaches to drug addiction and describes fundamental, state-of-the-art advances and major gaps in the field, as well as those within the broader context of neuroscience. A host of molecular and cellular determinants, combined with environmental circum-

TABLE 1. Statistics on Neuropsychiatric Disorders

Disorder	Heritability (%)	Cost (billions)	Complex genetics (billions)
Addictions	0.40	544.11	212.20
Alzheimer's disease and dementias	0.53	170.86	87.45
Pain (with migraine)	0.40	150.80	58.81
Head or spinal cord injury	0.05	94.41	4.25
Anxiety disorders	0.30	82.63	23.96
Schizophrenia	0.70	57.08	38.81
Depressive illness	0.40	53.14	18.60
Developmental disorders	0.33	35.68	9.28
Stroke	0.10	27.03	2.16
Parkinson's disease	0.10	15.96	1.44
Multiple sclerosis	0.40	7.62	2.90
Seizures	0.60	1.04	0.61
Huntington's disease	1.00	0.23	0
Total		$1240.59	$459.69

Data from Uhl and Grow 2004.

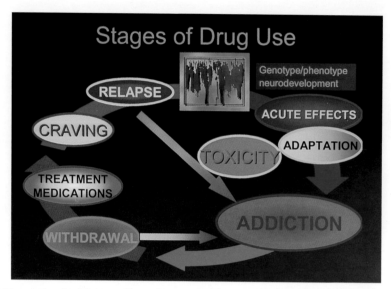

FIGURE 1. The cycle of addiction. The pathway to addiction is governed by many factors including the genotype and phenotype of the user, subject-specific response to the acute effects of a drug, neuroadaptation, progression to increased frequency of use, binge use, and addiction. Upon drug cessation from the adapted brain, physical or psychological withdrawal symptoms become prominent, and after these symptoms dissipate, drug craving can set in during abstinence. Treatment can be initiated at any point in the cycle, with interventions proving effective during the initial period of use or after frequent use, as well as during the addictive phase, withdrawal, and abstinence phase. Relapse to use can be triggered by drug-associated cues, stress, and other factors.

stances, contribute to the progression from use to abuse to addiction, with associated physical or psychological dependence (Fig. 1). Genetics can influence the pharmacokinetics of drugs, blood–brain barrier function, response of the direct targets of drugs (receptors, transporters, intracellular signaling pathways), neuroadaptive processes, and neurodevelopmental or neuropsychatric disorders that confer susceptibility to use and abuse. Even in the absence of genetic factors, drugs can generate anomalous signaling cascades that forge neuroadaptive changes to engender physical or psychological dependence and drug-seeking behavior (see the section entitled Genetics [Chapters 2–7]). The section Cell Biology and Pharmacology (Chapters 9–17) provides a detailed drug-specific overview of the most widely used drugs (alcohol, nicotinē, cannabinoids, cocaine, opioids, stimulants) and insight into the molecular targets of these drugs.

Until the advent of genomics/proteomics, candidate genes and proteins that mediate or bolster adaptive responses were scrutinized one at a time. It is now feasible to develop a comprehensive view of genes and their encoded proteins that fundamentally alter neuronal properties at various phases of drug use, addiction, withdrawal, craving, relapse to drug use, and therapeutic response. This strategy, described in the sections entitled Synaptic Plasticity and Addiction (Chapters 18–21) and Systems Analysis of Drug Abuse (Chapters 22 and 23), can provide a powerful molecular context to the concept of addiction as a disease and approaches for prevention, with relevance to other disciplines of neuroscience and neuropsychiatric disorders.

This volume also stresses the merits of integrating molecular approaches with behavioral and brain-imaging approaches (see Chapters 2–7). During the past 30 years, preclinical behavioral research generated a vast base of information on drug-induced spontaneous and schedule-controlled responses in various animal species. Despite exquisite refinements in behavioral analysis and models, amalgamation of behavioral with cellular and molecular approaches languished for decades. Fusion of these disciplines is accelerating, with increasing recognition that the significance of each discipline is immensely fortified by constructing behavioral models on a foundation of mechanism-based cellular models. The inclusion of human brain-imaging research intentionally provides another critical perspective and highlights another gap, in this case between preclinical drug-induced molecular changes and validation in human brain. Conversely, noninvasive imaging of the human brain exposed to drugs has revealed an array of changes in proteins or function, which warrants mechanistic resolution at the preclinical level. Preclinical research paradigms can also control for complex and confounding factors, polypharmacy, impure substances, underlying psychiatric diseases, genetic predisposition, or environmental factors, all of which frequently challenge clinical research. Bench-to-bedside translational research should be viewed as a priority. Without the ultimate objectives of corroborating preclinical findings in humans, clarifying mechanisms underlying changed behavior, and using this information to alleviate the consequences associated with drug abuse and addiction in human populations, justification for the research is enervated.

In addition, this timely book provides the reader with some fundamentals of neurodevelopment (see the section Development and Drug Abuse [Chapter 8]), an important factor in the interplay among genetics, development, and neural plasticity. It is increasingly apparent that drug-induced neuroadaptation may differ in the developing brain compared with the adult brain (e.g., Slotkin et al. 2004). Age-dependent responses to drugs and medicines, to addictive processes, and to learning, judgment, and reward are core to the field, but incompletely understood. The convergence of these methodological approaches is likely to accelerate progress in alleviating the consequences of this most costly of neuropsychiatric disorders. If this book accomplishes one or more of the following goals for the reader, we will be profoundly gratified:

1. Become familiar with different research approaches, brain imaging, genomics, genotyping/phenotyping, receptor/transporter trafficking, neuroadaptation and neurodevelopment, and neural activation mapping, and contemplate strategies for correlating these parameters with behavior.

2. Cultivate an appreciation of multidisciplinary addiction research and promote collaborations that meld highly focused in vitro and in vivo approaches.

3. Critique the strengths and limitations of cell and molecular biological approaches to addiction and enhance their validity.

4. Consider the cell and molecular biology of addiction as a venue for clarifying core brain functions of learning, memory, reward, and volitional behavior.

5. Destigmatize addiction research and recruit enthusiastic investigators into this highly exciting area of neuroscience research.

6. Develop creative strategies for validating preclinical discoveries in human subjects.

7. Develop translational research programs that lead to improved treatments.

Beyond the plane of pure science, this book can enhance the readers' capacity to make informed decisions about public policy on the basis of knowledge of the neurobiological consequences of drug use and addiction. Prospective molecular and cellular scientists must also bear in mind the enormous medical, public health, social, and behavioral consequences of addictive drugs. Although anomalous in the domain of a basic biology textbook, the following section provides the reader with a human context to the complex, costly, and daunting challenge of drugs to societies worldwide.

THE MAGNITUDE OF THE PROBLEM

Substance abuse problems are heavy burdens for individuals, their families, and society. No one initiates drug use with the intention of becoming addicted. From the addict's perspective, addiction may be an inadvertent, incidental, uncalculated, unanticipated, unforeseen, unintentional, unplanned, and random consequence of drug use. With the possible exception of obesity, the prevalence of substance abuse/addiction disorders is greater than any other preventable disease and more prevalent than many other chronic diseases. Drug abuse and addiction account for 500,000 or 25% of the 2 million annual deaths in the United States (Schneider Institute for Health Policy 2001). Of all the neuropsychiatric disorders, substance abuse/addiction disorders rank first in prevalence costs, surpassing Alzheimer's disease, depression, spinal cord injury, developmental disorders, and other devastating nervous system maladies (Uhl et al. 2002; Uhl and Grow 2004). In 1995, health care costs for abuse of alcohol, tobacco, and other drugs were estimated at $114 billion (Schneider Institute for Health Policy 2001). This number is partly attributable to the more than 500,000 annual visits that drug users make to costly emergency rooms for drug-related problems; a much higher estimate of $544.11 billion was recently estimated (Uhl and Grow 2004; Table 1).

Substance abuse can affect people from in utero to old age. Prenatal exposure to drugs is linked to low birth weight and to developmental disorders (Huestis and Choo 2002). Parental use of drugs can result in profound neglect and abuse of children. In adolescence and adulthood, drug use can be associated with poor school performance, accidents, unplanned sexual activity, pregnancy, violence, and criminal activity (Physician Leadership on National Drug Policy 2002). The work performance of illicit drug users is also problematic, with characteristic absenteeism, health problems, workplace injuries, high turnover, and lost productivity (Schneider Institute for Health Policy 2001). Older persons do not escape the long-term effects of drugs, because addiction, compromised health, and interrupted educational and social development take their toll. Whereas a high proportion of health problems and deaths are associated with alcohol and tobacco use, the use of illicit drugs is a significant contributor to medical, social, school work, and criminal justice issues. If illicit drugs were legalized or decriminalized, the contribution of Schedule I or II drugs to the spectrum of problems would likely increase (CASA Report 1995). On the basis of these considerations, it is not surprising to learn that in the year 2000, 67% of polled Americans ranked drug abuse among the top two or three problems facing American teenagers, far ahead of violence/crime/gun control (15%), education (9%), or teen pregnancy (8%). In Americans' views of serious health problems, drug abuse ranked above cancer (Blendon 2000; Physician Leadership on National Drug Policy 2002).

DRUG USE, ABUSE, AND ADDICTION

The convergence of environmental, individual, and drug-based factors in vulnerable individuals can lead to drug use, abuse, and addiction. Multiple factors can either promote or attenuate drug use.

Environmental Factors

The environment is a major influence on whether youth or adults will experiment with drugs or develop an aversion to them. Drug use can be encouraged by parents' indifference or acceptance of drug use in their children, access to drugs, chaotic homes with ineffective parents, lack of bonding with parents, sexual and physical abuse, the need for social acceptability, peer pressure, and the media (CASA Report 2002; Schneider Institute for Health Policy 2001). Countering these influences are the influence of friends, negative attitudes toward drugs and rules issued by parents, religious authorities, the media, the criminal justice system, affiliations with social and religious organizations, and public education messages on potential consequences of drugs.

Individual Risk Factors

Genetic predisposition (Uhl et al. 2002; Uhl and Grow 2004), psychiatric comorbidity (Volkow 2001; Armstrong and Costello 2002; Carey 2002; Drake et al. 2002), personality disorders (Sher and Trull 2002), poor school performance, inappropriate school behavior, and early drug use can all promote drug abuse problems (CASA Report 2002). Drug use and addiction has also been viewed as a form of self-medication to relieve psychiatric problems (Harris and Edlund 2005). Severe emotional shocks such as child abuse, the death of a parent, and alcoholic or drug-addicted parents can spur adolescents or young adults into drug use.

Drug-related Risks

Drugs exert powerful but capricious effects on the brain, with major variables being dose, cost, accessibility, route of administration (intravenous, smoking, insufflation, oral), chemical composition, and user response. The biological imprint of a drug in the brain can be as unpredictable as a New England storm, leaving little trace of passage or a trail of carnage that impact brain function, personality, and behavior.

THE CYCLE OF ADDICTION

The factors that trigger the transition from controlled use to addiction are not fully understood but are known to be drug specific and affect between 8% and 32% of users (Anthony and Petronis 1995; Wagner and Anthony 2002; Chen et al. 2005; O'Brien and Anthony 2005). When use progresses to the addictive stage (Fig. 1), the drug becomes a primary source of positive sensations, judgment deteriorates, and adverse consequences escalate. At this stage, abstinence from the drug is invariably manifest by physiological and/or psychological withdrawal. With prolonged abstinence, the addicted brain can generate intense craving and suppress efforts to control compulsive drug-seeking. Craving usually leads to relapse, and the addict can rotate through compulsive use, withdrawal, and relapse several times until the cycle can finally be broken. For different stages of drug use, adaptive changes can be observed in the brain—changes that are neither fully predictable nor necessarily reversible.

How Do Drugs Affect the Brain?

During the last decade, remarkable progress has been made in clarifying biological changes elicited by drugs. Researchers have identified the immediate targets of most drugs of abuse (e.g., see Madras et al. 1989; Bunzow et al. 2001; Nichols 2004; Piomelli 2004; Snyder and Pasternak 2004; Miller et al. 2005; Chapter 14, this volume), cloned the genes encoding these targets, and documented a prodigious array of intricate molecular and cellular adaptive changes generated by drugs (Kalivas 2004; Nestler 2004a,b; Kalivas et al. 2005). In experimental models, repeated administration of a drug leads to an excess of, and diminished production of, multiple transcribed genes in a single brain region, in conjunction with abnormal production of proteins (Loguinov et al. 2001; Freeman et al. 2002; Toyooka et al. 2002; Funada et al. 2004; Bahi and Dreyer 2005; Rhodes and Crabbe 2005; Thibault et al. 2005; Yamamoto et al. 2005; Yuferov et al. 2005). The findings provide powerful clues for clarifying how the brain adapts to drug-induced activity and sensations, as well as leads for designing drug therapies to treat addiction. The complexity of these findings can be summarized by a few core principles. Brain communication at the level of neurotransmission is a key to understanding how drugs trigger adaptation. At least 15% of the human genome encodes proteins involved in cellular communication. The human brain manufactures more than 100 different neurotransmitters and modulators for synaptic communication. Neurotransmitters and neurotransmission are exquisitely regulated because their function is key to survival. Drug structures resemble, but are not identical to, endogenous transmitters. The "imposters," cocaine, amphetamine, and ecstasy, have structural similarities to dopamine, serotonin, and norepinephrine; THC (Δ^9-tetrahydrocannabinol), an active ingredient in marijuana, resembles the brain's anandamide (Fig. 2A) and 2-arachidonylglycerol; heroin shares structural overlap with endorphin and enkephalins; LSD resembles the neurotransmitter serotonin (Fig. 2B). Because drugs are structurally similar to neurotransmitters, they influence neurotransmission, but the similarities do not portend that drugs fully imitate endogenous transmitter-mediated neurotransmission. Tightly regulated receptor-mediated signaling and synaptic communication are geared for endogenous transmitters, not the drug. Neurons generally do not process or regulate drugs with the same exquisite precision involved in controlling endogenous compounds. With the exception of amphetamines, most psychoactive drugs (e.g., LSD, morphine, cocaine, marijuana) are not substrates for transporters and are not sequestered away from the synapse if they trigger a burst of receptor activity. Drugs may activate receptors but are not necessarily able to trigger the receptor trafficking and processing required to recycle active receptors. Morphine can activate the μ-opioid receptor but is far less effective in promoting opioid receptor internalization than are endogenous peptides, a process implicated in the recycling of a

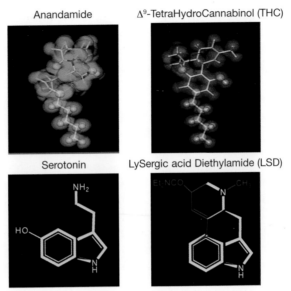

FIGURE 2. (*Top*) The endogenous neurotransmitter anandamide has structural similarities to Δ⁹-tetrahydrocannabinol (THC), an active constituent of marijuana. (Chemical model courtesy of Wayne Mascarella, incorporated into a Museum of Science exhibit, Boston [B.K. Madras, project director] and CD-ROM [B.K. Madras and L. Sato], entitled "Changing Your Mind: Drugs in the Brain.") Accumulating evidence indicates that both compounds act at the cannabinoid receptor (CB1) in the brain. (*Bottom*) The endogenous neurotransmitter serotonin has structural similarities to the hallucinogen lysergic acid diethylamide (LSD). Accumulating evidence indicates that both compounds act at the serotonin receptor (5-HT2C) in the brain.

functional opioid receptor (see Chapter 11, this volume). Both serotonin and LSD activate the serotonin 5-HT2A receptor but trigger different signal transduction cascades and unique and different patterns of gene induction (Almaula et al. 1996; Backstrom et al. 1999; Gonzalez-Maeso et al. 2003). Drugs may induce signals of anomalous strength and duration, and promote signaling cascades or trafficking that conceivably translate into euphoria, delusions, hallucinations, anger, and a host of other psychological and behavioral manifestations.

Drug-induced mnemonic effects may share components of other forms of learning and memory, and drug-seeking behavior may be prompted by these memories (Dong et al. 2004). With repeated and frequent use, the brain may adapt to and compensate for abnormal signals by altering production of gene copies, protein levels, and signaling networks to engender neuroadaptation of unknown reversibility. Drugs can alter glucose metabolism (London et al. 1990; Volkow et al. 1991), signal transduction cascades (Nestler 2004a,b), gene function expression (Nestler 2004a,b; Bahi and Dreyer 2005), protein production, cell morphology (Robinson and Kolb 2004), neural networks (Saka et al. 2004), and activity in specific brain regions, as concisely summarized in Chapters 6, 8, 13, and 21 of this volume. Some drugs are toxic to the brain: Cocaine promotes vasospasm and loss of normal blood flow in human brain (Kaufman et al. 1998). Frequent exposure to amphetamine, methamphetamine, and ecstasy produce cell-specific toxicity and damage (McCann and Ricaurte 1995; Villemagne et al. 1998; McCann et al. 2000;

Volkow et al. 2001). Heavy alcohol or inhalant use (e.g., toluene) can produce profound irreversible toxic effects that are manifest as shrinkage of brain gray matter.

At the psychological level, addiction is characterized by a persistent compulsion to use a drug(s), loss of control over drug use, a reduction in life-sustaining activities and social pursuits, medical and other adverse consequences, and possibly drug tolerance. Long-term users of specific drugs may experience pervasive changes in brain function and behavior. For example, ecstasy users can display dose-/time-dependent deficits in memory and report depression, anxiety, and sleep problems (Bolla et al. 1998; Parrott 2001). Both objective and self-reporting measures suggest numerous negative features associated with long-term heavy cannabis use. Heavy users themselves reported significantly lower educational attainment and income, and the majority of users rated the subjective effects of cannabis as engendering a negative effect on their cognition, memory, career, social life, physical and mental health, and levels of satisfaction (Gruber et al. 2003). Molecular and cellular research on the long-term consequences of drug exposure is needed to integrate abnormal molecular, metabolic, and functional changes with addictive behavior and compromised brain function. This gap is likely to close with an emphasis on translational, multidisciplinary research.

On withdrawal from a drug, the adapted brain is hyper-reactive compared with the pre-drug state, requiring the drug for "normal function." In withdrawal, anxiety, irritability, dysphoria, stress, and other psychological or physical discomforts (tremors, flu-like symptoms) emerge. Abrupt withdrawal from drugs such as alcohol and heroin unmasks the adapted brain and produces wrenching physical symptoms. Withdrawal from marijuana, nicotine, cocaine, and amphetamine generates a spectrum of symptoms including anxiety, irritability, dysphoria, insomnia or hypersomnia, aches, craving, and other drug-specific effects.

THE DISEASE MODEL OF ADDICTION

How does addiction conform to a definition of disease? Dictionaries define disease as "an alteration in the state of the body or of some of its organs, interrupting or disturbing the performance of the vital functions, and causing or threatening pain and weakness and characterized by an identifiable group of signs or symptoms"; "a pathological condition of a part, organ, or system of an organism resulting from various causes, such as infection, genetic defect, or environmental stress"; "a malady, affection, illness, sickness, disorder, applied figuratively to the mind, to the moral character and habits." Drug addiction can be viewed as a chronic, relapsing disease, characterized by compulsive, uncontrollable use despite adverse consequences. The disease model of addiction offers a framework to facilitate treatment research, reduce stigmatization by professionals, and focus on problem solving. The neurobiological changes, however, should not detract from the role of personal responsibility in propagating this behavior (Committee on Addictions 2002). Within the construct of a disease model, patients are urged to assume responsibility for compliance with treatment, as with other diseases. Compliance is comparable to patients with asthma, hypertension, and diabetes.

PREVENTION, INTERVENTION, AND TREATMENT OF ADDICTION

The goals of prevention are to stop initiation to drug use; the goals of intervention and treatment are to prevent progression to addiction or to reverse the behavioral patterns of the addicted state. Treatment can prevent progression to addiction, improve health and parenting, reduce medical costs, reduce transmission of infectious diseases such as AIDS, reduce crime, and restore a person to a functional role in society.

Problem drinkers and drug users do not necessarily suffer multiple serious consequences but are at risk for immediate consequences (see above) if the behaviors persist. Paradoxically, the health-care system does not address this issue aggressively (Lewis 1991; Saunders and Roche 1991; Dove 1999; Fleming et al. 1999; Wallace 2000; Blumenthal et al. 2001; Miller et al. 2001; Barnes 2002; Spangler et al. 2002; Abrams Weintraub et al. 2003), which leads to escalating health-care costs. Substantial evidence indicates that brief behavioral interventions are effective when used by clinicians who are not specialists in substance abuse treatment, especially when enhancing entry to more intensive substance abuse treatment (Saitz et al. 2000; Davis et al. 2003; Schermer et al. 2003). Equally relevant is the fact that a significant proportion of drug users (as many as 50%) have underlying psychiatric problems (Armstrong and Costello 2002; Carey 2002; Drake et al. 2002; Sher and Trull 2002; Harris and Edland 2005) that should be diagnosed and treated concurrently with drug treatment. For adolescents, brief interventions that include feedback on risks, an emphasis on personal responsibility, and alternatives for change have proven effective (Knight 2001; Brown 2004).

Treatment effectiveness can be measured by a range of outcome measures including reduced drug use, improved physical and mental health, employment, family relationships, reduced mortality, crime, and diminished medical, legal, social services, employment, and school costs to society (Blendon 2000; CASA Report 2002; Davis et al. 2003; Brown 2004; Nunes and Levin 2004). With regard to all of these measures, numerous studies have shown positive outcomes of treatment. A variety of treatment approaches is necessary to accommodate the needs of substance abusers who present with varying social skills, economic status, underlying psychiatric disorders, criminal activity, age, and family support systems. The majority of United States residents who had an alcohol or illicit drug problem in 2002 did not receive treatment for their problem (The 2002 National Survey on Drug Use and Health [http://www.samhsa.gov/oas/nhsda.htm]). Of the estimated 7.7 million individuals who needed treatment for an illicit drug problem in the past year, 18% received treatment (1.4 million). Similarly, of 18.6 million individuals in need of alcohol treatment, 8.8% received treatment. In a recent study, more than 75% of a population that needed treatment elected not to approach treatment providers for various reasons. Of the few who felt they needed treatment, 24% of illicit drug users made an effort to seek treatment. This may partially reflect the loss of judgment that gradually prevents drug users from understanding their predicament and it supports the need for external intervention. Outpatient drug-free, outpatient methadone, long-term residential, and short-term inpatient programs reduce drug use, including cocaine and heroin. The cost-effectiveness for treating compared with the cost of not treating is variable. Some reports indicate a cost savings equivalent to a 1:3 ratio, whereas others report much higher ratios. For example, an Institute of Medicine Report (1996) indicated that an untreated addict costs society $43,200 annually, compared with incarceration ($39,000), probation ($16,691), residential treatment ($12,467), methadone maintenance ($3,500), or outpatient treatment for cocaine ($2,722). These figures clearly indicate that medicines (e.g., methadone) for treating addiction (e.g., heroin) are cost-effective and much needed for other addictions. The cell biological consequences of drugs are critical for identifying creative, novel approaches to medications development.

CHALLENGES AND OPPORTUNITIES FOR CELL BIOLOGICAL RESEARCH

Note Added in Proof by Mark von Zastrow

In addition to its profound clinical and societal importance, addiction research provides a unique window for investigating fundamental questions of neural plasticity and behavior. How can the ingestion of a simple chemical, over time and with many repetitions, distort the personality of an addict almost beyond recognition? How do addicts adapt to escalating doses of drugs that are often in excess of those that would be lethal if administered to drug-naïve individuals? Why do addicts often chose drug ingestion over fundamental natural rewards such as food or sexual activity, despite (often horrendous) adverse consequences? These are fascinating questions that, in rapidly developing fields, have reached an unprecedented level of experimental accessibility. Indeed, the excitement and promise of addiction research have never been greater.

A main thesis of the following chapters is that addiction research must now transcend traditional barriers between molecular and systems-level thinking, and do so while striving for a high level of mechanistic rigor. Cell biology, with its traditional emphasis on linking biochemical mechanisms to integrated tissue function, seems well suited to this challenge. The workshops held at Cold Spring Harbor Laboratory have attempted to foster a "cell biology of addiction" and encourage promising junior scientists—as well as more senior scientists already accomplished in other areas—into this burgeoning field. This book attempts to synthesize and disseminate our progress in this effort. We hope that this book serves as a useful introduction to less experienced investigators and a focal point for additional discussion and debate among more seasoned members of the addiction research community.

REFERENCES

Abrams Weintraub T., Saitz R., and Samet J.H. 2003. Education of preventive medicine residents. Alcohol, tobacco, and other drug abuse. *Am. J. Prev. Med.* **24:** 101–105.

Almaula N., Ebersole B.J., Zhang D., Weinstein H., and Sealfon S.C. 1996. Mapping the binding site pocket of the serotonin 5-hydroxytryptamine2A receptor. Ser3.36(159) provides a second interaction site for the protonated amine of serotonin but not of lysergic acid diethylamide or bufotenin. *J. Biol. Chem.* **271:** 14672–14675.

Anthony J.C. and Petronis K.R. 1995. Early-onset drug use and risk of later drug problems. *Drug Alcohol Depend.* **40:** 9–15.

Armstrong T.D. and Costello E.J. 2002. Community studies on adolescent substance use, abuse, or dependence and psychiatric comorbidity. *J. Consult. Clin. Psychol.* **70:** 1224–1239.

Backstrom J.R., Chang M.S., Chu H., Niswender C.M., and Sanders-Bush E. 1999. Agonist-directed signaling of serotonin 5-HT2C receptors: Differences between serotonin and lysergic acid diethylamide (LSD). *Neuropsychopharmacology* (suppl. 2) **21:** 77S–81S.

Bahi A. and Dreyer J.L. 2005. Cocaine-induced expression changes of axon guidance molecules in the adult rat brain.

Mol. Cell. Neurosci. **28:** 275–291.

Barnes H.N. 2002. Creating meaning and value in substance abuse education. *Subst. Abuse* **23:** 203–209.

Blendon R. 2000. Public attitudes towards illegal drug use and drug treatment. Harvard School of Public Health and The Robert Wood Johnson Foundation Report.

Blumenthal D., Gokhale M., Campbell E.G., and Weissman J.S. 2001. Preparedness for clinical practice: Reports of graduating residents at academic health centers. *J. Am. Med. Assoc.* **286:** 1027–1034.

Bolla K.I., McCann U.D., and Ricaurte G.A. 1998. Memory impairment in abstinent MDMA ("Ecstasy") users. *Neurology* **51:** 1532–1537.

Brown S.A. 2004. Measuring youth outcomes from alcohol and drug treatment. *Addiction* (suppl. 2) **99:** 38–46.

Bunzow J.R., Sonders M.S., Arttamangkul S., Harrison L.M., Zhang G., Quigley D.I., Darland T., Suchland K.L., Pasumamula S., Kennedy J.L., Olson S.B., Magenis R.E., Amara S.G., and Grandy D.K. 2001. Amphetamine, 3,4-methylenedioxymethamphetamine, lysergic acid diethylamide, and metabolites of the catecholamine neurotransmitters are agonists of a rat trace amine receptor. *Mol. Pharmacol.* **60:** 1181–1188.

Carey K.B. 2002. Clinically useful assessments: Substance use and comorbid psychiatric disorders. *Behav. Res. Ther.* **40:** 1345–1361.

CASA Report. 1995. Legalization: Panacea or Pandora's box? The National Center on Addiction and Substance Abuse at Columbia University Report.

——. 2002. The National Center on Addiction and Substance Abuse at Columbia University Report, parts I–VII.

Chen C.Y., O'Brien M.S., and Anthony J.C. 2005. Who becomes cannabis dependent soon after onset of use? Epidemiological evidence from the United States: 2000–2001. *Drug Alcohol Depend.* **79:** 11–22.

Committee on Addictions (Committee on Addictions of the Group for the Advancement of Psychiatry). 2002. Responsibility and choice in addiction. *Psychiatr. Serv.* **53:** 707–713.

Davis T.M., Baer J.S., Saxon A.J., and Kivlahan D.R. 2003. Brief motivational feedback improves post-incarceration treatment contact among veterans with substance use disorders. *Drug Alcohol Depend.* **69:** 197–203.

Dong Y., Saal D., Thomas M., Faust R., Bonci A., Robinson T., and Malenka R.C. 2004. Cocaine-induced potentiation of synaptic strength in dopamine neurons: Behavioral correlates in GluRA(–/–) mice. *Proc. Natl. Acad. Sci.* **101:** 14282–14287.

Dove H.W. 1999. Postgraduate education and training in addiction disorders: Defining core competencies. *Psychiatr. Clin. North Am.* **22:** 481–488.

Drake R.E., Wallach M.A., Alverson H.S., and Mueser K.T. 2002. Psychosocial aspects of substance abuse by clients with severe mental illness. *J. Nerv. Ment. Dis.* **190:** 100–106.

Fleming M.F., Manwell L.B., Kraus M., Isaacson J.H., Kahn R., and Stauffacher E.A. 1999. Who teaches residents about the prevention and treatment of substance use disorders? A national survey. *J. Fam. Pract.* **48:** 725–729.

Freeman W.M., Dougherty K.E., Vacca S.E., and Vrana K.E. 2002. An interactive database of cocaine-responsive gene expression. *Scientific World Journal* **2:** 701–706.

Funada M., Zhou X., Satoh M., and Wada K. 2004. Profiling of methamphetamine-induced modifications of gene expression patterns in the mouse brain. *Ann. N.Y. Acad. Sci.* **1025:** 76–83.

Gonzalez-Maeso J., Yuen T., Ebersole B.J., Wurmbach E., Lira A., Zhou M., Weisstaub N., Hen R., Gingrich J.A., and Sealfon S.C. 2003. Transcriptome fingerprints distinguish hallucinogenic and nonhallucinogenic 5-hydroxytryptamine 2A receptor agonist effects in mouse somatosensory cortex. *J. Neurosci.* **23:** 8836–8843.

Gruber A.J., Pope H.G., Hudson J.I., and Yurgelun-Todd D. 2003. Attributes of long-term heavy cannabis users: A case-control study. *Psychol. Med.* **33:** 1415–1422.

Harris K.M. and Edlund M.J. 2005. Self-medication of mental health problems: New evidence from a national survey. *Health Serv. Res.* **40:** 117–134.

Huestis M.A. and Choo R.E. 2002. Drug abuse's smallest victims: In utero drug exposure. *Forensic Sci. Int.* **128:** 20–30.

Kalivas P.W. 2004. Recent understanding in the mechanisms of addiction. *Curr. Psychiatry Rep.* **6:** 347–351.

Kalivas P.W., Volkow N., and Seamans J. 2005. Unmanageable motivation in addiction: A pathology in prefrontal-accumbens glutamate transmission. *Neuron* **45:** 647–650.

Kaufman M.J., Levin J.M., Ross M.H., Lange N., Rose S.L., Kukes T.J., Mendelson J.H., Lukas S.E., Cohen B.M., and Renshaw P.F. 1998. Cocaine-induced cerebral vasoconstriction detected in humans with magnetic resonance angiography. *J. Am. Med. Assoc.* **279:** 376–380.

Knight J.R. 2001. The role of the primary care provider in preventing and treating alcohol problems in adolescents. *Ambul. Pediatr.* **1:** 150–161.

Lewis D.C. 1991. Recent advances in health professional training on substance abuse disorders. *Drug Alcohol Rev.* **10:** 45–53.

Loguinov A.V., Anderson L.M., Crosby G.J., and Yukhananov R.Y. 2001. Gene expression following acute morphine administration. *Physiol. Genomics* **6:** 169–181.

London E.D., Cascella N.G., Wong D.F., Phillips R.L., Dannals R.F., Links J.M., Herning R., Grayson R., Jaffe J.H., and Wagner H.N., Jr. 1990. Cocaine-induced reduction of glucose utilization in human brain. A study using positron emission tomography and [fluorine 18]-fluorodeoxyglucose. *Arch. Gen. Psychiatry* **47:** 567–574.

Madras B.K., Fahey M.A., Bergman J., Canfield D.R., and Spealman R.D. 1989. Effects of cocaine and related drugs in nonhuman primates. I. [3H]cocaine binding sites in caudate-putamen. *J. Pharmacol. Exp. Ther.* **251:** 131–141.

McCann U.D. and Ricaurte G.A. 1995. On the neurotoxicity of MDMA and related amphetamine derivatives. *J. Clin. Psychopharmacol.* **15:** 295–296.

McCann U.D, Eligulashvili V., and Ricaurte G.A. 2000. (+/–)3,4-Methylenedioxymethamphetamine ('Ecstasy')-induced serotonin neurotoxicity: Clinical studies. *Neuropsychobiology* **42:** 11–16.

Miller G.M., Verrico C.D., Jassen A., Konar M., Yang H., Panas H., Bahn M., Johnson R., and Madras B.K. 2005. Primate trace amine receptor 1 modulation by the dopamine transporter. *J. Pharmacol. Exp. Ther.* **313:** 983–994.

Miller N.S., Sheppard L.M., Colenda C.C., and Magen J. 2001. Why physicians are unprepared to treat patients who have alcohol- and drug-related disorders. *Acad. Med.* **76:** 410–418.

Nestler E.J. 2004a. Molecular mechanisms of drug addiction. *Neuropharmacology* (suppl. 1) **47:** 24–32.

——. 2004b. Historical review: Molecular and cellular mechanisms of opiate and cocaine addiction. *Trends Pharmacol. Sci.* **25:** 210–218.

Nichols D.E. 2004. Hallucinogens. *Pharmacol. Ther.* **101:** 131–181.

Nunes E.V. and Levin F.R. 2004. Treatment of depression in patients with alcohol or other drug dependence: A meta-analysis. *J. Am. Med. Assoc.* **291:** 1887–1896.

O'Brien M.S. and Anthony J.C. 2005. Risk of becoming cocaine dependent: Epidemiological estimates for the United States, 2000–2001. *Neuropsychopharmacology* **30:** 1006–1018.

Parrott A.C. 2001. Human psychopharmacology of Ecstasy

(MDMA): A review of 15 years of empirical research. *Hum. Psychopharmacol.* **16:** 557–577.

Physician Leadership on National Drug Policy. 2002. Adolescent Substance Abuse: A public health priority, pp. 1–102 (David Lewis, Project Director, www.PLNDP.org).

Piomelli D. 2004. The endogenous cannabinoid system and the treatment of marijuana dependence. *Neuropharmacology* (suppl. 1) **47:** 359–367.

Rhodes J.S. and Crabbe J.C. 2005. Gene expression induced by drugs of abuse. *Curr. Opin. Pharmacol.* **5:** 26–33.

Robinson T.E. and Kolb B. 2004. Structural plasticity associated with exposure to drugs of abuse. *Neuropharmacology* (suppl. 1) **47:** 33–46.

Saitz R., Sullivan L.M., and Samet J.H. 2000. Training community-based clinicians in screening and brief intervention for substance abuse problems: Translating evidence into practice. *Subst. Abuse* **21:** 21–31.

Saitz R., Friedmann P.D., Sullivan L.M., Winter M.R., Lloyd-Travaglini C., Moskowitz M.A., and Samet J.H. 2002. Professional satisfaction experienced when caring for substance-abusing patients: Faculty and resident physician perspectives. *J. Gen. Intern. Med.* **17:** 373–376.

Saka E., Goodrich C., Harlan P., Madras B.K., and Graybiel A.M. 2004. Repetitive behaviors in monkeys are linked to specific striatal activation patterns. *J. Neurosci.* **24:** 7557–7565.

Saunders J.B. and Roche A.M. 1991. Medical education in substance disorders. *Drug Alcohol Rev.* **10:** 263–265.

Schermer C.R., Bloomfield L.A., Lu S.W., and Demarest G.B. 2003. Trauma patient willingness to participate in alcohol screening and intervention. *J. Trauma* **54:** 701–706.

Schneider Institute for Health Policy. 2001. Substance abuse: The Nation's number one health problem. Brandeis University for the Robert Wood Johnson Foundation, Princeton, New Jersey.

Sher K.J. and Trull T.J. 2002. Substance use disorder and personality disorder. *Curr. Psychiatry Rep.* **4:** 25–29.

Slotkin T.A., Cousins M.M., and Seidler F.J. 2004. Administration of nicotine to adolescent rats evokes regionally selective upregulation of CNS alpha 7 nicotinic acetylcholine receptors. *Brain Res.* **1030:** 159–163.

Snyder S.H. and Pasternak G.W. 2003. Historical review: Opioid receptors. *Trends Pharmacol. Sci.* **24:** 198–205.

Spangler J.G., George G., Foley K.L., and Crandall S.J. 2002. Tobacco intervention training: Current efforts and gaps in US medical schools. *J. Am. Med. Assoc.* **288:** 1102–1109.

Thibault C., Hassan S., and Miles M. 2005. Using in vitro models for expression profiling studies on ethanol and drugs of abuse. *Addict. Biol.* **10:** 53–62.

Toyooka K., Usui M., Washiyama K., Kumanishi T., and Takahashi Y. 2002. Gene expression profiles in the brain from phencyclidine-treated mouse by using DNA microarray. *Ann. N.Y. Acad. Sci.* **965:** 10–20.

Uhl G.R. and Grow R.W. 2004. The burden of complex genetics in brain disorders. *Arch. Gen. Psychiatry* **61:** 223–229.

Uhl G.R., Liu Q.R., and Naiman D. 2002. Substance abuse vulnerability loci: Converging genome scanning data. *Trends Genet.* **18:** 420–425.

Villemagne V., Yuan J., Wong D.F., Dannals R.F., Hatzidimitriou G., Mathews W.B., Ravert H.T., Musachio J., McCann U.D., and Ricaurte G.A. 1998. Brain dopamine neurotoxicity in baboons treated with doses of methamphetamine comparable to those recreationally abused by humans: Evidence from [11C]WIN-35,428 positron emission tomography studies and direct in vitro determinations. *J. Neurosci.* **18:** 419–427.

Volkow N.D. 2001. Drug abuse and mental illness: Progress in understanding comorbidity. *Am. J. Psychiatry* **158:** 1181–1183.

Volkow N.D., Fowler J.S., Wolf A.P., Hitzemann R., Dewey S., Bendriem B., Alpert R., and Hoff A. 1991. Changes in brain glucose metabolism in cocaine dependence and withdrawal. *Am. J. Psychiatry* **148:** 621–626.

Volkow N.D., Chang L., Wang G.J., Fowler J.S., Leonido-Yee M., Franceschi D., Sedler M.J., Gatley S.J., Hitzemann R., Ding Y.S., Logan J., Wong C., and Miller E.N. 2001. Association of dopamine transporter reduction with psychomotor impairment in methamphetamine abusers. *Am. J. Psychiatry* **158:** 377–382.

Wagner F.A. and Anthony J.C. 2002. From first drug use to drug dependence: Developmental periods of risk for dependence upon marijuana, cocaine, and alcohol. *Neuropsychopharmacology* **26:** 479–488.

Wallace P. 2000. Medical students, drugs and alcohol: Time for medical schools to take the issue seriously. *Med. Educ.* **34:** 86–87.

Yamamoto H., Imai K., Takamatsu Y., Kamegaya E., Kishida M., Hagino Y., Hara Y., Shimada K., Yamamoto T., Sora I., Koga H., and Ikeda K. 2005. Methamphetamine modulation of gene expression in the brain: Analysis using customized cDNA microarray system with the mouse homologues of KIAA genes. *Brain Res. Mol. Brain Res.* **137:** 40–46.

Yuferov V., Nielsen D., Butelman E., and Kreek M.J. 2005. Microarray studies of psychostimulant-induced changes in gene expression. *Addict. Biol.* **10:** 101–118.

PART 1

Genetics

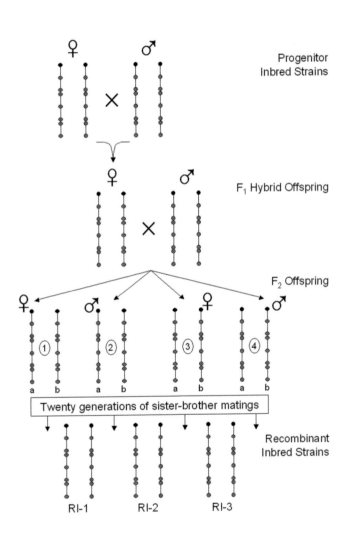

2 | Addiction Genetics and Genomics

George R. Uhl

Molecular Neurobiology Branch, National Institute on Drug Abuse–Intramural Research Program, National Institutes of Health, Baltimore, Maryland 21224

ABSTRACT

Classical genetic studies document (1) strong complex genetic contributions to substance abuse, (2) common genetic vulnerability to abuse of different abused substances, as well as more substance-specific genetic vulnerabilities, and (3) more prominent genetic influences for later phases of individuals' progression toward substance dependence. Increasing understanding of the human genome and its variants supports the idea that there are approximately three to four common haplotypes at each of the approximately two to three blocks of restricted haplotype diversity at each of about 30,000 human genes. If we assume that most addiction vulnerability variants lie within genes, current efforts in addiction molecular genetics and genomics should aim to understand (1) which of these approximately 250,000 haplotype variants contribute to the human individual differences in addiction vulnerability that can be observed in classical genetic studies, (2) how these variations contribute to addiction vulnerability, and (3) how this understanding can better match individuals with the prevention and treatment strategies most likely to work for them. Tools for this understanding include association and linkage-based genome-scanning methods, candidate gene studies, studies of influences of variants in gene structure, and regulation in vitro and in vivo, including positron emission tomography (PET) studies, pharmacogenetic and pharmacogenomic studies in human research volunteers, studies in human population samples, studies in postmortem human brains, and studies in genetically manipulated mice. A current attractive model for the molecular genetic architecture of human addiction vulnerability posits polygenic effects of common allelic variants, most influencing gene regulation. Many of these variants appear likely to influence mnemonic aspects of addiction and to contribute to differences in the brains of individuals with differential vulnerabilities to addictions. Other allelic variations could also produce effects on addiction vulnerability phenotypes by routes that include (1) altering drug metabolism and/or biodistribution, (2) changing drugs' rewarding properties, (3) acting through influences on the traits manifest by addicts, and (4) acting through influences on the psychiatric comorbidities manifest by addicts. Our understanding of addictions, addiction prevention, and addiction treatment is likely to be revolutionized by findings from human molecular genetic studies of addictions.

INTRODUCTION

Heritability of Addictions

Our understanding of the magnitude of genetic influences on addiction comes from classical genetic family, adoption, and (especially) twin studies. These studies also inform our thinking about the nature of the heritable components of addiction vulnerability.

Drug abuse vulnerability runs in families (for review, see Uhl et al.1995), although not in the manner that single-gene diseases move through families. Environment has a large role. Addictions are thus likely to be "complex disorders" and to receive contributions from allelic variations in a number of genes. Twin studies tell us that drug abuse vulnerability is 40–60% heritable (Pickens et al. 1991; Tsuang et al. 1996, 1998, 1999; Kendler et al. 1997, 1999, 2000; Kendler and Prescott 1998; True et al. 1999a,b; Uhl 1999; Uhl et al. 2002). Twin data support the idea that much of the genetic vulnerability to the abuse of legal and illegal addictive substances is shared (Tsuang et al. 1998; Kendler et al. 1999). Although parts of addiction vulnerability are likely to be specific for specific substances, most genetic influences are likely to be common to different addictive substances.

The phenotype that is best documented to be heritable is drug dependence, diagnosed using diagnostic and statistical manual (DSM) criteria, which states the following: "Twin studies also suggest that genes exert different effects on different aspects of individuals' progression from experimentation with drugs to use to regular use of drugs to DSM diagnoses of drug dependence." Only modest genetic influences impact initiation of drug use (Kendler et al. 1999; Tsuang et al. 1999). Despite this, some evidence does support genetic mediation of individual differences in acute drug responses. Individuals who have elevated genetic risk for developing alcoholism can display different responses to acute ethanol administration, for example (Schuckit et al. 2000). Only moderate genetic influences are likely to influence progression from initial use to regular use. The largest genetic heritable determinants appear to influence progression from regular use to dependence (Kendler et al. 1999; Tsuang et al. 1999). The major genetic influences on human addiction vulnerability are thus found in the later stages of the progression toward drug dependence.

Our understanding of the genetic architecture likely to underlie human addiction vulnerability is informed by understanding the ways in which patterns of DNA variations are identified in humans, by concepts about the likely frequencies of allelic variations that underlie common human disorders, and by results from molecular genetic studies of addiction.

Reviewing some of the aspects of our increasing understanding of the human genome and the variations found in the human genome is important to set the stage for studies of addiction-specific molecular genetics and genomics. We now know that the approximately 3.2 billion base pairs of the human genome contain millions of variations. The most common are single nucleotide polymorphisms (SNPs). The genome also contains insertion/deletion polymorphisms, short ("microsatellite") polymorphic repeat sequences, and long (variable number of tandem repeats [VNTR]) polymorphic repeat sequences. The number of possible combinations in which variation could occur, in theory, is dauntingly large.

As it turns out, however, each of the approximately 30,000 human genes may have perhaps three to four common haplotypes at each of perhaps two to three blocks of restricted haplotype diversity. These blocks of restricted haplotype diversity typically encompass dozens of variations that are thus present in many fewer combined patterns than we would find if all possible combinations were found frequently in the human genome. One view of addiction molecular genetics and genomics could thus address the questions: Which of these blocks of haplotype variants contributes to the human individual differences in addiction vulnerability and how do these variations contribute to addiction vulnerability?

One way to consider how these variants contribute to addiction vulnerability in the population is to consider contrasting "genetic architectures" for addiction. The more commonly considered genetic architecture is supported by the "common disease–common allele" hypothesis and by the high frequency of addictions in the population (Reich and Lander 2001). This hypothesis is based on assumptions of relatively modest genetic heterogeneity for human addiction vulnerability. It posits that many of the same addiction-vulnerability-enhancing allelic variants could contribute to vulnerability to drug abuse in many of the addicted individuals in the population. Such a model thus presumes low genetic heterogeneity and predicts fewer vulnerability alleles, higher frequencies of these alleles in the population, and less potent effects of each single vulnerability allele in individuals who harbor them than contrasting models (Fig. 1).

Contrasting models for the genetic architecture of human addiction vulnerability posit higher genetic heterogeneity. Within these models, rarer addiction-vulnerability-enhancing alleles identified in one addict would be found in fewer other addicts. Models with higher heterogeneity would generally predict larger numbers of vulnerability alleles, lower frequencies of these alleles in the population, and more potent effects of single vulnerability alleles within individuals and their families.

GOALS OF THE CHAPTER

This chapter describes the current status of rapidly advancing efforts to identify those genes and allelic variations likely to contribute to the molecular genetic bases of human addiction vulnerability. To benefit from the greater statistical certainty afforded by converging evidence with the least a priori bias about which genes "should" be involved (termed a "candidate gene" approach), the chapter focuses on the results of genome-scanning approaches. To best describe the work available to date, data are used from both linkage- and association-based genome-scanning approaches. The chapter attempts to describe integration of these observations with those from candidate gene strategies in human brain and (especially) transgenic and knockout mice.

One hypothesis explored here is that we are rapidly approaching one of the first major landmarks of drug abuse molecular genetics—the identification of molecular markers for many of the chromosomal regions, genes, and haplotypes that provide the genetic influences on drug abuse vulnerability that are documented in classical genetic studies. A second hypothesis is that the initial results of molecular genetic studies obtained to date support the assumption that polygenic genetic architecture underlies human addiction vulnerability without overwhelming genetic heterogeneity. The common disease–common allele hypothesis is thus supported by current data. A remarkable finding that emerges from these studies is the possibility that much addiction genetics may well reveal differences between the brains of individuals who are predisposed to addiction and those who are not, in ways that are likely to be relevant for memory-like aspects of human addiction vulnerability (Uhl 2004a,b).

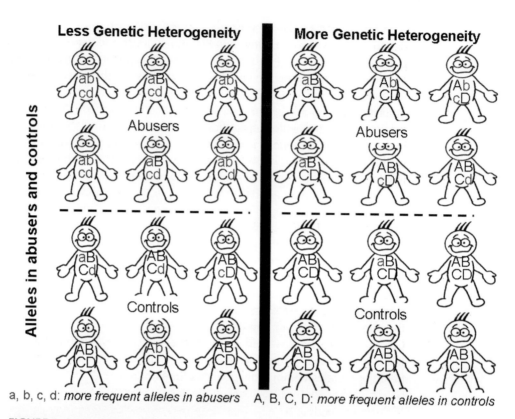

FIGURE 1. Contrasting models that might underlie low (*left panels*) and high (*right panels*) genetic heterogeneity in addictions. Alleles shown in lowercase letters are addiction predisposing. Alternate alleles displayed in uppercase letters were not addiction predisposing. With low genetic heterogeneity, "Abusers" (*top panels*) would express more addiction vulnerability alleles that were present at higher frequency in addicts (and in nonaddicts). With higher genetic heterogeneity, fewer addiction-predisposing alleles would be found in each abuser and addiction vulnerability allele frequencies would be lower.

MOLECULAR GENETIC GENOME SCANS

Remarkable Convergence of Results

One way to present the current state of the art for addiction molecular genetics is to examine data obtained from genome scans using linkage and association scans, with attention to the number of times that they have been reported. Linkage-based studies examine DNA from members of families, studying the ways in which allelic DNA markers and disease move through pedigrees. Association approaches examine the ways in which allelic DNA markers and disease move through unrelated members of the population. Association studies compare frequencies at genomic markers in affected individuals with those found in unaffected individuals.

We can best test the fit between prior and subsequent study results when we compare data from the largest possible number of reasonably powered studies of addictions. The greatest power for such approaches thus comes from studies of addictions to multiple addictive substances, both legal and illegal. In this setting, convergent molecular genetic findings are likely to be found at loci that harbor variants that influence (1) core vulnerabilities to (many) addictions and (2) actions of substances whose early use may provide gateways to use of other substances. It is important to note that searches for convergence in the way that I undertake in this chapter may well underestimate the frequencies of gene variants that provide specific influences on vulnerabilities to just one substance, especially when dependence on that substance is not a gateway to use of others.

Genome-scanning Studies (1998–2000)

Linkage-based genome scans for legal addictions were reported by Long et al. (1998; 172 sibpairs, 517 markers), Reich et al. (1998; 225 sibpairs, 291 markers), Foroud et al. (2000; 266 additional sibpairs, 351 markers), and Straub et al. (1999; 391 sibpairs; 451 markers). These studies used affected sibpairs from U.S. and New Zealand samples of several discrete ethnicities, including pedigrees collected at several U.S. sites as parts of the Collaborative Study on the Genetics of Alcoholism (COGA). They identified a number of loci with nominally significant log of odds ratio (LOD) scores. Unfortunately, there was little apparent agreement in the nominal results that these investigators reported. Even the two samples that Straub and colleagues reported, taken from New Zealand and Richmond, Virginia (U.S.), appeared to manifest little agreement. None of these data sets appeared to support substantial convergence of human addiction vulnerability gene-finding efforts. None appeared to support the concept that there were any major gene effects, or even moderate gene effects, in addiction vulnerability.

Association-based Study and Initial Convergent Data (2001–2002)

We reported the first association-based scan for illegal addictions in two different samples of unrelated abusers and controls (Uhl et al. 2001, 2002). We used 1494 SNPs and more than 1000 research volunteers with relatively extreme substance use/dependence phenotypes. When we modeled the power of this approach, we found that it was modest and likely to detect some, but not most, of the allelic variants that might underlie addiction.

We identified 41 markers for allelic variants that were associated with substance abuse vulnerability in both Caucasian and African-American research volunteer samples (Uhl et al. 2001). When we examined the chromosomal distribution of the 104 markers that included these 41 SNPs and the chromosomal markers linked to substance abuse vulnerability in one of the previously reported linkage studies, we found that 34 of these nominally positive markers clustered into 15 relatively small (+/– 5 million base pairs [Mb]) chromosomal regions (Uhl et al. 2001). Put another way, each of these rSA (reproducible substance abuse vulnerability) regions displayed nominally significant linkage and association findings from multiple, different studies. Modeling studies suggested that this clustering was unlikely to be the result of chance alone (Uhl et al. 2002); these rSAs thus represent convergent genome-scanning results.

Linkage and Genome-scanning Data (2002–2005)

The results of these early studies can be compared in turn with data from subsequent linkage and association studies. Stallings and co-workers reported linkage studies of polysubstance abuse phenotypes in affected sibpairs in which the proband had conduct disorder (Stallings et al. 2002). Gelernter and colleagues reported linkage to smoking in pedigrees in which the proband was identified by anxiety or panic disorder phenotypes (Gelernter 2002). Li and colleagues and Bergen and colleagues studied linkage to smoking (smoking rate) and alcohol (alcohol intake or maximal numbers of drinks) phenotypes in members of the Framingham Heart Study whose data were made available through the Genetics Analysis Workshop (GAW) (Bergen et al. 2003; Li et al. 2003; Ma et al. 2003). Several COGA investigators extended initial COGA linkage results to include COGA Wave II samples obtained from smaller pedigrees (Beirut et al. 2004). In addition, COGA investigators have reported association results from studies of members of COGA pedigrees (Porjesz et al. 2005).

These newer observations displayed substantial convergence with previously reported data. The strongest finding of Stallings et al. (2003) linkage at D3S1614 fits with the previously defined rSA2 region on chromosome 3. Four of the six strongest findings from this study lie within rSA regions; another is close. The strongest linkage finding of Gelernter et al. (2004) to the simple sequence length polymorphism D11S4046 fits with the previously defined rSA10 region of chromosome 11. Three of seven strongest findings from this report lie in rSA regions, and an additional observation is close. Five of the seven strongest findings of Wang et al. (2005) for smoking rate are within rSA regions, with one other finding close. Two of three strongest findings of Ma et al. (2003) for alcohol intake in the period 1970–1971 fit with rSA regions, with one other finding close. Eight of 16 strongest findings of Bergen et al. (2003) examining maximal rate of alcohol consumption fit with rSA regions.

Association Genome-scanning Data (2002–2005)

Results from 11,500 SNP arrays that provide the equivalent of about one-half billion person genotypes appear to buttress and extend these results. SNPs that lie in several of the previously defined rSA regions again display abuser-control differences anticipated if allelic variants that alter substance abuse vulnerability lie on blocks of restricted haplotype diversity marked by the SNPs that we have used in both initial and follow-up studies (Uhl et al. 2001; Liu et al. 2005). These studies identify 38 SNPs whose allelic frequencies differ between heavy polysubstance abusers and control individuals in both European- and African-American samples using even more stringent statistical criteria than those used for the 1494 SNP array analyses. When compared with data from other linkage studies, these results help identify a total of 36 rSA loci with support from multiple linkage and association studies (Table 1 and references therein).

The increasing resolution of association genome scanning with increasingly higher densities of markers allows some reasonable guesses about genes near the reproducibly

TABLE 1. Convergent Data That Forms rSA "Reproducible Substance Abuse Vulnerability" Chromosomal Regions Defined by Markers Linked or Associated with Addiction Vulnerability

rSA	Marker	Chromosome	Location (kbp)	rSA	Marker	Chromosome	Location (kbp)
rSA21	rs3012	1	34743		D9S910	9	88000
	D1S1598	1	39755	rSA31	D9S287	9	93845
rSA22	D1S1588	1	91686		D9S910	9	96500
	rs4492636	1	94541		D9S1690	9	99480
rSA23	GGAA5F09	1	159449	rSA32	D9S261	9	106215
	rs1553695	1	160359		D9S1677	9	107317
rSA24	D1S549	1	216032	rSA9	D10S1239	10	102860
	rs0664574	1	216096		D10S2469	10	106310
rSA25	rs0934993	2	4450		rs17078	10	106317
	D2S319	2	5036	rSA10	D11S1984	11	1533
rSA16	rs2026	2	234667		D11S4046	11	1927
	D2S338	2	237522		D11S4146	11	3706
rSA1	rs17207	3	26492	rSA19	D11S902	11	17452
	rs3278	3	27450		D11S2368	11	19245
	D3S2432	3	32137	rSA11	rs3459	11	29083
rSA17	D3S1279	3	152346		D11S1392	11	34604
	D3S1746	3	153050	rSA33	rs1792525	11	56492
rSA2	D3S1763	3	168560		D11S1985	11	58271
	D3S1614	3	169531		D11S4191	11	59775
	rs2917	3	172361	rSA13	D13S762	13	88722
	D3S3053	3	172808		rs1414291	13	89236
rSA26	rs2215267	4	17470		rs16931	13	90538
	D4S2633	4	18521	rSA14	D13S895	13	109100
rSA3	D4S2632	4	35601		D13S285	13	110743
	D4S2382	4	39947		rs2042	13	111559
	D4S174	4	40749	rSA34	D15S217	15	25676
	D4S1627	4	44094		D15S165	15	28976
	rs2709	4	44622		rs0735612	15	31792
rSA4	D4S1645	4	61986	rSA35	D15S816	15	92749
	D4S244	4	65494		D15S642	15	96307
rSA6	D4S1647	4	99893	rSA36	D16S675	16	8101
	rs17092	4	101541		GATA5H07	16	10300
	rs1395475	4	102573	rSA37	D16S539	16	77153
	ADH3	4	103210		D16S684	16	78074
	D4S2457	4	107037		D16S422	16	82691
	D4S3256	4	108344		D16S539	16	86167
	D4S3240	4	110253	rSA20	rs1391	17	64532
rSA27	D7S1793	7	48296		D17S2059	17	69098
	D7S1818	7	49133		rs2019877	17	72357
	rs1917288	7	51319		D17S1535	17	73159
rSA28	D7S2204	7	77772	rSA38	D18S844	18	72481
	rs1406159	7	78481		rs1942578	18	73319
rSA29	D7S1804	7	131247	rSA39	D20S107	20	39567
	rs1362985	7	133448		D21S2055	20	40111
rSA8	rs3410	9	27316		D20S481	20	44453
	D9S319	9	29549	rSA40	D22S689	22	25843
rSA30	D9S1120	9	83548		rs0471216	22	25935
	D9S257	9	85747	rSA15	rs0726396	X	148841
	D9S283	9	87905		DXS8061	X	150639
					rs3256	X	151048

Repositioning markers eliminates the previously defined rSA5, whereas linking rSA6 and rSA18 eliminates the previously defined rSA18 (for details, see Uhl et al. 2002; Liu et al. 2005).

positive SNPs that are associated with addiction vulnerability. These nearby genes include nerve cell adhesion molecule genes, an alcohol/acetaldehyde dehydrogenase gene cluster, and genes implicated in cellular signaling, gene regulation, development, and Mendelian disorders.

Association in the chromosome 4 alcohol/acetaldehyde dehydrogenase gene cluster is of particular interest. The allelic variations that have been most strongly associated with differential vulnerability to addiction are the striking variants in the alcohol (ADH) and acetaldehyde dehydrogenase (ALDH) genes that produce flushing when alcohol is consumed. Individuals with these striking ADH/ALDH variants are protected against alcoholism, especially in the Asian populations in which these variants are common (Yamashita et al. 1990). However, the findings from this study are also consistent with the idea that more subtle allelic variants in these same genes and in nearby genes in this same chromosomal region may produce the convergent association and linkage findings identified at this chromosomal locus in both European- and African-American samples.

Current Status

Identifying the genomic regions that contain addiction vulnerability alleles is an ongoing task. It is nevertheless gratifying that more recent studies have provided confirmation of several previously nominated loci. It is also gratifying that several previously identified nominally positive markers are now positioned closer to one another as the precision of the genomic backbone on which they have been localized has improved.

METHODOLOGICAL CONSIDERATIONS: CAVEATS AND LIMITATIONS

Although it is unlikely that *all* of the convergent data presented here are based on chance, it is also quite likely that these results include both false-positive and false-negative results. The simulation studies that we performed in 2001 indicated that the chance that at least one, two, three, or four of the 15 initially defined rSA regions were identified by chance were 0.78, 0.42, 0.17, 0.06, and 0.01, respectively, for example (Uhl et al. 2002). False-negative results are also quite likely, because no study performed to date has power sufficient to detect all of the allelic variants that contribute to addiction if we assume underlying polygenic genetic architecture and at least some genetic heterogeneity.

Comparing the results from "top down" genome-scanning approaches with data from "bottom up" candidate gene association studies also provides evidence for likely false-negative results from genome-scanning studies. For example, we, and a number of other investigators, have identified association and linkage of allelic variants of the catechol-*O*-methyltransferase (COMT) gene with addiction vulnerability (Smith et al. 1992; Persico et al. 1996; Vandenbergh et al. 1997, 2000). Although the COMT gene is located on chromosome 22 between 18.30 and 18.33 Mb, 10K array SNP markers that span this region do not display association signals with addiction vulnerability. Conceivably, the gaps between these markers explain this apparently false-negative association result, although they do not explain the failure of linkage-based genome-scanning studies to identify this locus. Ongoing studies with higher marker densities will help to improve the power of association-based genome scanning and reduce false-negative results of this sort.

WORKING HYPOTHESES CONCERNING THE GENETIC ARCHITECTURE OF SUBSTANCE ABUSE

Assembling the markers now linked to or associated with vulnerability to substance abuse now provides substantial support for polygenic inheritance for substance abuse vulnerability. This assembly of data also supports the idea that common allelic variants likely to contribute to vulnerability to abuse of several substances can be found in individuals from several different racial and ethnic groups, supporting the idea that many addiction vulnerability variants may have arisen relatively early in the process of evolving modern human genomes.

Although common allelic variants could thus have significant roles in vulnerabilities to abuse of multiple substances, each allelic variant could provide a distinctive pattern of influences and manifest effects on different aspects of drug action and/or addiction vulnerability. A number of such variants appear likely to also alter vulnerability to the other traits and behavioral disorders that often cooccur with substance disorders. However, other variants might be more selective. Alleles that alter drug metabolism, for example, might produce a straightforward shift of the dose-effect relationships for that drug without substantially altering brain chemistries or vulnerabilities to comorbidity conditions. Additional studies will be able to better identify the ways in which each allelic variant changes specific features of addiction vulnerability. Such studies can also help us to better understand the ways in which genetic and environmental influences interact.

FINE-MAPPING STUDIES

Increased understanding of addictions can come from molecular genetic studies that find addiction-linked and -associated genomic markers. However, better understanding of addictions requires identification of specific genes and their effects on addiction vulnerability mechanisms and behaviors.

Fine mapping is the process by which we can move from a linked or associated genomic region to a specific addiction-associated gene and specific addiction-associated haplotype. Subsequent molecular and neurobiological studies can provide validation of the molecular genetic observations and support specific cellular and behavioral mechanisms whereby the gene's variants can influence addiction phenotypes.

We performed fine-mapping studies and sought ancillary evidence for the validity of these molecular genetic observations at a number of chromosomal loci. Perhaps the best current example of such a progression from molecular genetic to molecular-biological and behavioral evidence comes from studies of the neuronal cell adhesion molecule (NrCAM) locus on the middle of chromosome 7.

We initially identified NrCAM through positional cloning studies that sought to follow up in a chromosome 7 region in which our initial association findings converged with those from linkage studies of COGA samples (Ishiguro et al. 2005). In the initial stages of this fine mapping, we were led to focus on the NrCAM gene because other investigators in this laboratory had previously cloned this gene using a subtractive hybridization/PCR (polymerase chain reaction) differential display paradigm based on regulation of its striatal expression levels after morphine administration. When we examined NrCAM expression patterns in brain, we also found that they were richly expressed in dopaminergic neurons that modulate reward and memory-related brain areas including striatum and hippocampus.

We identified increasingly stronger evidence for association between NrCAM genomic haplotypes and addiction vulnerability in four different samples, with the best evidence for haplotypes at the 5′ end of this gene. These haplotypes appeared to correlate well with the level of NrCAM expression in postmortem human brain samples. We thus thought that NrCAM knockout (KO) mice, especially heterozygote mice that express half of wild-type levels of NrCAM, might be good models for the human haplotype variants that we had cloned. When Grumet et al. (1997) provided us with these heterozygous KO mice, we were able to identify a striking reduction in the conditioned place preference for opiates and psychostimulants (Ishiguro et al. 2005). Because mouse-conditioned place preference for drugs is almost invariably a good predictor of human abuse liability, these mouse observations provide strong support for the idea that reduced NrCAM expression in humans is likely to lead to altered drug preference and presumably to altered addiction vulnerability.

APPLICATIONS TO SUBSTANCE ABUSING POPULATIONS

We recently estimated the large size of complex genetic contributions to addictions in U.S. society (Uhl and Grow 2004). This sizable impact should motivate efforts to identify allelic variants that contribute to addiction vulnerability. Conversely, results obtained to date indicate that no single allelic variant is likely to contribute a large fraction of this vulnerability in individuals from several different populations.

We believe that knowledge about which addiction vulnerability allelic variants are present in vulnerable individuals will improve our ability to individualize addiction treatments, bringing the power of pharmacogenetics to the complex and difficult task of treating these complex problems. We believe that such knowledge could even help to improve matching between individuals and the prevention strategies most likely to work for them. Individual and societal suffering could be substantially relieved by better understanding the complex human processes of human addiction through careful application of complex human genetic approaches. However, as we note below, application of complex genetics to population samples requires careful attention to minimizing risks of genotyping individuals, confidentiality, costs, genetic discrimination, and other attendant issues that are beyond the scope of this chapter.

SUMMARY AND CONCLUSIONS

Molecular genetic findings obtained to date are quite consistent with the basic messages that come from classical genetic studies of addictions—that there are complex genetic contributions to substance abuse that include factors that influence abuse of multiple different abused substances in individuals from several different underlying populations. Some of the variants that appear to predispose to addiction vulnerability lie at regions that may well have been predicted in advance on the basis of prior understanding of addictions, such as the chromosome 4 AHD/ALDH gene cluster. Most of the loci that appear to harbor reproducible substance abuser vulnerability loci, however, appear to harbor genes that would not have been readily predicted to contain addiction vulnerability alleles a priori. "Cell adhesion" molecule genes, in particular, appear to be dramatically overrepresented at these loci, in comparison to their frequency in the genome. Results to date support a model for the molecular genetic architecture of human addiction vulnerability that posits polygenic effects of common allelic variants with many effects on gene regulation.

GAPS AND FUTURE DIRECTIONS

The rSA regions documented here are an incomplete list of the addiction vulnerability loci that will be eventually elucidated. Only those allelic variants that provide enough influence on abuse of so many different substances in so many populations that they can be jointly detected in each of several studies that have used different methodologies and different clinical samples are likely to be listed here. However, both the brisk pace of ongoing work in a number of laboratories and the continuing progress in techniques for association-based genome scanning suggest that increasingly more addiction vulnerability haplotypes will be elucidated in the near term.

The number of studies that are currently in progress and the large numbers of comparisons being made in attempts to identify true addiction vulnerability loci, each with modest effect, also appear to suggest the need for caution against the large number of false-positive results likely to emerge, at least initially, from such efforts. More independent replications and positive findings in animal models in which single-gene effects can be studied in isolation each add to confidence in positive results.

Substance abusers have diagnoses of depression, antisocial personality disorder/conduct disorder, and adult residua of attention deficit/hyperactivity disorder more frequently than control populations (Regier et al. 1990; Rowland et al. 2002). Each of these diagnoses is likely to display heritable components. Each also appears to be comorbid with substance abuse more than with the other comorbid diagnoses. Thus, examination of the distribution of addiction vulnerability alleles in individuals with these diagnoses appears likely to help the understanding of these comorbid complex disorders as well.

These kinds of efforts may also help to elucidate heterogeneities in addicts. Genetic contributions to addiction in individuals with antisocial personality disorder may well differ from those found in addicts with depression, for example. Groups of individuals with apparently similar addictions but with different psychiatric comorbidities and with different constellations of associated genetic markers would provide quite plausible evidence for clinical molecular genetic heterogeneity of substance abuse.

Personality types provide another perspective. There are substantial heritabilities for several of the principal currently accepted personality domains (Ball 2001; Bouchard and Loehlin 2001; Livesley 2005). Individuals who score at the extremes of several tests of heritable personality types display especially enhanced risks of addiction. High "neuroticism" scores are found in substance abusers (Ball 2001; Carter et al. 2001), whereas this trait shows substantial heritability in twin studies (Jang et al. 1996). Some of the same allelic variants that predispose to substance abuse vulnerability are thus likely to predispose to neuroticism and perhaps to other personality types.

Individual differences in performance on memory tests and on tests of cognitive function that include executive function are substantially heritable (Wright et al. 2001; Fein et al. 2002a,b; Sim et al. 2002; Simon et al. 2002). The volumes of several brain structures important for these mnemonic and higher-order functions are also substantially heritable (Sullivan et al. 2001; Carmelli et al. 2002; Geschwind et al. 2002). Substance abusers that differ from one another on the basis of executive or other cognitive functions provide yet another way in which genetically heterogeneous addict subgroups might be defined. Furthermore, because the most heritable phenomena that accompany later stages of progression toward drug dependence appear likely to involve mnemonic processes, drug memory systems are thus especially strong candidates for genetically mediated individual differences in substance abuse behaviors. It appears likely that some addiction vulnerability alleles will also contribute to individual differences in cognitive functions, including executive functions and the kinds of habitual memory systems that appear to be engaged in drug-dependent individuals.

The future direction of addiction vulnerability genetics may thus be to inform thinking about addiction, to inform decisions about prevention and treatment matching for the appropriate individuals who, based on their genetic makeup, will be most likely to respond to the specific prevention or treatment interventions, and to enhance understanding of psychiatric comorbidities, personality types, mnemonic capacities, and cognitive strategies including executive functions.

ACKNOWLEDGMENTS

I thank the following investigators for help with personal studies cited in the chapter: Qing-Rong Liu, Daniel Naiman, Judith Hess, Bruce O'Hara, Antonio Persico, Donna Walther, Fely Carillo, Brenda Campbell, Linda Kahler, Fred Snyder, Carlo Contoreggi, Larry Rodriguez, Robert Grow, David Gorelick, Zhicheng Lin, Andrew Shapiro, and Leslie Cope and for financial support from the National Institute on Drug Abuse–Intramural Research Program.

REFERENCES

Ball S.A. 2001. The big five, alternative five, and seven personality dimensions: Validity in substance-dependent patients. In *Personality disorders and the five-factor model of personality disorder*, 2nd edition (ed. P.T. Costa and T.A. Widiger). American Psychological Assoiciation, Washington, D.C.

Bergen A.W., Yang X.R., Bai Y., Beerman M.B., Goldstein A.M., and Goldin L.R. 2003. Genomic regions linked to alcohol consumption in the Framingham Heart Study. *BMC Genet.* (suppl. 1) **4:** S101.

Bierut L.J., Rice J.P., Goate A., Hinrichs A.L., Saccone N.L., Foroud T., Edenberg H.J., Cloninger C.R., Begleiter H., Conneally P.M., Crowe R.R., Hesselbrock V., Li T.K., Nurnberger J.I., Jr., Porjesz B., Schuckit M.A., and Reich T. 2004. A genomic scan for habitual smoking in families of alcoholics: Common and specific genetic factors in substance dependence. *Am. J. Med. Genet. A* **124:** 19–27.

Bouchard T.J. and Loehlin J.C. 2001. Genes, evolution, and personality. *Behav. Genet.* **31:** 243–273.

Carmelli D., Swan G.E., DeCarli C., and Reed T. 2002. Quantitative genetic modeling of regional brain volumes and cognitive performance in older male twins. *Biol. Psychol.* **61:** 139–155.

Carter J.A., Herbst J.H., Stoller K.B., King V.L., Kidorf M.S., Costa P.T., Jr., and Brooner R.K. 2001. Short-term stability of NEO-PI-R personality trait scores in opioid-dependent outpatients. *Psychol. Addict. Behav.* **15:** 255–260.

Fein G., Di Sclafani V., and Meyerhoff D.J. 2002a. Prefrontal cortical volume reduction associated with frontal cortex function deficit in 6-week abstinent crack-cocaine dependent men. *Drug Alcohol Depend.* **68:** 87–93.

Fein G., Di Sclafani V., Cardenas V.A., Goldmann H., Tolou-Shams M., and Meyerhoff D.J. 2002b. Cortical gray matter loss in treatment-naive alcohol dependent individuals. *Alcohol: Clin. Exp. Res.* **26:** 558–564.

Foroud T., Edenberg H.J., Goate A., Rice J., Flury L., Koller D.L., Bierut L.J., Conneally P.M., Nurnberger J.I., Bucholz K.K., Li T.K., Hesselbrock V., Crowe R., Schuckit M., Porjesz B., Begleiter H., and Reich T. 2000. Alcoholism susceptibility loci: Confirmation studies in a replicate sample and further mapping. *Alcohol: Clin. Exp. Res.* **24:** 933–945.

Gelernter J. 2002. A genome scan for nicotine addiction in panic disorder clinic attendees. In *American Society for Human Genetics* satellite.

Gelernter J., Liu X., Hesselbrock V., Page G.P., Goddard A., and Zhang H. 2004. Results of a genomewide linkage scan: Support for chromosomes 9 and 11 loci increasing risk for cigarette smoking. *Am. J. Med. Genet. B Neuropsychiatr. Genet.* **128:** 94–101.

Geschwind D.H., Miller B.L., DeCarli C., and Carmelli D. 2002. Heritability of lobar brain volumes in twins supports genetic models of cerebral laterality and handedness. *Proc. Natl. Acad. Sci.* **99:** 3176–3181.

Grumet M. 1997. Nr-CAM: A cell adhesion molecule with ligand and receptor functions. *Cell Tissue Res.* **290:** 423–428.

Ishiguro H., Liu Q.-R., Gong J.-P., Hall S., Ujike H., Morales M., Sakurai T., Grumet M., and Uhl G.R. 2005. NrCAM in addiction vulnerability: Positional cloning, drug-regulation, haplotype-specific-expression and altered drug reward in knockout mice. *Neuropharmacology* (in press).

Jang K.L., Livesley W.J., and Vernon P.A. 1996. Heritability of the big five personality dimensions and their facets: A twin study. *J. Pers.* **64:** 577–591.

Kendler K.S. and Prescott C.A. 1998. Cocaine use, abuse and dependence in a population-based sample of female twins. *Br. J. Psychiatry* **173:** 345–350.

Kendler K.S., Thornton L.M., and Pedersen N.L. 2000. Tobacco consumption in Swedish twins reared apart and reared together. *Arch. Gen. Psychiatry* **57:** 886–892.

Kendler K.S., Prescott C.A., Neale M.C., and Pedersen N.L. 1997. Temperance board registration for alcohol abuse in a national sample of Swedish male twins, born 1902 to 1949. *Arch. Gen. Psychiatry* **54:** 178–184.

Kendler K.S., Karkowski L.M., Corey L.A., Prescott C.A., and Neale M.C. 1999. Genetic and environmental risk factors in the aetiology of illicit drug initiation and subsequent misuse in women. *Br. J. Psychiatry* **175:** 351–356.

Li M.D., Ma J.Z., Cheng R., Dupont R.T., Williams N.J., Crews K.M., Payne T.J., and Elston R.C. 2003. A genome-wide scan to identify loci for smoking rate in the Framingham Heart Study population. *BMC Genet.* (suppl. 1) **4:** S103.

Liu Q.-R., Drgon T., Walther D., Johnson C., Poleskaya O., Hess J., and Uhl G.R. 2005. Pooled association genome scanning: Validation and use to identify addiction vulnerability loci in two samples. *Proc. Natl. Acad. Sci.* **102:** 11864–11869.

Livesley W.J. 2005. Behavioral and molecular genetic contributions to a dimensional classification of personality disorder. *J. Pers. Disord.* **19:** 131–155.

Long J.C., Knowler W.C., Hanson R.L., Robin R.W., Urbanek M., Moore E., Bennett P.H., and Goldman D. 1998. Evidence for genetic linkage to alcohol dependence on chromosomes 4 and 11 from an autosome-wide scan in an American Indian population. *Am. J. Med. Genet.* **81:** 216–221.

Ma J.Z., Zhang D., Dupont R.T., Dockter M., Elston R.C., and Li M.D. 2003. Mapping susceptibility loci for alcohol consumption using number of grams of alcohol consumed per day as a phenotype measure. *BMC Genet.* (suppl. 1) **4:** S104.

Persico A.M., Bird G., Gabbay F.H., and Uhl G.R. 1996. D2 dopamine receptor gene TaqI A1 and B1 restriction fragment length polymorphisms: Enhanced frequencies in psychostimulant-preferring polysubstance abusers. *Biol. Psychiatry* **40:** 776–784.

Pickens R.W., Svikis D.S., McGue M., Lykken D.T., Heston L.L., and Clayton P.J. 1991. Heterogeneity in the inheritance of alcoholism. A study of male and female twins. *Arch. Gen. Psychiatry* **48:** 19–28.

Porjesz B., Rangaswamy M., Kamarajan C., Jones K.A., Padmanabhapillai A., and Begleiter H. 2005. The utility of neurophysiological markers in the study of alcoholism. *Clin. Neurophysiol.* **116:** 993–1018.

Regier D.A., Farmer M.E., Rae D.S., Locke B.Z., Keith S.J., Judd L.L., and Goodwin F.K. 1990. Comorbidity of mental disorders with alcohol and other drug abuse. Results from the Epidemiologic Catchment Area (ECA) Study. *J. Am Med. Assoc.* **264:** 2511–2518.

Reich D.E. and Lander E.S. 2001. On the allelic spectrum of human disease. *Trends Genet.* **17:** 502–510.

Reich T., Edenberg H.J., Goate A., Williams J.T., Rice J.P., Van Eerdewegh P., Foroud T, Hesselbrock V., Schuckit M.A., Bucholz K., Porjesz B., Li T.K., Conneally P.M., Nurnberger J.I., Jr., Tischfield J.A., Crowe R.R., Cloninger C.R., Wu W., Shears S., Carr K., Crose C., Willig C., and Begleiter H. 1998. Genome-wide search for genes affecting the risk for alcohol dependence. *Am. J. Med. Genet.* **81:** 207–215.

Rowland A.S., Lesesne C.A., and Abramowitz A.J. 2002. The epidemiology of attention-deficit/hyperactivity disorder (ADHD): A public health view. *Ment. Retard. Dev. Disabil. Res. Rev.* **8:** 162–170.

Schuckit M.A., Smith T.L., Kalmijn J., Tsuang J., Hesselbrock V., and Bucholz K. 2000. Response to alcohol in daughters of alcoholics: A pilot study and a comparison with sons of alcoholics. *Alcohol Alcohol.* **35:** 42–248.

Sim T., Simon S.L., Domier C.P., Richardson K., Rawson R.A., and Ling W. 2002. Cognitive deficits among methamphetamine users with attention deficit hyperactivity disorder symptomatology. *J. Addict. Dis.* **21:** 75–89.

Simon S.L., Domier C.P., Sim T., Richardson K., Rawson R.A., and Ling W. 2002. Cognitive performance of current methamphetamine and cocaine abusers. *J. Addict. Dis.* **21:** 61–74.

Smith S.S., O'Hara B.F., Persico A.M., Gorelick D.A., Newlin D.B., Vlahov D., Solomon L., Pickens R., and Uhl G.R. 1992. Genetic vulnerability to drug abuse. The D2 dopamine receptor Taq I B1 restriction fragment length polymorphism appears more frequently in polysubstance abusers. *Arch. Gen. Psychiatry* **49:** 723–727.

Stallings M.C., Hewitt J.K., Krauter K.S., Lessem J.M., Mikulich S.K., Rhee S.H., Smolen A., Young S.E., and Crowley T.J. 2002. A genome-wide search for QTLs influencing substance dependence vulnerabity. In the College on Problems of Drug Dependence.

Stallings M.C., Corley R.P., Hewitt J.K., Krauter K.S., Lessem J.M., Mikulich S.K., Rhee S.H., Smolen A., Young S.E., and Crowley T.J. 2003. A genome-wide search for quantitative trait loci influencing substance dependence vulnerability in adolescence. *Drug Alcohol Depend.* **70:** 295–307.

Straub R.E., Sullivan P.F., Ma Y., Myakishev M.V., Harris-Kerr C., Wormley B., Kadambi B., Sadek H., Silverman M.A., Webb B.T., Neale M.C., Bulik C.M., Joyce P.R., and Kendler K.S. 1999. Susceptibility genes for nicotine dependence: A genome scan and followup in an independent sample suggest that regions on chromosomes 2, 4, 10, 16, 17 and 18 merit further study. *Mol. Psychiatry* **4:** 129–144.

Sullivan E.V., Pfefferbaum A., Swan G.E., and Carmelli D. 2001. Heritability of hippocampal size in elderly twin men: Equivalent influence from genes and environment. *Hippocampus* **11:** 754–762.

True W.R., Heath A.C., Scherrer J..F, Xian H., Lin N., Eisen S.A., Lyons M.J., Goldberg J., and Tsuang M.T. 1999a. Interrelationship of genetic and environmental influences on conduct disorder and alcohol and marijuana dependence symptoms. *Am. J. Med. Genet.* **88:** 391–397.

True W.R., Xian H., Scherrer J.F., Madden P.A., Bucholz K.K., Heath A.C., Eisen S.A., Lyons M.J., Goldberg J., and Tsuang M. 1999b. Common genetic vulnerability for nicotine and alcohol dependence in men. *Arch. Gen. Psychiatry* **56:** 655–661.

Tsuang M.T., Lyons M.J., Harley R.M., Xian H., Eisen S., Goldberg J., True W.R., and Faraone S.V. 1999. Genetic and environmental influences on transitions in drug use. *Behav. Genet.* **29:** 473–479.

Tsuang M.T., Lyons M.J., Eisen S.A., Goldberg J., True W., Lin N., Meyer J.M., Toomey R., Faraone S.V., and Eaves L. 1996. Genetic influences on DSM-III-R drug abuse and dependence: A study of 3,372 twin pairs. *Am. J. Med. Genet.* **67:** 473–477.

Tsuang M.T., Lyons M.J., Meyer J.M., Doyle T., Eisen S.A., Goldberg J., True W., Lin N., Toomey R., and Eaves L. 1998. Co-occurrence of abuse of different drugs in men: The role of drug-specific and shared vulnerabilities. *Arch. Gen. Psychiatry* **55:** 967–972.

Uhl G.R. 1999. Molecular genetics of substance abuse vulnerability: A current approach. *Neuropsychopharmacology* **20:** 3–9.

———. 2004a. Molecular genetics of substance abuse vulnerability: Remarkable recent convergence of genome scan results. *Ann. N.Y. Acad. Sci.* **1025:** 1–13.

———. 2004b. Molecular genetic underpinnings of human substance abuse vulnerability: Likely contributions to understanding addiction as a mnemonic process. *Neuropharmacology* (suppl. 1) **47:** 140–147.

Uhl G.R. and Grow R.W. 2004. The burden of complex genetics in brain disorders. *Arch. Gen. Psychiatry* **61:** 223–229.

Uhl G.R., Liu Q.-R., and Naiman D. 2002. Substance abuse vulnerability loci: Converging genome scanning data. *Trends Genet.* **18:** 420–425.

Uhl G.R., Elmer G.I., Labuda M.C., and Pickens R.W. 1995. Genetic influences in drug abuse. In *Psychopharmacology: The fourth generation of progress* (ed. F.E. Bloom and D.J. Kupfer), pp. 1793–1806. Raven Press, New York.

Uhl G.R., Liu Q.-R., Walther D., Hess J., and Naiman D. 2001. Polysubstance abuse-vulnerability genes: Genome scans for association, using 1,004 subjects and 1,494 single-nucleotide polymorphisms. *Am. J. Hum. Genet.* **69:** 1290–1300.

Vandenbergh D.J., Rodriguez L.A., Miller I.T., Uhl G.R., and Lachman H.M. 1997. High-activity catechol-*O*-methyltransferase allele is more prevalent in polysubstance abusers. *Am. J. Med. Genet.* **74:** 439–442.

Vandenbergh D.J., Rodriguez L.A., Hivert E., Schiller J.H., Villareal G., Pugh E.W., Lachman H., and Uhl G.R. 2000. Long forms of the dopamine receptor (DRD4) gene VNTR are more prevalent in substance abusers: No interaction with functional alleles of the catechol-*O*-methyltransferase (COMT) gene. *Am. J. Med. Genet.* **96:** 678–683.

Wang D., Ma J.Z., and Li M.D. 2005. Mapping and verification of susceptibility loci for smoking quantity using permutation linkage analysis. *Pharmacogenomics J.* **5:** 166–172.

Wright M., De Geus E., Ando J., Luciano M., Posthuma D., Ono Y., Hansell N., Van Baal C., Hiraishi K., Hasegawa T., Smith G., Geffen G., Geffen L., Kanba S., Miyake A., Martin N., and Boomsma D. 2001. Genetics of cognition: Outline of a collaborative twin study. *Twin Res.* **4:** 48–56.

Yamashita I., Ohmori T., Koyama T., Mori H., Boyadjive S., Kielholz P., Gastpar M., Moussaoui D., Bouzekraoui M., and Sethi B.B. 1990. Biological study of alcohol dependence syndrome with reference to ethnic difference: Report of a WHO collaborative study. *Jpn. J. Psychiatry Neurol.* **44:** 79–84.

Catechol-*O*-Methyltransferase Genotype, Intermediate Phenotype, and Psychiatric Disorders

3

Ke Xu and David Goldman

Laboratory of Neurogenetics, National Institute on Alcohol Abuse and Alcoholism, National Institutes of Health, Rockville, Maryland 20852

ABSTRACT

Catechol-*O*-methyltransferase (COMT) metabolizes dopamine and other catecholamine neurotransmitters in the human brain. Variation in *COMT* function appears to have a critical role in cognitive function, especially function of the prefrontal cortex and in emotionality. A functional polymorphism, a valine-to-methionine substitution at codon 158, alters enzyme stability and activity. The *Val158Met* polymorphism is abundant across different populations worldwide. In addition, a *COMT* haplotype is associated with variation in COMT expression at the mRNA level. Two opposite-allele-configuration (yin-yang) haplotypes, one containing *Met* and another containing *Val*, are found across different human populations worldwide, providing evidence for balancing selection. The *Val* allele and haplotype have been linked to lower prefrontal cognitive function and schizophrenia, whereas the *Met* allele contributes to anxiety, diminished pain, and stress resiliency. Linkage of both *Val* and *Met* to addiction has been found. These bidirectional linkages may be congruent with the etiologic heterogeneity of addiction because there is substantial variation among clinical populations in the number of addicted patients with behavioral dyscontrol and the number of patients whose premorbid vulnerability was in the domain of anxiety and stress resiliency.

INTRODUCTION

Alcoholism is a chronic relapsing disease that is influenced by genetic and environmental factors. Only approximately one third of patients not using alcohol remain abstinent, whereas another one third of these patients will have fully relapsed within a year after alcohol withdrawal; the remaining one third will have suffered partial relapse. Prevention and treatment of alcoholism are currently based on diagnoses that rely on signs, symptoms, and clinical histories, rather than on understanding the disorder's underlying origins. What does contemporary research show about the neurobiology of addiction that can serve as a preliminary guide in the search for genes and specific factors that set the stage for this disease?

Like other drug abuse, alcoholism is moderately to highly heritable. Twin, family, and adoption studies have revealed that the genetic contribution to vulnerability in alcoholism is 40–60%. The inherited vulnerability is both substance-specific and nonspecific (Goldman and Bergen 1998). The challenge for the genetics of alcoholism is to identify functional genetic variants that confer vulnerability or protection. However, efforts to identify genetic variation for alcoholism have been hampered by phenotypic heterogeneity and limited understanding of neurobiological mechanism (Enoch et al. 2003a). Intermediate phenotypes, which are likely to lie closer to the fundamental origins of alcoholism, offer more power to identify genetic variants. Intermediate phenotypes themselves may be composites of different etiology factors, but are simpler than the entire complex disease of alcoholism. Intermediate phenotypes for alcoholism include differences in frontal cortical function and behavioral inhibition, individual differences in how people experience rewards, and large differences in anxiety, dysphoria at baseline, and response to stress. Therefore, two clinical domains in alcoholism—cognitive function/behavior dyscontrol and anxiety/stress response—serve as excellent intermediate phenotypes for genetic study of alcoholism.

Evidence from neurobiological, pharmacological, and behavioral studies have demonstrated that dopamine involves the mechanism of reward and reinforcement. Dopamine activity is also associated with executive cognitive function and emotionality. The genetic variations of dopamine receptors, transporters, and enzymes on the dopamine pathway determine dopamine activity in the additive circuitry and alter reward and reinforcement function. Thus, variation of dopamine function among individuals may affect vulnerability to addiction through influencing experience of reward, cognitive function, or anxiety/stress response. Therefore, the genes on the dopamine pathway are candidate genes for genetic study of alcoholism.

The *COMT* gene, discovered by Axelrod and Vesell (1970), metabolizes dopamine and other catecholamine neurotransmitters in human brain. A functional polymorphism in *COMT*, a valine-to-methionine substitution at codon 158 (*Val158Met*), alters enzyme stability and activity (Lachman et al. 1996). This *COMT* functional variant appears to have a critical role in cognitive function in the prefrontal cortex (PFC) (Egan et al. 2001; Malhotra et al. 2002) and anxiety/stress response (Enoch et al. 2003b), which are two intermediate domains of alcoholism. Counterbalancing effects of *Val158* and *Met158* alleles have been observed on cognition and anxiety. The *Val* allele has been linked to lower prefrontal cognitive function and schizophrenia, whereas the *Met* allele contributes to anxiety, diminished pain, and stress resiliency. Both *Val* and *Met* have been linked to addictions.

GOALS OF THE CHAPTER

This chapter provides an overview of *COMT* function and protein structures, functional polymorphism, gene expression in the human brain, and association with behavior, as derived from both animal and human studies. It describes *COMT* linkages to two intermediate phenotypes for alcoholism including prefrontal cognitive function, schizophrenia, and antipsychotic treatment response, as well as its potential role in anxiety and stress resiliency, and points to mechanisms for *COMT* involvement in alcoholism and addiction. The bidirectional linkages of *COMT* to two intermediate phenotypic domains may be congruent with the etiologic heterogeneity of addiction and provide evidence for the "Warriors versus Worriers" model in alcoholism and addiction.

COMT Function, Isoforms, and Protein Structure

In brain tissue, COMT catalyzes transfer of a methyl group from *S*-adenosylmethionine (SAM) to a catechol hydroxyl. COMT converts dopamine to 3-methoxytyramine, norepinephrine to normetanephrine, and epinephrine to metanephrine. COMT also metabolizes catechol estrogens and several neuroactive drugs, including L-Dopa, alphamethyl Dopa, and isoproterenol.

COMT is found in two forms: soluble (S-COMT; 25 kD) (Tilgmann and Kalkkinen 1991) and membrane bound (M-COMT; 30 kD) (Bertocci et al. 1991). The K_m value of S-COMT for dopamine is 10–100 times higher (lower affinity) than that of M-COMT, indicating that M-COMT accounts for most of the dopamine metabolism at the dopamine concentrations found in the mammalian brain (Roth 1992). S-COMT and M-COMT isoforms are encoded by different promoters and differ only in their amino termini (Fig. 1). S-COMT is 50 amino acids shorter than M-COMT in humans and 43 amino acids shorter in rats. S-COMT predominates in peripheral tissues including liver, erythrocytes, leukocytes, and kidney, but M-COMT predominates in the human central nervous system (CNS). In frontal cortex, M-COMT has a major role in degrading dopamine because expression of the dopamine transporter (DAT) is low. However, M-COMT is also expressed elsewhere in brain, for example, in striatum, where it is found in postsynaptic neurons and may act together with the dopamine transporter to terminate the action of dopamine in the synapse.

COMT is Mg^{2+} dependent. The active site of COMT includes the coenzyme-binding motif (SAM) and the catalytic site situated in the vicinity of the Mg^{2+} ion. The methyl transfer from the SAM sulfur to the catechol oxygen is a direct bimolecular transfer. The structure of rat M-COMT has been directly determined by X-ray crystallography, and it has

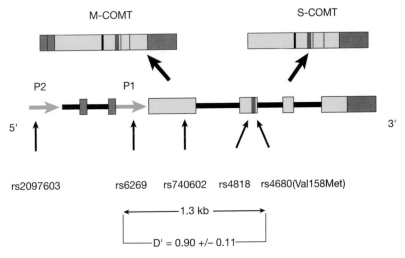

FIGURE 1. *COMT* gene structure with the location of five SNPs for estimation of linkage disequilibrium. Two promoters, P1 and P2, initiate two mRNAs that encode M-COMT and S-COMT isoforms. Coding exons are shown in *green*. A functional polymorphism, a valine-to-methionine substitution at codon 158 (rs4680), alters enzyme stability and activity. Linkage disequilibrium is extensive across the entire gene region and strong from SNP rs6269 to rs4680 (Mean D′ = 0.90 +/– 0.11).

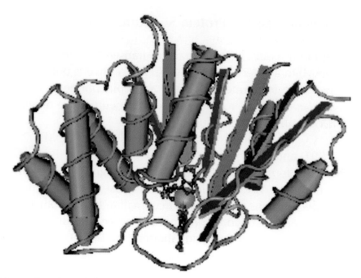

FIGURE 2. Schematic of rat COMT protein structure based on X-ray crystal structure. Seven β strands are shown as *brown* arrows showing directionality (N to C) and seven α helices are shown as *green* cylinders with a cone showing directionality. Random coil structure is indicated by the irregular *blue* line. Magnesium bound to the active site is shown as a *gray* sphere in the center of the structure. In humans, COMT consists of eight α helices and seven β strands. (Adapted from NCBI Entrez's Molecular Modeling Database [MMDB].)

seven α helices and seven β strands, as shown in Figure 2. In the human, M-COMT is composed of 271 amino acids and consists of eight α helices and seven β strands that are probably arranged in parallel (Vidgren et al. 1994).

Metabolism of Dopamine and Noradrenaline Is Drastically Altered in COMT Knockout Mice

In mice lacking *COMT*, increased levels of dopamine were observed in the frontal cortex of male mice but not female mice (Gogos et al. 1998; Huotari et al. 2002). CNS levels of norepinephrine were not altered in these animals. The male-female differences and failure of the *COMT* knockout to lead to consistent increases in catecholamine levels may be a result of compensatory changes in catecholamine function or neuronal architecture in response to the chronic knockout and can also reflect sexual dimorphism in brain function. In a model of anxiety, increased anxiety was seen in female but not male *COMT* knockout animals, whereas increased levels of aggression were observed in (+/–) males but not (–/–) males, data that parallel the observation of increased aggression in humans homozygous for the low-activity *COMT Met158* allele (Gogos et al. 1998). The main finding in the *COMT* knockout mice is that metabolism of both dopamine and noradrenaline is drastically altered in several brain regions: striatum, cortex, and hypothalamus. For noradrenaline, MHPG (3-methoxy 4-hydroxyphenylethylamine glycol) levels drop to undetectable, and DHPG (dihydroxyphenylethylamine glycol) levels dramatically increase in all these regions. For dopamine, HVA (homovanillic acid) levels drop to unde-tectable, and DOPAC (3,4-dihydroxyphenylacetic acid) levels dramatically increase (Huotari et al. 2002). Yet, as noted, the overall effects of the knockout on neurotransmit-ter level are modest and mixed. Overall, these results indicate that COMT maintains steady-state levels of catecholamines in a region-specific fashion and implicates *COMT* in normal emotional and social behaviors, but static measures of neurotransmitter levels do not effectively capture the functional effects of *COMT* genetic variation.

COMT Gene Structure, Polymorphisms, and Functional SNPs

In the human, the *COMT* gene is located on chromosome 22q11.21-q11.23. The gene spans 27 kb and includes six exons (Grossman et al. 1992). Two different promoters ini-tiate synthesis of mRNA as for M-COMT and S-COMT. Genetic variation at *COMT* is high. In addition to *COMT Val158Met*, approximately 60 single nucleotide polymor-phisms (SNPs) have been identified, based on the National Center for Biotechnology Information (NCBI) database. Seven SNPs are potential coding region variants ("mis-sense" SNPs) that encode a different amino acid at the appropriate position in the polypeptide. Eight SNPs are "synonymous" SNPs that do not change the encoded residue at a particular site and are not expected to be functional. Finally, 45 SNPs have been located within introns or noncoding regions.

In 1977, Weinshilboum and Raymond (1977) observed that erythrocyte COMT activity had a bimodal distribution in a randomly selected population. A subgroup of about 25% of the population had lower enzyme activity and a more thermolabile enzyme, indicating that a structural variation in the COMT enzyme was inherited. In 1996, Lachman et al. (1996) identified a G-to-A transition at codon 158 of M-COMT and at codon 108 of S-COMT, resulting in a valine-to-methionine substitution (SNP designation *Val158Met*, SNP database reference rs4680). It was recognized that the *Val* and *Met* alleles are codominant in action; enzyme activity of *Val/Met* heterozygotes is midway between the two homozy-gotes (Weinshilboum et al. 1999). The *Val* allele is more thermostable at body temperature and more active (Scanlon et al. 1979; Spielman and Weinshilboum 1981; Weinshilboum and Dunnette 1981; Lotta et al. 1995; Chen et al. 2004). *Val*-COMT-specific activity is approximately 40% higher than *Met*-COMT enzyme activity at normal body temperature in the dorsolateral prefrontal cortex (DLPFC), a brain region critical for cognitive function. Although the *Val/Val, Val/Met,* and *Met/Met* genotypes predict differences in synaptic cat-echolamine levels, so far, these predicted differences have not been empirically verified in vivo. In addition, as seen in *COMT* knockout mice, neuroadaptive changes could obscure direct effects of C*OMT* on neurotransmitter levels (Huotari et al. 2002).

Val158 is one of 24 hydrophobic amino-acid residues on the surface of *COMT* that anchor the enzyme to the membrane, leading to greater enzyme stability. Other mam-mals have leucine at residue 158. *Leu* is more hydrophobic than either *Val* or *Met*. *Leu*-COMT is more stable, about 60% more active than *Val*-COMT, and 100% more active than *Met*-COMT (Chen et al. 2004). Interestingly, in primate lineage, the *Val*- and *Met*-COMT substitutions may have been driven by selective pressure to slow the metabolism of dopamine and other neurotransmitters in brain (Chen et al. 2004). The *Val* and *Met* alleles may be maintained at high frequencies in human populations worldwide by the

TABLE 1. Frequencies of Two Complementary Yin-Yang Haplotypes of *COMT* in Five Populations

Haplotype	African-American	Chinese	Finnish	German	Plains Indians
Yin: 22112	0.09	0.20	0.13	0.24	0.11
Yang: 11221	0.08	0.24	0.23	0.28	0.22

Five loci of *COMT* correspond to SNPs shown in Figure 1. For each locus, allele 1 represents the abundant allele, and allele 2 represents the less frequent allele. Two haplotypes have opposite-allele configurations (yin-yang) at each locus. The haplotype frequencies are abundant across five populations, reflecting their ancient origin.

mechanisms of balancing selection or time- or niche-specific selection, as discussed below, that *Met* allele carriers benefit by better cognitive function, whereas *Val* allele carriers have a lower anxiety level and are less sensitive to stress stimuli.

In addition to the *Val158Met* polymorphism, other functional loci appear to exist at *COMT* because alleles at *Val158Met* are themselves differentially expressed at the RNA level, as are alleles at a transcribed loci in strong linkage disequilibrium with *Val158Met* (Zhu et al. 2004). Alleles at both *Val158Met* and rs4818, located only 64 bp apart, were observed to be differentially expressed, both in lymphoblasts and in postmortem brain. The higher-enzyme-activity *Val158* allele is generally underexpressed at the RNA level relative to *Met158* (Bray et al. 2003; Zhu et al. 2004). *Met158* and the very tightly linked rs4818 allele were observed to be overexpressed 1.7-fold relative to *Val158* and its corresponding allele (Zhu et al. 2004); Bray et al. (2003) saw that *Val* expression was 13–22% lower.

The *Val158Met* polymorphism is an ancient polymorphism, present in populations worldwide (Palmatier et al. 1999), with the frequency of the lower-activity allele, *Met158*, varying significantly from 0.01 to 0.62. The *Val* allele is an ancestral allele because it is also found in nonhuman primates. In addition, the *Val* and *Met* alleles are frequently found on opposite-configuration (yin-yang) haplotypes (Table 1), indicating ancient origin.

COMT mRNA Expression in Human Brain

COMT is widely expressed in human brain regions including the amygdala, thalamus, occipital pole, hippocampus, subthalamic nucleus, caudate nucleus, substantia nigra, frontal lobe, corpus callosum, putamen, temporal lobe, cerebral cortex, medulla, cerebellum, and spinal cord (Hong et al. 1998). Thus, *COMT* genetic variation could conceivably have pleiotropic effects resulting from the actions of the enzyme on different neurotransmitters in various brain regions. In both humans and rodents, *COMT* mRNA is expressed postsynaptically in large pyramidal and smaller neurons in all cortical layers of the prefrontal cortex (PFC), as well as in medium and large neurons in the striatum. *COMT* mRNA levels are significantly higher in neurons than in glia (Matsumoto et al. 2003). *COMT* mRNA levels in the PFC are higher than in the striatum in both rats and humans, despite the very high levels of dopamine in the striatum. The ability of the *Val158Met* genotype to predict prefrontal function appears to point to a particularly important role for *COMT* in that brain region (Egan et al. 2001). Altered prefrontal activity could also exert effects in other regions. It is postulated that increased tyrosine hydroxylase (TH) mRNA in

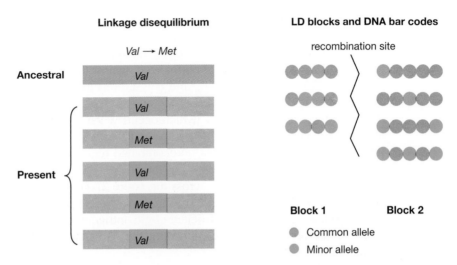

FIGURE 3. Schematic diagram describing linkage disequilibrium (LD) and haplotype block structure. (*Left*) Linkage disequilibrium: Each horizontal bar represents a chromosome with a region highlighted in yellow indicating the ancestral sequence. The *gray* region within each chromosome depicts the present chromosome brought about through recombination. The remaining *yellow* region defining the *COMT Val158Met* locus represents the region of LD. The *Val* allele is the ancestral allele and the *Met* allele is the more recent allele. During human evolution, recombination occurred, and *Val* and *Met* alleles distributed to different chromosomes in present-day populations. (*Right*) LD blocks and haplotypes. Recombination breaks up chromosomal segments (shown as the zigzag line). However, within an LD block, recombination rarely occurs. Therefore, allele combinations remain stable on the same chromosome. The arrangement of alleles is shown as beads on a string, where the common allele is depicted in *gray*, and the minor allele is shown in *yellow*. This allelic arrangement is called a haplotype. Only three to five haplotypes are present within each LD block.

mesencephalic dopamine neurons of *Val158* genotype individuals is caused by up-regulation resulting from diminished activity of prefrontal cortical dopamine neurons projecting to the mesencephalon (Akil et al. 2003).

Using Linkage Disequilibrium and Haplotypes to Locate the Functional Locus of *COMT*

Linkage disequilibrium (LD) is the co-occurrence of alleles, at neighboring loci in populations, that are observed more often than would be expected by chance. In the absence of knowledge of the actual functional locus, markers in LD or combinations of such markers (haplotypes) can be used to narrow the region containing the functional locus (Fig. 3). Although a series of loci in strong LD can theoretically generate numerous haplotypes, usually only a few are common in populations. These haplotypes can be followed, enabling tracking of the history and effects of ancient segments of DNA that may harbor alleles affecting function, and also enabling more general studies of population structure and history (Palmatier et al. 2004).

Several measures of LD are available. Three pairwise (locus-locus) measures are frequently used: D, D', and r^2. The term $D = P_{AB} - P_A * P_B$, where P_{AB} is the expected haplotype frequency and P_A and P_B are observed frequencies for allele combinations at two loci: A and B. The term D' is D normalized against the maximum value of D possible, given allele frequencies P_A and P_B, and r^2 is a direct measure of the statistical correlation of genotypes at genetic markers. Both D' and r^2 range from 0 to 1; however, r^2 can only reach 1 if there is allelic identity (which occurs when alleles at the two loci are equal in allele frequency). Thus, r^2 is a better measure for information content, whereas D' is more informative for detecting ancient unrecombined regions (haplotype blocks) of chromosomes as would be used to scan across the genome.

The variable D' is extensive across the entire 27-kb region of *COMT*. The two most abundant five-site *COMT* haplotypes are yin-yang (opposite) in configuration and are found worldwide, reflecting an ancient origin. Intriguingly, these yin-yang haplotypes differ at the *Val/Met* locus (Table 1), indicating the ancient presence of both alleles in human populations.

In brain, but not lymphoblast, a three-locus *COMT* haplotype further predicted *COMT* expression. The *Met158* allele was always found on the higher-expression haplotype. Although haplotype may predict *COMT* mRNA expression, the loci or mechanisms responsible are unknown. The genotype of rs2097603, located in the promoter 2 region and incorporated in the three-locus haplotype found to be predictive of *COMT* mRNA expression by Zhu et al. (2004), also predicted *COMT* enzyme activity. Again, the effect on expression is likely to be attributable to LD with an unknown functional locus. Incorporation of the *COMT* haplotype in linkage studies (Shifman et al. 2002) and further identification of other functional *COMT* loci appear to be warranted.

COMT Is Linked to Prefrontal Cognitive Function, Schizophrenia, and Antipsychotic Treatment Response

COMT has a critical role in PFC function because of the paucity of dopamine transporters in this region and consequent effect of *COMT* on dopamine levels in this region (Lewis et al. 2001; Mazei et al. 2002). In *COMT* knockout mice, dopamine level increased only in the PFC (Gogos et al. 1998), and even in this region, the effect was only observed in males. In the PFC, levels of norepinephrine are not altered, perhaps because of the abundance of norepinephrine transporters in this region. Disruption of PFC function is associated with deficits in a variety of measures of cognitive performance; for example, task switching and working memory. Dopamine modulates prefrontal cortical activity and performance (Sawaguchi and Goldman-Rakic 1991; Mattay et al. 1996; Seamans et al. 1998), and under many circumstances, performance is improved by augmenting dopamine neurotransmission. Working memory was enhanced in *COMT* knockout mice (Gogos et al. 1998).

The higher activity of the *COMT Val158* allele, as compared to the *Met158* allele, predicted lower levels of dopamine in *Val* allele carriers and consequent poorer performance on tasks testing PFC function, as was initially observed (Egan et al. 2001). In total, 10 of 12 studies report that *Val158* is linked to poorer PFC cognitive function. The effect of the polymorphism appears to be robust across a variety of clinical groups including healthy controls (Malhotra et al. 2002), schizophrenia patients (Bilder et al. 2002; Joober et al. 2002; Goldberg et al. 2003; Nolan et al. 2004), siblings of schizophrenics (Egan et al. 2001; Rosa et al. 2004), velocardiofacial syndrome patients with the 22q11.2 deletion syndrome (Bearden et al. 2004), and patients with brain injury (Lipsky et al. 2005). Ho et al. (2005) were unable to replicate the association of *Val158* to PFC cognitive performance in schizophrenics, although they did find a greater activation of the PFC during a working memory task, as previously observed (Egan et al. 2001). No effect of *COMT* genotype was observed for the Wisconsin Card Sort Test (WCST) error scores in children with attention deficit hyperactivity disorder (ADHD) (Taerk et al. 2004).

A consistent effect of *Val158Met* is seen across a variety of measures of PFC function: WCST perseverative errors (Egan et al. 2001), working memory tests such as the N-Back (Goldberg et al. 2003), WISC arithmetic, WISC digit span, and verbal fluency. *Val158Met* was thus the first polymorphism shown to modulate human cognitive performance, accounting for a modest 4% of variance in performance on the WCST (Egan et al. 2001). However, *Met158* is not always associated with better performance on measures of cognitive function. In one study, *Met/Met* homozygotes were able to more rapidly acquire an imitation rule but had more difficulty alternating between imitation and reversal than *Val158* carriers, indicating that the *Met158* allele was associated with better cognitive stability but poorer cognitive flexibility (Nolan et al. 2004). This could explain associations of *Met158* to aggressive behaviors (Strous et al. 1997; Lachman et al. 1998), which otherwise do not seem consistent with the general effect of *Met158* to improve frontal cognitive performance. Alternatively, association of *Met158* to aggression may be mediated through its effects on anxiety and emotionality. In addition, the failure of *Val158Met* to predict cognitive function in children could reflect developmental differences, for example, children may rely less on the *COMT* catabolic pathway than adults (Taerk et al. 2004).

Because of the inability to directly access the human PFC, functional imaging studies have been key to linking the *COMT Val158Met* locus to variation in information processing by this region. *Val158* heterozygotes and homozygotes activate the PFC more strongly to produce equivalent performance on the N-back working memory task, indicating cortical inefficiency (Egan et al. 2001). The degree of PFC activation was correlated with *Val158* allele dosage. This result was replicated in a study that also used amphetamine to enhance attention and cognitive function in the PFC (Mattay et al. 1996). Amphetamine significantly enhanced performance of *Val/Val* individuals in a working memory task performance but had no beneficial effect in *Met/Met* individuals. However, when task difficulty was increased, *Met/Met* individuals given amphetamine had a higher risk for adverse responses and deterioration of performance, resulting from increasing anxiety and agitation. These results appear to reflect a U-shaped dose-response curve for dopamine in PFC. Thus, too much or too little dopamine may impair frontal cognitive performance.

Deficits in PFC function are thought to be etiologic in schizophrenia, but the direct relationship of *COMT Val158Met* to this disease is considerably more obscure than the relationship of the locus to cognition. In schizophrenics, levels of *COMT* mRNA are relatively lower in the superficial (II/III) layers and higher in the intermediate/deep (IV/V) layers as compared to controls, in whom expression was homogeneous across layers. Velocardiofacial syndrome, caused by interstitial deletion of the chromosome 22q11 where *COMT* is located, is associated with a variety of psychiatric disorders including schizophrenia. Direct association of *Val158* to schizophrenia has been observed several times, but findings are inconsistent. Three meta-analyses encompassing 14–27 independent studies found that *Val158* was not significantly associated with schizophrenia (Glatt et al. 2003; Fan et al. 2005; Munafo et al. 2005). However, results from family-based association studies provide evidence that *Val158* is a risk factor for schizophrenia in patients of European ancestry (odds ratio = 1.5; 95% confidence interval = 0.9–2.4), but not for schizophrenia patients of Asian ancestry (Glatt et al. 2003). The family-based studies seem more compelling because case-control studies may be confounded by population admixture, obscuring linkage as well as leading to false positives. Furthermore, meta-analyses may not detect small risk effects obscured by cross-study heterogeneity, as may occur for schizophrenia, for which ascertainment criteria vary. On the basis of the 40–50% monozygotic twin concordance in schizophrenia, it can be concluded that the effects of risk alleles in schizophrenia are modified by environmental influences. A recent gene–environment study showed that adolescent-onset cannabis users carrying *Val158* exhibited more psychotic symptoms and were more likely to develop schizophreniform disorder (Caspi et al. 2005). Another possible origin of cross-study heterogeneity, especially when different ethnic populations are involved, is that other modifying loci may differ. Haplotype-based studies have consistently detected association of *COMT* to schizophrenia, as well as immediate adjunct genes to velocardiofacial syndrome (Shifman et al. 2002; Chen et al. 2004; Sanders et al. 2005). The risk haplotype for schizophrenia contains *Val158* (Shifman et al. 2002), but haplotype also appears to predict *COMT* mRNA expression, as already discussed.

In addition, evidence now shows that *Val158Met* is linked to treatment response in schizophrenics (Inada et al. 2003; Bertolino et al. 2004; Weickert et al. 2004). The *Val158Met* genotype predicted the effect of olanzapine on prefontal cortical function (Bertolino et al. 2004). The *Met* allele predicted more improvement in working memory and negative symptoms after 8 weeks of treatment with olanzapine. In the two-back test, *Met/Met* homozygous patients improved 20%, but *Val/Val* patients improved only 5%. A similar association of the *Met* allele to improvement in working memory improvement was seen in response to other antipsychotic medications (Weickert et al. 2004). In contrast, *Val* genotype patients appeared to be more resistant to antipsychotic treatment. The daily antipsychotic dosage in *Val/Val* patients was 79% more than in *Val/Met* patients and 68% more than in *Met/Met* patients (Inada et al. 2003).

COMT Has a Potential Role in Anxiety and Stress Resiliency

The role of the *Val158* allele in poorer cognitive performance and its possible role as a risk factor in schizophrenia raised the question of why it has been maintained at high frequencies in populations worldwide, especially because the *Val* and *Met* alleles tend to reside on ancient yin-yang haplotypes. The answer appears to be that the *Met* allele, although associated with better cognitive performance, is linked to diminished stress resiliency and higher levels of anxiety. A Warrior (*Val* allele) versus Worrier (*Met* allele) selectionist model can account for the maintenance of both *COMT* alleles via balancing selection or frequency-dependent selection.

One clue to the association of the *Met158* allele to anxiety was its higher frequency in obsessive-compulsive disorder (OCD). OCD is characterized by anxiety-producing intrusive thoughts and performance of anxiety-reducing rituals. *Met158* is more abundant in OCD, particularly in male patients (Karayiorgou et al. 1997, 1999). A family-based study also found that OCD patients were more likely to be *Met158/Met158* homozygotes (Schindler et al. 2000). More directly to the point, *Met158/Met158* homozygosity was associated with higher levels of dimensionally measured anxiety among women in two populations (Enoch et al. 2003a,b).

Anxiety and dysphoric mood are frequently triggered by stressful life experiences, and people who are anxious and dysphoric are less tolerant of stress and pain. It was therefore predicted that the *Met* allele would be associated with lower pain thresholds and to differences in brain opioid responses to painful, stressful stimuli that correlate with pain threshold (Zubieta et al. 2002, 2003). In an imaging genetics study using the μ-opioid receptor ligand C^{11} carfentanil, it was shown that *Met158/Met158* homozygous individuals are not able to further activate endorphin release after a painful stimulus and thus do not displace the exogenous radioligand from binding sites in the thalamus and various locations in the limbic system (Zubieta et al. 2003). There was a *Met158* allele dosage-effect, and a correlative effect on pain threshold and negative affective response to pain. In a replication study in a larger data set of women, a *Met*-containing four-locus *COMT* haplotype was again correlated with a lower pain threshold (Diatchenko et al. 2005). Recently, it was found that the effect of *COMT* genotype on the processing of unpleasant stimuli is most pronounced in the ventrolateral PFC (Smolka et al. 2005)—the same region previously implicated in the effect of *COMT* on cognition. The number of *Met* alleles was positively correlated with BOLD (blood oxygen level-dependent) response in the prefrontal region. *Met/Met* individuals showed the greatest BOLD changes, and *Val/Val* individuals had the smallest responses.

The *Val158Met* genotype explains up to 38% of the interindividual variance in PFC response to unpleasant stimuli. The *Met158* allele also predicts stronger activations by unpleasant stimuli of different parts of the limbic system including the hippocampus, amygdala, and thalamus (Smolka et al. 2005). Therefore, it is reasonable to state that *COMT* genetic variation may alter prefontal cortical function, leading to evolutionarily counterbalancing effects on emotion and cognitive function, or alternatively, this polymorphism could be exerting some of its primary effects on cognition or emotion elsewhere in the brain.

COMT Is Linked to Addiction

Three neurobiological mechanisms that appear to be critical in forming genetic vulnerability to addiction are reward, cognitive function, and stress/anxiety resiliency. As just discussed, substantial evidence shows that two of these, cognition and stress/anxiety resiliency, are modulated by *COMT Val158Met* and that they are modulated in opposite directions by the *Val* and *Met* alleles, as described in the "Warrior versus Worrier" model. It is also possible that *COMT* could modulate reward either through a direct effect on meso-limbic dopamine or via interactions between frontal cortex and the rest of the mesolimbic system. In this regard, the *COMT* genotype predicted both baseline μ-opioid receptor binding and binding modulated by pain-induced endomorphin release in the nucleus accumbens (Zubieta et al. 2003), a region rich in μ-opioid receptor binding. This effect of the *COMT* genotype on nucleus accumbens function could be a result of the direct action of *Val158Met* on catecholamine metabolism in the accumbens itself or elsewhere in brain. In animals, reduced dopamine signaling in the PFC leads to increased response of subcortically projecting midbrain dopamine neurons to stimuli such as stress. In humans, the *COMT* genotype directly predicts reduced dopamine synthesis in midbrain and affects the interaction with the PFC (Meyer-Lindenberg et al. 2005). *Met/Met* homozygotes that have relatively greater prefrontal synaptic dopamine were observed to have lower tyrosine hydroxylase levels in midbrain, indicating that dopamine levels in PFC and the striatum may be inversely modulated by *Val158Met* (Akil et al. 2003).

COMT variation may thus contribute to vulnerability to addiction via a variety of mechanisms, raising the possibility of variation in the strength and direction of the linkage of this gene to addictions across different clinical addiction groups in which these causes of addiction vary in importance. *COMT* linkage to addiction, but with seemingly puzzling contradictions, is what has in fact been observed. The *Val* allele was higher in frequency in DSM-III-R diagnosed polysubstance abusers, and relative risk was 2.8 (95% confidence interval = 1.3–6.1) (Vandenbergh et al. 1997).

This result supported a family-based association study of heroin addiction (Horowitz et al. 2000) in which the *Val* allele was transmitted in excess to heroin addicts. The *Met158 allele* has also been linked to addictions—to alcoholism in two case-control studies and a family-based study (Tiihonen et al. 1999; Hallikainen et al. 2000; Kauhanen et al. 2000; Wang et al. 2001). All of these studies were conducted on late-onset alcoholics, who tend to be characterized by higher levels of anxiety, to which the *Met158* allele may contribute as described.

There was no association of *Met158* in Finns with early-onset alcoholism (Hallikainen et al. 2000), in which factors other than anxiety appear to have more of a role. However, the early-/late-onset dichotomy cannot be extended too far. *Met158* was associated with early-onset Type 2 alcoholism in a family-based study in a German population (Wang et al. 2001). Effects of *COMT* on reward could also come into play.

Because alcohol-induced euphoria—and perhaps more saliently, anticipation—is associated with the rapid release of dopamine in limbic areas, the *Met158* low-activity allele appears to lead to lower dopamine levels in midbrain (Meyer-Lindenberg et al. 2005), and *Met158* carriers could have a more profound experience of reward. *Met158/Met158* homozygous controls reported 27% higher weekly alcohol consumption compared with other genotype groups (Kauhanen et al. 2000).

RESEARCH OPPORTUNITIES OF *COMT* IN ALCOHOLISM: GAPS AND FUTURE DIRECTIONS

As discussed above, *COMT* has an important role in addiction through different mechanisms. The functional *Val158Met* SNP has been extensively studied in linkage studies with a variety of clinical phenotypes including alcoholism. Haplotype-based linkage and *COMT* expression suggested that additional functional loci at *COMT* may exist. Identifying functional loci and understanding how genetic variants affect gene function are warranted for the near future. Increased understanding of *COMT* and environmental interaction in alcoholism and other psychiatric disorders can provide more biological guidance for diagnosis, prevention, and treatment of alcoholism.

REFERENCES

Akil M., Kolachana B.S., Rothmond D.A., Hyde T.M., Weinberger D.R., and Kleinman J.E. 2003. Catechol-O-methyltransferase genotype and dopamine regulation in the human brain. *J. Neurosci.* **23:** 2008–2013.

Axelrod J. and Vesell E.S. 1970. Heterogeneity of N- and O-methyltransferases. *Mol. Pharmacol.* **6:** 78–84.

Bearden C.E., Jawad A.F., Lynch D.R., Sokol S., Kanes S.J., McDonald-McGinn D.M., Saitta S.C., Harris S.E., Moss E., Wang P.P., Zackai E., Emanuel B.S., and Simon T.J. 2004. Effects of a functional COMT polymorphism on prefrontal cognitive function in patients with 22q11.2 deletion syndrome. *Am. J. Psychiatry* **161:** 1700–1702.

Bertocci B., Miggiano V., Da Prada M., Dembic Z., Lahm H.W., and Malherbe P. 1991. Human catechol-O-methyltransferase: Cloning and expression of the membrane-associated form. *Proc. Natl. Acad. Sci.* **88:** 1416–1420.

Bertolino A., Caforio G., Blasi G., De Candia M., Latorre V., Petruzzella V., Altamura M., Nappi G., Papa S., Callicott J.H., Mattay V.S., Bellomo A., Scarabino T., Weinberger D.R., and Nardini M. 2004. Interaction of COMT Val108/158Met genotype and olanzapine treatment on prefrontal cortical function in patients with schizophrenia. *Am. J. Psychiatry* **161:** 1798–1805.

Bilder R.M., Volavka J., Czobor P., Malhotra A.K., Kennedy J.L., Ni X., Goldman R.S., Hoptman M.J., Sheitman B., Lindenmayer J.P., Citrome L., McEvoy J.P., Kunz M., Chakos M., Cooper T.B., and Lieberman J.A. 2002. Neurocognitive correlates of the COMT Val158Met polymorphism in chronic schizophrenia. *Biol. Psychiatry* **52:** 701–707.

Bray N.J., Buckland P.R., Williams N.M., Williams H.J., Norton N., Owen M.J., and O'Donovan M.C. 2003. A haplotype implicated in schizophrenia susceptibility is associated with reduced COMT expression in human brain. *Am. J. Hum. Genet.* **73:** 152–161.

Caspi A., Moffitt T.E., Cannon M., McClay J., Murray R., Harrington H., Taylor A., Arseneault L., Williams B., Braithwaite A., Poulton R., and Craig I.W. 2005. Moderation of the effect of adolescent-onset cannabis use on adult psychosis by a functional polymorphism in the catechol-O-methyltransferase gene: Longitudinal evidence of a gene X environment interaction. *Biol. Psychiatry* **57:** 1117–1127.

Chen J., Lipska B.K., Halim N., Ma Q.D., Matsumoto M., Melhem S., Kolachana B.S., Hyde T.M., Herman M.M., Apud J., Egan M.F., Kleinman J.E., and Weinberger D.R. 2004. Functional analysis of genetic variation in catechol-O-methyltransferase (*COMT*): Effects on mRNA, protein, and enzyme activity in postmortem human brain. *Am. J. Hum. Genet.* **75:** 807–821.

Diatchenko L., Slade G.D., Nackley A.G., Bhalang K., Sigurdsson A., Belfer I., Goldman D., Xu K., Shabalina S.A., Shagin D., Max M.B., Makarov S.S., and Maixner W. 2005. Genetic basis for individual variations in pain perception

and the development of a chronic pain condition. *Hum. Mol. Genet.* **14:** 135–143.

Egan M.F., Goldberg T.E., Kolachana B.S., Callicott J.H., Mazzanti C.M., Straub R.E., Goldman D., and Weinberger D.R. 2001. Effect of COMT Val108/158 Met genotype on frontal lobe function and risk for schizophrenia. *Proc. Natl. Acad. Sci.* **98:** 6917–6922.

Enoch M.A., Schuckit M.A., Johnson B.A., and Goldman D. 2003a. Genetics of alcoholism using intermediate phenotypes. *Alcohol Clin. Exp. Res.* **27:** 169–176.

Enoch M.A., Xu K., Ferro E., Harris C.R., and Goldman D. 2003b. Genetic origins of anxiety in women: A role for a functional catechol-O-methyltransferase polymorphism. *Psychiatr. Genet.* **13:** 33–41.

Fan J.B., Zhang C.S., Gu N.F., Li X.W., Sun W.W., Wang H.Y., Feng G.Y., St Clair D., and He L. 2005. Catechol-O-methyltransferase gene Val/Met functional polymorphism and risk of schizophrenia: A large-scale association study plus meta-analysis. *Biol. Psychiatry* **57:** 139–144.

Glatt S.J., Faraone S.V., and Tsuang M.T. 2003. Association between a functional catechol O-methyltransferase gene polymorphism and schizophrenia: Meta-analysis of case-control and family-based studies. *Am. J. Psychiatry* **160:** 469–476.

Gogos J.A., Morgan M., Luine V., Santha M., Ogawa S., Pfaff D., and Karayiorgou M. 1998. Catechol-O-methyltransferase-deficient mice exhibit sexually dimorphic changes in catecholamine levels and behavior. *Proc. Natl. Acad. Sci.* **95:** 9991–9996.

Goldberg T.E., Egan M.F., Gscheidle T., Coppola R., Weickert T., Kolachana B.S., Goldman D., and Weinberger D.R. 2003. Executive subprocesses in working memory: Relationship to catechol-O-methyltransferase Val158Met genotype and schizophrenia. *Arch. Gen. Psychiatry* **60:** 889–896.

Goldman D. and Bergen A. 1998. General and specific inheritance of substance abuse and alcoholism. *Arch. Gen. Psychiatry* **55:** 964–965.

Grossman M.H., Emanuel B.S., and Budarf M.L. 1992. Chromosomal mapping of the human catechol-O-methyltransferase gene to 22q11.1–q11.2. *Genomics* **12:** 822–825.

Hallikainen T., Lachman H., Saito T., Volavka J., Kauhanen J., Salonen J.T., Ryynanen O.P., Koulu M., Karvonen M.K., Pohjalainen T., Syvalahti E., Hietala J., and Tiihonen J. 2000. Lack of association between the functional variant of the catechol-O-methyltransferase (COMT) gene and early-onset alcoholism associated with severe antisocial behavior. *Am. J. Med. Genet.* **96:** 348–352.

Ho B.C., Wassink T.H., O'Leary D.S., Sheffield V.C., and Andreasen N.C. 2005. Catechol-O-methyl transferase Val158Met gene polymorphism in schizophrenia: Working memory, frontal lobe MRI morphology and frontal cerebral blood flow. *Mol. Psychiatry* **10:** 287–298.

Hong J., Shu-Leong H., Tao X., and Lap-Ping Y. 1998.

Distribution of catechol-O-methyltransferase expression in human central nervous system. *Neuroreport* **9:** 2861–2864.

Horowitz R., Kotler M., Shufman E., Aharoni S., Kremer I., Cohen H., and Ebstein R.P. 2000. Confirmation of an excess of the high enzyme activity COMT *val* allele in heroin addicts in a family-based haplotype relative risk study. *Am. J. Med. Genet.* **96:** 599–603.

Huotari M., Gogos J.A., Karayiorgou M., Koponen O., Forsberg M., Raasmaja A., Hyttinen J., and Mannisto P.T. 2002. Brain catecholamine metabolism in catechol-*O*-methyltransferase (COMT)-deficient mice. *Eur. J. Neurosci.* **15:** 246–256.

Inada T., Nakamura A., and Iijima Y. 2003. Relationship between catechol-O-methyltransferase polymorphism and treatment-resistant schizophrenia. *Am. J. Med. Genet. B Neuropsychiatr. Genet.* **120:** 35–39.

Joober R., Gauthier J., Lal S., Bloom D., Lalonde P., Rouleau G., Benkelfat C., and Labelle A. 2002. Catechol-O-methyltransferase *Val-108/158-Met* gene variants associated with performance on the Wisconsin Card Sorting Test. *Arch. Gen. Psychiatry* **59:** 662–663.

Karayiorgou M., Altemus M., Galke B.L., Goldman D., Murphy D.L., Ott J., and Gogos J.A. 1997. Genotype determining low catechol-*O*-methyltransferase activity as a risk factor for obsessive-compulsive disorder. *Proc. Natl. Acad. Sci.* **94:** 4572–4575.

Karayiorgou M., Sobin C., Blundell M.L., Galke B.L., Malinova L., Goldberg P., Ott J., and Gogos J.A. 1999. Family-based association studies support a sexually dimorphic effect of *COMT* and *MAOA* on genetic susceptibility to obsessive-compulsive disorder. *Biol. Psychiatry* **45:** 1178–1189.

Kauhanen J., Hallikainen T., Tuomainen T.P., Koulu M., Karvonen M.K., Salonen J.T., and Tiihonen J. 2000. Association between the functional polymorphism of catechol-O-methyltransferase gene and alcohol consumption among social drinkers. *Alcohol Clin. Exp. Res.* **24:** 135–139.

Lachman H.M., Nolan K.A, Mohr P., Saito T., and Volavka J. 1998. Association between catechol O-methyltransferase genotype and violence in schizophrenia and schizoaffective disorder. *Am. J. Psychiatry* **155:** 835–837.

Lachman H.M., Papolos D.F., Saito T., Yu Y.M., Szumlanski C.L., and Weinshilboum R.M. 1996. Human catechol-O-methyltransferase pharmacogenetics: Description of a functional polymorphism and its potential application to neuropsychiatric disorders. *Pharmacogenetics* **6:** 243–250.

Lewis D.A., Melchitzky D.S., Sesack S.R., Whitehead R.E., Auh S., and Sampson A. 2001. Dopamine transporter immunoreactivity in monkey cerebral cortex: Regional, laminar, and ultrastructural localization. *J. Comp. Neurol.* **432:** 119–136.

Lipsky R.H., Sparling M.B., Ryan L.M., Xu K., Salazar A., Goldman D., and Warden D.L. 2005. Association of COMT Val158Met polymorphism with executive functioning following traumatic brain injury. *J. Neuropsych. Clin. Neurosci.* (in press).

Lotta T., Vidgren J., Tilgmann C., Ulmanen I., Melen K., Julkunen I., and Taskinen J. 1995. Kinetics of human soluble and membrane-bound catechol O-methyltransferase: A revised mechanism and description of the thermolabile variant of the enzyme. *Biochemistry* **34:** 4202–4210.

Malhotra A.K., Kestler L.J., Mazzanti C., Bates J.A., Goldberg T., and Goldman D. 2002. A functional polymorphism in the COMT gene and performance on a test of prefrontal cognition. *Am. J. Psychiatry* **159:** 652–654.

Matsumoto M., Weickert C.S., Akil M., Lipska B.K., Hyde T.M., Herman M.M., Kleinman J.E., and Weinberger D.R. 2003. Catechol *O*-methyltransferase mRNA expression in human and rat brain: Evidence for a role in cortical neuronal function. *Neuroscience* **116:** 127–137.

Mattay V.S., Frank J.A., Santha A.K., Pekar J.J., Duyn J.H., McLaughlin A.C., and Weinberger D.R. 1996. Whole-brain functional mapping with isotropic MR imaging. *Radiology* **201:** 399–404.

Mazei M.S., Pluto C.P., Kirkbride B., and Pehek E.A. 2002. Effects of catecholamine uptake blockers in the caudate-putamen and subregions of the medial prefrontal cortex of the rat. *Brain Res.* **936:** 58–67.

Meyer-Lindenberg A., Kohn P.D., Kolachana B., Kolachana B., Kippenhan S., McInerney-Leo A., Nussbaum R., Weinberger D.R., and Berman K.F. 2005. Midbrain dopamine and prefrontal function in humans: Interaction and modulation by *COMT* genotype. *Nat. Neurosci.* **8:** 594–596.

Munafo M.R., Bowes L., Clark T.G., and Flint J. 2005. Lack of association of the COMT (Val(158/108) Met) gene and schizophrenia: A meta-analysis of case-control studies. *Mol. Psychiatry.* **10:** 765–770.

Nolan K.A., Bilder R.M., Lachman H.M., and Volavka J. 2004. Catechol O-methyltransferase Val158Met polymorphism in schizophrenia: Differential effects of Val and Met alleles on cognitive stability and flexibility. *Am. J. Psychiatry* **161:** 359–361.

Palmatier M.A., Kang A.M., and Kidd K.K. 1999. Global variation in the frequencies of functionally different catechol-O-methyltransferase alleles. *Biol. Psychiatry* **46:** 557–567.

Palmatier M.A., Pakstis A.J., Speed W., Paschou P., Goldman D., Odunsi A., Okonofua F., Kajuna S., Karoma N., Kungulilo S., Grigorenko E., Zhukova O.V., Bonne-Tamir B., Lu R.B., Parnas J., Kidd J.R., DeMille M.M., and Kidd K.K. 2004. COMT haplotypes suggest P2 promoter region relevance for schizophrenia. *Mol. Psychiatry* **9:** 859–870.

Rosa A., Peralta V., Cuesta M.J., Zarzuela A., Serrano F., Martinez-Larrea A., and Fananas L. 2004. New evidence of association between COMT gene and prefrontal neurocognitive function in healthy individuals from sibling pairs discordant for psychosis. *Am. J. Psychiatry* **161:** 1110–1112.

Roth J.A. 1992. Membrane-bound catechol-O-methyltransferase: A reevaluation of its role in the O-methylation of the catecholamine neurotransmitters. *Rev. Physiol. Biochem. Pharmacol.* **120:** 1–29.

Sanders A.R., Rusu I., Duan J., Vander Molen J.E., Hou C., Schwab S.G., Wildenauer D.B., Martinez M., and Gejman P.V. 2005. Haplotypic association spanning the 22q11.21

genes COMT and ARVCF with schizophrenia. *Mol. Psychiatry* **10:** 353–365.

Sawaguchi T. and Goldman-Rakic P.S. 1991. D1 dopamine receptors in prefrontal cortex: Involvement in working memory. *Science* **251:** 947–950.

Scanlon P.D., Raymond F.A., and Weinshilboum R.M. 1979. Catechol-O-methyltransferase: Thermolabile enzyme in erythrocytes of subjects homozygous for allele for low activity. *Science* **203:** 63–65.

Schindler K.M., Richter M.A., Kennedy J.L., Pato M.T., and Pato C.N. 2000. Association between homozygosity at the COMT gene locus and obsessive compulsive disorder. *Am. J. Med. Genet.* **96:** 721–724.

Seamans J.K., Floresco S.B., and Phillips A.G. 1998. D1 receptor modulation of hippocampal-prefrontal cortical circuits integrating spatial memory with executive functions in the rat. *J. Neurosci.* **18:** 1613–1621.

Shifman S., Bronstein M., Sternfeld M., Pisante-Shalom A., Lev-Lehman E., Weizman A., Reznik I., Spivak B., Grisaru N., Karp L., Schiffer R., Kotler M., Strous R.D., Swartz-Vanetik M., Knobler H.Y., Shinar E., Beckmann J.S., Yakir B., Risch N., Zak N.B., and Darvasi A. 2002. A highly significant association between a COMT haplotype and schizophrenia. *Am. J. Hum. Genet.* **71:** 1296–1302.

Smolka M.N., Schumann G., Wrase J., Grusser S.M., Flor H., Mann K., Braus D.F., Goldman D., Buchel C., and Heinz A. 2005. Catechol-*O*-methyltransferase *val158met* genotype affects processing of emotional stimuli in the amygdala and prefrontal cortex. *J. Neurosci.* **25:** 836–842.

Spielman R.S. and Weinshilboum R.M. 1981. Genetics of red cell COMT activity: Analysis of thermal stability and family data. *Am. J. Med. Genet.* **10:** 279–290.

Strous R.D., Bark N., Parsia S.S., Volavka J., and Lachman H.M. 1997. Analysis of a functional catechol-O-methyltransferase gene polymorphism in schizophrenia: Evidence for association with aggressive and antisocial behavior. *Psychiatry Res.* **69:** 71–77.

Taerk E., Grizenko N., Ben Amor L., Lageix P., Mbekou V., Deguzman R., Torkaman-Zehi A., Ter Stepanian M., Baron C., and Joober R. 2004. *Catechol-O-methyltransferase (COMT) Val108/158 Met* polymorphism does not modulate executive function in children with ADHD. *BMC Med. Genet.* **5:** 30.

Tiihonen J., Hallikainen T., Lachman H., Saito T., Volavka J., Kauhanen J., Salonen J.T., Ryynanen O.P., Koulu M., Karvonen M.K., Pohjalainen T., Syvalahti E., and Hietala J. 1999. Association between the functional variant of the cat-echol-O-methyltransferase (COMT) gene and type 1 alcoholism. *Mol. Psychiatry* **4:** 286–289.

Tilgmann C. and Kalkkinen N. 1991. Purification and partial sequence analysis of the soluble catechol-O-methyltransferase from human placenta: Comparison to the rat liver enzyme. *Biochem. Biophys. Res. Commun.* **174:** 995–1002.

Vandenbergh D.J., Rodriguez L.A., Miller I.T., Uhl G.R., and Lachman H.M. 1997. High-activity catechol-O-methyltransferase allele is more prevalent in polysubstance abusers. *Am. J. Med. Genet.* **74:** 439–442.

Vidgren J., Svensson L.A., and Liljas A. 1994. Crystal structure of catechol O-methyltransferase. *Nature* **368:** 354–358.

Wang T., Franke P., Neidt H., Cichon S., Knapp M., Lichtermann D., Maier W., Propping P., and Nothen M.M. 2001. Association study of the low-activity allele of catechol-O-methyltransferase and alcoholism using a family-based approach. *Mol. Psychiatry* **6:** 109–111.

Weickert T.W., Goldberg T.E., Mishara A., Apud J.A., Kolachana B.S., Egan M.F., and Weinberger D.R. 2004. Catechol-O-methyltransferase *val108/158met* genotype predicts working memory response to antipsychotic medications. *Biol. Psychiatry* **56:** 677–682.

Weinshilboum R. and Dunnette J. 1981. Thermal stability and the biochemical genetics of erythrocyte catechol-O-methyltransferase and plasma dopamine-beta-hydroxylase. *Clin. Genet.* **19:** 426–437.

Weinshilboum R. and Raymond F. 1977. Variations in catechol-O-methyltransferase activity in inbred strains of rats. *Neuropharmacology* **16:** 703–706.

Weinshilboum R.M., Otterness D.M., and Szumlanski C.L. 1999. Methylation pharmacogenetics: Catechol O-methyltransferase, thiopurine methyltransferase, and histamine N-methyltransferase. *Annu. Rev. Pharmacol. Toxicol.* **39:** 19–52.

Zhu G., Lipsky R.H., Xu K., Ali S., Hyde T., Kleinman J., Akhtar L.A., Mash D.C., and Goldman D. 2004. Differential expression of human *COMT* alleles in brain and lymphoblasts detected by RT-coupled 5′ nuclease assay. *Psychopharmacology* **177:** 178–184.

Zubieta J.K., Heitzeg M.M., Smith Y.R., Bueller J.A., Xu K., Xu Y., Koeppe R.A., Stohler C.S., and Goldman D. 2003. COMT *val158met* genotype affects μ-opioid neurotransmitter responses to a pain stressor. *Science* **299:** 1240–1243.

Zubieta J.K., Smith Y.R., Bueller J.A., Xu Y., Kilbourn M.R., Jewett D.M., Meyer C.R., Koeppe R.A., and Stohler C.S. 2002. μ-opioid receptor-mediated antinociceptive responses differ in men and women. *J. Neurosci.* **22:** 5100–5107.

4 | Identifying Genes Affecting Addiction Risk in Animal Models

John C. Crabbe, Ph.D.

*Portland Alcohol Research Center, VA Medical Center, Portland, Oregon 97239;
Department of Behavioral Neuroscience, Oregon Health & Science University,
Portland, Oregon 97239*

ABSTRACT

Individual differences in susceptibility to addiction are due in part to differences in sequence or expression of many genes. Because individual gene effects on risk are small, it is difficult to identify them. Animal models, particularly the laboratory mouse, offer attractive systems with which to work toward this goal. Much effort is devoted to transgenic manipulations of individual candidate genes' function and, more recently, to "omics" approaches (genomics, proteomics, metabolomics) based on gene expression. This chapter focuses on a third approach, gene mapping. Here, the strategy is to first localize each such influential gene in the genome (at which point it is termed a quantitative trait locus, or QTL) and then map its location with increasing precision, until only one viable candidate remains in the QTL, and the quantitative trait gene (QTG) is discovered. An example of success is the pursuit of QTLs affecting acute withdrawal severity from alcohol and pentobarbital, leading to the demonstration that the QTG is *Mpdz*, which codes for a multiple PDZ (Mpdz) domain protein. Currently, QTL mapping projects in humans and laboratory animals are incorporating information about gene expression as well as gene sequence differences in the search for QTGs relevant to addiction risk.

INTRODUCTION

How does one identify risk factors for susceptibility to alcoholism or drug dependence? We know that roughly 50% of individual differences in the complex behaviors that earn a psychiatric diagnosis as a substance or alcohol use disorder can be traced to the influence of genes. The other half of the risk landscape is environmental, including the influence of family members, peer groups, and work situations (Enoch and Goldman 2001; Kreek 2001; Crabbe 2002). From decades of psychosocial research, we know a great deal about many important environmental risk factors that both exacerbate and moderate risk. We also know that these addictions are not disorders, such as Huntington's disease, which reflects the influence of a single gene whose protein product exerts a binary effect on diagnosis that is reflected in a strictly Mendelian pattern of inheritance (Quaid et al. 1996). Also, individuals do not inherit risk, they inherit specific genes relevant to risk.

Addiction diagnoses are examples of complex behavioral traits whose presence and/or severity reflects the influence of many individual genes, each of which increases or decreases risk to a modest degree. Thus, the task of identifying specific genes relevant to risk is formidable.

Signal, Noise, and Complexity in Gene Identification. In addition to the small signal that any single gene can emit, the larger challenge facing those studying gene identification is the fact that genes act only in contexts, or environments. The interaction of specific genes and specific environmental pressures is a focus of the many interesting attempts to identify individual risk genes for psychiatric disorders (van den Bree et al. 1998; Caspi et al. 2003; Gunzerath and Goldman 2003). Someone with a very poor genetic prognosis may still avoid alcoholism or drug dependence with favorable environmental support, whereas a very challenging environment may influence another individual to become dependent even if that individual is genetically low risk.

Attempts to unravel the complexities of genes affecting risk in humans have generally come from family studies across multiple generations. By identifying as initial targets (probands) individuals with a specific diagnosis and then comparing concordance (possession of the same diagnosis) in relatives, the extent of genetic contributions can be estimated. In the addictions, monozygotic (identical) cotwins of those addicted have about twice the concordance as dizygotic (fraternal) cotwins, even when reared apart from the proband. Risk among siblings tends to resemble that between dizygotic twins, and both fraternal twins and siblings share 50% of identity of their genomes (Whitfield et al. 2004). Yet, such approaches cannot easily move beyond the estimation of proportion of overall risk that is genetic (heritability) to begin to identify the specific genes responsible.

Three other complexities should also be considered. Genes must be expressed to be relevant. Although same-sex monozygotic twins are exact genetic duplicates with respect to the sequence and protein product of each of their genes, they do not always share diagnoses. Usually, half or fewer of monozygotic cotwins of diagnosed substance abusers are also abusers, and one possible difference between them is how and when their identical genomes' genes are expressed and actively working (Kendler et al. 2003). A second complexity is that part of the environment relevant for understanding the influences of specific genes is other genes. That is, genes interact (a phenomenon called epistasis), such that the end result of having the same specific genotype at gene A may differ among individuals who have genotype X versus genotype Y at gene B (Wade 2001; Carlborg and Haley 2004; Marchini et al. 2005).

A third complexity is that behaviors interact as well. Here is a simple example, taken from research with animal models. The "elevated plus maze" is an apparatus used to detect an anxiety-like state in rodents (Lister 1987; Flint 2001). It is comprised of two arms, crossed in the middle to form a plus shape, and elevated above the floor. One arm has high walls, whereas the other is simply a plank exposed to the room. An animal placed on this apparatus will explore both open and closed arms, and the amount of time it spends in the open arms is measured to indicate its level of anxiety. An anxious animal will avoid the open arms. This attribution is based on extensive studies with drugs, which have found that drugs that are effective anxiolytics in the clinic increase open-arm time in the plus maze in rodents, whereas ineffective drugs do not (Rodgers and Dalvi 1997). It is frequently stated that after mice or rats are made physically dependent on alcohol and the drug is then withdrawn, they become anxious for a while, much like humans.

Many studies have shown that the percent of time in (or entries into) open arms of an elevated plus maze decreases during withdrawal.

However, a recent review of this literature uncovered something quite different. In most studies, the principal behavior that animals demonstrated during withdrawal was a profound reduction in overall activity. Because they did not explore the apparatus during the short test period, they spent much more time in a "safe" alley. But the reduction of locomotor activity competed with the exploratory drive that is necessary to support enough exploration to make this a valid test of anxiety. In other words, there was competition for the allocation of behaviors, rather than a simple decrease in "anxiety-like" behavior (Kliethermes 2005).

GOALS OF THE CHAPTER

In this chapter, I discuss the strengths and weaknesses of animal models for gene-chasing experiments. I briefly mention gene targeting and gene-expression-based approaches, but focus on QTL gene mapping approaches. The example of a drug withdrawal locus is used to illustrate several of the methods that can be used to locate a quantitative trait gene. Many groups are actively seeking risk genes in human populations. The challenges are formidable for many reasons that are described in more detail elsewhere, and we do not consider the human literature here, but rather refer the reader to other reviews (Jacob et al. 2001; Hurley et al. 2002; Schuckit et al. 2004; Uhl 2004). The remainder of this chapter explores the attempts to model aspects of the addictions in laboratory animals and the use of these models for identifying and studying specific genes of importance.

FINDING RISK GENES: ANIMAL MODELS

It is expected that this chapter will give the reader an understanding of the value and limits of rodent models for studying complex traits. Furthermore, he or she should understand the logic of the pursuit of a gene, exemplified by a successful search. Animal models offer both advantages and disadvantages over studying psychiatric disease directly in humans. For example, more tools are available to the scientist who uses animal models, and the underlying biology can be probed more invasively and systematically. Genes can be introduced, deleted, mutated, or modulated. In addition, many features of the environment can be regulated much more closely by the experimenter, facilitating quite sophisticated studies of gene–environment interaction. However, the cost of using animal models is high. Animals cannot report directly on the hedonic effects of drugs, nor on any other aspect of their motivational or emotional experiences. Attempts to model certain features of the environment in laboratory animals have little face validity (it is hard to imagine a relevant model for the intense conditions usually lumped under the heading "peer pressures," but we know they are a crucial influence for adolescents). Given that the psychiatric diseases that include substance abuse disorders are largely behaviorally defined, it seems clear that currently no complete and convincing animal model of alcoholism or cocaine dependence exists. Rather, a good animal model selects certain features of the targeted disease and attempts to develop a partial model of those features (McClearn 1979). The whole picture, therefore, must ultimately be carefully (and cautiously) reassembled from the partial images.

Approaches to Gene Discovery

Three broad approaches to establishing a gene–behavior link are available to the animal researcher. Each is introduced below along with its basic strengths and weaknesses, but our focus is on only one method, that of gene mapping.

1. Gene Manipulation. Currently, the most popular method in the neuroscience of addictions is that of manipulating a gene directly and studying the consequences. In the past, neuroscientists have lesioned a brain nucleus and studied the ensuing behavioral differences, a technique used for many years in rats (Harker and Whishaw 2004). They have also studied mutants created accidentally by nature, a method most often used in mice, an animal on which geneticists have long focused their attention (Reith and Bernstein 1991). Today, molecular biologists can replace the gene for any protein of interest with a dysfunctional variant in an embryo (i.e., make a null mutant, or "knockout") and study function if the animal proves to be viable when it develops. The many variations on this theme include overexpression transgenics (mutants in which single bases in a gene's DNA are mutated) and most recently, conditional mutants (where the embryonic introduction of the mutant gene includes sequences that allow the experimenter to turn gene expression in the adult animal on or off at will) (Muller 1999; Homanics 2002). Predecessors of conditional mutants can be found in studies that used various methods to depress brain function temporarily, such as the application of reversible cooling to a brain area (Lomber 1999).

Hundreds of studies now report behavioral and other neurobiological differences between null mutants and wild-type animals. A reasonable case can be made for the potential importance of the manipulated gene for understanding many different drugs of abuse (Homanics 2002; Picciotto and Corrigall 2002; Stephens et al. 2002; Crabbe and Phillips 2004). Null mutants have their own limitations, of course, many of which resemble those faced when using any other lesioning technique for understanding complex functional networks (Gerlai 1996; Bowers et al. 2000).

2. The "Omics" Approaches. Much energy in neuroscience at present, including addictions, is focused on attempts to study more than one gene at a time. These studies are looking for differences in the expression of genes with hopes of identifying groups of genes that, when considered together, predict sensitivity to a drug's effects, or result from them (Plomin and Crabbe 2000). Such functional genomics studies often use some variant of microarray expression analyses, in which thousands of genes (or pieces of them, or of their cDNA) can be arrayed on a chip. When exposed to brain tissue from an animal, only those genes actively being expressed in that tissue will be captured by the chip's embedded probes and provide a signal in the analysis. Groups and families of genes are affected by drug treatment or its withdrawal, and examination of their protein products allows the experimenter to deduce which biochemical pathways are affected by the drug. Genomics then evolves into proteomics, in which analysis of thousands of proteins is substituted for the gene expression arrays. The analyses of the groups of proteins' data with the goal of identifying pathways has been termed metabolomics (Klose et al. 2002).

The "omics" approaches have their antecedents in studies of gene or protein expression or localization that were conducted one by one (Reilly et al. 2001). The advantages of multiplex simultaneous analysis are obvious. Notably, they include the ability to discover the involvement of a gene or protein whose role was not previously suspected (Kerns et al. 2005). As with any new method, the disadvantages are also surfacing. Among those are cost and a host of highly technical difficulties, but perhaps more sobering is the sheer mass of data that results from such experiments. These analyses have forced the accompanying development of bioinformatics approaches to attempt to organize the data. Studies are increasingly moving beyond simply providing a long list of genes differentially expressed. On the plus side, the widespread application of genomics and informatics is responsible for the development of an entirely new field of study (Fischer et al. 2003).

3. Gene Mapping. The third broad approach to gene finding, gene mapping, is the one with the longest history in both animal and human genetics. Each gene is known to exist in multiple variants, due to differences in DNA sequence that arise for various reasons. The alternative alleles at a gene (which are sometimes confusingly called "genes") may not be significant, or they may lead to the synthesis of different variants of the protein and have a profound effect on gene function. In addition, many variations in DNA sequence lie outside the physical range of what we understand to be the genes' regulatory and decoding regions. These sequence variations are inherited as rigorously as genes and serve as markers for genetic analyses. The sequence variations can be short, repeated motifs (as in microsatellite repeats) or simply single base substitutions (single nucleotide polymorphisms, or SNPs). We now have a high-fidelity map of the location in the genome for each such polymorphism, in multiple species. Therefore, studies of many individuals who show individual differences in a drug-related response attempt to establish statistically reliable co-occurrence between a specific polymorphism and a specific behavior. If such a correspondence is found, it is thought that a linked gene (one physically very close to the polymorphic marker) must exert influence on the trait being mapped (Mackay 2001). Interest in gene mapping in animal models is high because of the great similarity between mouse and human genomes. Because we share an evolutionary ancestor with mice from not too long ago (on an evolutionary timescale), more than 80% of the sequence (and genes) in the mouse genome can be traced directly to a location in the human genome (Silver 1995).

Attempts to map genes relevant for addiction are faced with trying to detect weak signals in multiple genomic locations simultaneously. Because the traits mapped are not all or none, they have been termed quantitative traits, and the initial mapping signals, or chromosomal regions where a signal is detected, are called quantitative trait loci (QTL). There are many ways to approach QTL mapping in animal models, explained in detail elsewhere (Silver 1995; Mackay 2001; Palmer and Phillips 2002). A large proportion of such studies targeting the addictions have occurred with mice, because they are mammals and their genome is so much better explored than that of the rat. QTL mapping studies for behavioral traits have been ongoing since the early 1990s. The process is iterative and cumulative, so that a broad region of a chromosome is slowly but surely narrowed until only a few genes are left in the remaining QTL. Each remaining gene in turn is treated as a candidate and studied to determine whether the trait being mapped is affected by the candidate. Eventually, all candidates but one are eliminated, and the remaining gene is termed a quantitative trait gene (QTG) (Mackay 2001). A recent review describes more than 100 QTL for behavioral responses that have been mapped in laboratory mice (Flint

2003). These include many for responses to drugs of abuse including alcohol, cocaine, and several others (Crabbe et al. 1994). Although progress was initially slow because of the primitive tools available and the sketchy articulation of the mouse map of genomic markers, it has accelerated during the past few years as the technology has improved. More than 25 genes affecting complex traits in mammalian species are now generally accepted as proven QTGs (Korstanje and Paigen 2002; Shirley et al. 2004; Yalcin et al. 2004). Identifying QTGs in animal models has usually involved positional cloning, using a strategy that eliminated all other possibilities by narrowing the region and comparing the sequences and expression levels of the genes in the progenitor strains. We have discussed elsewhere how various sources may be used for final proof that a candidate gene is the true QTG (Crabbe et al. 1999; Belknap et al. 2001).

Nonetheless, gene mapping is a slow business. To date, a single QTG has been discovered in addiction research (Shirley et al. 2004). We describe the process in some detail because it is not too different conceptually from many other methods. Using a combination of data from standard inbred strains, recombinant inbred strains, an F_2 segregating population, short-term selectively bred lines, congenic strains, and, finally, interval-specific congenic strains (see below), Kari Buck and her colleagues (1997) gradually isolated a small region of mouse chromosome 4 that contained a gene whose product affected the severity of acute alcohol and pentobarbital withdrawal. The QTL region initially contained several hundred genes, then more than 40, then 5 (Fehr et al. 2002). Finally, the gene *Mpdz*, which codes for a multiple PDZ domain protein, was shown to be the only remaining gene in the region that affected the response (Shirley et al. 2004).

Alcohol and Pentobarbital Withdrawal Severity

The method of Buck et al. starts with two inbred strains of mice. Each individual within an inbred strain is essentially a clone, an identical twin of all of the other individuals within that strain. These strains are very stable over time, so studies done 40 or 50 years ago with inbred strains can be directly compared with studies done today. For example, C57BL/6 inbred mice were reported to prefer to drink ethanol solutions (and DBA/2 mice to avoid them) more than 45 years ago (McClearn and Rodgers 1959), and this finding has been replicated many times since. Across a number of mice from a number of inbred strains, characterized for a trait related to drug sensitivity or drug dependence, there will be individual differences in response. To the extent that differences among the average responses of multiple strains are greater than the average within-strain individual differences (which are the nongenetic or environmental individual differences), evidence for a genetic contribution to the trait is provided (Hegmann and Possidente 1981). A community effort called the Mouse Phenome Project, curated by The Jackson Laboratory, has cumulated more than 500 neurobiological traits across 8–40 inbred strains in a database for public access (Bogue 2003).

When manipulated by the tail, mice exhibit a characteristic, mild convulsion called the handling-induced convulsion (HIC). This sign of central nervous system excitability was discovered in the 1970s by Dora Goldstein at Stanford. It is quite reliable, varies quantitatively, and its severity is exacerbated during withdrawal from alcohol dependence (Goldstein and Pal 1971; Goldstein 1972). The two different inbred strains of mice mentioned above, C57BL/6J and DBA/2J, show a very modest basal level of expression of this convulsion sign. The administration of 4 g/kg of ethanol, which is an anesthetic dose, suppresses this little bit of behavioral output for the 2–6 hours that it takes for alcohol to be metabolized and eliminated. During the next several hours, the HIC sign waxes and then wanes in DBA/2J mice, presumably because their central nervous system becomes excitable during alcohol withdrawal. In contrast, C57BL/6J mice show a very modest acute withdrawal reaction (Roberts et al. 1992). The area under the curve describing HIC scores over time is taken as an index of the severity of withdrawal. Following prolonged exposure to alcohol (for example, exposing mice to alcohol vapor), the same HIC index can be used to monitor chronic withdrawal severity, which is much more marked. When many different inbred mouse strains were compared for alcohol withdrawal severity, there were large strain differences, indicating a substantial genetic contribution to this trait (Crabbe et al. 1983; Metten and Crabbe 2005). Finally, the HIC also can be used to index withdrawal from barbiturates, benzodiazepines, or virtually any other drug with sedative effects on the brain (Crabbe et al. 1991). Studies of several sorts indicate that the genetic contributions to acute and chronic withdrawal from ethanol, barbiturates, and benzodiazepines involve substantially, but not completely, the same genes (Metten and Crabbe 1999).

Nominating QTL Regions Using Inbred and Recombinant Inbred Strains

Crossing two inbred strains begins to take advantage of the recombination that occurs during meiosis, where genetic material from one homolog of a chromosome is exchanged for the analogous chromosomal segment of the other homolog (see Fig. 1). With two inbred strains, each strain has by definition two copies of only one allele for each gene—that is, they are obligate homozygotes for all genes. The F_1 hybrid cross animal inherits one chromosome homolog for each chromosome from each progenitor strain. Although all F_1 mice are genetically identical, they are heterozygous for each gene where the progenitor strains differ. When the F_2 generation of a cross between two F_1 mice is created, the chromosomes will experience a number of crossover events (i.e., recombinations) during meiosis. On each chromosome, each F_2 mouse will therefore possess a unique mosaic of chromosomal material (see Fig. 1; Silver 1995; Broman 2005).

If a single pair of F_2 mice is mated, and their offspring are then mated brother to sister, the animals begin to form a new inbred strain. Each generation, the animals become 50% more genetically similar, and after 20 generations of brother-sister matings, there would be essentially no cases in which the offspring do not possess two copies of the same allele at each gene. This new inbred strain is called a recombinant inbred (RI) strain because the end result is to capture one random shuffling of the genetic recombination deck in the F_2 population and render the specific patterns of recombinations replicable from then on. Donald Bailey realized more than 30 years ago that such new RI strains would be extremely valuable for genetic mapping (Bailey 1971). From the F_2 cross of C57BL/6J and DBA/2J, Benjamin Taylor created 26 RI strains at The Jackson Laboratory. These are called BXD RI strains, and several other sets of RI strains are available from The Jackson Laboratory, each starting with a different pair of inbred strains as progenitors.

Each BXD RI strain has also been genotyped, and each has a characteristic pattern of genetic markers, where the allele at each marker is either of C57BL/6J or DBA/2J origin (e.g., see Crabbe and Belknap 1992; Silver 1995). More than 1500 DNA microsatellite markers that have been genotyped in all of the BXD RI strains are systematically scattered

FIGURE 1. Generation of new recombinant inbred (RI) strains from two progenitor inbred strains. A single chromosome is shown schematically. The first row shows mice from two different progenitor inbred mouse strains. Each mouse has two copies of the chromosome, one inherited from each of its parents. Because its parents were inbred, the two homologs are identical. The black dot at the top is the centromere. The blue or red dots represent genes for which the two inbred strains are polymorphic (possess different alleles). The thin black lines represent the rest of the chromosomal DNA, including many genes for which the two inbred progenitors have the same allele. These regions of the chromosome are uninformative for gene mapping. The second row depicts two F_1 hybrid offspring of a cross between the progenitor strains. All F_1 mice are genetically identical (except for the X and Y chromosomes, which are not depicted). However, each is heterozygous at any gene for which the progenitor strains differ. The third row depicts four F_2 offspring of the cross of two F_1 mice. It is in this generation that information relevant for mapping begins to appear. At some point during meiosis, the chromosome homologs may recombine and exchange chromosomal material. This happens on average once per meiosis, so that the average F_2 mouse inherits chromosomal homologs carrying two different crossovers, one from each parent. In the figure, mouse 1 has inherited zero and one crossover on her a and b homologs, respec-

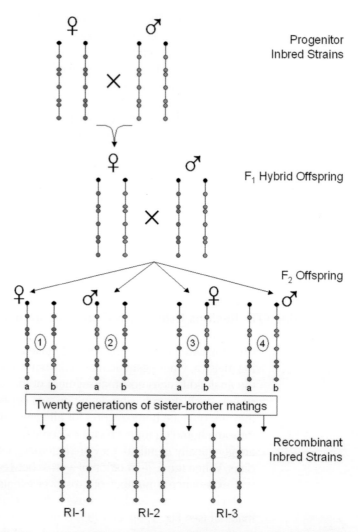

tively. Mice 2 and 3 have each inherited a single crossover per homolog. Mouse 4 has inherited one crossover on his a homolog, but two on his b homolog. Brother-sister pairs of F_2 offspring are then systematically inbred (i.e., to each other) for 20 generations to form new inbred strains. Crossovers continue to occur, but each generation of inbreeding doubles the probability that the two alleles at a gene are identical. By the time new combinations are fixed in the homozygous state in new recombinant inbred strains (row 4), an average of three or four crossovers has been fixed in each new RI strain (for RI strains 1–3, two, three, and four recombinations are depicted, respectively). (Adapted from Grisel and Crabbe 1995, 1996.)

throughout the chromosomes. Thus, we have a relatively dense genetic map or fingerprint for each RI strain.

The utility of RI strains for mapping genes is that the individual strains can be characterized for their average response on a trait to be mapped. Their stability gives this phenotyping the ability to occur over time, using different animals. Because the genotype of each RI strain essentially never changes, one can now calculate the association between a particular RI strain's phenotypic score and whether the genome is C57BL/6J-like or DBA/2J-like in a particular location (Crabbe and Belknap 1992; Palmer and Phillips 2002).

We tested the 21 available BXD RI strains for acute ethanol withdrawal severity using the HIC as described (Belknap et al. 1993; Buck et al. 1997). The strains differed a great deal, with DBA/2J being the most severely withdrawing strain and C57BL/6J among the least. When we then correlated the genotypic score at each marker (zero or one, where a DBA/2J allele was arbitrarily given the value of 1) with the phenotypic (withdrawal) score, we found seven regions of the genome at which there was a tendency toward an association between genotype and phenotype. For example, we found a number of markers in the midregion of mouse chromosome 4 that were very highly correlated with the severity of acute alcohol withdrawal. The BXD RI strains with the DBA/2J genotype in this region had high withdrawal and those with the C57BL/6J genotype had low withdrawal. The markers themselves were generally not functional genes. However, the locations at which a significant association between a marker and the phenotype could be found suggested that a gene of importance for the withdrawal response lay so near the marker that they had not been separated by recombination during the generations of forming the RI strains (Silver 1995).

The advantage of this approach is that it is not necessary to do any genotyping to map a trait—the BXD RI strains have all been genotyped. One limitation is that there were only 21 genotypes. The analogy would be to attempt a genetic association study in 21 humans, and this number of data points is simply insufficient to find a signal from most genes. One of the reasons for this low statistical power is that one is forced to correct for multiple comparisons; the question of whether there was gene nearby this marker was asked more than 1500 times, and at a nominal level of $p < .05$, one might expect to find 60 cases of a false-positive answer of "yes." Even if one looks, instead, for groups of linked markers that give the same answer, 21 genotypes is just not enough to map a gene.

Additional Mapping in B6D2F$_2$ Mice and Short-term Selected Lines

What was necessary, then, was to look at additional populations, do the mapping study again in other ways, and combine the data until the statistical power was great enough to sort the true regions of QTL association from those that were false. From the many available ways to do this, Dr. Buck chose to examine 400 B6D2 F$_2$ animals, where she first ascertained for each animal an individual acute alcohol withdrawal score. She then genotyped these mice with a number of markers selected so that they bracketed the regions of the chromosomes nominally identified in the RI data. Selective genotyping allowed her to reduce the number of statistical tests performed. Because the experiment had been done using the same gene pool (i.e., only C57BL/6J or DBA/2J alleles were available to assort in these populations), the statistical results could be combined. Still, there was insufficient power to prove that any specific region harbored a QTG, for reasons governing the level of statistical association required to rule out false positives (Lander and Kruglyak 1995).

At this stage, Dr. Buck turned to what is probably the oldest technique in behavior genetics, i.e., selective breeding. The method works according to the same reasoning that leads one to breed pointers for show points or select dam and sire to produce horses that run fast but do not fall over. She tested a number of F_2 mice for acute alcohol withdrawal and found a nearly normal distribution for the severity of withdrawal. She then selected the individual mice with the highest withdrawal scores and mated them. Similarly, she selected the lowest-scoring mice for mating for a separate line. She repeated this process for a few generations, and saw that gradually the high-scoring mice produced ever higher scoring offspring, whereas the low scorers showed low withdrawal. In essence, she was using natural selection—exerting selective pressure by choosing who was "fittest" and allowed them to mate.

Selective breeding concentrates alleles that affect the selected trait. In the line selected for high withdrawal (high alcohol withdrawal [HAW] mice), she predicted that all of the alleles that acted to increase withdrawal would increase in frequency over generations, and that the alleles that were protective against withdrawal would concentrate in the low alcohol withdrawal (LAW) line. Selection occurs rapidly under conditions like these, and within four generations, there were marked changes in the genetics of the population. Because she already knew from the RI and F_2 populations the likely QTLs, she did not need to genotype across the whole genome; rather, she selected several markers surrounding each provisional QTL region, which saved statistical power by reducing the number of associations computed.

In the starting F_2 population, the gene frequencies, on average, were .5 for each gene. That is, half of the alleles were from DBA/2J and half were from C57BL/6J. Mice were genotyped after the second and fourth generations of selective breeding. In Figure 2A, the HAW and LAW lines rapidly diverged in their withdrawal severity across generations. In Figure 2B, one can see the gene frequencies for a marker (*Tryp1*) in the middle region of chromosome 4, an area that had been provisionally identified in the BXD RI and the B6D2F_2 results. The frequency of the DBA/2J allele at this marker increased over generations in the HAW line, whereas it decreased in the LAW line (Buck et al. 1997). By comparing the rate at which the allele frequencies changed with what could have occurred due to chance if there had been no selection in matings, she determined that the change in allele frequency was not an accident and represented a true QTL association (Belknap et al. 1997).

Together, these results identified three significant QTL, on chromosomes 1, 4, and 11, and two others, both on chromosome 2, that were suggestive of the presence of a QTL (Buck et al. 1997). The problem then was to find the QTG within the QTL. Mapping is an approximate technique, and at this stage, we knew that each QTL region that had been mapped contained hundreds of genes, any one of which could have been responsible for the QTL association.

Fine Mapping with Congenic and Interval-specific Congenic Lines

The real trick, then, is to isolate a much smaller region of DNA, where the association still holds true. The usual method for doing that is to produce a congenic strain—an inbred strain that differs only in a very, very small proportion of the genome from a background inbred strain. To create a congenic, one starts with an F_1, in our case, the B6XD2 F_1. We elected to create a congenic strain by repetitive backcrossing of C57BL/6J genetic material onto the background of the DBA/2J genotype (see Fig. 3).

FIGURE 2. Short-term selective breeding and QTL mapping. *A* shows the severity of acute alcohol withdrawal plotted versus generations of selective breeding. Starting with the F_2 population created from intercrosses of C57BL/6J x DBA/2J mice (generation zero), the highest-scoring mice were mated to produce high alcohol withdrawal (HAW) mice. The lowest-scoring mice were mated to form the low alcohol withdrawal (LAW) line. Testing and selection were practiced each generation. Over succeeding generations, withdrawal scores in offspring of HAW mice increased, whereas those of LAW mice became slightly smaller. *B* shows the frequency of the DBA/2J-derived allele at the gene *Tryp1*. This coat color locus (previously called "b") was DBA-like 50% of the time in the F_2 generation, and the frequency of DBA/2J alleles increased to nearly 100% as withdrawal severity increased. Because the *Tryp1* gene is in the provisional QTL interval on mouse chromosome 4, this provided further evidence that a linked gene nearby affected withdrawal severity. Means ± S.E. shown. (Reprinted, with permission, from Buck et al. 1997, ©Society for Neuroscience.)

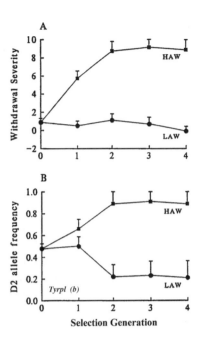

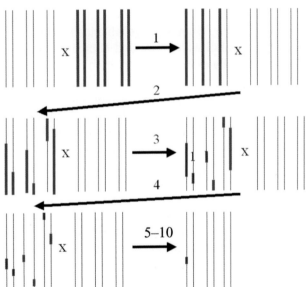

FIGURE 3. Creation of a congenic strain. For the example shown, only three chromosomes are depicted. The color represents two inbred strains, and the chromosomes are shown as homolog pairs, as though the strains were polymorphic at every locus on all chromosomes. The blue (donor) inbred strain is backcrossed to the gray (recipient) strain at step 1. Resulting offspring are F_1 hybrids, which are backcrossed to the recipient strain (step 2). The offspring now have 50% less chromosomal material from the donor strain. Two additional backcrosses (steps 3 and 4) dilute the blue contribution further, and after ten generations (steps 5–10), the result is a congenic line that possesses DNA from the donor strain at less than 1% of its genome. Mating two members of the congenic line shown in the lower right corner will produce one quarter of offspring that are homozygous for the introgressed region of the chromosome (not shown), at which point the stock is called a congenic strain. Breeders are selected at each generation of backcrossing by examining polymorphic markers in the region of the chromosome targeted by the QTL analysis—in this case, a region midway through the leftmost chromosome—and selecting only individuals with donor alleles at those markers.

Consider only the chromosome 4 region surrounding the provisional QTL. When an F_1 animal is backcrossed to a pure DBA/2J inbred, the offspring pick up only the DBA/2J genotype on one chromosome 4 homolog from their pure DBA/2J parent. From their F_1 parent, they pick up a chromosome 4 homolog that may be either the C57BL/6J or DBA/2J genotype. Now, the process is repeated. With the first subsequent backcross generation, the donor chromosome homolog is more likely to have a recombination that makes it, on average, 50% of each genotype. Thus, the offspring of this backcross will now be 75% like DBA/2J and only 25% like C57BL/6J. The process is repeated for another generation, and now the backcross offspring are, on average, 87.5% DBA/2J and 12.5% C57BL/6J. The only chromosome region we care about is that surrounding the chromosome 4 QTL, and offspring are genotyped for several markers in the region to determine what is happening there. After about ten generations, by selecting animals to perpetuate the backcross line that have retained C57BL/6J markers surrounding the QTL, one has produced a congenic strain that is now more than 99% DBA/2J, but possesses only a very small region of C57BL/6J genotype. The conventional approach to producing such a congenic strain requires many hundreds of mice and at least 3 years.

In fact, Buck and her co-workers created a number of these interval-specific congenic lines (ISCL), each of which possessed a different small piece of DNA from C57BL/6J near the QTL. Figure 4 shows this schematically. She then tested each of these congenic lines for acute alcohol withdrawal and compared their severity scores with the pure DBA/2J inbred strain. By examining the pattern of results, it was evident that line ISCL5 possessed the QTG, because these mice showed significantly lower withdrawal than DBA/2J (Fehr et al. 2002). This proved that she had captured the gene in that congenic line. In ISCL5, rather than hundreds of genes, there were only about 30 genes from which to choose. Furthermore, this ISCL affected not only acute ethanol but also acute pentobarbital withdrawal severity.

Continuing to backcross and test mice from the ISCL5 congenic strain, she was able to find more recombinant animals that still showed lower withdrawal scores. What remained after this further analysis was a region containing only 18 genes or gene-related sequences, but which still showed the phenotypic difference on withdrawal. From the dozen genes that could be considered candidate QTGs at this point, there was evidence of differential protein coding as a result of the C57BL/6J versus DBA/2J genotype for only one. This gene, *Mpdz*, coded for a zinc finger protein. A final piece of evidence was obtained by examining the haplotypes and protein variants in Mpdz in 15 standard inbred strains. These inbred strains had all been characterized for the severity of acute alcohol and pentobarbital withdrawal, using exactly the same phenotypic measurement of HIC severity (Metten and Crabbe 1994). They differed markedly in acute alcohol and pentobarbital withdrawal, and the six different haplotypes obtained by examining 18 SNPs were shown to result in three different protein variants of Mpdz. Protein variant correlated strongly with withdrawal severity across the inbred strains (Fehr et al. 2002).

Mpdz, a Quantitative Trait Gene for Acute Withdrawal

No consensus exists for exactly what constitutes proof that a specific gene is the QTG underlying a QTL. Our belief is that the best evidence is to perform a variety of different experiments designed to disprove the candidate QTG. One attempt has been to study the

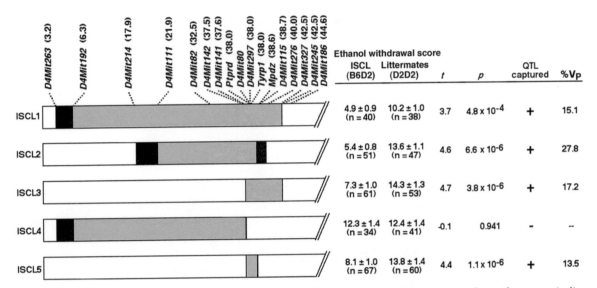

FIGURE 4. Use of congenic lines for QTL fine mapping. (*Left*) Five interval-specific congenic lines (ISCL1–ISCL5) were developed and tested to map the region surrounding a QTL on mouse chromosome 4. MIT microsatellite markers and candidate genes are shown across the top, with their chromosomal position (in centimorgans from the centromere) in parentheses. For each ISCL, the region introgressed from the C57BL/6J donor strain is shown in *gray* (i.e., for all markers in this region, mice of that ISCL had two alleles from C57BL/6J). For regions shown in *white*, all markers had the DBA/2J allele. The boundary regions (*black*) were lacking in an informative marker, so they could have been of either C57BL/6J or DBA/2J origin. (*Right*) Acute ethanol withdrawal scores (mean ± S.E.) and statistical outcomes for the heterozygous ISCL and littermate DBA/2J inbred mice. Alcohol withdrawal was less severe in all ISCL lines except ISCL4. This demonstrated that the introgressed segment from the low withdrawal C57BL/6J strain was able to influence the phenotype, reflecting the effect of a gene captured somewhere in the gray regions of ISCL1, ISCL2, ISCL3, and ISCL5. The very small region remaining in ISCL5 allowed the inference that only genes in this region needed to be further considered as candidates. The percent of the variance in withdrawal accounted for by the gene(s) in each ISCL line was estimated and is also given. These candidates included the gene *Mpdz*. (Reprinted, with permission, from Fehr et al. 2002, ©Society for Neuroscience.)

pattern of gene expression during withdrawal from ethanol in DBA/2J and C57BL/6J mice. These studies found that *Mpdz* was the only gene known to map within the QTL interval on chromosome 4 that was differentially expressed in these strains (Daniels and Buck 2002). Further analyses with reverse transcriptase–polymerase chain reaction (RT-PCR) and in situ hybridization confirm the differential brain-regional expression of *Mpdz* in these strains as well as in the congenic versus background strains. Western analyses of protein levels are also consistent (Shirley et al. 2004).

Mpdz encodes a multiple PDZ (Mpdz) domain protein with 13 PDZ domains. PDZ domain proteins act as intracellular scaffolding proteins that facilitate coupling of binding partners in signal transduction pathways. One of Mpdz's binding partners is 5-HT2C receptors. Serotonin is important for many traits related to alcoholism and alcohol responses (LeMarquand et al. 1994a,b), including the development of tolerance to ethanol (Khanna et al. 1981). Thus, further studies are exploring the possible serotonergic mediation of *Mpdz*-related effects on ethanol responses.

Another approach is to create bacterial artificial chromosome (BAC) transgenics. Here, C57BL/6J-derived BACs containing the *Mpdz* gene sequence have been microinjected into DBA/2J embyros, and for those embryos that incorporate the C57BL/6J *Mpdz* gene, the prediction is that adult mice will show lower withdrawal scores. In the end, "proof" that *Mpdz* is the chromosome 4 QTG will rely on the accumulation of evidence from many sources (Belknap et al. 2001).

SUMMARY

Importance of Animal Models in Behavioral Genomics

Both gene manipulation and "omics" share the feature that they focus on a single end point, either a single gene or a single behavior, respectively. We have argued that we are now able to undertake something that we have called behavioral genomics (Plomin and Crabbe 2000). This perspective attempts to relate multiple genes' interactive and collective effects on the whole organism's behavior. It recognizes that behaviors are themselves complex, may compete, and do not occur in isolation. In addition, it places the focus of analysis on the behaviors themselves rather than allowing a single behavior to represent the complexity of addiction liability. We know that the effects of QTLs affecting withdrawal severity from both alcohol and pentobarbital display epistasis. That is, the efficacy of a QTL (actually, of its as yet unidentified QTG) to affect withdrawal depends on the genotype at other chromosomal locations (Hood et al. 2001; Bergeson et al. 2003).

For gene mapping studies, the newest focus of interest is on a combination of gene expression and gene sequence-based sources of genetic influence. The classic QTL mapping approach just described, which ends with positional cloning of the candidate QTG, relies almost entirely on differences in gene sequence, as they are played out in different protein variants with different neurobiological functions. However, frequently the QTL signals identified from marker-based mapping appear to derive ultimately from differential expression of the underlying QTG producing the signal (Schadt et al. 2003). New methods that wed sequence and expression orientations are proving to be highly successful at identifying the QTGs for complex traits (Jansen and Nap 2001); the rationale for these methods is discussed elsewhere (Hitzemann et al. 2004). A series of papers providing elegant demonstrations of the power of this approach has recently appeared (Bystrykh et al. 2005; Chesler et al. 2005; Hubner et al. 2005). And, many approaches to finding the relevant genes exist, as discussed in a recent review (Flint et al. 2005).

CONCLUSIONS, GAPS, AND OPPORTUNITIES

Translating complex traits into their constituent genetic influences is a difficult task. Although this review has concentrated on the complexities, it should not be concluded that the task is insurmountable. The pace of development of the relevant technologies is accelerating, and what was barely possible 10 years ago is a routine undergraduate laboratory skill today. As genetic data accumulate, more and more gene finding analyses can be performed completely in silico (Grupe et al. 2001; Klein et al. 2004; Liao et al. 2004). However, proof of function will always require the whole human or laboratory animal.

There is increasing hope that knowledge of statistical risk for addiction can gradually be replaced with knowledge about specific risk- or protection-conferring genes.

The *Mpdz* story is just beginning. Studies are underway to determine whether a polymorphism in, near, or affecting the expression of the homologous human gene, *Mpdz*, is present in populations of alcohol or drug abusers. Whereas *Mpdz* may play a pleiotropic role in promoting many kinds of seizures in mice (Fehr et al. 2004), the functional effector system through which *Mpdz* influences withdrawal severity is unknown. One of the receptors for which the protein effects coupling is the serotonin 2C subtype, and studies are actively pursuing this possible pathway. Regardless of how *Mpdz* is found to affect risk, it will explain only part of the story, and other QTLs will need to be promoted to QTGs before we will be able to understand this gene–behavior system more fully.

ACKNOWLEDGMENTS

Preparation of this chapter was supported by grants from the Department of Veterans Affairs, the National Institute on Drug Abuse, and the National Institute on Alcohol Abuse and Alcoholism of the National Institutes of Health. Thanks to Mark Rutledge-Gorman for help with the preparation of this manuscript.

REFERENCES

Bailey D.W. 1971. Recombinant-inbred strains. An aid to finding identity, linkage, and function of histocompatibility and other genes. *Transplantation* **11:** 325–327.

Belknap J.K., Richards S.P., O'Toole L.A., Helms M.L., and Phillips T.J. 1997. Short-term selective breeding as a tool for QTL mapping: Ethanol preference drinking in mice. *Behav. Genet.* **27:** 55–66.

Belknap J.K., Hitzemann R., Crabbe J.C., Phillips T.J., Buck K.J., and Williams R.W. 2001. QTL analysis and genomewide mutagenesis in mice: Complementary genetic approaches to the dissection of complex traits. *Behav. Genet.* **31:** 5–15.

Belknap J.K., Metten P., Helms M.L., O'Toole L.A., Angeli-Gade S., Crabbe J.C., and Phillips T.J. 1993. Quantitative trait loci (QTL) applications to substances of abuse: Physical dependence studies with nitrous oxide and ethanol in BXD mice. *Behav. Genet.* **23:** 213–221.

Bergeson S.E., Warren R.K., Crabbe J.C., Metten P., Erwin V.G., and Belknap J.K. 2003. Chromosomal loci influencing chronic alcohol withdrawal severity. *Mammal. Genome* **14:** 454–463.

Bogue M. 2003. Mouse Phenome Project: Understanding human biology through mouse genetics and genomics. *J. Appl. Physiol.* **95:** 1335–1337.

Bowers B.J., Collins A.C., and Wehner J.M. 2000. Background genotype modulates the effects of γ-PKC on the development of rapid tolerance to ethanol-induced hypothermia. *Addict. Biol.* **5:** 47–58.

Broman K.W. 2005. Mapping expression in randomized rodent genomes. *Nat. Genet.* **37:** 209–210.

Buck K.J., Metten P., Belknap J.K., and Crabbe J.C. 1997. Quantitative trait loci involved in genetic predisposition to acute alcohol withdrawal in mice. *J. Neurosci.* **17:** 3946–3955.

Bystrykh L., Weersing E., Dontje B., Sutton S., Pletcher M.T., Wiltshire T., Su A.I., Vellenga E., Wang J., Manly K.F., Lu L., Chesler E.J., Alberts R., Jansen R.C., Williams R.W., Cooke M.P., and de Haan G. 2005. Uncovering regulatory pathways that affect hematopoietic stem cell function using "genetical genomics." *Nat. Genet.* **37:** 225–232.

Carlborg O. and Haley C.S. 2004. Epistasis: Too often neglected in complex trait studies? *Nat. Rev. Genet.* **5:** 618–625.

Caspi A., Sugden K., Moffitt T.E., Taylor A., Craig I.W., Harrington H., McClay J., Mill J., Martin J., Braithwaite A., and Poulton R. 2003. Influence of life stress on depression: Moderation by a polymorphism in the 5-HTT gene. *Science* **301:** 386–389.

Chesler E.J., Lu L., Shou S., Qu Y., Gu J., Wang J., Hsu H.C., Mountz J.D., Baldwin N.E., Langston M.A., Threadgill D.W., Manly K.F., and Williams R.W. 2005. Complex trait analysis of gene expression uncovers polygenic and pleiotropic networks that modulate nervous system function. *Nat. Genet.* **37:** 233–242.

Crabbe J.C. 2002. Genetic contributions to addiction. *Annu. Rev. Psychol.* **53:** 435–462.

Crabbe J.C. and Belknap J.K. 1992. Genetic approaches to drug dependence. *Trends Pharmacol. Sci.* **13:** 212–219.

Crabbe J.C. and Phillips T.J. 2004. Pharmacogenetic studies of alcohol self-administration and withdrawal. *Psychopharma-*

cology **174:** 539–560.

Crabbe J.C., Belknap J.K., and Buck K.J. 1994. Genetic animal models of alcohol and drug abuse. *Science* **264:** 1715–1723.

Crabbe J.C., Merrill C.D., and Belknap J.K. 1991. Acute dependence on depressant drugs is determined by common genes in mice. *J. Pharmacol. Exp. Ther.* **257:** 663–667.

Crabbe J.C., Young E.R., and Kosobud A. 1983. Genetic correlations with ethanol withdrawal severity. *Pharmacol. Biochem. Behav.* (suppl.1) **18:** 541–547.

Crabbe J.C., Phillips T.J., Buck K.J., Cunningham C.L., and Belknap J.K. 1999. Identifying genes for alcohol and drug sensitivity: Recent progress and future directions. *Trends Neurosci.* **22:** 173–179.

Daniels G.M. and Buck K.J. 2002. Expression profiling identifies strain-specific changes associated with ethanol withdrawal in mice. *Genes Brain Behav.* **1:** 35–45.

Enoch M.A. and Goldman D. 2001. The genetics of alcoholism and alcohol abuse. *Curr. Psychiat. Rep.* **3:** 144–151.

Fehr C., Shirley R.L., Belknap J.K., Crabbe J.C., and Buck K.J. 2002. Congenic mapping of alcohol and pentobarbital withdrawal liability loci to a <1 centimorgan interval of murine chromosome 4: Identification of *Mpdz* as a candidate gene. *J. Neurosci.* **22:** 3730–3738.

Fehr C., Shirley R.L., Metten P., Kosobud A.E., Belknap J.K., Crabbe J.C., and Buck K.J. 2004. Potential pleiotropic effects of *Mpdz* on vulnerability to seizures. *Genes Brain Behav.* **3:** 8–19.

Fischer G., Ibrahim S.M., Brockmann G.A., Pahnke J., Bartocci E., Thiesen H.J., Serrano-Fernandez P., and Moller S. 2003. Expressionview: Visualization of quantitative trait loci and gene-expression data in Ensembl. *Genome Biol.* **4:** R77.

Flint J. 2001. Is this mouse anxious? The difficulties of interpreting the effects of genetic action. *Behav. Pharmacol.* **12:** 461–465.

———. 2003. Analysis of quantitative trait loci that influence animal behavior. *J. Neurobiol.* **54:** 46–77.

Flint J., Valdar W., Shifman S., and Mott R. 2005. Strategies for mapping and cloning quantitative trait genes in rodents. *Nat. Rev. Genet.* **6:** 271–286.

Gerlai R. 1996. Gene targeting in neuroscience: The systemic approach. *Trends Neurosci.* **19:** 188–189.

Goldstein D.B. 1972. Relationship of alcohol dose to intensity of withdrawal signs in mice. *J. Pharmacol. Exp. Ther.* **180:** 203–215.

Goldstein D.B. and Pal N. 1971. Alcohol dependence produced in mice by inhalation of ethanol: Grading the withdrawal reaction. *Science* **172:** 288–290.

Grisel J.E. and Crabbe J.C. 1995. Quantitative trait loci mapping. *Alcohol Health Res. World* **19:** 220–227.

———. 1996. Genetic basis of alcoholism and drug addiction. In *Encyclopedia of molecular biology and molecular medicine* (ed. R.A. Meyers), pp. 146–155. VCH Publishers, New York.

Grupe A., Germer S., Usuka J., Aud D., Belknap J.K., Klein R.F., Ahluwalia M.K., Higuchi R., and Peltz G. 2001. In silico mapping of complex disease-related traits in mice. *Science* **292:** 1915–1918.

Gunzerath L. and Goldman D. 2003. G x E: A NIAAA workshop on gene-environment interactions. *Alcohol. Clin. Exp. Res.*

27: 540–562.

Harker K.T. and Whishaw I.Q. 2004. A reaffirmation of the retrosplenial contribution to rodent navigation: Reviewing the influences of lesion, strain, and task. *Neurosci. Biobehav. Rev.* **28:** 485–496.

Hegmann J.P. and Possidente B. 1981. Estimating genetic correlations from inbred strains. *Behav. Genet.* **11:** 103–114.

Hitzemann R., Reed C., Malmanger B., Lawler M., Hitzemann B., Cunningham B., McWeeney S., Belknap J., Harrington C., Buck K., Phillips T., and Crabbe J. 2004. On the integration of alcohol-related quantitative trait loci and gene expression analyses. *Alcohol. Clin. Exper. Res.* **28:** 1437–1448.

Homanics G.E. 2002. Knockout and knockin mice. In *Methods in alcohol-related neuroscience research* (ed. Y. Liu et al.), pp. 31–61. CRC Press, Boca Raton, Florida.

Hood H.M., Belknap J.K., Crabbe J.C., and Buck K.J. 2001. Genomewide search for epistasis in a complex trait: Pentobarbital withdrawal convulsions in mice. *Behav. Genet.* **31:** 93–100.

Hubner N., Wallace C.A., Zimdahl H., Petretto E., Schulz H., Maciver F., Mueller M., Hummel O., Monti J., Zidek V., Musilova A., Kren V., Causton H., Game L., Born G., Schmidt S., Muller A., Cook S.A., Kurtz T.W., Whittaker J., Pravenec M., and Aitman T.J. 2005. Integrated transcriptional profiling and linkage analysis for identification of genes underlying disease. *Nat. Genet.* **37:** 243–253.

Hurley T.D., Edenberg H.J., and Li T.-K. 2002. Pharmacogenomics of alcoholism. In *Pharmacogenomics: The search for individualized therapies* (ed. J. Licinio et al.), pp. 417–439. Wiley-VCH, Weinheim, Germany.

Jacob T., Sher K.J., Bucholz K.K., True W.T., Sirevaag E.J., Rohrbaugh J., Nelson E., Neuman R.J., Todd R.D., Slutske W.S., Whitfield J.B., Kirk K.M., Martin N.G., Madden P.A., and Heath A.C. 2001. An integrative approach for studying the etiology of alcoholism and other addictions. *Twin Res.* **4:** 103–118.

Jansen R.C. and Nap J.P. 2001. Genetical genomics: The added value from segregation. *Trends Genet.* **17:** 388–391.

Kendler K.S., Jacobson K.C., Prescott C.A., and Neale M.C. 2003. Specificity of genetic and environmental risk factors for use and abuse/dependence of cannabis, cocaine, hallucinogens, sedatives, stimulants, and opiates in male twins. *Am. J. Psychiat.* **160:** 687–695.

Kerns R.T.R.A., Hassan S., Cage M.P., York T., Sikela J.M., Williams R.W., and Miles M.F. 2005. Ethanol-responsive brain region expression networks: Implications for behavioral responses to acute ethanol in DBA/2J versus C57BL/6J mice. *J. Neurosci.* **25:** 2255–2266.

Khanna J.M., Kalant H., Le A.D., and LeBlanc A.E. 1981. Role of serotonergic and adrenergic systems in alcohol tolerance. *Prog. Neuro-Psychopharmacol.* **5:** 459–465.

Klein R.F., Allard J., Avnur Z., Nikolcheva T., Rotstein D., Carlos A.S., Shea M., Waters R.V., Belknap J.K., Peltz G., and Orwoll E.S. 2004. Regulation of bone mass in mice by the lipoxygenase gene Alox15. *Science* **303:** 229–232.

Kliethermes C.L. 2005. Anxiety-like behaviors following chronic ethanol exposure. *Neurosci. Biobehav. Rev.* **28:** 837–850.

Klose J., Nock C., Herrmann M., Stuhler K., Marcus K., Bluggel M., Krause E., Schalkwyk L.C., Rastan S., Brown S.D., Bussow K., Himmelbauer H., and Lehrach H. 2002. Genetic analysis of the mouse brain proteome. *Nat. Genet.* **30:** 385–393.

Korstanje R. and Paigen B. 2002. From QTL to gene: The harvest begins. *Nat. Genet.* **31:** 235–236.

Kreek M.J. 2001. Drug addictions. Molecular and cellular endpoints. *Ann. N.Y. Acad. Sci.* **937:** 27–49.

Lander E. and Kruglyak L. 1995. Genetic dissection of complex traits: Guidelines for interpreting and reporting linkage results. *Nat. Genet.* **11:** 241–247.

LeMarquand D., Pihl R.O., and Benkelfat C. 1994a. Serotonin and alcohol intake, abuse, and dependence: Findings of animal studies. *Biol. Psychiat.* **36:** 395–421.

———. 1994b. Serotonin and alcohol intake, abuse, and dependence: Clinical evidence. *Biol. Psychiat.* **36:** 326–337.

Liao G., Wang J., Guo J., Allard J., Cheng J., Ng A., Shafer S., Puech A., McPherson J.D., Foernzler D., Peltz G., and Usuka J. 2004. In silico genetics: Identification of a functional element regulating H2-Ealpha gene expression. *Science* **306:** 690–695.

Lister R.G. 1987. The use of a plus-maze to measure anxiety in the mouse. *Psychopharmacology* **92:** 180–185.

Lomber S.G. 1999. The advantages and limitations of permanent or reversible deactivation techniques in the assessment of neural function. *J. Neurosci. Methods* **86:** 109–117.

Mackay T.F. 2001. The genetic architecture of quantitative traits. *Annu. Rev. Genet.* **35:** 303–339.

Marchini J., Donnelly P., and Cardon L.R. 2005. Genome-wide strategies for detecting multiple loci that influence complex diseases. *Nat. Genet.* **37:** 413–417.

McClearn G.E. 1979. Genetics and alcoholism simulacra. *Alcohol. Clin. Exp. Res.* **3:** 255–258.

McClearn G.E. and Rodgers D.A. 1959. Differences in alcohol preference among inbred strains of mice. *Quart. J. Stud. Alcohol* **20:** 691–695.

Metten P. and Crabbe J.C. 1994. Common genetic determinants of severity of acute withdrawal from ethanol, pentobarbital and diazepam in inbred mice. *Behav. Pharmacol.* **5:** 533–547.

———. 1999. Genetic determinants of severity of acute withdrawal from diazepam in mice: Commonality with ethanol and pentobarbital. *Pharmacol. Biochem. Behav.* **63:** 473–479.

———. 2005. Alcohol withdrawal severity in inbred mouse strains. *Behav. Neurosci.* (in press).

Muller U. 1999. Ten years of gene targeting: Targeted mouse mutants, from vector design to phenotype analysis. *Mech. Dev.* **82:** 3–21.

Palmer A.A. and Phillips T.J. 2002. Quantitative trait locus (QTL) mapping in mice. In *Methods in alcohol-related neuroscience research* (ed. Y. Liu et al.), pp. 1–30. CRC Press, Boca Raton, Florida.

Picciotto M.R. and Corrigall W.A. 2002. Neuronal systems underlying behaviors related to nicotine addiction: Neural

circuits and molecular genetics. *J. Neurosci.* **22:** 3338–3341.

Plomin R. and Crabbe J. 2000. DNA. *Psychol. Bull.* **126:** 806–828.

Quaid K.A., Dinwiddie S.H., Conneally P.M., and Nurnberger J.I., Jr. 1996. Issues in genetic testing for susceptibility to alcoholism: Lessons from Alzheimer's disease and Huntington's disease. *Alcohol. Clin. Exp. Res.* **20:** 1430–1437.

Reilly M.T., Fehr C., and Buck K.J. 2001. Alcohol and gene expression in the central nervous system. In *Nutrient-gene interactions in health and disease* (ed. N. Moussa-Moustaid et al.), pp. 131–162. CRC Press, Boca Raton, Florida.

Reith A.D. and Bernstein A. 1991. Molecular basis of mouse developmental mutants. *Genes Dev.* **5:** 1115–1123.

Roberts A.J., Crabbe J.C., and Keith L.D. 1992. Genetic differences in hypothalamic-pituitary-adrenal axis responsiveness to acute ethanol and acute ethanol withdrawal. *Brain Res.* **579:** 296–302.

Rodgers R.J. and Dalvi A. 1997. Anxiety, defence and the elevated plus-maze. *Neurosci. Biobehav. Rev.* **21:** 801–810.

Schadt E.E., Monks S.A., Drake T.A., Lusis A.J., Che N., Colinayo V., Ruff T.G., Milligan S.B., Lamb J.R., Cavet G., Linsley P.S., Mao M., Stoughton R.B., and Friend S.H. 2003. Genetics of gene expression surveyed in maize, mouse and man. *Nature* **422:** 297–302.

Schuckit M.A., Smith T.L., and Kalmijn J. 2004. The search for genes contributing to the low level of response to alcohol: Patterns of findings across studies. *Alcohol. Clin. Exp. Res.* **28:** 1449–1458.

Shirley R.L., Walter N.A., Reilly M.T., Fehr C., and Buck K.J. 2004. *Mpdz* is a quantitative trait gene for drug withdrawal seizures. *Nat. Neurosci.* **7:** 699–700.

Silver LM. 1995. *Mouse genetics.* Oxford University Press, New York.

Stephens D.N., Mead A.N., and Ripley T.L. 2002. Studying the neurobiology of stimulant and alcohol abuse and dependence in genetically manipulated mice. *Behav. Pharmacol.* **13:** 327–345.

Uhl G.R. 2004. Molecular genetics of substance abuse vulnerability: Remarkable recent convergence of genome scan results. *Ann. N.Y. Acad. Sci.* **1025:** 1–13.

van den Bree M.B., Johnson E.O., Neale M.C., and Pickens R.W. 1998. Genetic and environmental influences on drug use and abuse/dependence in male and female twins. *Drug Alcohol Depend.* **52:** 231–241.

Wade M.J. 2001. Epistasis, complex traits, and mapping genes. *Genetica* **112–113:** 59–69.

Whitfield J.B., Zhu G., Madden P.A., Neale M.C., Heath A.C., and Martin N.G. 2004. The genetics of alcohol intake and of alcohol dependence. *Alcohol. Clin. Exp. Res.* **28:** 1153–1160.

Yalcin B., Willis-Owen S.A., Fullerton J., Meesaq A., Deacon R.M., Rawlins J.N., Copley R.R., Morris A.P., Flint J., and Mott R. 2004. Genetic dissection of a behavioral quantitative trait locus shows that Rgs2 modulates anxiety in mice. *Nat. Genet.* **36:** 1197–1202.

5 | Endorphins, Gene Polymorphisms, Stress Responsivity, and Specific Addictions: Selected Topics

Mary Jeanne Kreek

Laboratory of the Biology of Addictive Diseases, The Rockefeller University,
The Rockefeller University Hospital, New York, New York 10021

ABSTRACT

This chapter reviews the work of my Laboratory of the Biology of Addictive Diseases at The Rockefeller University on selected topics related to genetics, stress responsivity, and the role of the endogenous opioid system in specific addictions. Summarized here are some aspects of epidemiology as well as our earliest work leading to the first (and still most) effective pharmacotherapy for an addiction, i.e., methadone maintenance treatment for heroin addiction. I also highlight evidence that addictions are metabolic disorders with a genetic component resulting, in part, from abnormalities in the hypothalamic-pituitary–axis response to stress.

INTRODUCTION

Prevalence of Addiction to Alcohol, Heroin, and Cocaine Use

The prevalence of specific drug abuse and addiction problems and the vulnerability to develop addictions should preface any discussion on the relevance of basic and clinical research findings to the addictions. Both "drug abuse" and "drug addiction" have been defined by the American Psychiatric Association in their specific diagnostic criteria (the most recent of which is called DSM-IV). These criteria were developed primarily for outreach in identifying persons who need psychiatric intervention or care, as well as for related reimbursement purposes. However, they were not developed for research use. One of the most widely accepted definitions of addiction, which is operationally probably the best to use, is one that is adapted here from the World Health Organization: "Drug addiction is compulsive drug-seeking behavior and drug self-administration without regard for negative consequences to self and others." A more stringent diagnosis, developed by the United States government in the Federal Regulations for entry into opioid agonist treatment with methadone or levo-alpha acetylmethadol (LAAM), is the definition of "multiple daily self-administrations of illicit opiates for one year or more with development of

This chapter is based on and excerpted from the Cold Spring Harbor Laboratory course lecture of August 5, 2003.

tolerance and physical dependence and without regard to potential harm to self, others, or society," as cited and discussed in the National Academy of Sciences Institute of Medicine Report of 1995, on consideration of revising the Federal Guidelines for opioid agonist treatment (Rettig and Yarmolinsky 1995).

We have recently developed and validated a further assessment of drug abuse and drug addiction that is extremely useful for research purposes, because it yields a quantitative assessment of magnitude, duration, and timing of any drug abuse; the first of these scales was developed for use for lifetime, and we now, in addition, use this scale a second time to ascertain drug use in the last 30 days (Kellogg et al. 2005).

It is now estimated that whereas more than 177 million persons over the age of 16 in the United States have used alcohol, approximately 15 million are alcoholics or have the disease of alcoholism. More than 26 million are estimated to have used cocaine at some time, and 2–3 million are estimated to be cocaine-addicted, meeting DSM-IV criteria. Various studies sponsored by the National Institutes of Health as well as other government agencies have also estimated that 3.6–3.7 million individuals in the United States have used heroin at some time (http://www.oas.samhsa.gov/nhsda/2k3tabs/Sect1peTabs1to66. htm#tab1.26a) and that approximately 1 million are addicted to heroin.

Incidence of Illicit Use of Prescription Medicines

During the last 3 years, there has been increasing recognition of the very widespread illicit use of prescription opiates, sometimes illicit use from the start of self-administration and, in other cases, continuing use of pain medications prescribed originally for a limited time course only. It is estimated now by the National Institute on Drug Abuse that between 4 and 8 million have illicitly used prescription opiates at some time (http://www. oas.samhsa.gov/nhsda/2k3tabs/Sect1peTabs1to66.htm#tab1.26a). However, the resultant numbers of persons addicted to specific opiate medicines have eluded definition and may be very difficult to determine because of various issues of confidentiality, along with reticence to report such problems on the part of both patients and their physicians.

Incidence of Abusers Who Become Addicted

In looking at these data from a different standpoint, it is of interest that according to our own meta-analyses of numerous epidemiological surveys during the last 20 years, approximately 1 in 8 to 1 in 15 of those who ever self-administer alcohol or cocaine will become addicted to those agents; in sharp contrast, it is of considerable interest and potential importance that 1 in 3 to 1 in 5 who ever self-administer heroin will become addicted to opiates. Our analyses were made using several surveys during several years, including the government and government-funded National Household Survey on Drug Abuse (now called the "National Survey on Drug Use and Health," http://www.icpsr.umich.edu/SAMHDA and http://www.oas.samhsa.gov/nhsda/2k3tabs/toc.htm), The DAWN Network Reports of Emergency Room Visits (http://dawninfo.samhsa.gov), the National Comorbidity Survey (http://www.hcp.med.harvard.edu/ncs), and the Monitoring the Future Reports of the University of Michigan (http://monitoringthefuture.org). Our estimates are very similar to those reported by Anthony et al. (1994) in their more formal meta-analysis.

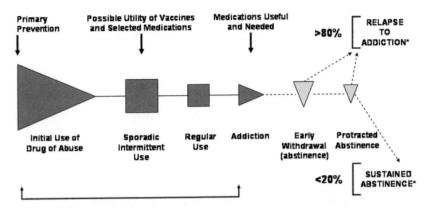

SHORT ACTING OPIOIDS: 1 in 3 to 1 in 5 who ever self-administer progress to addiction
COCAINE: 1 in 8 to 1 in 15 who ever self-administer progress to addiction
ALCOHOL: 1 in 8 to 1 in 15 who ever self-administer progress to addiction

*with no medications

FIGURE 1. This diagram depicts the progression from any initial self-administration of a potential drug of abuse and the progression toward regular use and addiction, as well as the outcome of attempts at recovery by abstinence-based approaches alone. Data show that less than 20% of those who reach accepted criteria for addiction are able to remain in sustained complete abstinence with no illicit use of any drug or alcohol; this is in contrast to findings made when appropriate, targeted, effective pharmacotherapies are combined with behavioral treatments, in which case, sustained abstinence from any illicit or excessive use of drugs or alcohol can be achieved in more than 80% of subjects. Estimates based on 50 years of diverse epidemiologic studies from the United States and Europe of the numbers of individuals who progress to addiction after self-exposure are also provided. (Modified, with permission, from Kreek et al. 2002.)

Natural History of Drug Abuse and Addiction

The natural history of drug abuse and addiction is also worthy of consideration (see Fig. 1). Self-administration of a drug of abuse must, by definition, precede abuse or addiction. Unfortunately, it is only before initial self-administration that primary prevention (or "just say no") has been shown to be effective. Sporadic intermittent drug use will proceed in some individuals. Once addicted, very few persons are able to be treated without targeted medications, i.e., by abstinence-based approaches alone, irrespective of how effective and humane those approaches may be. Studies have shown that less than 20% of long-term heroin addicts are able to remain in a heroin- or illicit-opiate-free state on a permanent, long-term basis. More than 80% will relapse to opiate abuse and addiction within 1 year. For other addictions, similar data have been forthcoming. We are very fortunate now to have three effective pharmacotherapies for opiate addiction: long-term methadone maintenance treatment, long-term buprenorphine maintenance treatment, and although now used to a very limited extent because of some possible medical adverse effects in a few individuals, long-term maintenance treatment with LAAM (Kreek 2000a; Kreek and Vocci 2002). In addition, it has been repeatedly shown that behavioral treatment must accompany such pharmacotherapy and that adequate doses of both the opiate agonist or partial agonist, as well as behavioral treatment, must be used (McLellan et al. 1993; Kreek 2000a; Kreek and Vocci 2002; Kreek et al. 2002).

Development of Methadone Maintenance Treatment

Our own interest in studying addictive diseases goes back now more than 40 years (Dole et al. 1966a,b; Kreek 1972, 1973a,b, 1987, 2000a; Kreek and Vocci 2002; Kreek et al. 2002). At that time, I had the privilege to be recruited by Dr. Vincent Dole. At the same time, he recruited the late Dr. Marie Nyswander. Dr. Nyswander was a psychiatrist who had worked for many years in the field and had written a book on the problems of heroin addicts seeking treatment. As for myself, I was a young clinical investigator who had had considerable clinical and laboratory research experience, despite my level of training, because of many summers and some winters spent working in the laboratory of the late Frederic C. Bartter at the National Institutes of Health, as well as while a medical student at Columbia University College of Physicians & Surgeons in the laboratory of the late Donald Tapley (e.g., see Kreek et al. 1963).

We joined together in the initial work that led to the development of methadone maintenance treatment, including clinical research studies at The Rockefeller University Hospital that documented our hypothesized mechanism of action of this long-acting, orally effective synthetic opioid in humans, i.e., through the mechanism of tolerance and cross-tolerance (Dole et al. 1966a,b). We showed that methadone, which we found to be long-acting in humans ($t_{1/2}$ hr) as compared with heroin or morphine, which are very short-acting (3 min and 4 hr, respectively), administered in steady moderate-to-high doses (80–120 mg/d) would prevent withdrawal symptoms and also reduce or eliminate drug craving; it would also prevent the subjective perception or objective measurements of any superimposed short-acting narcotic, such as heroin, of an "affordable dose" amount, and thus with a dose-dependent effectiveness in cross-tolerance (see Fig. 2) (Dole et al. 1966a,b).

Of even greater conceptual importance was our initial hypothesis that, in retrospect, was frame-shifting for the entire field of neurobiological research pertaining to the addictive diseases. We hypothesized that (heroin) addiction is a disease—a metabolic disease—of the brain with resultant behaviors of "drug hunger" and drug self-administration, despite negative consequences to self and others and, furthermore, that heroin addiction is not simply a criminal behavior or attributable alone to antisocial personality or some other personality disorder (Table 1).

Methadone maintenance treatment for opiate (primarily heroin) addiction has turned out to be extremely effective, both in the United States and worldwide (Table 2). In addition, other opioid agonist and partial agonist treatments derived from the original research with methadone maintenance have proven to be effective, including treatment with LAAM and also the partial agonist buprenorphine, especially when combined with naloxone to prevent abuse liability (Table 2) (Kreek et al. 2002). However, as we hypothesized, opioid antagonist treatment has not been effective in the treatment of unselected heroin addicts, except in a limited number of cases where severe alternatives of negative contingencies exist, such as parolees who will be sent back to jail if any relapse occurs, in those states where, unfortunately, opioid agonist treatment with methadone or buprenorphine is not allowed (see Table 2).

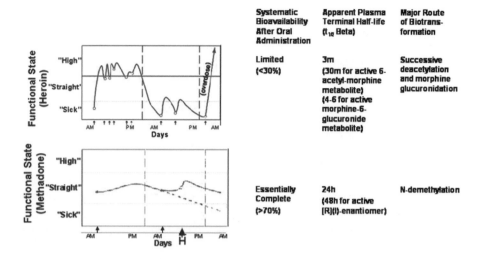

Systematic Bioavailability After Oral Administration	Apparent Plasma Terminal Half-life (t₁β Beta)	Major Route of Biotrans- formation
Limited (<30%)	3m (30m for active 6-acetyl-morphine metabolite) (4-6 for active morphine-6-glucuronide metabolite)	Successive deacetylation and morphine glucuronidation
Essentially Complete (>70%)	24h (48h for active [R](l)-enantiomer)	N-demethylation

FIGURE 2. The left side shows a pictorial representation of clinical observations made of the pharmacodynamics of the short-acting opioid heroin, as contrasted to the long-acting orally effective opioid, methadone. These observations were made at The Rockefeller University Hospital in 1964 by the original research team pioneering the use of long-term pharmacotherapy for the management of heroin addiction (Dole et al. 1966a,b). In 1964, no analytical technologies were available to determine plasma levels or even urine levels of opiates and opioids. On the right is a summary of the early and repeatedly validated pharmacokinetic studies done 8–10 years later by the independent groups of Inturrisi and Kreek concerning the pharmacokinetics of biotransformation of heroin following intravenous administration and methadone following oral administration in human subjects. The heroin pharmacokinetic studies were conducted in pain patients receiving morphine and morphine-equivalent analgesia on a long-term basis. The methadone pharmacokinetic studies were conducted in well-stabilized, long-term methadone maintenance patients. The stable isotope studies were also conducted in long-term, stabilized, methadone-maintained patients using intravenous administration of one of two deuterated species of methadone D₃ and D₅, coupled with the use of a third deuterated species, D₈, with gas chromatography and chemical ionization mass spectrometry with selected ion monitoring for plasma measurements. (Graphs reprinted, with permission, from Dole et al. 1966a [©American Medical Association]; data from Inturrisi and Verebely 1972a,b; Kreek 1973c; Kreek et al. 1976; Inturrisi et al. 1984.)

GOALS OF THE CHAPTER

The goals of this chapter are to highlight the current prevalence of alcohol, heroin, and cocaine use and addiction; to provide an overview of the natural history of drug abuse and addiction; to highlight the earliest research leading to the development of methadone maintenance treatment; and to present the 1964 hypothesis that addictions are metabolic disorders of the brain with behavioral manifestations. In addition, a very brief overview is presented on the role of the stress-responsive hypothalamic-pituitary-adrenal (HPA) axis and the role of specific components of the endogenous opioid system, including the interrelationships between stress responsivity and the opioid system, in the development of

TABLE 1. Development of Methadone Maintenance Treatment (1964 Onward)

Initial clinical research on mechanisms and treatment using methadone maintenance pharmacotherapy at The Rockefeller Hospital of The Rockefeller Institute for Medical Research (by the mid-1960s, The Rockefeller University) performed by the team of

Vincent P. Dole, Jr., M.D., *Professor and Head of the Laboratory of Physiology and Metabolism (now Professor Emeritus).*

Marie Nyswander, M.D., *Guest Investigator. Joined Dole Lab in winter 1964 (now deceased).*

Mary Jeanne Kreek, M.D., *Guest Investigator. Joined Dole Lab in winter 1964 (now Professor and Head of the Lab).*

First publications describing methadone maintenance treatment research

- **1964:** Initial clinical research on mechanisms and treatment using methadone maintenance pharmacotherapy performed at The Rockefeller Hospital of The Rockefeller Institute for Medical Research:

  Dole et al. 1966a (also recorded in the Association of American Physicians meeting transcription of discussion).

- **1964:** Translational applied clinical research performed at Manhattan General Hospital:

  Dole and Nyswander 1965.

TABLE 2. Methadone Maintenance Treatment for Opiate (Heroin) Addiction and Overall Opiate Addiction Treatment Outcome

Number of patients in treatment: 212,000 (United States), >500,000 (worldwide)

Efficacy in "good" treatment programs using adequate methadone doses (80–150 mg/d) and adequate behavioral treatment:

Voluntary retention in treatment (1 year or more)	50–80%
Continuing use of illicit heroin	5–20%

Actions of methadone treatment:
- Prevents withdrawal symptoms and "drug hunger"
- Blocks euphoric effects of short-acting narcotics
- Allows normalization of disrupted physiology

Mechanism of action:

Long-acting narcotic provides steady levels of opioid at specific μ receptor sites (in recent studies, methadone was found to be a full μ-opioid receptor agonist, which internalizes like endorphins and also has modest NMDA receptor complex antagonism)

Opiate addiction treatment outcome[a]

Methadone maintenance	50–80%
Buprenorphine-naloxone maintenance	40–50%[b]
LAAM[c] maintenance	50–80%
Naltrexone maintenance	10–20%
"Drug free" (nonpharmacotherapeutic)	5–30%
Short-term detoxification (any mode)	5–20%

Data from Kreek (1996a, 2000a, 2002); Kreek and Vocci (2002); Kreek et al. (2002).

[a]One-year retention in treatment and/or follow-up with significant reduction or elimination of illicit use of opiates.

[b]Maximum effective sublingual dose (24–32 mg sl) equivalent to about 60–70 mg/d oral methadone. (Data based on 6-month follow-up only.)

[c]Use curtailed because of limited manufacture in the United States (2003); use stopped in much of Europe because of concern about QTc-interval prolongation in some patients.

specific addictive diseases. Finally, some very recent findings are described, primarily from our laboratory, on specific polymorphisms of the endogenous opioid system genes, which include both functional variants and, in a few cases, have been found to have a very significant association with specific addictive diseases.

ADDICTION DISRUPTS MOLECULAR NEUROBIOLOGY

Rapid Increases and Decreases in Opiate Levels Disrupt Normal Homeostasis

Our work in humans (and very strongly supported by our work in animal models) has shown that disruption of levels of gene expression, peptides, receptor-mediated events (including signal transduction systems), neurochemistry, physiology, and thus behavior occurs with the very short-acting "on/off" effects of drugs of abuse, administered or self-administered chronically. However, steady-state exposure to the long-acting opioid agonist methadone—for which pharmacokinetics render it long-acting in humans (24-hr half-life for the racemic mixture; 36-hr half-life for the active l[R] enantiomer) or in rats and mice where, because of the short-acting half-life of methadone in rodents (90 min for rats, 60 min for mice), steady-state pump infusion is essential—results in normalization of disrupted events (such as changes in gene expression and levels of hormones of the important stress-responsive HPA axis) or no disruption if only steady-state administration is ever used (see Fig. 2) (Kreek 1973c, 1979, 2000a; Burstein et al. 1980; Zhou et al. 1996c; Kreek and Vocci 2002).

"On/Off" Effects of Drugs of Abuse Cause Atypical Responsivity to Stress, Contributing to Drug Abuse and Addiction

For more than 20 years, our laboratory has conducted bidirectional translational research. In addition to studying in humans, when possible, perturbations of specific physiologic functions and using novel challenge compounds or potential therapeutic agents, whenever possible, based on findings made in laboratory studies, we have also used clinical observations of addicts, including especially the patterns and modes of drug self-administration during addiction to develop novel animal models (see Table 3).

We have used these novel models for our studies of both stress responsivity and the role of the endogenous opioid system components in specific addictive diseases. In our studies in both humans and animal models, we have shown that stress responsivity, involving the HPA axis, has a major role in specific addictive diseases (Fig. 3, left panel).

In the late 1960s, we postulated—and started to incorporate into our prospective and one-point-in-time initial studies, conducted to determine the physiological effects and medical safety of methadone maintenance treatment—that an atypical responsivity to stressors caused by chronic use of drugs of abuse may, in part, contribute to the persistence of and relapse to self-administration of drugs of abuse and addiction (see Table 4).

Furthermore, we hypothesized that such atypical responsivity to stress and stressors in some individuals may exist before the use of addictive drugs on a genetic or acquired basis and may lead to the development of drug addiction (Table 4) (Kreek 1972, 1973a,b, 1987, 1992, 1996a,b, 1997a,b, 2000b; Kreek et al. 1981, 1983, 1984, 2002; Kreek and Hartman 1982).

TABLE 3. Novel and Conventional Animal Models

- **"Binge" pattern cocaine administration model:**
  Constant or ascending dose
  Mimics most common pattern of human use in addiction

- **Intermittent morphine (heroin) administration model:**
  Constant or ascending dose
  Mimics most common pattern of human use in addiction

- **"Binge" pattern oral ethanol administration model:**
  Mimics common pattern of human excessive use

- **Pump methadone administration model:**
  Converts short-acting pharmacokinetic properties of opioid agonist in rodent to long-acting human
  pharmacokinetic profile

- **Extended access self-administration with or without high-dose drug**
  (cocaine or opiate)

Laboratory of the Biology of Addictive Diseases (1987–2005).

We have documented that such atypical stress responsivity occurs in early or late abstinence in human heroin, cocaine, and alcohol addicts (see Table 5). We have shown in innumerable studies in animal models using "binge" pattern cocaine, intermittent administration of morphine, and also bolus and "binge" administration of alcohol, as well as self-administration paradigms, that specific components of the HPA stress-responsive axis are profoundly altered by the intermittent effects of a drug of abuse (e.g., see Fig. 3) (Zhou et al. 1996a,b, 1998, 1999a,b, 2000, 2002, 2003a,b; Spangler et al. 1997a; Mantsch et al. 2000, 2001, 2003; Yuferov et al. 2001).

In studies conducted during the last 40 years, we have found profound neuroendocrine effects of each of three major drugs of abuse (opiates, cocaine, and alcohol) on the HPA stress-responsive axis in humans (Table 5) (Kreek 1972, 1973a,b, 1978; Kreek et al. 1981, 1983, 1984; Kreek and Hartman 1982; Kosten et al. 1986, 1987; Kennedy et al. 1990; Culpepper-Morgan et al. 1992; Rosen et al. 1996; Culpepper-Morgan and Kreek 1997; King et al. 1997, 2002; Schluger et al. 1998, 2001, 2003; Jacobsen et al. 2001; O'Malley et al. 2002; Borg et al. 2002; Sinha et al. 2003).

Exogenous Opiates Acutely, but not Chronically, Activate HPA Axis in Rodents but in Humans, HPA Axis Is Always Suppressed by Exogenous Opiates

We have learned that during self-administration of cocaine, using our extended (10-hr) access model, with high, medium, and low doses of cocaine per infusion, there is activation and disruption of the circadian pattern of HPA axis hormones and also blunting of hormone levels and circadian pattern after 1 and 4 days of withdrawal from cocaine (Fig. 3, right panel) (Mantsch et al. 2000, 2001, 2003). We and other investigators have found that acutely, and subacutely, all exogenous opioids will cause activation of the hormones of the HPA axis in rats and in mice. However, during chronic opiate administration to rodents, suppression of this stress-responsive HPA axis is found. In contrast, in humans, acutely all opiates suppress the HPA axis, and this suppression persists during chronic intermittent exposure to short-acting opioids seen in humans during

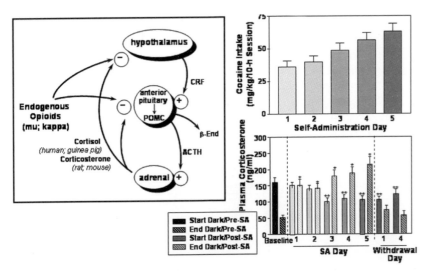

FIGURE 3. Effects of cocaine self-administration and extended access (10 hr) on plasma corticosterone in rats. The left depicts the HPA axis in humans and other mammals. On the top right, the total amount of cocaine self-administered over a 10-hour session, expressed in mg/kg, are presented for each of the 5 days of self-administration. On the bottom right, the plasma levels of corticosterone, the stress-responsive hormone of the HPA axis cascade, are shown before any cocaine administration, at the start of the dark period (beginning of activity) and the end of the dark period (end of activity), both for one baseline day and then at the beginning and end of the dark period for each of 5 days of cocaine self-administration over a 10-hour period. In addition, the levels of corticosterone on cocaine withdrawal on day 1 and 4 are shown, again with levels determined at the beginning and end of the dark period. It should be noted that the first day of cocaine self-administration resulted in the flattening of the normal circadian rhythm of corticosterone levels. By day 3, enhanced corticosterone levels were observed at the end of cocaine self-administration, and this pattern persisted. At the beginning of the withdrawal period, there is a persistent and significant reduction of the preactivity beginning of dark-period corticosterone levels compared with baseline levels and a partially flattened circadian rhythm. These data show a failure to return to a normal pattern of the stress-responsive hormone for up to 4 days after cessation of cocaine self-exposure. (Reprinted, with permission, from Mantsch et al. 2000.)

chronic cycles of heroin addiction. Of major importance, we have repeatedly documented that each component of both the basal and heroin-challenged components of the stress-responsive HPA axis becomes normalized in humans receiving long-term, steady-dose treatment with the long-acting opioid methadone, in which setting there is a steady state of plasma levels of methadone during most of the 24-hour dosing interval

TABLE 4. Hypothesis—Atypical Responsivity to Stressors: A Possible Etiology of Addictions

Atypical responsivity to stress and stressors may, in part, contribute to the persistence of, and relapse to, self-administration of drugs of abuse and addictions.

Such atypical stress responsivity in some individuals may exist before the use of addictive drugs on a genetic or acquired basis, and lead to the acquisition of drug addiction.

Genetic, environmental, and direct drug effects may each contribute to this atypical stress responsivity.

Data from Kreek (1972, 1973a,b, 1978, 1987, 1992) and Kreek et al. (2002).

TABLE 5. Neuroendocrine Effects of Opiates, Cocaine, and Alcohol in Humans: Hormones Involved in Stress Response

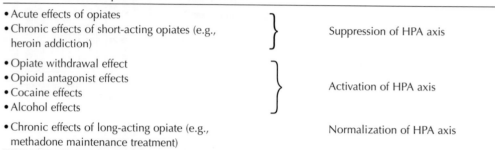

• Acute effects of opiates • Chronic effects of short-acting opiates (e.g., heroin addiction)	Suppression of HPA axis
• Opiate withdrawal effect • Opioid antagonist effects • Cocaine effects • Alcohol effects	Activation of HPA axis
• Chronic effects of long-acting opiate (e.g., methadone maintenance treatment)	Normalization of HPA axis

Data from Kreek (1972, 1973a,b, 1978, 1987, 1992; Kreek et al. (1983, 1984, 1996a, 2000a, 2002). (HPA) Hypothalamic-pituitary-adrenal axis (involved in stress response).

(see Table 5). We found in a rigorous positron emission tomography (PET) study of healthy normal volunteers and stabilized long-term methadone-maintained former heroin addicts that only 20–30% of μ-opioid receptors in all specific brain regions studied are occupied by methadone as used in maintenance treatment at adequate doses (Kling et al. 2000).

Opiate Withdrawal, Cocaine, and Alcohol Activate HPA Axis

Our group, and our group in collaboration with the group of Kosten, have shown profound activation of the HPA axis in opiate withdrawal (see Table 5). Furthermore, our group has shown in healthy volunteers, as well as volunteers with specific addictive diseases, that there is a profound activation of the HPA axis when an opiate antagonist is administered to any human and potentially exaggerated if given to an opiate-dependent individual (Kreek 1978; Kosten et al. 1987; Culpepper-Morgan et al. 1992; Rosen et al. 1996; Culpepper-Morgan and Kreek 1997; Schluger et al. 1998; King et al. 2002). Thus, we know that the HPA axis is, in healthy humans, in tonic modulation (inhibition) by the μ endogenous opioid system and that this system becomes deranged, as well documented by our group and other investigators, during cycles of opiate addiction (Fig. 3, left panel). In addition, our group and other investigators have shown in humans, as well as in rodent models, that both cocaine and alcohol significantly activate the stress-responsive HPA axis (Table 5).

κ Receptor Modulation of HPA Axis

In further studies, we have shown that both μ and κ receptor-directed endogenous opioids have a role in modulating the stress-responsive HPA axis. This was documented by the intravenous bolus administration of high doses of two different opiate antagonists that can be administered in humans. One, naloxone, is primarily directed to the μ-opioid receptor in humans and nonhuman primates, whereas another, nalmefene, is directed in humans and nonhuman primates at the μ- and κ-opioid receptor systems (Schluger et al. 1998; Bart et al. 2005a). In other studies, our group and many other investigators have shown that the stress-responsive systems of other parts of the brain have a profound impact on specific components of the forebrain and midbrain stress-responsive systems (e.g., see Kreek and Koob 1998; Zhou et al. 1996a, 2003a).

DRUGS OF ABUSE PRODUCE PROFOUND CHANGES IN THE CNS

Drug Seeking due to Changes in Brain that Develop during Chronic Exposure to Drugs of Abuse

Not only do drugs of abuse have a profound effect on the function of the HPA axis, but they also produce profound changes in the entire central nervous system (CNS). For pursuing our laboratory-based neurobiological work related to drug abuse and addiction, we have developed several novel, as well as used conventional, animal models (see Table 3). To summarize more than 15 years of basic laboratory-based molecular neurobiological work (including many studies in which we have used the novel models), we have established that the "drug-craving" or "drug hunger" that contributes to relapse to self-administration and addiction to a drug of abuse is, in part, dependent on disruption and changes in the brain because of neuroplasticity involving specific neurotransmitter and neuropeptide receptor systems, and alterations in specific components of signal-transduction systems, and, thus, in gene expression, related proteomics to these symptoms, overall neurochemistry, as well as synaptogenesis, potentially neurogenesis, and resultant behaviors (Kreek 1997b, 2002; Kreek et al. 2002, 2004b). Although we recognize that there will be many other not-yet-identified specific neurotransmitters, neuropeptides, and receptor systems that have a role, there is a likelihood that those systems already identified as having a major role will continue to be thus characterized.

We have focused on the endogenous opioid system and have shown that the μ-opioid receptor is central to reward and also has a major role in the modulation of the stress-responsive HPA axis. We have shown that the κ-opioid receptor system, with its ligands, has a critical countermodulatory role on the reward and reinforcing effects of drugs of abuse, targeted in part on the dopaminergic system, and also that the κ system modulates specific components of neuroendocrine function, including stress responsivity (see other chapters in this volume).

Chronic "Binge" Pattern Cocaine Administration Leads to Decreased Basal and Attenuated Cocaine-induced Increases in Dopamine Levels in Rodents

Studies have shown that dopamine acting at D1 and D2-like receptors (D2, D3, and possibly D4) clearly has a major role in the immediate rewarding and reinforcing effects of drugs of abuse. However, many of our studies, including our microdialysis studies, have shown that with repeated "binge" pattern cocaine administered to a rat or to a mouse, there is progressive reduction in basal dopamine levels, a highly significant reduction in cocaine-induced surges in dopamine levels in the extracellular fluid of both the caudate putamen and the nucleus accumbens, and progressive and significant reduction in the binding potential of both D1 and D2 dopaminergic receptors, as measured by PET, with all findings made in our laboratory using the "binge" pattern model (Fig. 4) (Maisonneuve

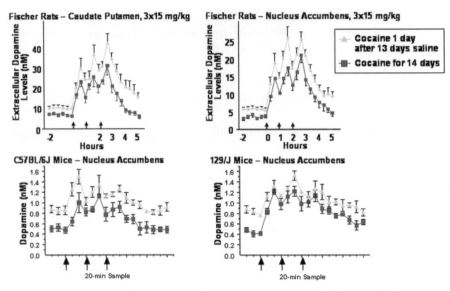

FIGURE 4. The upper two panels show the effects of acute (1-day) and chronic (14-day) cocaine on extracellular dopamine levels in the caudate putamen (*left*) and nucleus accumbens (*right*) in awake freely moving Fischer rats. The lower panels show the effects of acute (1-day) and chronic (14-day) cocaine on dopamine levels in the nucleus accumbens of awake freely moving mice (C57BL/6J mice, *left*; 129J mice, *right*). All dopamine levels are expressed in nanomolars. In each case, the dose of cocaine used was 15 mg/kg x 3 doses 1 hour apart, with no further cocaine for 24 hours. (Upper right: reprinted, with permission, from Maisonneuve et al. 1995; lower right: reprinted, with permission, from Zhang et al. 2003 [©John Wiley & Sons].)

and Kreek 1994; Maisonneuve et al. 1995; Tsukada et al. 1996; Maggos et al. 1998; Zhang et al. 2001, 2003).

"Binge" Pattern Cocaine Leads to Up-regulation of μ- and κ-opioid Receptors, and κ-, but not μ-, Directed Endogenous Opioid Peptides: Impact on Reward and Countermodulation of Reward

We have found profound receptor density and molecular-biological changes in the endogenous opioid system evoked, as we hypothesized in the mid-1980s, by cocaine, and in particular, after "binge" pattern exposure or self-administration of cocaine, as well as after opioid agonist and antagonist administration (Unterwald et al. 1992, 1993, 1994, 1995, 2001; Spangler et al. 1993, 1996a,b, 1997a,b; Wang et al. 1999; Yuferov et al. 1999, 2001, 2005b). These include changes in a second major component of reward as related to drugs of abuse (including, as we now know, opiates, cocaine, alcohol, and probably also amphetamine and, to a lesser extent, nicotine and other drugs of abuse), i.e., the μ-opioid receptor system. We have found that, for instance, acute μ-opioid receptor gene expression is altered by "binge" cocaine; then progressively and with time, there is a highly significant up-regulation of μ-opioid receptor density, an up-regulation that persists long after withdrawal from chronic cocaine, which may reflect, in part, alterations in trafficking, including endocytosis and reappearance on the cell membrane of the μ-

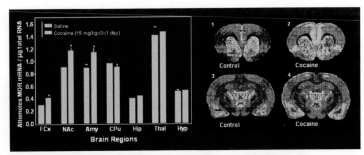

FIGURE 5. The impact of acute "binge" pattern cocaine on μ-opioid receptor mRNA levels, quantitatively measured by modified solution hybridization RNase protection assay are on the left panel. The right panel shows the impact of chronic 14-day binge pattern cocaine on increasing μ-opioid receptor density in selected brain regions, including the caudate putamen, nucleus accumbens, anterior cingulate, and amygdala. (Left: reprinted, with permission, from Yuferov et al. 1999 [©Elsevier]; right: reprinted, with permission, from Unterwald et al. 1992 [©Elsevier].)

opioid receptor, as well as potential alterations in synthesis of receptor (Fig. 5) (Unterwald et al. 1992, 1993, 2001; Wang et al. 1999; Yuferov et al. 1999; Bailey et al. 2005a,b). However, there is no increase in gene expression or peptide levels for the critical potentially "rewarding" endogenous opioid peptides, including enkephalins and β-endorphin, which are the endogenous ligands binding to the μ receptor. Thus, a "relative endorphin deficiency" develops.

Furthermore, we have recently found in a model of morphine self-administration, with extended (18-hr) sessions and with significant dose self-escalation, a highly significant reduction in GTPSγS binding in the amygdala of rats, an important area for modulating rewards and other emotions, such as pleasure, fear, and stress; this reduction implies an attenuation of responsivity of the μ system (Kruzich et al. 2003).

From 1992 onward, our group and other investigators have also learned that chronic "binge" pattern cocaine, or self-administration of cocaine, on an acute or chronic basis results in a persistent and recurrent elevation of gene expression of the dynorphin gene that codes for the κ-opioid-receptor-directed peptides; other investigators have shown that dynorphin peptides are, indeed, increased in that setting (e.g., see Daunais et al. 1993; Hurd and Herkenham 1993; Spangler et al. 1993, 1996a,b, 1997b). We have also shown that "binge" pattern cocaine on a chronic basis results in increased density of κ-opioid receptors (see Fig. 6). Thus, with increase both in the κ-opioid receptors and in the endogenous peptide ligand of those receptors, there is increased activity of the κ-opioid receptor system, in contrast to the relative imbalance of the μ-opioid receptor system with relative μ endorphin deficiency (Unterwald et al. 1994, 2001). We suggest that this increased dynorphin activity of the κ-opioid receptor system acts in "countermodulatory" role to the μ-opioid and to dopamine-related "reward."

Activation of κ-opioid Receptors Attenuate Extracellular Dopamine Levels Increased by Cocaine Countermodulation

We have shown that the natural dynorphin peptides, and as well as synthetic κ-opioid receptor ligands (studied by our group and other investigators), acting at the κ-opioid recep-

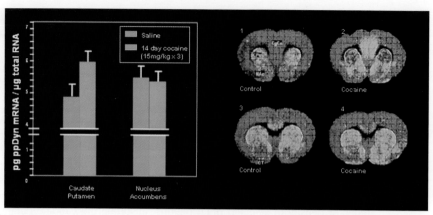

FIGURE 6. The impact of chronic "binge" pattern cocaine on dynorphin mRNA levels is shown on the left panel; increased levels occur after acute (1-day) and subacute (3-day) "binge" pattern cocaine as well as after 14 days as shown here, and, specifically in the caudate putamen, with dynorphin mRNA levels measured by modified solution hybridization RNase protection assay. The right panel shows the impact of "binge" pattern cocaine on increasing κ-opioid receptor density in the caudate putamen, nucleus accumbens, anterior cingulate, and olfactory tubercle. (Data from Spangler et al. 1996a; reprinted, with permission, from Unterwald et al. 1992 [©Elsevier].)

tors have a counterregulatory role of both modulating basal dopamine levels in the extracellular fluid of rodent caudate putamen and nucleus accumbens and modulating cocaine-induced dopaminergic surges. We have shown both in rat and now in mouse that natural dynorphin peptide, dynorphin A_{1-17} causes a dose-dependent progressive reduction in basal dopaminergic tone (Zhang et al. 2004a,b, 2005). Furthermore, we have shown that dynorphin A_{1-17} attenuates or may even prevent cocaine-induced surges in dopaminergic levels. Therefore, there is a direct countermodulation of dopamine extracellular levels by dynorphin. The studies from our laboratory, as well as the studies of other investigators, have shown that both intermittent morphine and "binge" alcohol enhance levels of prepro-dynorphin gene expression and increase in message of the κ-opioid receptor (e.g., see Wang et al. 1999). Thus, κ receptor-dynorphin up-regulation seems to be a general countermodulatory event to respond to dopaminergic surges. Work from several laboratories has described the mechanism of this increase in dynorphin gene expression through now well-defined mechanisms of signal-transduction-mediated enhancement of transcription factors triggered by dopamine acting at postsynaptic D_1- or D_2-like receptors.

Dynorphin Acts to Lower Dopamine Tone in the Tuberoinfundibular Dopamine System in Humans

We have shown in whole humans (staying within the constraints of measurements that can be made in living humans) that dynorphin peptide administered peripherally will reach the hypothalamic sites regulating prolactin release, which are largely outside of the blood–brain barrier in humans and nonhuman primates (although not in rodents). There, dynorphin peptide acting primarily at the κ-opioid receptor acts to lower the dopaminergic tone of the tuberoinfundibular dopaminergic system natural sequence (Butelman et al. 1999; Kreek et al. 1999; Bart et al. 2003). The impact of this lowering of dopaminergic tone can be read out by measurements of peripheral serum prolactin levels, because prolactin in humans is directly under tonic inhibition of tuberoinfundibular dopamine and thus can serve as a "biomarker" of this hypothalamic dopaminergic system. We have found an elevation of prolactin levels in a classical dose response when increasing amounts of natural sequence, but shortened dynorphin A_{1-13} peptide, is administered intravenously to healthy, normal, human volunteers. Furthermore, we have shown that in long-term, stabilized methadone-maintained patients, although a dose-dependent response is still seen, we find significantly attenuated serum levels of prolactin compared with those seen in healthy volunteers studied under the same clinical research paradigm (Bart et al. 2003).

HYPOTHESIS: GENETIC POLYMORPHISMS IN THE μ-OPIOID RECEPTOR UNDERLIE ADDICTION

Overview

The work of our laboratory in human molecular genetics related to the biology of addictive diseases began in 1994. In our initial work, we defined multiple variants of the μ-opioid receptor, two of which were common by definition, and both of which we found had profoundly different allelic frequencies in different ethnic/cultural groups. Furthermore, our early studies documented that one of these very common variants, A118G, is a functional variant with significantly tighter binding of the longest of the endogenous opioids, β-endorphin, as well as greater signal transduction through G-protein-coupled potassium, inwardly rectifying channels when β-endorphin binds to that receptor.

We hypothesized that the A118G variant, and possibly other variants or haplotypes of the μ-opioid receptor, would be strongly associated with opiate addiction, because opiate addiction, and also all effective approaches to treatment of opiate addiction, have been targeted very specifically at the μ-opioid receptor. We also suggested that this functional variant would be associated with alcoholism, because, as stated immediately above, one of the two effective treatments for alcoholism is an opioid antagonist, such as naltrexone or nalmefene, again primarily targeted at the μ-opioid receptor system. By studying a population with a minimal ethnic/cultural admixture, central Sweden, where more than 70% of study subjects are Swedes with Swedish ancestry, we were recently able to show a very highly significant association of the A118G variant with opiate addiction and, further, in separates studies, a very strong association of this A118G variant with alcoholism. In addition, the attributable risk, as determined by Professor Jurg Ott, for the A118G variant with opiate addiction was 21% in Swedes with Swedish ancestry and 18% in all the central Swedish population studied, and the attributable risk of developing alcoholism because of this one A118G variant of the μ-opioid receptor for developing alcoholism was 11% in the central Swedish population.

In our initial publication in the Proceedings of the National Academy of Sciences in 1998, we hypothesized that this A118G variant, which had both different binding characteristics and function in molecular-cellular constructs, would be found to significantly alter normal human physiology in those systems under significant modulation by the μ-opioid receptor systems. The most important system for the addictive diseases that we hypothesized would be altered was stress responsivity, because our group had clearly documented that the μ-opioid receptor agonist tonically inhibits both the hypothalamic and pituitary sites of the HPA axis. We also coined a term, "physiogenetics," analogous to the long-standing term, "pharmacogenetics," meaning differences in response to a medicine as a result of genetic factors, but with our work implying differences in normal physiologic functioning because of some genetic alteration. Furthermore, we hypothesized that such physiogenetics could lead to differences in pharmacodynamics, i.e., the physiological response to some medicine or pharmacological agent. Proof of principle of those findings was initially forthcoming from the group of Wand and then the group of Kranzler (Wand et al. 2002; Hernandez-Avila et al. 2003). We also hypothesized that pharmacogenetics therefore would pertain and specifically that alcoholics with the A118G variant would give a more positive response to treatment with an opiate antagonist such as naltrexone or nalmefene, because our group has shown (and based on an earlier hypothesis with proof of principle from our laboratory) that many alcoholics are, in part, seeking activation of the stress-responsive axis, which should be facilitated by removal of the endogenous opioids from the μ-opioid receptors. Proof of principle was provided by the combined groups of Volpicelli, O'Brien, and Kranzler (Oslin et al. 2003).

Details of Human Genetics Studies in a Historical Sequence and Perspective

In 1994, we joined forces with Dr. Lei Yu (then of the University of Indiana, now of the University of Cincinnati), the scientist who first, along with, independently, G.R. Uhl, cloned the μ-opioid receptor in rodents and then in humans, building on the elegant work of Keiffer et al. and Evans et al., who cloned a first opioid receptor (the δ-opioid receptor) from a cell line with abundant δ receptor expression (Evans et al. 1992; Kieffer et al. 1992; Chen et al. 1993; Eppler et al. 1993; Wang et al. 1993, 1994; Mestek et al. 1995). We hypothesized that some of the individual variability to the susceptibility to the development of, persistence of, or relapse to opiate addiction may be because of polymorphisms in the μ-opioid receptor (Bond et al. 1998; Kreek 2000b). Furthermore, after a few years of work, we also hypothesized that individual differences in response to one's own endogenous opioids, "physiogenetics," might be because of the presence of variants, just as the dynamic response to opioid-directed pharmacotherapies ("pharmacogenetics") might be mediated through variant forms of the μ receptor (Kreek 2000b; LaForge et al. 2000a). In our first joint paper in 1998, we first reported the concomitant presence of multiple variants of the μ-opioid receptor in the coding region, specifically, five single nucleotide polymorphisms (SNPs) (and subsequently several more rare SNPs) (Bond et al. 1998). We also reported that two of these SNPs were, by definition, not rare, i.e., 1% or greater allelic frequency across different ethnicities (Table 6). Both of these variants are in the amino terminus of the μ-opioid receptor, a site involved in the initial binding of the opiate receptor and directly leading to the resultant signal transduction or activation.

Furthermore, we defined these two variants (A118G and C17T) as having very high allelic frequency across ethnic groups (10.6 and 6.6 allelic frequency across groups then studied, respectively), although with potentially very important highly and significantly different allelic frequencies across multiple ethnic/cultural groups; both of these variants also resulted in amino-acid changes in the μ-opioid receptor peptide. The most common

TABLE 6. SNPs in Moderate Allelic Frequency in the Coding Region of the Human μ-opioid Receptor Gene

Variant (nucleotide position)[a]	Exon location	Protein domain	Corresponding amino-acid change	Allele frequency
A118G	1	amino terminus	Asn-4 Asp (N40D)	10.5% (26 heterozygous; 3 homozygous)
C17T	1	amino terminus	Ala-6 Val (A6V)	6.6% (14 heterozygous; 3 homozygous)

Data from Bond et al. (1998).

[a]Nucleotide position 1 is first base of the start codon.

of these, the variant located at A118G, resulted in an amino-acid change (asparagine to aspartic acid) at a putative glycosylation site (Fig. 7).

We developed a novel and simple-to-use "SNP Chip." This technique allows rapid identification of these two most common variants using a technique different from most microarrays and different from any standard DNA sequencing technique. It is much more suitable for potential widespread use, such as in academic or commmunity hospitals or in patient-care-oriented laboratories that we then perceived (and now know) could be of great importance in understanding differences in human physiology, as well as for some specific disorders (LaForge et al. 2000b).

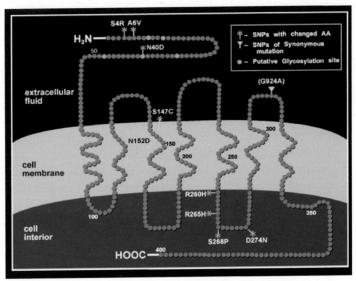

FIGURE 7. Diagram of all amino-acid residues of the seven *trans*-membrane G-protein-coupled μ-opioid receptor indicating the sites of each amino acid in which there is an SNP in the code for that amino acid. Both sites and the impact of SNPs that have either changed or not changed an amino acid (synonymous mutation) are noted. Sites of putative glycosylation are also noted.

C17T μ-opioid Receptor Variant Not Significantly Associated with Opiate Dependence

We went on to ask whether either the A118G or C17T variants were associated with severe opiate (heroin) dependence. Looking at all ethnic groups taken together, we found no association of the A118G variant with opiate dependence (Table 7). However, when we looked at the C17T variant in all ethnic groups together, there was an almost significant level of potential association ($p = 0.054$), but clearly not reaching significant levels (Table 7) (Bond et al. 1998). This was of interest, because one earlier paper, which had identified only this one variant and not the other more common variant, A118G (which, parenthetically, was the sole major variant identified in the only other earlier report, and which probably was not identified in the report that identified the C17T variant for both technical sequencing, as well as patient population ethnic/cultural background reasons), also reported a close, but not significant, association of C17T with opiate dependence (Table 7) (Bergen et al. 1997; Berrettini et al. 1997).

Of interest, and importance for subsequent studies, we also found that there were highly significant differences in the allelic frequency in the three different ethnic populations studied for both the A118G and C17T variants; this was further confirmed in our later studies (Tables 8 and 9).

Functional Significance in Molecular-cellular Constructs and Healthy Humans of A118G Variant of μ-opioid Receptor

We had decided a priori to study the potential function of any of the *most* common variants found in the coding region of the μ receptor. We therefore conducted a series of very rigorous molecular and cell biological studies comparing the prototype and resulting variant receptor. We first studied the relative binding of a variety of endogenous and exogenous opioids and opiates to both the prototype μ receptor and the altered receptor resulting from the A118G variant (Bond et al. 1998). What we learned is that with one endogenous μ-opioid receptor ligand only, the longest of the endogenous opioids, the 31-residue β-endorphin—made in the pituitary, as well as in many regions of the brain, and peripherally in the gastrointestinal tract, the immune system, and possibly elsewhere—there was a threefold greater binding (as shown by a significant shift to the left of the binding curves) with a threefold greater binding affinity of β-endorphin to the variant receptor than to the prototype (Bond et al. 1998). Furthermore, in molecular-cellular construct studies of one of the most common pathways of signal transduction, the G-protein-coupled inwardly rectifying potassium channels (GIRKs), we found similarly a threefold increased activation of the variant

TABLE 7. Allelic Frequency Associations with Opioid Dependence: A118G and C17T

	C	T	A	G	Total
Dependent	207	19	206	20	226
	(0.916)	(0.084)	(0.912)	(0.088)	
Nondependent	77	1	66	12	78
	(0.987)	(0.013)	(0.846)	(0.154)	
Yate's corrected χ^2	3.70 ($p = 0.054$)[a]		1.98 ($p = 0.159$)		

Data from Bond et al. (1998).

[a]This finding is similar to that obtained by Berrettini et al. (1997): $\chi^2 = 4.1$ ($p = 0.05$). (Yate's corrected $\chi^2_{(1)}$ = 8.22 [$p = 0.0041$]).

TABLE 8. Allelic Frequencies of the Variant Allele of the A118G SNP of the Human μ-opioid Receptor Gene in Diverse Populations

Ethnicity or population	Bergen et al. (1997)	Bond et al. (1998)	Gelernter et al. (1999)	Szeto et al. (2001)	Tan et al. (2003)	Bart et al. (2004)
Caucasian						
European-American	0.105 (100)	0.115 (52)	0.141 (543)			
Finnish-Caucasian	0.122 (324)					
Swedish-Caucasian						0.107 (187)
Indian					0.442 (137)	
Asian						
Japanese			0.485 (34)			
Han-Chinese				0.362 (297)		
Chinese					0.351 (208)	
Thai					0.438 (56)	
Malay					0.446 (156)	
Hispanic		0.142 (67)	0.117 (47)			
African-American		0.016 (31)	0.028 (144)			
Southwest Native American			0.163 (367)			
Other						
Ethiopian			0.170 (49)			
Bedouin			0.080 (43)			
Ashkenazi			0.210 (93)			

Data from LaForge et al. (2000a; pers. comm. 2005).

Allele frequency for the variant allele is shown for various study populations. Numbers in parentheses are the numbers of subjects whose genotype was ascertained in each study. A study of Han-Chinese found the 118G allele at a frequency of 0.321 and no occurrence of the 17T allele in 540 subjects (Li et al. 2000).

TABLE 9. Allelic Frequencies of the Variant Allele of the C17T SNP of the Human μ-opioid Receptor Gene in Diverse Populations

Ethnicity or population	Bergen et al. (1997)	Bond et al. (1998)	Gelernter et al. (1999)	Szeto et al. (2001)	Tan et al. (2003)	Bart et al. (2004)
Asian						
Japanese			0.000 (35)			
Han-Chinese						
Chinese						
Thai						
Malay						
Indian						
Southwest Native American						
Caucasian						
European-American		0.019 (52)	0.008 (470)			
Finnish-Caucasian						
Swedish-Caucasian						0.000 (309)
Hispanic		0.037 (66)	0.011 (46)			
African-American		0.210 (31)	0.140 (143)			
Other						
Ethiopian			0.080 (49)			
Bedouin			0.050 (43)			
Ashkenazi			0.016 (93)			

Data from LaForge et al. (2000a; pers. comm. 2005).

Allele frequency for the variant allele is shown for various study populations. Numbers in parentheses are the numbers of subjects whose genotype was ascertained in each study. A study of Han-Chinese found the 118G allele at a frequency of 0.321 and no occurrence of the 17T allele in 540 subjects (Li et al. 2000).

receptor resulting from the A118G polymorphism when β-endorphin is the binding ligand, as compared with the prototype (Bond et al. 1998). However, none of the other endogenous opioids or exogenous opiates resulted in any change of binding or activation of the variant receptor. These findings led us to predict that individuals with one copy of this variant would have differences in the physiological responses of any system modulated by the μ-opioid receptor. We specifically predicted that stress responsivity, which we had studied in both healthy volunteers and individuals with specific addictive diseases for several years, would be found to be different in individuals with one copy of this variant (Bond et al. 1998; LaForge et al. 2000a). Work first reported by Wand and colleagues at Johns Hopkins University and later by Kranzler and colleagues at the University of Connecticut showed that this hypothesis was correct and thus provided proof of the principle of "physiogenetics" (Wand et al. 2002; Hernandez-Avila et al. 2003). They found that cortisol levels after incremental intravenous naloxone administration, or naltrexone in a second study, were elevated to a significantly greater extent in individuals with one copy of the A118G variant than in healthy normal volunteer subjects. Subsequent work by other investigators has also provided support for our early hypothesis with "proof of principle" that "physiogenetics" is a term correctly applied to the functional differences of the μ-opioid receptor resulting from the A118G gene variant.

Genetic Polymorphism in μ-opioid Receptor Predicts Positive Response to Opiate Antagonist Treatment of Alcoholism

We also predicted that the response to a specific opiate antagonist used in treatment of alcoholism, such as naltrexone (or nalmefene), would be different in individuals with one or two copies of the A118G variant because of different binding of β-endorphin, which has been shown to enhance stress responsivity with this variant (O'Malley et al. 1992; Volpicelli et al. 1992; Mason et al. 1994; Kreek 2000b; LaForge et al. 2000a,b). In earlier studies, we had found that nonalcoholic adults with or without a positive family history for alcoholism gave profoundly different responses to orally administered naltrexone (King et al. 2002). We showed that those with a family history of alcoholism had a much greater response to naltrexone with respect to HPA axis activation as measured by greater elevation in plasma levels of stress-response hormones ACTH and cortisol than did those with a negative family history. Our research findings further supported our hypothesis that alcoholics with this variant may have a better response to naltrexone treatment.

In other extensive studies conducted in our collaboration with the group of Dr. Stephanie O'Malley at Yale, we postulated and documented with "proof of the principle" that at least some alcoholics are seeking (not avoiding) the stimulatory effect of alcohol on the HPA axis and will drink increasing amounts of alcohol to achieve that effect (O'Malley et al. 2002). In this laboratory-based study (which had undergone thorough ethical review), a group of alcoholics were randomized to either placebo or naltrexone treatment for 1 week; they were then studied 6 hours after that day's oral dose of naltrexone or placebo (Table 10). At the 6-hour timepoint, as many of our own previous studies have shown, the anticipated acute naltrexone-induced activation of the HPA axis through the mechanism of endogenous opioid disinhibition had subsided and hormones had returned to normal levels. Just as had been found in the outpatient treatment research setting, as shown in the seminal clinical trials work of Volpicelli, O'Brien, and O'Malley (and later by other investigators), subjects in our clinical laboratory study receiving naltrexone drank many

TABLE 10. Alcoholic Drinks Consumed 6 Hours after Administration of Oral Naltrexone or Placebo and following a Priming Dose during Each of Two 2-hour Consecutive Sessions

	Number of drinks		Total number of drinks during 2-hour session
	First hour of choice	Second hour of choice	
Naltrexone	1.25 +/– 0.48	0.65 +/– 0.45	1.9 +/– 0.72
Placebo	2.20 +/– 0.04	2.40 +/– 0.41	4.6 +/– 0.85

Data from O'Malley et al. (2002).

fewer alcoholic drinks during the next two 1-hour sessions during which up to four drinks, or, alternatively, three dollars per drink, would be given to the subjects, as compared with the placebo-treated subjects. Specifically, by the end of the 2-hour test period, those receiving naltrexone had elected to consume 1.9 drinks, whereas those on placebo had consumed a mean of 4.6 drinks (Table 10) (O'Malley et al. 2002).

As had been hypothesized by us, in the setting of disinhibition of the HPA axis by naltrexone, individuals receiving naltrexone had a prompt and much greater alcohol-induced activation of the HPA axis, as reflected by higher plasma levels of ACTH and cortisol, than those receiving placebo treatment, despite the fact that blood alcohol levels were much higher in the placebo-treated group, because of the almost threefold greater number of drinks consumed (O'Malley et al. 2002).

Furthermore, and very excitingly now supporting our hypotheses, the urge to drink alcohol or "craving" was significantly lower during and also immediately following the 2-hour drinking session in those receiving naltrexone, and was related to the elevated levels of plasma cortisol. Those on placebo wished to consume more alcohol, we postulate, in an attempt to get modest activation of the hormones of the HPA axis, including CRF, ACTH, and cortisol (with the precise relative roles of each component hormone yet to be fully elucidated). Those on naltrexone, because of opioid antagonist disinhibition of the endogenous opioid system, achieved a much larger activation response to modest amounts of alcohol with respect to HPA axis activation, which resulted in a reduction of their craving for further drinking and, undoubtedly, contributed to their decision to consume less than two drinks during the sessions (O'Malley et al. 2002).

The groups of O'Brien and Volpicelli, working with the group of Kranzler, were able to recontact and invite to participate those volunteer subject alcoholics who had completed their three earlier naltrexone treatment clinical trials. Approximately one sixth of the subjects agreed to consent for and participate in genetics studies (Oslin et al. 2003). After genotyping was performed, it was found that most of those individuals who responded "positively" to naltrexone treatment, with decreased numbers of days drinking alcohol and decreased amounts of alcohol consumed, were, in fact, mostly individuals heterozygous or homozygous for the functional A118G variant of the μ-opioid receptor (Oslin et al. 2003).

A118G Polymorphism of μ-opioid Receptor Is Highly Significantly Associated Both with Opiate Dependence and Alcoholism, and with High Attributable Risk

In very recent studies (ongoing at the time of the Cold Spring Harbor Laboratory course in August of 2003, but completed and reported in the 18 months thereafter), we found a highly significant association between the functional A118G polymorphism of the μ-opioid receptor gene and heroin addiction in central Sweden studies conducted in collaboration with Dr. Markus Heilig. We performed these studies using subjects in central Sweden because of the very minimal admixture of the Swedish population with non-Swedes. Although since the beginning of World War II there has been an influx of foreign nationals, the admixture is only now beginning to occur. Therefore, as predicted, we found that of our unselected volunteer and opiate-dependent subjects meeting the criteria for entry into methadone maintenance treatment in the United States (1 year or more of multiple daily self-administrations of opiates), or possibly even the then-current criteria in Sweden (4 years of regular opiate use), Swedes with Swedish parents comprise most (more than 70%) of the population. Furthermore, we found that whether we analyzed the data for the entire population or for only Swedes with Swedish heritage, there was a highly significant association of the A118G variant with heroin addiction. With rigorous statistical genetics analysis by Professor Jurg Ott, also at The Rockefeller University, an unexpectedly high attributable risk of the genetic component of opiate addiction to this single A118G variant in the overall central Swedish population, with an 18% attributable risk, was found to be because of this single µ-opioid receptor variant, and, in Swedes with Swedish parentage, 21% of the attributable risk was a result of the A118G variant (Table 11) (Bart et al. 2004).

In a similar, even more recent study conducted in central Sweden, we also found a highly significant association, as hypothesized, of this A118G polymorphism of the µ-opioid receptor gene, with its resultant functional receptor variant, with alcoholism. Again, in overall population or in the Swedish population with Swedish parents, a substantial attributable genetic risk was found to be because of this variant in analysis by Ott, with 11% of the attributable risk a result of this single variant (Table 12) (Bart et al. 2005b).

Dynorphin Gene Potentially Functional Variants May be Associated with Cocaine Dependence

In other human genetics studies, we examined the potential role of a repeat element in a putative AP-1 consensus site in the 5´ promoter region of the dynorphin gene with cocaine addiction. All humans have one, two, three, or four copies of this 68-bp repeat. In studies of this variant by Zimprich, in the laboratory of the first discoverer of this variant, V. Höllt, and using a reporter construct, it has been suggested that the presence of three or four copies of this variant yields greater activation and thus, by assumption, greater amounts of dynorphin than one or two copies (Zimprich et al. 2000). One can hypothesize that this enhancement can either protect, by allowing increased modulation and thus protection of dopaminergic surge, or increase vulnerability by enhancing potential dysphoria resulting from basal and stimulated levels of dynorphin peptide. In an initial study of a group of cocaine-abusing or -dependent individuals, we found an association with cocaine addiction (Chen et al. 2002).

κ Receptor SNP and Specific Haplotype Associated with Opiate Dependence

In other very recent studies, all completed and reported subsequent to the second Cold Spring Harbor Laboratory course, we redefined the structure of the κ-opioid receptor

TABLE 11. Association between a Functional Polymorphism in the μ-opioid Receptor Gene and Opiate Addiction in Central Sweden

	All subjects		Swedish with both parents Swedish	
Genotype	Controls (n = 170)	Opiate dependent (n = 139)	Controls (n = 120)	Opiate dependent (n = 67)
A/A	147 (0.865)	98 (0.705)	104 (0.867)	46 (0.687)
A/G	21 (0.123)	39 (0.281)	15 (0.125)	19 (0.283)
G/G	2 (0.012)	2 (0.014)	1 (0.008)	2 (0.030)
	RR = 2.86	$\chi^2(1)$ = 13.403 p = 0.00025	RR = 2.97	$\chi^2(1)$ = 8.740 p = 0.0031

	Opiate dependent (n = 139)	Control (n = 170)
G/G; A/G	41	23
A/A	98	147
118G allele frequency	0.155	0.074

Data from Bart et al. (2004).

In the entire study group in this central Swedish population, attributable risk due to genotypes with a G allele in this population is 18%, and attributable risk due to genotypes with a G allele in Swedes with Swedish parents is 21% (with confidence interval ranging from 8 to 28%).

TABLE 12. Association between a Functional Polymorphism in the μ-opioid Receptor Gene and Alcoholism in Central Sweden

	Swedish with two Swedish parents		Non-Swedish without Swedish parents	
	Alcohol dependent (n = 193)	control (n = 120)	Alcohol dependent (n = 196)	Control (n = 50)
A118	158	104	141	43
A118G, G118G	35	16	55	7
		OR = 1.92 $\chi^2(1)$ = 7.18, p = 0.0074		

	Alcohol dependent (n = 389)	Control (n = 170)
G/G; A/G	90	23
A/A	299	147
118G allele frequency	0.125	0.074
	Overall 118G allele frequency = 0.109	

Data from Bart et al. (2005b).

In the entire study group in this central Swedish population, attributable risk due to genotypes with a G allele is 11.1% (with confidence interval ranging from 3.6 to 18%).

gene, elucidating the presence of a fourth exon (Fig. 8) (Yuferov et al. 2004). Furthermore, we identified 12 polymorphisms in the coding exonic, intronic, or the nearby 5′ or adjacent regions (LaForge et al. 2000c; Yuferov et al. 2004). One of these variants, G36T, has been shown to have pointwise association with opiate addiction. In collaboration with Ott, we also defined the potentially significant haplotypes in each of the three ethnic groups studied and defined the most common haplotype in all groups (Yuferov et al. 2004). In addition, in the Hispanic group, we defined one set of haplotypes very strongly associated with opiate addiction (Yuferov et al. 2004). Other studies ongoing to date in our laboratory have suggested an association of the enkephalin gene with a specific addictive disease.

CONCLUSION

In summary, to date, we have shown that four of the six endogenous opioid family genes—the μ-opioid receptor, the κ-opioid receptor, the dynorphin peptide gene, and the enkephalin peptide gene—have an association of specific variants or haplotypes with a specific addictive disease yielding either increased vulnerability or increased protection to develop addiction following self-exposure. All of these specific genes were selected on the basis of our basic clinical and laboratory molecular and neurobiological work, i.e., our bidirectional translational research, and provide further support for the role of the endogenous opioid system in each of three specific addictive diseases (Kreek et al. 2002, 2004a,b, 2005; Yuferov et al. 2005a,b).

GAPS AND OPPORTUNITIES

Only highlights of the role of the μ-opioid receptor system in the reward mechanisms of each of the major drugs of abuse, as well as the role of the κ-opioid receptor system in countermodulation of that reward, primarily through targeting dopamine (which has been well documented to be part of reward mechanisms), are presented herein. Much work not covered in this brief review has been done by our laboratory and other investigators. However, further work is needed concerning the specific, as well as more abundant non-specific, signal-transduction mechanisms involved in the already discovered pathways of reward and countermodulation. Also yet to be explored further is the relative role of other neurotransmitters, neuropeptides, and their specific receptors or sites of action, both in reward and in countermodulation thereof. Our laboratory and many others are defining which drug-induced abnormalities, attributable to intrinsic neuroplasticity of the brain, persist for extended periods of time (and may, in some cases, be permanent changes), in contrast to other changes, which may undergo slow restoration to normal status with time or may reverse immediately on drug withdrawal.

Further studies of human molecular genetics, with studies of both genes identified through targeted molecular-neurobiological studies, as well as whole-genome scans, are in progress and ongoing in many laboratories, including our own.

Possibly the greatest need at this time is to significantly increase the amount and quality of research conducted in both healthy humans and humans with specific addictive diseases, to determine the relative role and importance of the findings made in the various

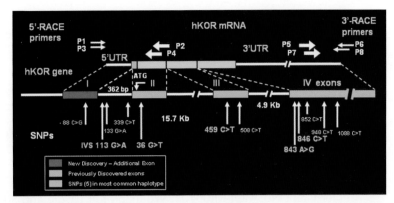

FIGURE 8. The human κ-opioid receptor gene, including its recently elucidated structure, and 12 recently identified SNPs, including an indication of the SNPs in the most common haplotype. It should be noted that the SNP 36G>T is associated with a pointwise association with opiate addiction. (Reprinted, with permission, from Yuferov et al. 2004 [©Lippincott, Williams & Wilkins].)

animal and molecular models, to the extent feasible and ethical, and to identify realistic targets for medicines, as well as behavioral treatment. Similarly, early applied clinical research is needed to determine the success of those targeted medicines, as well as applied field clinical research to determine the efficacy of any pharmacotherapy-behavioral treatment combination thereof in "real world" settings.

Above all, and not discussed at all in this chapter, is the need to enhance knowledge of scientists, physicians, and other health-care providers and policy makers about specif-

TABLE 13. Haplotypes in *OPRK1* in Hispanic Group Significantly Associated with Opiate Addiction and Most Common Haplotype in All Ethnic Groups

Significant haplotypes in *OPRK1* in Hispanic group					Frequencies	
					Controls	Cases
IVS1 113G>A	36G>T	459C>T	843A>G	846C>T	(*n* = 25)	(*n* = 35)
Haplotypes						
A	T	C	G	T	0.060	0
G	G	T	G	C	0	0.071
Sum					1	1

Most common haplotype in *OPRK1* in all groups					Frequencies		
					Controls	Cases	
IVS1 113G>A	36G>T[a]	459C>T	843A>G	846C>T	Controls	Cases	
Haplotype					0.6571	0.8119	(Caucasians)
					(*n* = 64)	(*n* = 65)	
G	G	C	A	C	0.2416	0.3779	(African-
					(*n* = 34)	(*n* = 36)	Americans)
					0.5890	0.7286	(Hispanics)
					(*n* = 25)	(*n* = 35)	

Data from Yuferov et al. (2004).
[a]Significant pointwise association with opiate addiction in all groups.

ic addictive diseases and exciting discoveries in molecular neurobiology and genetics related to these diseases, with the ultimate goal (or at least hope) of eradicating stigma and prejudice through knowledge and information.

ACKNOWLEDGMENTS

This work was supported in part by the National Institutes of Health–National Institute of Drug Abuse (NIH-NIDA) Research Center Grant Nos. P60-DA05130 and R01-DA12848; NIH-NIDA Research Scientist Award Grant No. K05-DA00049; the New York State Office of Alcoholism and Substance Abuse Services; and the NIH-General Clinical Research Center Grant No. MOI-RR00102. We thank Kitt Lavoie and Susan Russo for their assistance in the preparation of this manuscript.

REFERENCES

Adelson M.O., Hayward R., Bodner G., Bleich A., Gelkopf M., and Kreek M.J. 2000. Replication of an effective opiate addiction pharmacotherapeutic treatment model: Minimal need for modification in a different country. *J. Maint. Addict.* **1:** 5–13.

Anthony J.C., Warner L.A., and Kessler R.C. 1994. Comparative epidemiology of dependence on tobacco, alcohol, controlled substances, and inhalants: Basic findings from the national comorbidity survey. *Exp. Clin. Psychopharm.* **2:** 244–268.

Bailey A., Gianotti R., Ho A., and Kreek M.J. 2005a. Persistent upregulation of mu-opioid but not adenosine receptors in brains of long-term withdrawn escalating dose "binge" cocaine-treated rats. *Synapse* **57:** 160–166.

Bailey A., Yuferov V., Bendor J., Schlussman S.D., Zhou Y., Ho A., and Kreek M.J. 2005b. Immediate withdrawal from chronic "binge" cocaine administration increases μ-opioid receptor mRNA levels in rat frontal cortex. *Mol. Brain Res.* **137:** 258–262.

Bart G., Borg L., Schluger J.H., Green M., Ho A., and Kreek M.J. 2003. Suppressed prolactin response to dynorphin A(1-13) in methadone maintained versus control subjects. *J. Pharmacol. Exp. Ther.* **306:** 581–587.

Bart G., Schluger J.H., Borg L., Ho A., Bidlack J., and Kreek M.J. 2005a. Nalmefene induced elevation in serum prolactin in normal human volunteers: Partial kappa opioid agonist activity? *Neuropsychopharmacology* (In press).

Bart G., Heilig M., LaForge K.S., Pollak L., Leal S.M., Ott J., and Kreek M.J. 2004. Substantial attributable risk related to a functional mu-opioid receptor gene polymorphism in association with heroin addiction in central Sweden. *Mol. Psychiatry* **9:** 547–549.

Bart G., Kreek M.J., Ott J., LaForge K.S., Proudnikov D., Pollak L., and Heilig M. 2005b. Increased attributable risk related to a functional mu-opioid receptor gene polymorphism in association with alcohol dependence in central Sweden. *Neuropsychopharmacology* **26:** 106–114.

Bergen A.W., Kokoszka J., Peterson R.J., Long J.C., Linnoila M., and Goldman D. 1997. Opioid receptor gene variants: Lack of association with alcohol dependence. *Mol. Psych.* **2:** 490–494.

Berrettini W.H., Hoehe M.R., Ferrada T.N., and Gottheil E. 1997. Human mu opioid receptor gene polymorphisms and vulnerability to substance abuse. *Addiction Biol.* **2:** 303–308.

Bond C., LaForge K.S., Tian M., Melia D., Zhang S., Borg L., Gong J., Schluger J., Strong J.A., Leal S.M., Tischfield J.A., Kreek M.J., and Yu L. 1998. Single nucleotide polymorphism in the human mu opioid receptor gene alters beta-endorphin binding and activity: Possible implications for opiate addiction. *Proc. Natl. Acad. Sci.* **95:** 9608–9613.

Borg L., Ho A., Wells A., Joseph H., Appel P., Moody D., and Kreek M.J. 2002. The use of levo-alpha-acetylmethadol (LAAM) in methadone patients who have not achieved heroin abstinence. *J. Addict. Dis.* **21:** 13–22.

Burstein Y., Grady R.W., Kreek M.J., Rausen A.R., and Peterson C.M. 1980. Thrombocytosis in the offspring of female mice receiving dl-methadone. *Proc. Soc. Exp. Biol. Med.* **164:** 275–279.

Butelman E.R., Harris T., Perez A., and Kreek M.J. 1999. Effects of systemically administered dynorphin A(1-17) in rhesus monkeys. *J. Pharmacol. Exp. Ther.* **290:** 678–686.

Chen A.C.H., LaForge K.S., Ho A., McHugh P.F., Bell K., Schluger R.P., Leal S.M., and Kreek M.J. 2002. A potentially functional polymorphism in the promoter region of prodynorphin gene may be associated with protection against cocaine dependence or abuse. *Am. J. Med. Genet.* **114:** 429–435.

Chen Y., Mestek A., Liu J., Hurley J.A., and Yu L. 1993. Molecular cloning and functional expression of a mu-opioid receptor from rat brain. *Mol. Pharmacol.* **44:** 8–12.

Culpepper-Morgan J.A. and Kreek M.J. 1997. HPA axis hypersensitivity to naloxone in opioid dependence: A case of naloxone induced withdrawal. *Metabolism* **46:** 130–134.

Culpepper-Morgan J.A., Inturrisi C.E., Portenoy R.K., Foley K., Houde R.W., Marsh F., and Kreek M.J. 1992. Treatment of opioid induced constipation with oral naloxone: A pilot study. *Clin. Pharmacol. Ther.* **23:** 90–95.

Daunais J.B., Roberts D.C., and McGinty J.F. 1993. Cocaine self-administration increases preprodynorphin, but not c-fos, mRNA in rat striatum. *Neuroreport* **4:** 543–546.

Dole V.P. and Nyswander M.E. 1965. A medical treatment for diacetylmorphine (heroin) addiction. *J. Am. Med. Assoc.* **193:** 646–650.

Dole V.P., Nyswander M.E., and Kreek M.J. 1966a. Narcotic blockade. *Arch. Intern. Med.* **118:** 304–309.

Dole V.P., Nyswander M.E., and Kreek M.J. 1966b. Narcotic blockade: A medical technique for stopping heroin use by addicts. *Trans. Assoc. Am. Phys.* **79:** 122–136.

Eppler C.M., Hulmes J.D., Wang J.B., Johnson B., Corbett M., Luthin D.R., Uhl G.R., and Linden J. 1993. Purification and partial amino acid sequence of a mu opioid receptor from rat brain. *J. Biol. Chem.* **268:** 26447–26451.

Evans C.J., Keith D.E., Jr., Morrison H., Magendzo K., and Edwards R.H. 1992. Cloning of a delta opioid receptor by functional expression. *Science* **258:** 1952–1955.

Gelernter J., Kranzler H., and Cubells J. 1999. Genetics of two mu opioid receptor gene (OPRM1) exon 1 polymorphisms: Population studies, and allele frequencies in alcohol- and drug-dependent subjects. *Mol. Psychiatry* **4:** 476–483.

Hernandez-Avila C., Wand G., Luo X., Gelernter J., and Kranzler H.R. 2003. Association between the cortisol response to opioid blockage and the Asn40Asn polymorphism at the mu-opioid receptor locus (OPRM1). *Am. J. Med. Genet. B Neuropsychiatr. Genet.* **118:** 60–65.

Hurd Y.L. and Herkenham M. 1993. Molecular alterations in the neostriatum of human cocaine addicts. *Synapse* **13:** 357–369.

Inturrisi C.E. and Verebely K. 1972a. Disposition of methadone in man after a single oral dose. *Clin. Pharmacol. Ther.* **13:** 923–930.

———. 1972b. The levels of methadone in the plasma in methadone maintenance. *Clin. Pharmacol. Ther.* **13:** 633–637.

Inturrisi C.E., Max M.B., Foley K.M., Schultz M., Shin S.-U., and Houde R.W. 1984. The pharmacokinetics of heroin in patients with chronic pain. *N. Engl. J. Med.* **310:** 1213–1217.

Jacobsen L.K., Giedd J.N., Kreek M.J., Gottschalk C., and Kosten T.R. 2001. Quantitative medial temporal lobe brain morphology and hypothalamic-pituitary-adrenal axis function in cocaine dependence: A preliminary report. *Drug Alcohol Depend.* **62:** 49–56.

Kellogg S.H., McHugh P.F., Belll K., Schluger J.H., Schluger R.P., LaForge K.S., Ho A., and Kreek M.J. 2003. The Kreek-McHugh-Schluger-Kellogg Scale: A new rapid method for quantifying substance abuse and its possible applications. *Drug Alcohol Depend.* **69:** 137–150.

Kennedy J.A., Hartman N., Sbriglio R., Khuri E., and Kreek M.J. 1990. Metyrapone-induced withdrawal symptoms. *Br. J. Addict.* **85:** 1133–1140.

Kieffer B.L., Befort K., Gaveriaux-Ruff C., and Hirth C.G. 1992. The delta-opioid receptor: Isolation of a cDNA by expression cloning and pharmacological characterization. *Proc. Natl.*

Acad. Sci. **89:** 12048–12052.

King A.C., Volpicelli J.R., Gunduz M., O'Brien C.P., and Kreek M.J. 1997. Naltrexone biotransformation and incidence of subjective side effects: A preliminary study. *Alcohol: Clin. Exp. Res.* **21:** 906–909.

King A.C., Schluger J., Gunduz M., Borg L., Perret G., Ho A., and Kreek M.J. 2002. Hypothalamic-pituitary-adrenocortical (HPA) axis response and biotransformation of oral naltrexone: Preliminary examination of relationship to family history of alcoholism. *Neuropsychopharmacology* **26:** 778–788.

Kling M.A., Carson R.E., Borg L., Zametkin A., Matochik J.A., Schluger J., Herscovitch P., Rice K.C., Ho A., Eckelman W.C., and Kreek M.J. 2000. Opioid receptor imaging with PET and [^{18}F]cyclofoxy in long-term methadone-treated former heroin addicts. *J. Pharmacol. Exp. Ther.* **295:** 1070–1076.

Kosten T.R., Kreek M.J., Raghunath J., and Kleber H.D. 1986. A preliminary study of beta-endorphin during chronic naltrexone maintenance treatment in ex-opiate addicts. *Life Sci.* **39:** 55–59.

Kosten T.R., Kreek M.J., Swift C., Carney M.K., and Ferdinands L. 1987. Beta-endorphin levels in CSF during methadone maintenance. *Life Sci.* **41:** 1071–1076.

Kreek M.J. 1972. Medical safety, side effects and toxicity of methadone. In *Proceedings of the Fourth National Conference on Methadone Treatment*, National Association for the Prevention of Addiction to Narcotics (NAPAN)-NIMH, pp. 171–174. Washington, D.C.

———. 1973a. Medical safety and side effects of methadone in tolerant individuals. *J. Am. Med. Assoc.* **223:** 665–668.

———. 1973b. Physiological implications of methadone treatment. In *Proceedings of the Fifth National Conference of Methadone Treatment*, National Association for the Prevention of Addiction to Narcotics (NAPAN) II-NIMH, pp. 824–836. Washington, D.C.

———. 1973c. Plasma and urine levels of methadone. *N.Y. State J. Med.* **73:** 2773–2777.

———. 1978. Medical complications in methadone patients. *Ann. N.Y. Acad. Sci.* **311:** 110–134.

———. 1979. Methadone disposition during the perinatal period in humans. *Pharmacol. Biochem. Behav.* (suppl.) **11:** 1–7.

———. 1987. Multiple drug abuse patterns and medical consequences. In *Psychopharmacology: The third generation of progress* (ed. H.Y. Meltzer), pp. 1597–1604. Raven Press, New York.

———. 1992. Rationale for maintenance pharmacotherapy of opiate dependence. In *Addictive states. Association for research in nervous and mental disease* (ed. C.P. O'Brien and J.H. Jaffe), pp. 205–230. Raven Press, New York.

———. 1996a. Opiates, opioids and addiction. *Mol. Psychiatry* **1:** 232–254.

———. 1996b. Opioid receptors: Some perspectives from early studies of their role in normal physiology, stress responsivity and in specific addictive diseases. *J. Neurochem. Res.* **21:** 1469–1488.

———. 1997a. Clinical update of opioid agonist and partial agonist medications for the maintenance treatment of opioid addiction. *Semin. Neurosci.* **9:** 140–157.

——. 1997b. Opiate and cocaine addictions: Challenge for pharmacotherapies. *Pharmacol. Biochem. Behav.* **57:** 551–569.

——. 2000a. Methadone-related opioid agonist pharmacotherapy for heroin addiction: History, recent molecular and neurochemical research and the future in mainstream medicine. *Ann. N.Y. Acad. Sci.* **909:** 186–216.

——. 2000b. Opiates, opioids, SNPs and the addictions: Nathan B. Eddy Memorial Award for lifetime excellence in drug abuse research lecture. *Natl. Inst. Drug Abuse Res. Monogr. Ser.* **180:** 3–22.

——. 2002. Molecular and cellular neurobiology and pathophysiology of opiate addiction. In *Neuropsychopharmacology: The Fifth generation of progress* (ed. K.L. Davis), pp. 1491–1506. Lippincott Williams and Wilkins, Philadelphia.

Kreek M.J. and Hartman N. 1982. Chronic use of opioids and antipsychotic drugs: Side effects, effects on endogenous opioids and toxicity. *Ann. N.Y. Acad. Sci.* **398:** 151–172.

Kreek M.J. and Koob G.F. 1998. Drug dependence: Stress and dysregulation of brain reward pathways. *Drug Alcohol Depend.* **51:** 23–47.

Kreek M.J. and Vocci F.J. 2002. History and current status of opioid maintenance treatments: Blending conference session. *J. Subst. Abuse Treat.* **23:** 93–105.

Kreek M.J., LaForge K.S., and Butelman E. 2002. Pharmacotherapy of addictions. *Nat. Rev. Drug Discov.* **1:** 710–726.

Kreek M.J., Nielsen D.A., and LaForge K.S. 2004a. Genes associated with addiction: Alcoholism, opiate and cocaine addiction. *Neuromolecular Med.* **5:** 85–108.

Kreek M.J., Guggenheim F.G., Ross J.E., and Tapley D.F. 1963. Glucuronide formation in the transport of testosterone and androstenedione by rat intestine. *Biochem. Biophys. Acta* **74:** 418–427.

Kreek M.J., Schecter A.J., Gutjahr C.L., and Hecht M. 1980. Methadone use in patients with chronic renal disease. *Drug Alcohol Depend.* **5:** 197–205.

Kreek M.J., Bart G., Lilly C., LaForge K.S., and Nielsen D.A. 2005. Pharmacogenetics and human molecular genetics of opiate and cocaine addictions and their treatments. *Pharmacol. Rev.* **57:** 1–26.

Kreek M.J., Gutjahr C.L., Garfield J.W., Bowen D.V., and Field F.H. 1976. Drug interactions with methadone. *Ann. N.Y. Acad. Sci.* **281:** 350–371.

Kreek M.J., Schluger J., Borg L., Gunduz M., and Ho A. 1999. Dynorphin A1-13 causes elevation of serum levels of prolactin through an opioid receptor mechanism in humans: Gender differences and implications for modulations of dopaminergic tone in the treatment of addictions. *J. Pharmacol. Exp. Ther.* **288:** 260–269.

Kreek M.J., Schlussman S.D., Bart G., LaForge K.S., and Butelman E.R. 2004b. Evolving perspectives on neurobiological research on the addictions: Celebration of the 30th anniversary of NIDA. *Neuropharmacology* **47:** 324–344.

Kreek M.J., Wardlaw S.L., Friedman J., Schneider B., and Frantz A.G. 1981. Effects of chronic exogenous opioid administration on levels of one endogenous opioid (beta-endorphin) in man. In *Advances in endogenous and exogenous opioids* (ed. E. Simon and H. Takagi), pp. 364–366. Kodansha Ltd. Publishers, Tokyo.

Kreek M.J., Ragunath J., Plevy S., Hamer D., Schneider B., and Hartman N. 1984. ACTH, cortisol and beta-endorphin response to metyrapone testing during chronic methadone maintenance treatment in humans. *Neuropeptides* **5:** 277–278.

Kreek M.J., Wardlaw S.L., Hartman N., Raghunath J., Friedman J., Schneider B., and Frantz A.G. 1983. Circadian rhythms and levels of beta-endorphin, ACTH, and cortisol during chronic methadone maintenance treatment in humans. *Life Sci.* **33:** 409–411.

Kruzich P., Chen A.C.H., Unterwald E.M., and Kreek M.J. 2003. Subject-regulated dosing alters morphine self-administration behavior and morphine-stimulated [^{35}S]GTPγS binding. *Synapse* **47:** 243–249.

LaForge K.S., Yuferov V., and Kreek M.J. 2000a. Opioid receptor and peptide gene polymorphisms: Potential implications for addictions. *Eur. J. Pharmacol.* **410:** 249–268.

LaForge K.S., Shick V., Spangler R., Proudnikov D., Yuferov V., Lysov Y., Mirzabekov A., and Kreek M.J. 2000b. Detection of single nucleotide polymorphisms of the human mu opioid receptor gene by hybridization or single nucleotide extension on custom oligonucleotide gelpad microchips: Potential in studies of addiction. *Am. J. Med. Genet. Neuropsychiatr. Genet.* **96:** 604–615.

LaForge K.S., Kreek M.J., Uhl G.R., Sora I., Yu L., Befort K., Filliol D., Favier V., Hoehe M., Kieffer B.L., and Höllt V. 2000c. Symposium XIII. Allelic polymorphism of human opioid receptors: Functional studies. Genetic contributions to protection from, or vulnerability to, addictive diseases. *Natl. Inst. Drug Abuse Res. Monogr. Ser.* **180:** 47–50.

Li T., Liu X., Zhu Z.H., Zhao J., Hu X., Sham P.C., and Collier D.A. 2000. Association analysis of polymorphisms in the *mu* opioid gene and heroin abuse in Chinese subjects. *Addict. Biol.* **5:** 181–186.

Maggos C.E., Tsukada H., Kakiuchi T., Nishiyama S., Myers J.E., Kreuter J., Schlussman S.D., Unterwald E.M., Ho A., and Kreek M.J. 1998. Sustained withdrawal allows normalization of *in vivo* [11C]N-methylspiperone dopamine D2 receptor binding after chronic "binge" cocaine: A positron emission tomography study in rats. *Neuropsychopharmacology* **19:** 146–153.

Maisonneuve I.M. and Kreek M.J. 1994. Acute tolerance to the dopamine response induced by a binge pattern of cocaine administration in male rats: An *in vivo* microdialysis study. *J. Pharmacol. Exp. Ther.* **268:** 916–921.

Maisonneuve I.M., Ho A., and Kreek M.J. 1995. Chronic administration of a cocaine "binge" alters basal dopamine extracellular levels in male rats: An *in vivo* microdialysis study. *J. Pharmacol. Exp. Ther.* **272:** 652–657.

Mantsch J.R., Ho A., Schlussman S.D., and Kreek M.J. 2001. Predictable individual differences in the initiation of cocaine self-administration by rats under extended-access conditions are dose-dependent. *Psychopharmacology* **157**: 31–39.

Mantsch J.R., Schlussman S.D., Ho A., and Kreek M.J. 2000. Effects of cocaine self-administration on plasma corticosterone and prolactin in rats. *J. Pharmacol. Exp. Ther.* **294**: 239–247.

Mantsch J.R., Yuferov V., Mathieu-Kia A.-M., Ho A., and Kreek M.J. 2003. Neuroendocrine alterations in a high-dose, extended-access rat self-administration model of escalating cocaine use. *Psychoneuroendocrinology* **28**: 836–862.

Mason B.J., Ritvo E.C., Morgan R.O., Salvato F.R., Goldberg G., Welch B., and Mantero-Atienza E. 1994. A double-blind, placebo-controlled pilot study to evaluate the efficacy and safety of oral nalmefene HCl for alcohol dependence. *Alcohol. Clin. Exp. Res.* **18**: 1162–1167.

McLellan A.T., Arndt I.O., Metzger D.S., Woody G.E., and O'Brien C.P. 1993. The effects of psychosocial services in substance treatment. *J. Am. Med. Assoc.* **269**: 1953–1959.

Mestek A., Hurley J.H., Bye L.S., Campbell A.D., Chen Y., Tian M., Liu J., Schulman H., and Yu L. 1995. The human mu opioid receptor: Modulation of functional desensitization by calcium/calmodulin-dependent protein kinase and protein kinase C. *J. Neurosci.* **15**: 2396–2406.

O'Malley S.S., Krishnan-Sarin S., Farren C., Sinha R., and Kreek M.J. 2002. Naltrexone decreases craving and alcohol self-administration in alcohol dependent subjects and activates the hypothalamo-pituitary-adrenocortical axis. *Psychopharmacology* **160**: 19–29.

O'Malley S.S., Jaffe A.J., Chang G., Schottenfeld R.S., Meyer R.E., and Rounsaville B. 1992. Naltrexone and coping skills therapy for alcohol dependence: A controlled study. *Arch. Gen. Psychiatry* **49**: 881–889.

Oslin D.W., Berrettini W., Kranzler H.R., Pettinati H., Gelernter J., Volpicelli J.R., and O'Brien C.P. 2003. A functional polymorphism of the mu-opioid receptor gene is associated with naltrexone response in alcohol-dependent patients. *Neuropsychopharmacology* **28**: 1546–1552.

Rettig R.A. and Yarmolinsky A., eds. 1995. *Federal regulation of methadone treatment*, pp. 37–60. National Academy of Sciences, National Academy Press, Washington, D.C.

Rosen M.I., McMahon T.J., Hameedi F.A., Pearsall H.R., Woods S.W., Kreek M.J., and Kosten T.R. 1996. Effect of clonidine pretreatment on naloxone-precipitated opiate withdrawal. *J. Pharmacol. Exp. Ther.* **276**: 1128–1135.

Schluger J.H., Borg L., Ho A., and Kreek M.J. 2001. Altered HPA axis responsivity to metyrapone testing in methadone maintained former heroin addicts with ongoing cocaine addiction. *Neuropsychopharmacology* **24**: 568–575.

Schluger J.H., Bart G., Green M., Ho A., and Kreek M.J. 2003. Corticotropin-releasing factor testing reveals a dose-dependent difference in methadone maintained vs control subjects. *Neuropsychopharmacology* **28**: 985–994.

Schluger J.H., Ho A., Borg L., Porter M., Maniar S., Gunduz M., Perret G., King A., and Kreek M.J. 1998. Nalmefene causes greater hypothalamic-pituitary-adrenal axis activation than naloxone in normal volunteers: Implications for the treatment of alcoholism. *Alcohol: Clin. Exp. Res.* **22**: 1430–1436.

Sinha R., Talih M., Malison R., Cooney N., Anderson G.M., and Kreek M.J. 2003. Hypothalamic-pituitary-adrenal axis and sympatho-adreno-medullary responses during stress-induced and drug cue-induced cocaine craving states. *Psychopharmacology* **170**: 62–72.

Spangler R., Unterwald E.M., and Kreek M.J. 1993. "Binge" cocaine administration induces a sustained increase of prodynorphin mRNA in rat caudate-putamen. *Mol. Brain Res.* **19**: 323–327.

Spangler R., Zhou Y., Schlussman S.D., Ho A., and Kreek M.J. 1997a. Behavioral stereotypes induced by "binge" cocaine administration are independent of drug-induced increases in corticosterone levels. *Behav. Brain Res.* **86**: 201–204.

Spangler R., Ho A., Zhou Y., Maggos C., Yuferov V., and Kreek M.J. 1996a. Regulation of kappa opioid receptor mRNA in the rat brain by "binge" pattern cocaine administration and correlation with preprodynorphin mRNA. *Mol. Brain Res.* **38**: 71–76.

Spangler R., Zhou Y., Maggos C.E., Schlussman S.D., Ho A., and Kreek M.J. 1997b. Prodynorphin, proenkephalin and κ opioid receptor mRNA responses to acute "binge" cocaine. *Mol. Brain Res.* **44**: 139–142.

Spangler R., Zhou Y., Maggos C.E., Zlobin A., Ho A., and Kreek M.J. 1996b. Dopamine antagonist and "binge" cocaine effects on rat opioid and dopamine transporter mRNAs. *Neuroreport* **7**: 2196–2200.

Szeto C.Y., Tang N.L., Lee D.T., and Stadlin A. 2001. Association between mu opioid receptor gene polymorphisms and Chinese heroin addicts. *Neuroreport* **12**: 1103–1106.

Tan E.C., Tan C.-H., Karupathivan U., and Yap E.P.H. 2003. Mu opioid receptor gene polymorphism and heroin dependence in Asian populations. *Neuroreport* **14**: 569–572.

Tsukada H., Kreuter J., Maggos C.E., Unterwald E., Kakiuchi T., Nishiyama S., Futatsubashi M., and Kreek M.J. 1996. Effects of "binge" pattern cocaine administration on dopamine D_1 and D_2 receptors in the rat brain: An *in vivo* study using positron emission tomography. *J. Neurosci.* **16**: 7670–7677.

Unterwald E.M., Fillmore J., and Kreek M.J. 1996. Chronic repeated cocaine administration increases dopamine D1 receptor-mediated signal transduction. *Eur. J. Pharmacol.* **318**: 31–35.

Unterwald E.M., Horne-King J., and Kreek M.J. 1992. Chronic cocaine alters brain μ opioid receptors. *Brain Res.* **584**: 314–318.

Unterwald E.M., Kreek M.J., and Cuntapay M. 2001. The frequency of cocaine administration impacts cocaine-induced receptor alterations. *Brain Res.* **900**: 103–109.

Unterwald E.M., Rubenfeld J.M., and Kreek M.J. 1994. Repeated cocaine administration upregulates κ and μ, but not δ, opioid receptors. *Neuroreport* **5**: 1613–1616.

Unterwald E.M., Cox B.M., Kreek M.J., Cote T.E., and Izenwasser S. 1993. Chronic repeated cocaine administration alters basal and opioid-regulated adenylyl cyclase activity. *Synapse* **15**: 33–38.

Unterwald E.M., Rubenfeld J.M., Imai Y., Wang J.-B., Uhl G.R., and Kreek M.J. 1995. Chronic opioid antagonist administration upregulates mu opioid receptor binding without altering mu opioid receptor mRNA levels. *Mol. Brain Res.* **33:** 351–355.

Volpicelli J.R., Alterman A.I., Hayashida M., and O'Brien C.P. 1992. Naltrexone in the treatment of alcohol dependence. *Arch. Gen. Psychiatry* **49:** 879–880.

Wand G.S., McCaul M., Yang X., Reynolds J., Gotjen D., Lee S., and Ali A. 2002. The mu-opioid receptor gene polymorphism (A118G) alters HPA axis activation induced by opioid receptor blockade. *Neuropsychopharmacology* **26:** 106–114.

Wang J.B., Imai Y., Eppler C.M., Gregor P., Spivak C.E., and Uhl G.R. 1993. Mu opiate receptor: cDNA cloning and expression. *Proc. Natl. Acad. Sci.* **90:** 10230–10234.

Wang J.B., Johnson P.S., Persico A.M., Hawkins A.L., Griffin C.A., and Uhl G.R. 1994. Human mu opiate receptor. cDNA and genomic clones, pharmacologic characterization and chromosomal assignment. *FEBS Lett.* **338:** 217–222.

Wang X.M., Zhou Y., Spangler R., Ho A., Han J.S., and Kreek M.J. 1999. Acute intermittent morphine increases preprodynorphin and mu-opioid receptor mRNA levels in the rat brain. *Mol. Brain Res.* **66:** 184–187.

Yuferov V., Bart G., and Kreek M.J. 2005a. Clock reset for alcoholism. *Nat. Med.* **11:** 23–24.

Yuferov V., Nielsen D.A., Butelman E.R., and Kreek M.J. 2005b. Microarray studies of psychostimulant-induced changes in gene expression. *Addict. Biol.* **10:** 101–118.

Yuferov V., Zhou Y., LaForge K.S., Spangler R., Ho A., and Kreek M.J. 2001. Elevation of guinea pig brain preprodynorphin mRNA expression and hypothalamic-pituitary-adrenal axis activity by "binge" pattern cocaine administration. *Brain Res. Bull.* **55:** 65–70.

Yuferov V., Zhou Y., Spangler R., Maggos C.E., Ho A., and Kreek M.J. 1999. Acute "binge" cocaine increases mu-opioid receptor mRNA levels in areas of the rat mesolimbic mesocortical dopamine system. *Brain Res. Bull.* **48:** 109–112.

Yuferov V., Fussell D., LaForge K.S., Nielsen D.A., Gordon D., Ho A., Leal S.M., Ott J., and Kreek M.J. 2004. Redefinition of the human kappa opioid receptor gene (OPRK1) structure and association of haplotypes with opiate addiction. *Pharmacogenetics* **14:** 793–804.

Zhang Y., Schlussman S.D., Ho A., and Kreek M.J. 2001. Effect of acute binge cocaine on levels of extracellular dopamine in the caudate putamen and nucleus accumbens in male C57BL/6J and 129/J mice. *Brain Res.* **921:** 172–177.

——. 2003. Effect of chronic "binge cocaine" on basal levels and cocaine-induced increases of dopamine in the caudate putamen and nucleus accumbens of C57BL/6J and 129/J mice. *Synapse* **50:** 191–199.

Zhang Y., Butelman E., Schlussman S.D., Ho A., and Kreek M.J. 2004a. Effect of the endogenous κ opioid agonist dynorphin A(1-17) on cocaine-evoked increases in striatal dopamine levels and cocaine-induced place preference in C57BL/6J mice. *Psychopharmacology* **172:** 422–429.

——. 2004b. Effect of the κ opioid agonist R-84760 on cocaine-induced increases in striatal dopamine levels and cocaine-induced place preference in C57BL/6J mice. *Psychopharmacology* **173:** 146–152.

——. 2005. The plant derived hallucinogen, salvinorin A, decreases basal dopamine levels in the caudate putamen and produces conditioned place aversion in mice by agonist actions on kappa opioid receptors. *Psychopharmacology* **179:** 551–558.

Zhou Y., Spangler R., Ho A., and Kreek M.J. 2003a. Increased CRH mRNA levels in the rat amygdala during short-term withdrawal from chronic "binge" cocaine. *Mol. Brain Res.* **114:** 73–79.

Zhou Y., Spangler R., Schlussman S.D., Ho A., and Kreek M.J. 2003b. Alterations in hypothalamic-pituitary-adrenal axis activity and in levels of POMC and CRH-R1 mRNAs in the pituitary and hypothalamus of the rat during chronic "binge" cocaine and withdrawal. *Brain Res.* **964:** 187–199.

Zhou Y., Franck J., Spangler R., Maggos C.E., Ho A., and Kreek M.J. 2000. Reduced hypothalamic POMC and anterior pituitary CRF1 receptor mRNA levels after acute, but not chronic, daily "binge" intragastric alcohol administration. *Alcohol: Clin. Exp. Res.* **24:** 1575–1582.

Zhou Y., Spangler R., LaForge K.S., Maggos C.E., Ho A., and Kreek M.J. 1996a. Corticotropin-releasing factor and CRF-R1 mRNAs in rat brain and pituitary during "binge" pattern cocaine administration and chronic withdrawal. *J. Pharmacol. Exp. Ther.* **279:** 351–358.

——. 1996b. Modulation of CRF-R1mRNA in rat anterior pituitary by dexamethasone: Correlation with POMC mRNA. *Peptides* **17:** 435–441.

Zhou Y., Spangler R., Maggos C.E., LaForge K.S., Ho A., and Kreek M.J. 1996c. Steady-state methadone in rats does not change mRNA levels of corticotropin-releasing factor, its pituitary receptor or proopiomelanocortin. *Eur. J. Pharmacol.* **315:** 31–35.

Zhou Y., Yuferov V.P., Spangler R., Maggos C.E., Ho A., and Kreek M.J. 1998. Effects of memantine alone and with acute "binge" cocaine on hypothalamic-pituitary-adrenal activity in the rat. *Eur. J. Pharmacol.* **352:** 65–71.

Zhou Y., Schlussman S.D., Ho A., Spangler R., Fienberg A.A., Greengard P., and Kreek M.J. 1999a. Effects of chronic 'binge' cocaine administration on plasma ACTH and corticosterone levels in mice deficient in DARPP-32. *Neuroendocrinology* **70:** 196–199.

Zhou Y., Spangler R., Maggos C.E., Wang X.M., Han J.S., Ho A., and Kreek M.J. 1999b. Hypothalamic-pituitary-adrenal activity and pro-opiomelanocortin mRNA levels in the hypothalamus and pituitary of the rat are differentially modulated by acute intermittent morphine with or without water restriction stress. *J. Endocrinol.* **163:** 261–267.

Zhou Y., Spangler R., Schlussman S.D., Yuferov V.P., Sora I., Ho A., Uhl G.R., and Kreek M.J. 2002. Effects of acute "binge" cocaine on preprodynorphin, preproenkephalin, pro-opiomelanocortin and corticotropin-releasing hormone receptor mRNA levels in the striatum and hypothalamic-pituitary-adrenal axis of mu-opioid receptor knockout mice. *Synapse* **45:** 220–229.

Zimprich A., Kraus J., Woltje M., Mayer P., Rauch E., and Höllt V. 2000. Anallelic variation in the human prodynorphin gene promoter alters stimulus-induced expression. *J. Neurochem.* **74:** 472–477.

6 Imaging the Addicted Brain

Nora D. Volkow,[1-3] Gene-Jack Wang ,[3] Joanna S. Fowler,[4] and Rita Z. Goldstein[3]

[1]National Institute on Drug Abuse, Bethesda, Maryland 20892; [2]National Institute on Alcohol Abuse and Alcoholism, Bethesda, Maryland 20892; [3]Medical Department, Brookhaven National Laboratory, Upton, New York 11973; [4]Chemistry Department, Brookhaven National Laboratory, Upton, New York 11973

ABSTRACT

Imaging technologies are giving researchers and clinicians a greater understanding of the underlying neurobiological basis of drug addiction and the differences in individual vulnerabilities to abuse and addiction. Unprecedented progress has been made in elucidating molecular and cellular mechanisms that, coupled with genetic and environmental factors, can lead to compulsive drug taking. This chapter describes some of the brain-imaging techniques currently used to study living human subjects, with a focus on the ways in which positron emission tomography (PET) has furthered our understanding of the long-term changes in behavior and brain function that take place in the addicted individual.

INTRODUCTION

When one considers the multitude of synthetic and natural compounds, it is surprising to note that few are actually addictive. What are the properties of drugs that cause them to induce compulsive intake despite known negative consequences? How do drugs interact with biological systems to produce the biochemical changes that occur with addiction? What characteristics of individuals make them vulnerable to addiction, whereas others seem "protected?" An increased understanding of how drugs of abuse affect the human brain is essential for developing effective prevention strategies and treatment approaches for individuals that do become addicted.

A puzzle also remains as to the extent that changes in the brains of those who are addicted to drugs can be attributed to chronic drug exposure, genes that predispose them to addiction, the effects of an environment at vulnerable periods that facilitates the development of addiction, or a combination of these. Researchers in the PET imaging group at the Brookhaven National Laboratory began studying the neurotransmitter dopamine (DA) two decades ago because drugs of abuse cause increases in brain DA in specific brain areas, and these increases appear crucial for a drug's reinforcing effects. What has remained elusive, however, is understanding the exact role of DA in addiction. Neuroimaging studies combined with neuropsychological tools are now providing

93

insights into the involvement of the brain DA system in the core underlying neurocognitive-behavioral mechanisms in drug addiction.

It is now recognized that drug addiction is a complex, chronic, and relapsing brain disorder, but the perception that addiction reflects moral weakness remains all too common. Although the precise mechanisms that underlie addiction and relapse remain to be fully elucidated, research is revealing that addiction is likely to be the result of an individual's unique combination of genetic and biological factors interacting with environmental influences and chronic drug use. Although this chapter focuses on the role of DA in drug addiction, it is clear that DA is only one of multiple neurotransmitters involved in the complex neurobehavioral changes that occur in the addicted individual.

GOALS OF THE CHAPTER

This chapter enables the reader to develop a basic understanding of (1) the specificities and limitations of commonly used brain-imaging technologies and their applications to the study of the neurobiological basis of addiction and (2) the involvement of dopamine in drug abuse and addiction.

METHODOLOGICAL CONSIDERATIONS

Brain-imaging Methods Used in Studies of Addiction

Imaging techniques are now essential to refine our knowledge about biochemical, physiological, and functional brain processes; brain anatomy and tissue composition; neurotransmitters; energy use and blood flow; and drug distribution and kinetics in the living human brain. Imaging techniques vary in their sensitivities, specificities, and limitations, giving each technique unique properties for studies of the brain. It is helpful to compare the most commonly used techniques with respect to spatial and temporal resolution and principal applications (Table 1). Imaging studies of drug users can include protocols with age-matched, nonaddicted individuals to compare the functionality of brain areas affected by the chronic use of drugs. Some techniques are more amenable for repeated imaging, for example, studying the same individual before and after periods of abstinence, to compare ways in which brain function changes with use and recovery in the same person.

Magnetic Resonance Imaging (MRI)

MRI has wide applications in clinical medicine for diagnosis and evaluation, and it has the highest spatial resolution of the noninvasive imaging technologies currently available for functional brain mapping. Because MRI does not require ionizing radiation, multiple studies can be performed in the same patient without radiation exposure or known long-term risk. Protons in the water molecules that comprise all brain tissue emit characteristic radio waves as they respond to an externally applied magnetic field, and this produces clinical MRI signals. Thus, the intensity of an MRI signal depends on the water content, and brain water content differs for white matter (myelinated fibers and tracts) and gray matter (neu-

TABLE 1. Brain-imaging Techniques Commonly Used in Drug Abuse Research

Imaging technique	Principal applications	Temporal resolution	Spatial resolution	Advantages	Limitations
Magnetic resonance imaging (MRI)	tissue morphology; composition; physiological function; biochemical and pharmacological processes; receptor-ligand interaction	sec–hr	1.0–1.5 mm	no ionizing radiation; noninvasive; maximizes tissue contrasts; high anatomical resolution	claustrophobia prohibits participation of some subjects
Magnetic resonance spectroscopy (MRS)	cerebral metabolism; specific chemicals in physiological processes; tissue composition; drug kinetics; distribution; metabolites	dependent on field strength; ~7–27 min	1 cm^3 or more with ^{1}H-MRS; decreases with higher field strengths	noninvasive; metabolic changes observable; can be performed as part of MRI	limited sensitivity and specificity; diagnostic and therapeutic roles not established; acquisition and interpretation not standardized
Functional magnetic resonance imaging (fMRI)	changes in brain chemical composition; fluid dynamics (e.g., glucose, blood); functional brain activation mapping	0.5–5 sec	~3 mm for BOLD; depends on image/signal noise and field strength acquisition speed	no ionizing radiation; noninvasive; hundreds of acquisitions possible during each session	motion can corrupt the signal, a real problem with these populations; signal loss in the most interesting regions (vmPFC and amygdala)
Positron emission tomography (PET)	detection and quantification of physiological function; metabolic, biochemical, and pharmacological processes; drug distribution and kinetics; receptor-ligand interaction; enzyme targeting	15–60 sec (limited by physiology, e.g., 90% of maximal change in regional cerebral blood flow [rCBF] occurs 5–8 sec after stimulus)	2–4 mm	high sensitivity; quantitative for rCBF, striatal glucose consumption, and regional cerebral metabolic rateof oxygen consumption	radiation exposure; indirectly measures neuronal activity; not widely available
Single-photon-emission computed tomography (SPECT)	receptor-ligand interaction; physiological function; biochemical and pharmacological processes	min (limited by physiology)	6–8 mm	quantitative; less expensive than PET; more widelyavailable; measurescerebral metabolism	radiation exposure; low temporal rCBF; low resolution; sensitive to rCBF changes

ronal tissue). Differences in signal intensities produce contrasts in the resulting MRI image.

MRI has provided invaluable information about changes in gross brain volume over time. For example, the dynamic anatomical development of human cortical gray matter from childhood through early adulthood has been documented with MRI (Gogtay et al. 2004). Indeed, the slow maturation of the prefrontal cortex (PFC), which is essential for higher-order decision making and self-monitoring, has been hypothesized to be one of the contributors to the greater vulnerability of adolescents to drug abuse and addiction. In studies of substance abuse, MRI has been used to compare the brains of addicted individuals with those of nondrug users. These studies have shown that individuals with a history of polysubstance use have smaller prefrontal lobes than the same brain regions in healthy nonusing volunteers (Liu et al. 1998). These findings are of particular importance to the study of drug addiction because patients with injuries to the PFC often have behavioral disruptions that can include aggression, poor judgment, and an inability to inhibit inappropriate responses (Eslinger et al. 1992), which are strikingly similar to those observed in many addicted individuals (Bechara et al. 1994; Bechara 2001). This has led to the hypothesis that PFC dysfunction has an important role in the neuropathological basis of drug addiction. Imaging studies have also been conducted to identify those individuals who are at higher risk for addiction to attempt to distinguish between the factors that may have preceded drug use and that may predispose an individual to an increased vulnerability to addiction from the changes that are secondary to chronic drug use.

Magnetic Resonance Spectroscopy (MRS)

When MRI is used to assess the chemical composition of the brain and to measure the brain concentration and kinetics of certain drugs, this mode of use is denoted as magnetic resonance spectroscopy (MRS). MRS is based on a chemical's unique resonance frequency when stimulated by a magnet. Results from MRS imaging are displayed graphically as a spectrum, with peaks that represent detected chemicals. Chemical compounds, including those endogenously found in the brain, must be present in a relatively large concentration, or they must provide a unique spectral "signature" for detection by MRS. Therefore, when MRS is used to image drug kinetics, only those that achieve relatively high brain concentrations, such as alcohol, can be detected.

During the last 20 years, MRS has been performed in patients in an effort to understand the pathological biochemistry of various neurological and psychiatric disorders, identify differences between diagnostic groups, and monitor brain changes associated with drug therapy. MRS studies have also helped to identify changes in brain metabolites associated with cognitive dysfunction.

N-acetyl aspartate (NAA), which provides a uniquely detectable MRS signal, is a naturally occurring amino acid that is found in the central nervous system (CNS). Measures of NAA are used as an index of neuronal viability, and MRS has shown that concentrations or ratios of NAA correlate with measures of cognitive function in healthy subjects (Rae et al. 1998). Reductions in NAA concentrations in basal ganglia and frontal white matter have been reported in methamphetamine abusers and interpreted as evidence of the neurotoxic effects of this drug to the human brain (Ernst et al. 2000). Reductions in gamma-aminobutyric acid (GABA) concentration, which is the main inhibitory neurotransmitter in the brain, have been documented by MRS studies both in alcoholics and in cocaine-addicted subjects (Volkow et al. 2000; Dick et al. 2004; Edenberg et al. 2004).

These findings could provide a potential underlying mechanism for the benefits reported with the use of pharmacological agents that increase brain GABA concentration in the treatment of addiction (Dewey et al. 1998; Brodie et al. 2003).

Functional Magnetic Resonance Imaging (fMRI)

MRI can also be used to measure the function of the human brain, and this is referred to as functional MRI (fMRI). When oxygenated blood flows into stimulated (activated) brain regions, the surrounding magnetic field is altered. fMRI detects and localizes these small magnetic variations. The change in signal intensity detected by fMRI is attributable to differences in the magnetic properties of oxygenated versus deoxygenated blood, known as BOLD (blood oxygen level-dependent) contrast (Springer 1999). During stimulation, an excess of arterial blood is delivered into the active area with concomitant changes in the ratio of deoxyhemoglobin to oxyhemoglobin, which then gives rise to the activation BOLD signal. fMRI has high spatial resolution (2–4 mm in human studies), does not require ionizing radiation, and is therefore amenable to repeated imaging sessions with the same subject. Further, the safety of this technique allows for investigations in subjects of all ages.

fMRI has been invaluable in helping to reveal how rewarding stimuli shape and affect human behavior. Changes in activation of specific brain regions in healthy subjects responding to behavioral stimuli and neuropsychological tasks can be measured and compared with those of individuals addicted to drugs. For example, viewing erotic films was associated with a larger extent of activation (both in volume and strength) in a distributed network of brain regions implicated in reward processing in healthy subjects as compared to cocaine-addicted subjects. In contrast, in cocaine-dependent subjects, a similar reward circuitry is activated by experimental cocaine infusions, and this has been associated with subjective reports of cocaine-induced euphoria and craving (Breiter et al. 1997; Breiter and Rosen 1999). These findings corroborate in addicted subjects the ability of a drug to recruit brain regions involved in processing natural rewards (see Fig. 1).

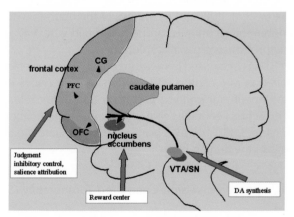

FIGURE 1. Brain dopamine system. Dopamine cell bodies synthesize dopamine in the ventral tegmental area (VTA) and substantia nigra (SN). The nucleus accumbens is often called the brain's reward center and is linked to the reinforcing effects of drugs of abuse. The frontal cortex includes the orbitofrontal cortex (OFC), involved in salience attribution, the prefrontal cortex (PFC), involved in judgment, planning, and executive control, and the cingulate gyrus (CG), associated with emotional processing and cognitive and behavioral control.

fMRI has thus identified the brain regions on which to focus molecular and neurochemical studies of addictions. The anterior cingulate gyrus (CG; associated with emotional processing, and cognitive and behavioral control) is activated in some cocaine-addicted individuals when viewing videotapes containing cocaine-associated cues, even if subjects do not report cocaine craving. fMRI has also shown activation in the frontal lobe of addicted individuals compared with healthy subjects during cocaine-cue tapes (Wexler et al. 2001). Reduced prefrontal activation, reflected behaviorally as impaired decision making, has been detected in methamphetamine-dependent subjects (Paulus et al. 2002), whereas significant hypoactivity in midline areas of the anterior CG has been shown in cocaine-abusing subjects (Kaufman et al. 2003).

Pharmacogenomic studies using fMRI have shown that variations in neurotransmitter expression may contribute to the vulnerability to abuse drugs, and to addiction, depression, and suicidal behavior. For example, a known polymorphism in the catechol-*O*-methyltransferase gene (COMT; the enzyme that catalyzes degradation of DA, epinephrine, and norepinephrine) has been shown to affect the brain's response to amphetamine in the PFC (Mattay et al. 2003).

Positron Emission Tomography

PET uses radiotracers labeled with short-lived positron-emitting isotopes to track biochemical transformations as well as the movement of drugs in the living human brain. The availability of positron emitters for natural elements (carbon-11, oxygen-15, nitrogen-13, and fluorine-18, which can be used to substitute for hydrogen) and the high sensitivity of PET (measures concentrations in the nanomolar–picomolar range) allow the flexibility to label compounds that are of pharmacological and physiological relevance, and the use of those radiotracers to measure neurochemical processes without perturbing them. The short half-lives of PET isotopes also make them relatively benign; subjects are exposed to limited amounts of radioactivity, often allowing multiple studies (important in addiction research) to be performed with the same subjects (Fowler et al. 2003; Kung et al. 2003; Volkow et al. 2003a). PET studies of the brain do not provide much anatomical information, but the technique can be combined with MRI when mapping chemical changes to maximize image spatial resolution.

PET has been used to measure brain glucose metabolism, cerebral blood flow; the concentration of receptors, transporters, and enzymes involved in the synthesis or metabolism of neurotransmitters; and neurotransmitter release (Reivich et al. 1979; Wang et al. 1999; Schreckenberger et al. 2004). These applications are illustrated for the DA synapse in Figure 2. PET has also been used to evaluate the distribution and pharmacokinetics of drugs in the human brain.

Single-photon-emission Computed Tomography (SPECT)

SPECT shares many technical similarities with PET and can also be used to measure receptors, transporters, and enzymes. SPECT uses single-photon (gamma)-emitting isotopes that typically have longer half-lives than those of PET. SPECT quantifies very small (nanomolar–picomolar) concentrations of cellular elements important to drug abuse research, but the resolution and sensitivity of SPECT are lower than for PET. Furthermore, SPECT isotopes cannot be incorporated into drugs because their constituent elements are

FIGURE 2. Dopamine synapse. (*A*) When a dopamine cell fires, dopamine (DA) is released into the synapse (gap between the cells) and interacts with a DA receptor. Excess DA is taken back up into the cell, via the dopamine transporter, where it is stored or inactivated by monoamine oxidase A (MAO A). (*B*) When cocaine is present, it binds to the dopamine transporter, preventing the reuptake of dopamine. This results in a buildup of DA in the synapse, causing an intense stimulation of the dopamine receptors on a neighboring cell.

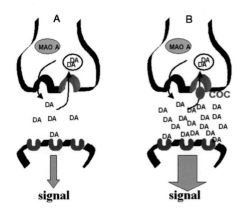

absent from pharmaceuticals (i.e., ^{99}Tc or ^{123}I). No SPECT counterpart to ^{18}FDG exists for measuring brain glucose metabolism.

APPLICATIONS OF PET FOR SUBSTANCE ABUSE STUDIES

Critical Role of Brain Imaging in Drug Addiction Studies

Brain imaging has identified regions, circuits, neurochemicals, and metabolic processes involved in intoxication, craving, withdrawal, and relapse. Imaging is also furthering our understanding of how environmental variables interact with brain circuits to affect the propensity to take drugs compulsively. For example, psychological stress is a known risk factor for maladaptive drug use, and imaging studies are beginning to assess the effects of polymorphisms of the COMT and SERT genes on the sensitivity of limbic brain regions to stress (Hariri et al. 2002).

Imaging Has Increased our Understanding of Dopamine's Involvement in Addiction

Most drugs of abuse increase the concentration of extracellular DA in limbic brain regions, and this effect has been linked with the drugs' ability to be reinforcing (Di Chiara and Imperato 1988; Koob and Bloom 1988). In humans, PET studies have corroborated the relevance of increased DA levels to the reinforcing effects and saliency of drugs, as well as the importance of the dynamic characteristics of DA increases. These studies have shown that it is not DA increases per se in striatum that are associated with the reinforcing effects of drugs, but rather, "fast" DA increases.

Levels of extracellular DA in the striatum are established by both tonic and phasic DA cell firing. Tonic firing maintains baseline steady-state levels and sets the overall responsiveness of the DA system, whereas phasic DA cell firing leads to fast DA changes that highlight the saliency of stimuli. Large, fast increases in brain DA that mimic but exceed the phasic DA increases in response to environmental events may be a way that DA encodes saliency (Grace 2000). The magnitude and duration of the phasic DA increases induced by drugs (particularly stimulants) are significantly greater and of longer duration than the DA increases induced by natural reinforcers. Moreover, drugs of abuse do not result in adaptation with repeated administration, whereas natural reinforcers lose the ability to increase DA in the nucleus accumbens (NAc) shell with repeated exposure (Di Chiara 2002). This lack of attenuation is believed to strengthen the reinforcing value of the drug.

Radiolabeled cocaine and methylphenidate (MP) have been used in PET studies to investigate how DA in the human brain is involved in the reinforcing effects of stimulant drugs. Cocaine is considered one of the most reinforcing of the drugs of abuse, increasing extracellular DA by blocking DA transporters (DATs). The stimulant drug MP also increases extracellular DA by blocking DATs. Cocaine and MP have similar in vitro affinities for inhibition of DA uptake, but MP has a much lower level of abuse and a well-accepted clinical profile for treating attention deficit hyperactivity disorder (ADHD) (Parran and Jasinski 1991).

We thus used PET to compare the regional brain distribution and pharmacokinetics of carbon-11 ($[^{11}C]$)-labeled cocaine and MP. Such studies have been used to assess the relationship between brain pharmacokinetics and temporal patterns of the self-reported "high" (Volkow et al. 1994). Intravenous MP and cocaine each enter the brain rapidly—4–6 minutes for cocaine and 8–10 minutes for MP—and have similar regional distribution. Both drugs show the highest uptake in the basal ganglia, where they compete with DA for binding to DATs. However, for MP, the rate of brain clearance is significantly slower (90-min half-life) than for cocaine (20-min half-life) (Fig. 3). This rate of clearance of both drugs (when administered intravenously) parallels the temporal changes for the perception of the "high." For cocaine, the reduction in "high" follows its fast clearance from the brain, whereas for MP, the "high" declined rapidly despite the persisting presence of high levels of MP remaining in the brain.

Thus, the initial fast uptake of MP and cocaine into the brain, rather than their steady-state presence, seems necessary for drug-induced reinforcement. Cocaine's rate of clearance from the brain corresponds well to the frequency of administration that is reported by abusers (every 20–30 min during smoked or intravenous cocaine binging). It appears that the slow rate of clearance of MP constrains the frequency of administration, because slow clearance might lead to DAT saturation and longer duration of (aversive) side effects with repeated administration.

Drugs, Dopamine, and Reinforcement

Although DAT blockade is the initial pharmacological effect of MP and cocaine, the subsequent increase in DA and resulting activation of DA receptors are responsible for their behavioral effects. $[^{11}C]$raclopride competes with endogenous DA for occupancy of the DA D2 receptors and can be used in PET studies to assess relative changes in synaptic DA. Two PET scans with $[^{11}C]$raclopride are required in such studies: one after pretreatment with placebo and one after pretreatment with the drug being evaluated. The difference between these conditions is relative to the changes in synaptic DA that are induced by the drug under study. Comparison of the DA changes induced by intravenous cocaine to those by intravenous MP revealed equipotent effects as assessed by the similar decreases in striatal $[^{11}C]$raclopride binding. In human, cocaine-addicted individuals, intravenous MP decreased $[^{11}C]$raclopride binding in a dose-dependent manner and induced a "high" in most subjects. The subjects that had the greatest DA increases (i.e., decreased $[^{11}C]$raclopride binding) were those that perceived the most intense "high." Subjects in whom MP did not increase striatal DA did not perceive a "high" (see Fig. 4). Findings from

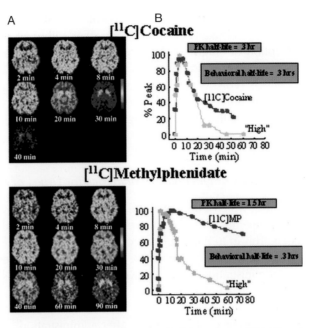

FIGURE 3. (*A*) Images for [¹¹C]cocaine and [¹¹C]methylphenidate ([¹¹C]MP) at the level of the basal ganglia at various times after radiotracer administration. Red represents the highest concentration of radiotracer. (*B*) Time-activity curves for [¹¹C]cocaine and [¹¹C]MP plotted with the corresponding temporal patterns for the subjective reports of "high" after pharmacological doses of intravenous cocaine or MP. The peak corresponds to the normalized maximum "high" reported by each subject. Both drugs have rapid uptake into the brain, and the initial uptake parallels the temporal changes for the perception of "high" for both. However, for MP, the rate of clearance is significantly slower than for cocaine. Reduction in "high" for cocaine follows its fast brain clearance, whereas for MP, the "high" declined rapidly despite the persisting presence of high brain levels. (Reprinted, with permission, from Volkow et al. 2004b [©Macmillan Publishers Ltd.].)

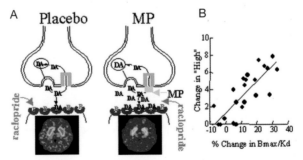

FIGURE 4. (*A*) PET-raclopride method for measuring changes in DA induced by administration of MP. Subjects are tested on two occasions: one after placebo and one after MP. MP blocks DA transporters, allowing DA to accumulate in the synapse. Because [¹¹C]raclopride can only bind to those DA D2 receptors that are not occupied by DA, the difference in its binding between placebo and MP reflects MP-induced increases in synaptic DA. (*B*) Correlation between MP-induced increases in DA (assessed as percent of change in Bmax/Kd, the measure of DA D2 receptor availability obtained with [¹¹C]raclopride) and self-reports of a "high" induced by MP. (Reprinted, with permission, from Volkow et al. 2004a [©Elsevier].)

MP/cocaine PET studies are consistent with other studies in the literature that correlate increases in striatal DA induced by other stimulant drugs (e.g., amphetamine) with the perception of "high" or of euphoria. Combined, these findings provide evidence that stimulant drugs cause increases in striatal DA that are associated with the drugs' rewarding effects (Volkow et al. 2004a). This is relevant because it is believed that at least for natural reinforcers, DA does not encode for reward itself, but rather, for prediction of reward. Thus, these findings highlight another potential difference between natural and drug reinforcers.

Importance of Brain Pharmacokinetics for Drugs of Abuse

The rate at which a drug of abuse enters the brain is a powerful determinant of its reinforcement properties; the faster the entry of the drug into the brain, the greater its rewarding effects. This in turn explains why routes of administration that lead to faster brain delivery (intravenous, smoking) are more reinforcing than routes that lead to slower brain uptake (oral administration). Thus, intravenous MP, which leads to very fast delivery of the drug to the brain (peak uptake is 10 min after administration), is very rewarding to cocaine-addicted individuals, whereas oral MP, which enters the brain very slowly (peak brain uptake does not occur until 60–90 min after administration), is much less rewarding. This difference occurred although striatal DA increases were comparable, and DAT blockade was equivalent (~70%) after a 20-mg oral dose of MP or a 0.5-mg/kg intravenous dose. However, oral MP did not induce increases in self-reports of "high," nor was there a correlation between the "high" and the DA increases. In contrast, intravenous MP induced significant increases in subjects' perceptions of reinforcement, which were correlated with the DA increases (Volkow et al. 2001b). These differences between intravenous and oral MP are therefore likely to reflect the differences in the functional consequences of fast versus slow DA increases in DA-regulated circuits.

The PET studies discussed in this section have focused on how stimulant drugs affect levels and kinetics of brain DA. However, the interactions of DA with other neurotransmitters, such as glutamate and GABA, and with nonpharmacological variables, such as those affecting learning and memory, likely have a role in modulating the magnitude of the DA response and contribute to a drug's reinforcing effects (Bell et al. 2000; Szumlinski et al. 2004). They are described in the following section.

Nonpharmacological Variables Modulate the Reinforcing Effects of Drugs

Neutral stimuli can acquire reinforcing properties and motivate salience through conditioned-incentive learning. Addiction may facilitate and contribute to another form of learning—habit learning—causing common stimuli to automatically elicit well-learned sequences of behavior. It has been suggested that in extreme cases, habit learning may be associated with perseveration and contribute to ritualistic behaviors that are linked with drug intake (Powell 1995). Declarative memory is also involved in associating affective conditions or circumstances with drug-taking experiences. Indeed, PET and fMRI have both shown that cue-elicited craving and intoxication activate brain regions known to be involved in memory, including the hippocampus and amygdala (Grant et al. 1996; Childress et al. 1999; Wang et al. 1999; Kilts et al. 2001).

These nonpharmacological variables, including conditioned responses, shape an individual's expectation of a drug's effects. Expectation, in turn, modulates the responses of the individual to that drug. For example, in drug abusers, subjective "pleasurable" responses to the drug are increased when subjects expect to receive the drug (Kirk et al. 1998). In laboratory animals, cocaine-induced DA increases in the NAc are larger when animals are given the drug in an environment in which they had previously received it, compared to a novel environment. Cocaine-induced changes in brain metabolism are also different when animals self-administer the drug than when administration is involuntary (Graham and Porrino 1995).

[18]FDG (2-[^{18}F]fluoro-2-deoxy-D-glucose) PET studies of the effects of drug expectation on brain metabolism in human cocaine abusers have used MP, taking advantage of the drug's long half-life. The increases in regional brain metabolism were about 50% larger when the drug was expected than when unexpected. The largest increases in metabolism were in the cerebellum and thalamus (Volkow et al. 2003c). When MP was not expected but received, metabolism was increased in the left lateral OFC. Increases in self-reports of "high" also increased by about 50% when subjects expected MP rather than when they did not, and these increases significantly correlated with the metabolic increases in the thalamus. These findings show the involvement of the OFC in unexpected rewards in humans and suggest an involvement of the thalamus in conditioned responses.

Disrupted Brain Circuits Are Involved in the Process of Addiction

We hypothesize that addiction results from adaptations to intermittent, drug-induced, supraphysiological increases in DA. In support of this hypothesis, imaging studies of addicted individuals have provided evidence that multiple brain DA circuits are disrupted, including those that underlie reward and saliency, motivation and drive, inhibitory control and disinhibition, and memory and conditioning. Direct and indirect glutamatergic and GABAergic innervations connect these circuits to each other, and any given brain region participates in more than one circuit. Although traditionally the neurobiology of addiction primarily implicated limbic brain circuits (amygdala, NAc), neuroimaging studies have implicated the involvement of prefrontal cortical regions (e.g., OFC, anterior CG) (for a review, see Goldstein and Volkow 2002) as well as the temporal insula (Franklin et al. 2002). Specifically, it has been postulated that disruption of the OFC and CG underlie compulsive drug intake as well as the poor impulse control regarding drug intake that characterizes addiction (Volkow and Fowler 2000). The role of the temporal insula in the addiction process is much less understood, but this brain region likely underlies the strong autonomic responses to the drug and drug stimuli that occur with conditioning.

Brain circuits involved in motivation and drive are intricately related to reward circuits. Primate studies have shown that the OFC and CG are involved in regulating the role of DA in motivation (Koob 1996; Hollerman and Schultz 1998). Thus, DA-mediated OFC disruptions would alter the saliency value of a stimulus as a function of its previous conditioning to the drug, whereas dysfunction of the CG would affect inhibitory control. Interestingly, there is *increased*, rather than decreased, activation of the OFC in addicted subjects when they are presented with drug-related stimuli or memories, or if given a drug. This activation has been associated with the intensity of desire for the drug, suggesting that drugs or drug-related stimuli in addicted subjects activate the OFC. This offers a potential explanation for an addict's enhanced motivation and drive to take the drug. OFC and anterior CG abnormalities have been reported in cocaine abusers, alcoholics, methamphetamine abusers, heroin abusers, and marijuana abusers, suggesting that this is a common abnormality across various types of drug addictions (Goldstein and Volkow 2002).

Several brain regions, including the anterior CG, the lateral OFC, and the dorsolateral PFC, regulate inhibitory control and affect processes involved in executive control (Royall et al. 2002). These areas are disrupted in many addicted individuals. Drug-induced changes in dendrites of the PFC have been documented in preclinical studies, with significant branching and increased spine density resulting from repeated drug self-administration in rats (Robinson et al. 2001). Neuronal changes and the resulting disruption of prefrontal brain activity in drug-addicted individuals could cause impairments in attribution of emotional/motivational salience, self-monitoring, and behavioral control, and have an important role in the cognitive changes that perpetuate drug self-administration. The specific role of each region in drug addiction and the differences between addiction and other disorders that share a dysfunction of the PFC remain to be fully characterized.

Differences in Individual Responses to Drugs Suggest Addiction Vulnerability

Why do some individuals become addicted while others do not? Using functional neuroimaging techniques, it is difficult to determine if the brain abnormalities of drug-addicted individuals were preexisting and contributed to drug abuse, or if dysfunction was a consequence of drug abuse. Vulnerability may result from the disrupted activity of inhibitory control circuits, a decreased sensitivity of reward circuits to natural reinforcers, an increased sensitivity to conditioned drug stimuli, and/or increased responses of motivation and drive circuits to drugs. Preclinical studies have provided important information about the involvement of DA, glutamate, opioids, and serotonin in modulating the predisposition to drug self-administration (Laakso et al. 2002).

In laboratory animals, DA function modulates the predisposition to drug self-administration, and genetic manipulations of DA D2 receptors markedly affect drug self-administration (Piazza et al. 1991; Thanos et al. 2001). In humans, imaging can be used to compare differences in DA D2 receptor levels in nonaddicted subjects with behavioral responses to drugs of abuse to gain further insight into the effects of drug "liking," especially if conducted during vulnerable periods (e.g., adolescence, stress).

Baseline levels of striatal DA D2 receptors in healthy, non-drug-abusing individuals vary in concentration. Behavioral responses of nonabusing subjects to intravenous MP were found to be negatively correlated with their baseline levels of DA D2 receptors (Volkow et al. 1999, 2002). Individuals with lower levels of DA D2 receptors (half of the study subjects) described MP as pleasant. The remainder described intravenous MP as unpleasant, and the disparity could not be attributed to differences in plasma MP concentration (Fig. 5). Thus, there may be an optimal range for DA D2 receptor stimulation such that too little is not reinforcing, whereas too much stimulation may become aversive. This suggests an inverted U-shaped curve to describe the relationship between drug-induced DA D2 receptor activation and its rewarding effects.

PET studies have shown that levels of DA D2 receptors in cocaine-abusing individuals are significantly reduced (Figs. 6 and 7). These changes persist after withdrawal, and cor-

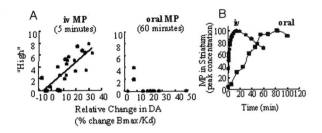

FIGURE 5. (*A*) Relationship between changes in DA as assessed by decreases in striatal [^{11}C]raclopride binding after intravenous and oral MP. The relationship was significant for intravenous but not for oral MP. Oral MP did not increase the self-reports of "high" except in one subject. (*B*) Time-activity curves for the uptake of [^{11}C]MP in the baboon brain after intravenous and oral administration. The curves were normalized to maximal uptake. (Reprinted, with permission, from Volkow et al. 2004b [©Macmillan Publishers Ltd.].)

FIGURE 6. In subjects with low DA D2 receptors (*closed star; bottom PET image*), the large MP-induced increases in DA result in optimal stimulation and perception of the effects of the drug as pleasant. In subjects with high D2 receptors (*open star; upper PET image*), the large MP-induced DA increases move them to the far right of an inverted U-shaped curve, resulting in overstimulation and perception of drug effects as unpleasant. (Reprinted, with permission, from Volkow et al. 2004a [©Elsevier].)

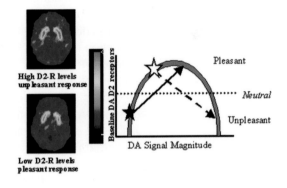

FIGURE 7. DA D2 receptors and brain glucose metabolism in control subjects and cocaine abusers. Drug-addicted subjects have decreases in DA D2 receptors in striatum and decreases in metabolism in the orbital frontal cortex (OFC). (Modified, with permission, from Volkow et al. 2004a [©Elsevier].)

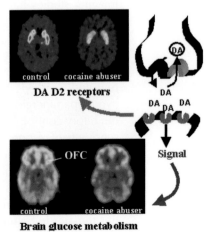

relate with self-ratings of dysphoria (Volkow et al. 1990, 1993). Studies using [18]FDG in the same cocaine-abusing subjects have revealed low metabolism in the OFC and the anterior CG (Volkow et al. 1993, 2003b). Additionally, low levels of DA D2 receptors have been documented in alcoholics (Volkow et al. 1996), heroin abusers (Wang et al. 1997; Volkow et al. 2001b), and methamphetamine abusers (Volkow et al. 2001b). The low DA receptor levels in methamphetamine abusers have been associated with low metabolism in the OFC.

If DA D2 receptors modulate sensitivity to physiological and natural rewards, low receptor levels could predispose an individual to use drugs as a way to compensate for the decreased activation of reward circuits. Alternatively, low DA D2 receptor levels could predispose that individual to psychostimulant abuse by favoring initial pleasant drug responses, whereas high receptor levels may be protective by favoring unpleasant responses. The fact that some nonaddicted subjects have been shown to have receptor levels comparable to those of drug-addicted subjects indicates that low DA D2 receptor levels alone are not sufficient to cause drug addiction, and that it is more likely that D2 receptors have a role in modulating vulnerability to addiction.

Information from human brain-imaging studies can be used to design preclinical experiments to investigate associated variables that may be causally related. For example, recent work has provided evidence that high levels of DA D2 receptors are causally related to a reduction in alcohol intake. An adenoviral vector was used to deliver the DA D2 receptor gene into the NAc of rats previously trained to self-administer alcohol. The resulting transient increase in receptor levels (50%) was within the physiological range and produced a significant reduction in alcohol intake. Previous levels of alcohol intake returned as the DA D2 receptors fell to baseline levels (Thanos et al. 2001). Because basal levels of DA D2 receptors are modulated by both genetic and environmental factors (Morgan et al. 2002), variations in receptor levels may be one neurobiological mechanism that underlies the predisposition to drug abuse and addiction. These results also open the possibility for developing treatments for drug abuse and addiction that involve increasing DA D2 receptor levels.

SUMMARY

Imaging studies have corroborated the role of DA in motivation and in the rewarding effects of abused drugs in humans. These studies have shown that phasic increases in DA and subsequent fast and marked activation of postsynaptic DA receptors, not the tonic level of DA per se, are relevant for drug reinforcement. Smooth DA increases have been associated with motivation and the attribution of saliency to drugs. Imaging studies have revealed marked disruptions of DA brain function in addicted individuals. This hypodopaminergic state may lead to dysregulation of reward processing, motivation, and inhibitory control. Disrupted reward circuits and OFC dysfunction may further impair the sensitivity of responsiveness to natural reinforcers, which could contribute to the compulsive drug seeking and self-administration that characterize individuals addicted to drugs. CG dysfunction could underlie the poor inhibitory control also displayed by addicts. Finally, research has shown that nonpharmacological variables associated with conditioned responses can modulate the reinforcing effects of drugs in addicted subjects, but this remains to be further investigated using functional neuroimaging tools.

FUTURE DIRECTIONS IN IMAGING STUDIES OF THE ADDICTED BRAIN

As we acquire basic knowledge about addiction-related brain circuits and how they are affected by environmental variables, dual approaches that pair behavioral interventions with medicines will likely offer new and effective treatments for addictions and the neurobiological changes caused by them. It is conceivable that strategies designed to "exercise" brain circuits using specific cognitive and behavioral activities could strengthen those circuits affected by chronic drug use, analogous to remediating reading and learning disabilities. For example, behavioral strategies might activate and strengthen inhibitory control circuits to increase the likelihood of successful abstinence from drug taking.

Continuing research in drug addiction must also seek to understand how genes predispose individuals to drug abuse. This may offer new targets for developing pharmacological as well as nonpharmacological interventions. Similarly, a better understanding of the interactions among genes, environment, and neurobiology will assist in the development of improved behavioral interventions to counteract the deleterious effects of the life stressors on the brain that facilitate drug abuse and addiction.

ACKNOWLEDGMENTS

The authors thank Dr. Cheryl Kassed for manuscript preparation and the agencies that helped fund the studies discussed in this chapter (Department of Education, National Institute on Drug Abuse, Office of National Drug Control Policy, and National Institute on Alcohol Abuse and Alcoholism).

REFERENCES

Bechara A. 2001. Neurobiology of decision-making: Risk and reward. *Semin. Clin. Neuropsychiatry* **6:** 205–216.

Bechara A., Damasio A.R., Damasio H., and Anderson S.W. 1994. Insensitivity to future consequences following damage to human prefrontal cortex. *Cognition* **50:** 7–15.

Bell K., Duffy P., and Kalivas P.W. 2000. Context-specific enhancement of glutamate transmission by cocaine. *Neuropsychopharmacology* **23:** 335–344.

Breiter H.C. and Rosen B.R. 1999. Functional magnetic resonance imaging of brain reward circuitry in the human. *Ann. N.Y. Acad. Sci.* **877:** 523–547.

Breiter H.C., Gollub R.L., Weisskoff R.M., Kennedy D.N., Makris N., Berke J.D., Goodman J.M., Kantor H.L., Gastfriend D.R., Riorden J.P., Mathew R.T., Rosen B.R., and Hyman S.E. 1997. Acute effects of cocaine on human brain activity and emotion. *Neuron* **19:** 591–611.

Brodie J.D., Figueroa E., and Dewey S.L. 2003. Treating cocaine addiction: From preclinical to clinical trial experience with gamma-vinyl GABA. *Synapse* **50:** 261–265.

Childress A.R., Mozley P.D., McElgin W., Fitzgerald J., Reivich M., and O'Brien C.P. 1999. Limbic activation during cue-induced cocaine craving. *Am. J. Psychiatry* **156:** 11–18.

Dewey S.L., Morgan A.E., Ashby C.R., Jr., Horan B., Kushner S.A., Logan J., Volkow N.D., Fowler J.S., Gardner E.L., and Brodie J.D. 1998. A novel strategy for the treatment of cocaine addiction. *Synapse* **30:** 119–129.

Di Chiara G. 2002. Nucleus accumbens shell and core dopamine: Differential role in behavior and addiction. *Behav. Brain Res.* **137:** 75–114.

Di Chiara G. and Imperato A. 1988. Drugs abused by humans preferentially increase synaptic dopamine concentrations in the mesolimbic system of freely moving rats. *Proc. Natl. Acad. Sci.* **85:** 5274–5278.

Dick D.M., Edenberg H.J., Xuei X., Goate A., Kuperman S., Schuckit M., Crowe R., Smith T.L., Porjesz B., Begleiter H., and Foroud T. 2004. Association of GABRG3 with alcohol dependence. *Alcohol Clin. Exp. Res.* **28:** 4–9.

Edenberg H.J., Dick D.M., Xuei X., Tian H., Almasy L., Bauer L.O., Crowe R.R., Goate A., Hesselbrock V., Jones K., Kwon J., Li T.K., Nurnberger J.I., Jr., O'Connor S.J., Reich T., Rice J., Schuckit M.A., Porjesz B., Foroud T., and Begleiter H. 2004. Variations in GABRA2, encoding the alpha 2 subunit of the GABA(A) receptor, are associated with alcohol dependence and with brain oscillations. *Am. J. Hum. Genet.* **74:** 705–714.

Ernst T., Chang L., Leonido-Yee M., and Speck O. 2000. Evidence for long-term neurotoxicity associated with methamphetamine abuse: A 1H MRS study. *Neurology* **54:** 1344–1349.

Eslinger P.J., Grattan L.M., Damasio H., and Damasio A.R. 1992. Developmental consequences of childhood frontal lobe damage. *Arch. Neurol.* **49:** 764–769.

Fowler J.S., Ding Y.S., and Volkow N.D. 2003. Radiotracers for positron emission tomography imaging. *Semin. Nucl. Med.* **33:** 14–27.

Franklin T.R., Acton P.D., Maldjian J.A., Gray J.D., Croft J.R., Dackis C.A., O'Brien C.P., and Childress A.R. 2002. Decreased gray matter concentration in the insular, orbitofrontal, cingulate, and temporal cortices of cocaine patients. *Biol. Psychiatry* **51:** 134–142.

Gogtay N., Giedd J.N., Lusk L., Hayashi K.M., Greenstein D., Vaituzis A.C., Nugent T.F., III, Herman D.H., Clasen L.S., Toga A.W., Rapoport J.L., and Thompson P.M. 2004. Dynamic mapping of human cortical development during childhood through early adulthood. *Proc. Natl. Acad. Sci.* **101:** 8174–8179.

Goldstein R.Z. and Volkow N.D. 2002. Drug addiction and its underlying neurobiological basis: Neuroimaging evidence for the involvement of the frontal cortex. *Am. J. Psychiatry* **159:** 1642–1652.

Grace A.A. 2000. The tonic/phasic model of dopamine system regulation and its implications for understanding alcohol and psychostimulant craving. *Addiction* (suppl. 2) **95:** S119–S128.

Graham J.H. and Porrino L.J. 1995. Neuroanatomical basis of cocaine self-administration. In *Neurobiology of cocaine: Cellular and molecular mechanisms* (ed. P. Ronald and J. Hammer), pp. 3–14. CRC Press, Boca Raton, Florida.

Grant S., London E.D., Newlin D.B., Villemagne V.L., Liu X., Contoreggi C., Phillips R.L., Kimes A.S., and Margolin A. 1996. Activation of memory circuits during cue-elicited cocaine craving. *Proc. Natl. Acad. Sci.* **93:** 12040–12045.

Hariri A.R., Mattay V.S., Tessitore A., Kolachana B., Fera F., Goldman D., Egan M.F., and Weinberger D.R. 2002. Serotonin transporter genetic variation and the response of the human amygdala. *Science* **297:** 400–403.

Hollerman J.R. and Schultz W. 1998. Dopamine neurons report an error in the temporal prediction of reward during learning. *Nat. Neurosci.* **1:** 304–309.

Kaufman J.N., Ross T.J., Stein E.A., and Garavan H. 2003. Cingulate hypoactivity in cocaine users during a GO-NOGO task as revealed by event-related functional magnetic resonance imaging. *J. Neurosci.* **23:** 7839–7843.

Kilts C.D., Schweitzer J.B., Quinn C.K., Gross R.E., Faber T.L., Muhammad F., Ely T.D., Hoffman J.M., and Drexler K.P. 2001. Neural activity related to drug craving in cocaine addiction. *Arch. Gen. Psychiatry* **58:** 334–341.

Kirk J.M., Doty P., and De Wit H. 1998. Effects of expectancies on subjective responses to oral delta9-tetrahydrocannabinol. *Pharmacol. Biochem. Behav.* **59:** 287–293.

Koob G.F. 1996. Hedonic valence, dopamine and motivation.

Mol. Psychiatry **1:** 186–189.

Koob G.F. and Bloom F.E. 1988. Cellular and molecular mechanisms of drug dependence. *Science* **242:** 715–723.

Kung H.F., Kung M.P., and Choi S.R. 2003. Radiopharmaceuticals for single-photon emission computed tomography brain imaging. *Semin. Nucl. Med.* **33:** 2–13.

Laakso A., Mohn A.R., Gainetdinov R.R., and Caron M.G. 2002. Experimental genetic approaches to addiction. *Neuron* **36:** 213–228.

Liu X., Matochik J.A., Cadet J.L., and London E.D. 1998. Smaller volume of prefrontal lobe in polysubstance abusers: A magnetic resonance imaging study. *Neuropsychopharmacology* **18:** 243–252.

Mattay V.S., Goldberg T.E., Fera F., Hariri A.R., Tessitore A., Egan M.F., Kolachana B., Callicott J.H., and Weinberger D.R. 2003. Catechol O-methyltransferase val[158]-met genotype and individual variation in the brain response to amphetamine. *Proc. Natl. Acad. Sci.* **100:** 6186–6191.

Morgan D., Grant K.A., Gage H.D., Mach R.H., Kaplan J.R., Prioleau O., Nader S.H., Buchheimer N., Ehrenkaufer R.L., and Nader M.A. 2002. Social dominance in monkeys: Dopamine D2 receptors and cocaine self-administration. *Nat. Neurosci.* **5:** 169–174.

Parran T.V., Jr., and Jasinski D.R. 1991. Intravenous methylphenidate abuse. Prototype for prescription drug abuse. *Arch. Intern. Med.* **151:** 781–783.

Paulus M.P., Hozack N.E., Zauscher B.E., Frank L., Brown G.G., Braff D.L., and Schuckit M.A. 2002. Behavioral and functional neuroimaging evidence for prefrontal dysfunction in methamphetamine-dependent subjects. *Neuropsychopharmacology* **26:** 53–63.

Piazza P.V., Rouge-Pont F., Deminiere J.M., Kharoubi M., Le Moal M., and Simon H. 1991. Dopaminergic activity is reduced in the prefrontal cortex and increased in the nucleus accumbens of rats predisposed to develop amphetamine self-administration. *Brain Res.* **567:** 169–174.

Powell J. 1995. Conditioned responses to drug-related stimuli: Is context crucial? *Addiction* **90:** 1089–1095.

Rae C., Karmiloff-Smith A., Lee M.A., Dixon R.M., Grant J., Blamire A.M., Thompson C.H., Styles P., and Radda G.K. 1998. Brain biochemistry in Williams syndrome: Evidence for a role of the cerebellum in cognition? *Neurology* **51:** 33–40.

Reivich M., Kuhl D., Wolf A., Greenberg J., Phelps M., Ido T., Casella V., Fowler J., Hoffman E., Alavi A., Som P., and Sokoloff L. 1979. The [18F]fluorodeoxyglucose method for the measurement of local cerebral glucose utilization in man. *Circ. Res.* **44:** 127–137.

Robinson T.E., Gorny G., Mitton E., and Kolb B. 2001. Cocaine self-administration alters the morphology of dendrites and dendritic spines in the nucleus accumbens and neocortex. *Synapse* **39:** 257–266.

Royall D.R., Lauterbach E.C., Cummings J.L., Reeve A., Rummans T.A., Kaufer D.I., LaFrance W.C., Jr., and Coffey C.E. 2002. Executive control function: A review of its promise and challenges for clinical research. A report from the

Committee on Research of the American Neuropsychiatric Association. *J. Neuropsychiatry Clin. Neurosci.* **14:** 377–405.

Schreckenberger M., Amberg R., Scheurich A., Lochmann M., Tichy W., Klega A., Siessmeier T., Grunder G., Buchholz H.G., Landvogt C., Stauss J., Mann K., Bartenstein P., and Urban R. 2004. Acute alcohol effects on neuronal and attentional processing: Striatal reward system and inhibitory sensory interactions under acute ethanol challenge. *Neuropsychopharmacology* **29:** 1527–1537.

Springer C.S. 1999. Principles of bulk magnetic susceptibility and the sign of the fMRI response. In *Functional magnetic resonance imaging* (ed. C.T.W. Moonen and P.A. Bandettini), pp. 91–102. Springer Verlag, New York.

Szumlinski K.K., Dehoff M.H., Kang S.H., Frys K.A., Lominac K.D., Klugmann M., Rohrer J., Griffin W., 3rd, Toda S., Champtiaux N.P., Berry T., Tu J.C., Shealy S.E., During M.J., Middaugh L.D., Worley P.F., and Kalivas P.W. 2004. Homer proteins regulate sensitivity to cocaine. *Neuron* **43:** 401–413.

Thanos P.K., Volkow N.D., Freimuth P., Umegaki H., Ikari H., Roth G., Ingram D.K., and Hitzemann R. 2001. Overexpression of dopamine D2 receptors reduces alcohol self-administration. *J. Neurochem.* **78:** 1094–1103.

Volkow N.D. and Fowler J.S. 2000. Addiction, a disease of compulsion and drive: Involvement of the orbitofrontal cortex. *Cereb. Cortex* **10:** 318–325.

Volkow N.D., Fowler J.S., and Wang G.J. 2003a. Positron emission tomography and single-photon emission computed tomography in substance abuse research. *Semin. Nucl. Med.* **33:** 114–128.

——. 2003b. The addicted human brain: Insights from imaging studies. *J. Clin. Invest.* **111:** 1444–1451.

——. 2004a. The addicted human brain viewed in the light of imaging studies: Brain circuits and treatment strategies. *Neuropharmacology* (suppl. 1) **47:** 1–13.

Volkow N.D., Fowler J.S., Wang G.J., and Swanson J.M. 2004b. Dopamine in drug abuse and addiction: Results from imaging studies and treatment implications. *Mol. Psychiatry* **9:** 557–569.

Volkow N.D., Fowler J.S., Wang G.J., Hitzemann R., Logan J., Schlyer D.J., Dewey S.L., and Wolf A.P. 1993. Decreased dopamine D2 receptor availability is associated with reduced frontal metabolism in cocaine abusers. *Synapse* **14:** 169–177.

Volkow N.D., Wang G.J., Fowler J.S., Logan J., Gatley S.J., Gifford A., Hitzemann R., Ding Y.S., and Pappas N. 1999. Prediction of reinforcing responses to psychostimulants in humans by brain dopamine D2 receptor levels. *Am. J. Psychiatry* **156:** 1440–1443.

Volkow N.D., Wang G.J., Fowler J.S., Logan J., Hitzemann R., Ding Y.S., Pappas N., Shea C., and Piscani K. 1996. Decreases in dopamine receptors but not in dopamine transporters in alcoholics. *Alcohol Clin. Exp. Res.* **20:** 1594–1598.

Volkow N.D., Fowler J.S., Wolf A.P., Schlyer D., Shiue C.Y., Alpert R., Dewey S.L., Logan J., Bendriem B., and Christman D. 1990. Effects of chronic cocaine abuse on postsynaptic dopamine receptors. *Am. J. Psychiatry* **147:** 719–724.

Volkow N.D., Wang G., Fowler J.S., Logan J., Gerasimov M., Maynard L., Ding Y., Gatley S.J., Gifford A., and Franceschi D. 2001a. Therapeutic doses of oral methylphenidate significantly increase extracellular dopamine in the human brain. *J. Neurosci.* **21:** RC121.

Volkow N.D., Wang G.J., Fowler J.S., Logan J., Schlyer D., Hitzemann R., Lieberman J., Angrist B., Pappas N., and MacGregor R. 1994. Imaging endogenous dopamine competition with [^{11}C]raclopride in the human brain. *Synapse* **16:** 255–262.

Volkow N.D., Wang G.J., Fowler J.S., Thanos P.P., Logan J., Gatley S.J., Gifford A., Ding Y.S., Wong C., Pappas N., and Thanos P. 2002. Brain DA D2 receptors predict reinforcing effects of stimulants in humans: Replication study. *Synapse* **46:** 79–82.

Volkow N.D., Wang G.J., Ma Y., Fowler J.S., Zhu W., Maynard L., Telang F., Vaska P., Ding Y.S., Wong C., and Swanson J.M. 2003c. Expectation enhances the regional brain metabolic and the reinforcing effects of stimulants in cocaine abusers. *J. Neurosci.* **23:** 11461–11468.

Volkow N.D., Chang L., Wang G.J., Fowler J.S,. Ding Y.S., Sedler M., Logan J., Franceschi D., Gatley J., Hitzemann R., Gifford A., Wong C., and Pappas N. 2001b. Low level of brain dopamine D2 receptors in methamphetamine abusers: Association with metabolism in the orbitofrontal cortex. *Am. J. Psychiatry* **158:** 2015–2021.

Volkow N.D., Wang G.J., Fowler J.S., Franceschi D., Thanos P.K., Wong C., Gatley S.J., Ding Y.S., Molina P., Schlyer D., Alexoff D., Hitzemann R., and Pappas N. 2000. Cocaine abusers show a blunted response to alcohol intoxication in limbic brain regions. *Life Sci.* **66:** L161–L167.

Wang G.J., Volkow N.D., Fowler J.S., Cervany P., Hitzemann R.J., Pappas N.R., Wong C.T., and Felder C. 1999. Regional brain metabolic activation during craving elicited by recall of previous drug experiences. *Life Sci.* **64:** 775–784.

Wang G.J., Volkow N.D., Fowler J.S., Logan J., Abumrad N.N., Hitzemann R.J,. Pappas N.S., and Pascani K. 1997. Dopamine D2 receptor availability in opiate-dependent subjects before and after naloxone-precipitated withdrawal. *Neuropsychopharmacology* **16:** 174–182.

Wexler B.E., Gottschalk C.H., Fulbright R.K., Prohovnik I., Lacadie C.M., Rounsaville B.J., and Gore J.C. 2001. Functional magnetic resonance imaging of cocaine craving. *Am. J. Psychiatry* **158:** 86–95.

7 | Neurotoxic Effects of Drugs of Abuse: Imaging and Mechanisms

Dean F. Wong

The Russell H. Morgan Department of Radiology and Radiological Science, Psychiatry and Environmental Health Sciences, Johns Hopkins University Medical School and Bloomberg School of Public Health, Baltimore, Maryland 21287

ABSTRACT

The investigation of neurotoxicity has evolved with the advent of neuroreceptor imaging by positron emission tomography/single-photon emission-computed tomography (PET/SPECT). It is now possible to directly image toxic pathophysiological processes as well as to examine the potential impact of treatment. Once neuroreceptors have been radiolabeled, it is possible to study them at multiple sites in the synapse. They can then be used to study various disorders and aid in the development of new drugs.

Imaging can be used as a biomarker for treatment of substance abuse. Studies have shown that methamphetamine induces a reduction in dopamine transporter in humans. Other studies have shown that cocaine craving can induce dopamine release. GBR-12909 (a monoamine transport inhibitor) has been shown to inhibit intravenous amphetamine-induced dopamine release.

This chapter reviews the role of neuroreceptor imaging (especially the dopamine system) in helping to characterize the pathophysiology and potential treatment for substance abuse. In particular, the role of pre-, post-, and intrasynaptic imaging in drug abuse, such as methamphetamine abuse or cocaine abuse, is reviewed. Some of the historical aspects of developing such dopamine radioligands are briefly reviewed and put in the context of substance abuse research.

An increasingly important role of imaging is in medicinal development. Thus, the role of neuroreceptor imaging in testing potential treatments for cocaine use—providing receptor dopamine transporter occupancies as a guide for dosing and examining mechanisms of actions—is described. Finally, a technical appendix provides more detail about imaging methodology for investigators more interested in some of the complexities and limitations associated with this work.

INTRODUCTION

The investigation of neurotoxicity in substance abuse research has traditionally been the subject of classical histochemistry and autoradiographic studies and more recently, molecular biology techniques. However, with the advent of neuroreceptor imaging by PET/SPECT, it has become possible to directly image the toxic pathophysiological processes as well as to examine the potential impact of treatment.

GOALS OF THE CHAPTER

This chapter describes the role of noninvasive brain imaging in neurotoxicity and other areas of substance abuse and describes how imaging has directly impacted our understanding of these disorders, their mechanisms, and potential therapies. PET or SPECT ligands can be devised that bind with relatively high affinity and selectivity to key brain proteins, ranging from extracellular transporters or receptors to enzymes, vesicular transporters, and other intracellular targets.

DOPAMINE SYSTEMS OVERVIEW

Biogenic amine receptors are the direct or indirect targets of a wide range of drugs, including the psychostimulants cocaine, amphetamines, and other drugs that release dopamine. Drugs abused by humans (opiates, ethanol, nicotine, amphetamine, and cocaine) increase extracellular dopamine concentrations in rodent striatum and nucleus accumbens, whereas drugs with aversive properties (e.g., κ-opioid agonists) reduce dopamine release in these brain regions. In contrast, drugs not abused by humans (e.g., imipramine, atropine, or diphenhydramine) are unable to modify synaptic dopamine concentrations, providing biochemical evidence that activation of dopamine transmission may be a fundamental property of abused drugs (Di Chiara et al. 1988). In large measure, these conclusions were drawn from rodent data, with sparse confirmation in humans. Accordingly, a noninvasive view of human brain dopamine systems can clarify the neuropharmacological and toxicological effects of drugs of abuse. Brain imaging of dopamine systems promises to provide a noninvasive means of documenting the effects of drugs in brain, when other approaches are not feasible.

Radiolabeling of D2 Dopamine Receptors Is a Preamble for PET/SPECT Imaging of Receptors

Imaging the D_2 receptor and other neuroreceptors is now a routine procedure. The field has become increasingly important for dopamine (DA), serotonin (5-HT), and norepinephrine receptors, and for intrasynaptic measures of reuptake sites. These studies were first performed with PET in humans in the early 1980s (Wagner et al. 1983; Wong et al. 1984) as well as with SPECT (Eckelman et al. 1984).

In addition to [^{18}F]fluorodopa studies first reported in *Nature* (Garnett et al. 1983), the Brookhaven group pioneered the development of MAO_a and MAO_b (monoamine oxydase) ligands (Fowler et al. 1993, 2001). The enzymatic breakdown, precursor generation, reuptake sites, and postsynaptic sites are known for D_1, D_2, and D_3, but not for D_4.

To a lesser extent, there is similar data for serotonin and, to an even lesser extent, for the cholinergic, γ-aminobutyric acid (GABA) and glutamate systems.

The Dopamine System Can Now Be Studied at Multiple Sites in the Synapse

Because DA has such an important role in reward and reinforcement in substance abuse, a historical description and brief summary of dopaminergic PET and SPECT ligands available for imaging is given below.

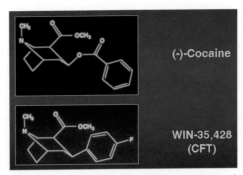

FIGURE 1. Chemical structure of anatomy: (–)-cocaine and WIN 35,428.

Initially, it was hoped that D_2 ligands would bind exclusively to postsynaptic receptors and reflect receptor availability. In fact, D_2 radioligands compete with extracellular levels of DA for D_2 receptors, particularly if the radioligand has modest D_2-receptor affinity (Seeman et al. 1989). Although this can be considered a hindrance to quantifying receptor density, the observation permitted indirect and noninvasive scrutiny of intrasynaptic DA. Notwithstanding this intriguing and serendipitous discovery, this approach is currently an area of great debate for imaging methodology, because changes in affinity state and other explanations have not been resolved. The prototype for these studies is the D_2-receptor ligand, raclopride, widely used in challenge studies, and it is more effective for viewing intrasynaptic changes in dopamine levels than for measuring receptor density (B_{max}). The very fact that the field has taken something that was initially considered to be a potential confound (e.g., DA) to routine measures is an important testimony to progress in the field.

Figure 1 illustrates a cocaine analog (Scheffel et al. 1991; Boja et al. 1992), WIN 35,428, that we have used. First introduced and recommended as a radiolabeled probe for PET or SPECT by the Madras laboratory (Madras et al. 1989; Canfield et al. 1990), this basic structure can and has been modified with substitutions including fluorine and other small carbon chain substitutions to develop improved PET ligands with suitable affinity and selectivity. The development of this series, along with others, is currently a continuing area of research. Parallel research is being conducted to improve the properties of these ligands, such as the Wagner et al. (1983) study with [11C]N-methylspiperone (NMSP), a derivative of the antipsychotic spiperone. In those early days, the idea of adding a methyl group to an amine, which simplifies radiolabeling but could cause a significant change in the pharmacology, was not attempted. Now N-methylation is routinely performed with the caveat that the pharmacology must be checked afterward to verify that the favorable properties have not declined. Currently, it is possible to image pre-, post-, and intrasynaptic DA (see Table 1).

TABLE 1. Imaging Analysis of Pre-, Post-, and Intrasynaptic DNA

Presynaptic	Postsynaptic	Intrasynaptic
Transporters/reuptake density (e.g., DAT) Enzyme activity (e.g., DDC) Precursors to neurotransmitters Vesicular transporters	Receptor density (e.g., D2) Receptor subtypes (e.g., D_1 or D_2) Receptor affinity	Endogenous neurotransmitter release 2° drugs (e.g., DA 2° amphetamine) Endogenous neurotransmitter basal concentration Enzyme activity (e.g., MAO)

METHODOLOGICAL CONSIDERATIONS

Despite the tremendous advances in radioligand imaging of neuroreceptors since the early 1980s, considerable challenges continue for radioligand development. First, a potential ligand must retain its pharmacologic specificity after labeling with PET or SPECT isotopes (e.g., ^{11}C or ^{123}I). In addition, the radioligands must pass the blood–brain barrier—they must have fairly low polarity to cross into the brain, but cannot be so lipophilic that the ligands exhibit too much nonspecific binding. If the ligand is too extensively metabolized in plasma or in the brain, or if the target background (i.e., signal-to-noise ratio) is too low, it is unusable. A potential ligand must also be dismissed if the kinetics in the brain do not allow for sufficient binding during the limited imaging time. Finally, the compound must have sufficient affinity and specificity for the receptors or transporters of interest. These are but a few of the methodological concerns that must be addressed when developing a successful radioligand.

Ligand Development

With respect to substance abuse, the most important development is the ability to image transporters or reuptake sites such as for DA, 5-HT, and, more recently, norepinephrine. Much of the work done by neuroreceptor imaging laboratories is very dependent on collaborations with investigators who develop new ligands, such as those working in preclinical pharmaceutical companies (Wong et al. 2002). One approach has been for universities to negotiate master agreements with large pharmaceutical companies for development of ligands that will be used in their early stages of drug development. With this kind of partnership, and with the understanding that eventually these compounds will be made available to the general public (just like the human genome), measurements of many other transmitter systems can be expanded. An example of the typical issues and algorithm for ligand development is provided by Wong and Pomper (2003).

Limitations of PET and SPECT

Unfortunately, there are a limited number of ligands with sufficient affinity and kinetic characteristics suitable for in vivo binding in brain. We are limited by the necessity to radiolabel them with isotopes that can be externally imaged with current instrumentation. For PET, this means being able to substitute an [^{11}C] (20-min half-life), [^{18}F] (110-min half-life), or a [^{99m}Tc] (6-hr half-life), [^{123}I] (13-hr half-life) with SPECT. Unfortunately, of the hundreds of thousands of compounds that may be available at a pharmaceutical company, a university, or in a government library of compounds and structures, only a fraction is suitable for potential imaging. Selection of potential imaging compounds is often determined by techniques such as in vivo tritium (^3H) distribution or, more recently, with sophisticated mass spectrometry methods. Ultimately, for external imaging and clinical studies, a PET or SPECT radioisotope is used. This reduces the number of potential compounds considerably but leads to ongoing essential research and development as new targets and discoveries made in postmortem tissues or cell culture systems require verification in living brain. This is an area of continuing interest on the part of the National

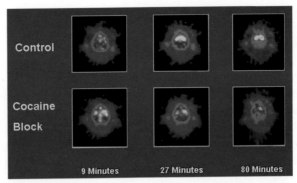

FIGURE 2. [^{11}C]WIN 35,428 PET study in baboons. (Reprinted, with permission, from Wong et al. 1993.)

Institutes of Health (NIH), the Food and Drug Administration (FDA), and academia as evidenced by recent workshops (Brady 2005).

After 10 years of working on D$_2$ receptors, the first study of dopamine transporters (beginning in 1991) was conducted with [^{11}C]WIN 35,428 (Wong et al. 1993). Bertha Madras had originally proposed and discovered [^{11}C]WIN as a potential PET or SPECT imaging agent for the DA transporter and as a marker for degeneration of DA neurons (Kaufman and Madras 1991). She has subsequently extended this to technetium imaging agents for SPECT imaging (Madras et al. 1998).

Our pharmacology collaborators, led by Mike Kuhar, proposed using [^{11}C]WIN 35,428 for PET studies. Figure 2 is an example of [^{11}C]WIN 35,428 in the control state, showing the binding in the baboon brain and the blockade that occurs after unlabeled (–) cocaine is given 5 minutes beforehand.

DA Transporter Radioligands Image: Differences in Human Diseases

The WIN 35,428 (WIN) compound was used for examining neurotoxicity, in collaboration with George Ricaurte (Department of Neurology, Johns Hopkins University). There was selective binding to the caudate putamen over 90 minutes, with most of the activity in the cortex being nonspecific (Villemagne et al. 1998a).

There are certainly 5-HT DA transporters (DATs) in the cortex, but WIN is fairly selective, although its affinity is not as high as some of the SPECT compounds that are available, but it is quite selective. It does not bind very well to 5-HT transporters (Wong et al. 1993).

DAT Abnormalities in Neuropsychiatric Disorders Show the Usefulness of PET-radiolabeled Cocaine Derivatives

WIN has been used to show that patients with Lesch-Nyhan syndrome showed no overlap of DAT density compared with controls (Wong et al. 1996). Lesch-Nyhan is a disease of hyperuricemia and self-injurious behavior.

The dramatic changes were predicted almost 15 years earlier, but they were based only on three postmortem studies that had been performed by investigators at the University of

Michigan (Lloyd et al. 1981). There was some evidence that DA would be decreased in Lesch-Nyhan syndrome, but it was not confirmed until we actually performed the study. Similar studies with fluorodopa have also shown this dramatic reduction (Ernst et al. 1999).

The next point illustrates the role animal models have in imaging. George Breese developed the neonatal 6-hydroxy DA lesion in neonatal rats, demonstrating that self-injurious behavior could be mimicked in neonate rodents, just like the Lesch-Nyhan syndrome, but not in adult rodents (Breese et al. 1995). This is a great example of how the study of animal models really led to a hypothesis that was confirmed with minimal post-mortem work. Our studies of DAT in Lesch-Nyhan syndrome and those with [18F]fluorodopa (Wong et al. 1993; Ernst et al. 1996) are examples of how imaging can be used to confirm a hypothesis that may be difficult to prove otherwise.

Lesch-Nyhan syndrome is a very rare disorder, so these studies were unique. Although these studies may not make a significant impact on our society, because Lesch-Nyhan is a rare disease, the scientific approaches that were used illustrate how translational investigations integrate with basic science.

Quantification of PET/SPECT Neuroreceptors Requires Understanding the Kinetics

PET/SPECT studies are performed on a routine basis that are similar to association and dissociation studies. In studies of PET receptors, an entire dynamic series is done. Because of their emphasis on rigorous data collection and methodology, Johns Hopkins University is very active in this field. For example, the majority of humans in Johns Hopkins PET studies get an arterial line for blood sampling. There is a special interest in quantitative modeling, which has simplified the technique. A complete dynamic study is performed. Typically, between 30 and 50 brain images are obtained per PET scan, of which each brain slice can contain between 15 and 64 or more (depending on the scanner) spatial slices in the brain. These are used to create a time-activity curve. An arterial plasma curve is measured simultaneously from the subject's radial artery. About six to eight discrete plasma measurements are taken to obtain high-performance liquid chromatography (HPLC) correction of the parent versus the metabolized radiotracer, because data from only the parent compound are desired. If there are metabolites, it must be known whether the metabolites are getting into brain and whether they are interfering with the image. Once we obtain more experience with the compound, this can be simplified, but most studies are done dynamically and analyzed by various models.

Carbon-11-based PET scans usually are conducted for 60–90 minutes. Steady-state imaging is preferred, but this is determined by the pharmacology of the ligand. Ideally, it would go to steady state or to pseudo-equilibrium. If a bolus injection is used, a true steady state can never be achieved, whereas if a continuous infusion is used, it is possible to approach equilibrium. Obviously, equilibrium conditions are ideal (since in vitro binding is typically done at equilibrium) because all of the binding transients and the equations are simplified. Unfortunately, some of the ligands with the highest affinity and that are the most selective, do not reach equilibrium during the 90–120 minutes of the scan. This is a limitation of carbon-11 or fluorine-18, because of their short half-lives. [123]I studies may continue for 6 hours or more (see Table 2).

The obvious logistical limitations to human imaging studies are primarily related to the availability of effective ligands. Some of the most successful ligands are those such as raclo-

TABLE 2. Typical Radionuclides Used in PET/SPECT and Physical Properties

Isotope	PET/SPECT	Decay mode (%)	Half-life (min)	Energy (MeV) (max/most abundant)	Production
^{11}C	PET	β^+ 100	20.1	0.97	C
^{18}F	PET	β^+ 97 EC3	109.8	0.64	C and R
^{15}O	PET	β^+ 100	2.04	1.74	C
^{13}N	PET	β^+ 100	9.996	1.20	C
^{62}Cu	PET	β^+ 97 EC3	9.7	2.92	G[^{62}Zn]
^{82}Rb	PET	β^+ 95 EC5	1.3	3.4	G[^{82}Sn]
^{68}Ga	PET	β^+ 89 EC11	68	1.899	G[^{68}Ge]
^{99m}Tc	SPECT	IT100	360	0.140	G[^{99}Mo]
^{123}I	SPECT	EC100	780	0.159	C
^{131}I	SPECT	β^- 100	8 (days)	0.192	C/R
Annihilation-γ	PET	–	–	0.511	(see above)

(EC) Electric capture; (IT) isomeric transition; (C) cyclotron; (R) reactor; (G) generator.

pride, for which imaging information can be achieved in the first hour; however, this is driven by the pharmacology, the on-and-off rates to the receptor, and other complex factors. Unfortunately, with imaging, we are limited by the half-life of the isotopes. For situations in which a radioligand does not achieve equilibrium within a reasonable window of time, other techniques have been developed that still allow receptor quantification (Wong et al. 1986a).

Ligands that bind almost irreversibly, and are often delivery dependent, are of limited use. An example is iodinated quinuclidinyl benzoate (QNB). Other tracers, such as [^{11}C]WIN 35,428, are slowly reversible, making modeling more challenging. Tracers that equilibrate too rapidly may include [^{11}C]nicotine. This is not only a result of poor binding to the receptor, but also a result of the tracer kinetics. Ideal ligand binding should equilibrate somewhere in between these extremes. Raclopride is intermediate; it peaks in about 25 minutes and imaging is completed by 1 hour.

The relevance of imaging for both the pathophysiology and treatment issues is that multiple timepoints can be monitored. With imaging, multiple treatments or drug perturbations can be performed and then eventually examined as a function of time.

PET and SPECT DAT Radioligands Have Various Trade-offs

For assessing neurotoxicity, the basic mechanisms of receptor localization, selectivity, and specificity are of concern. In 1983, the concern was whether spiperone was going to be selective enough, because it binds to 5-HT$_{2a}$, α-adrenergic receptors, as well as D$_2$ receptors. However, the radioligand, [^{11}C]NMSP, was the only viable D$_2$ radioligand available at the time. Compounds such as [^{11}C]raclopride became available several years later.

Many of the advantages of [^{11}C]methylphenidate, versus a very selective tropane such as [^{11}C]WIN 35,428 for imaging the DAT, were initially ignored because it binds to multiple transporter sites and in vitro has a higher level of nonspecific binding. But, as a number of groups have found, it is more convenient for quantification because of its reversible binding (Volkow et al. 1995). Because there is some anatomical separation between 5-HT and norepinephrine transporters and DATs in the striatum, methylphenidate may be a reasonable choice for measurement of the DAT. The University of Michigan, Brookhaven National Laboratories, and Johns Hopkins University use it routinely for imaging DAT. Because no ligand is perfect, and the criteria vary for each target and application, rational criteria for probes are not universally accepted, and the optimal approach to new ligand development is not always clear (Wong and Pomper 2003).

PET versus SPECT

PET and SPECT each have various advantages (Table 2). SPECT can detect radioactive emissions from ^{99m}Tc or ^{123}I. SPECT has the advantage of being available wherever there is a technetium generator, which is basically every nuclear medicine department. SPECT also has the option of using iodine (^{123}I), which, unfortunately, is now only available for North America from a cyclotron in Vancouver. Its relatively long half-life (13 hours) allows 24-hour studies. PET has the advantage of short-lived isotopes (^{11}C, 20 minutes; ^{18}F, 110 minutes).

Variability and Individual Differences

Another challenge for preclinical PET imaging research is the use of anesthesia for primate research. A few unique centers, such as Hammamatsu (Tsukada et al. 2004), have trained monkeys (not baboons) that remain still for a number of hours in an unusual upright PET scanner to address this challenge. The majority of studies use a within-subjects design for drug treatment studies, which can reduce individual variability in PET measures. Some variability in baboons might be a result of anesthesia issues, but other technical factors, such as the size of the brain and the partial volume effects, may have a role. Variability in humans is also observed, but the potential contributing factors, including genotype (Jönsson et al. 2003), age, latent disease states, personality, neuroadaptation to stress, drugs, or other factors, have not been routinely investigated in preclinical or clinical subjects. The Karolinska group investigated a number of psychological scores for factors and personality, such as introversion and social reclusiveness (Farde et al. 1997), and the Columbia University group has looked at the effects of social anxiety (Schneier et al. 2000). Abnormalities in D_2 receptors were associated with personality scores, which conceivably account for a portion of the variability. A problem with PET studies is sample size. With PET imaging, it is difficult to obtain the huge numbers that would be needed for a large genotyping study. Typical numbers in a human PET study are anywhere between 10 and 20. A sample size of 30 individuals is a rare luxury. A combined analysis was conducted on prior PET work by Wong et al. (1997a,b,c) on D_2 DA in schizophrenia from 1986 (Wong 1986b) through 1997; this study included 35 schizophrenic subjects, and similar work (Singer et al. 2000) with Tourette's syndrome included 30 subjects.

It is difficult to determine all of the factors affecting variability. As more imaging groups with different capabilities become involved, there will be a better appreciation of both genotyping and individual differences (which are sometimes manifest by phenotypes of different personalities). For example, the Wake Forest group studied social isolation in monkeys and its effect on D_2 DA as measured by PET (Morgan et al. 2002). These were acute experiments, and DA was measured in the brain after sacrificing the monkeys. The

DA levels for the controls went from 6.21 down to 1.58 for the socially isolated animals, a 75% reduction in DA. With PET imaging (using a DAT ligand), DA levels went down 39% (from 1.09 ± 0.06 to 0.66 ± 0.23), and when measured at postmortem, there was a 57% reduction (from 259.3 ± 1.7 to 111.6 ± 1.5).

SPECIFIC RELEVANCE TO SUBSTANCE ABUSE OR APPLICATIONS FOR SUBSTANCE ABUSE PROBLEMS

Given the role of dopamine in reinforcement and reward mechanisms, measurements of in vivo chemistry using imaging have been an important part of substance abuse research, beginning with the first imaging studies of dopamine receptors in the early 1980s. Brain imaging is especially important in substance abuse because it allows the direct measurement of the dopaminergic system within various abuse conditions, such as following drug washout and relapse. In addition, it allows for direct measurement of dopamine states in coordination with mood states such as euphoria, dysphoria during drug taking, or craving responses to cues. In the section below, we describe the role of brain imaging in the development of pharmacological treatments for substance abuse.

Imaging as a Biomarker for Treating Substance Abuse

There is a clearly a role for imaging in the progressive treatment of substance abuse. Until such in vivo imaging existed, traditional methods included in vitro receptor binding, in vitro and in vivo autoradiography, and, subsequently, studying animal models. With the advent of external imaging, such as PET and SPECT, it was possible to examine the acute and chronic dosing and mechanisms of action in correlation with behavior in nonhuman primates and in human studies, and more recently, in rodent studies. These latter studies have the advantage of being able to investigate knockout animals and specifically genetically modified rodents, and to provide in vivo imaging.

Methamphetamine Abuse

Methamphetamine (Meth) (also known as speed, crystal, crank, go, and ice) is the most widely distributed, illegally manufactured, and abused type of amphetamine. An estimated 4 million individuals in the United States have abused Meth at least once. The Monitoring of the Future Study (which assesses the extent of drug use among adolescents) found that in 1996, 4.4% of high school seniors had used crystal Meth at least once (2.7% in 1990). The Substance Abuse and Mental Health Services Administration's Drug Abuse Warning Network reports that from 1991 to 1994, the number of Meth-related visits to hospital emergency departments more than tripled. According to the Drug Enforcement Administration (DEA), Meth has been the most prevalent clandestinely produced controlled substance in the United States since 1979.

Meth and MDMA

Meth abuse is an increasingly serious problem, with its origins in the western United States and rapid movement eastward. There are a number of theories about why Meth might be toxic in one or two doses. Radical formation, 3-4 methylenedioxymethamphet-amine (MDMA), or toxic metabolites may be involved as the underlying mechanisms (Kraemer et al. 2002; Jiang et al. 2004).

Theories for Meth Toxicity

Methamphetamine is selectively toxic to DA and 5-HT nerve terminals in the central nervous system. PET imaging has provided evidence for the toxicity of methamphetamine to dopamine terminals, as shown by the reduced dopamine transporter density (McCann et al. 1998b; Villemagne et al. 1998b) measured in humans and baboons. Autoradiography studies in rodents have shown decreases in serotonin transporter, dopamine transporters, and vesicular monoamine transporter–type 2 (VMAT-2) after methamphetamine administration (Guilarte et al. 2003; Armstrong and Noguchi 2004).

Several theories exist with regard to the mechanism behind methamphetamine's toxic effects:

1. The excitatory feed-forward loop theory holds that repeated depolarization leads to Na^+ and Ca^{++} influx, leading to cell death (Seiden and Sabol 1996);

2. Meth toxicity is mediated by the DA system (Itzhak and Achat-Mendes 2004);

3. The N-methyl-D-asparate (NMDA) receptor has a role in toxicity (Sonsalla et al. 1998).

Neurotoxicity: Relevance to the Cell Biology of Drug Addiction

Meth is selective for both DA and 5-HT neuroterminals in the central nervous system (CNS), with long-lasting depletion of neurotransmitters (Guilarte et al. 2003), although even this finding is controversial. Meth neurotoxicity is detected in humans using PET imaging of the DAT (Villemagne et al. 1998b) and in rats using autoradiography (Armstrong and Noguchi 2004). The rodent results showed that Meth is most toxic in fore-brain regions to both 5-HT and DA terminals. Human PET imaging studies showed a reduction in presynaptic concentration in rate-limiting enzymes, and a reduction of DAT reuptake sites is observed (Volkow et al. 2000) qualitatively, especially after repeated Meth administration (Villemagne et al. 1998b).

When all of the doses from 0.5 to 2 mg/kg of the drug treatment are combined, the reduction is highly significant. In the baboon brain, the binding reduction measured by WIN 35,428 for ligand-binding potential (Fig. 3) was linear and dose dependent. Every individual baboon (used as its own control) that was studied had different slopes, but there was a monotonic decrease in their DAT binding, depending on dose.

Binding potential is an estimate of the available receptors divided by their affinity. This is now an outcome measure that is routinely used by investigators who do single measurements under a single condition. Unlike fMRI (functional magnetic resonance imaging), which allows for multiple determinations, with this study design, PET produces one outcome measured per scan. If duplicate or triplicate PETs are desired, the study can be repeated, either as test/retest in the same animal or repeated in a different animal, all under the same pharmacological conditions.

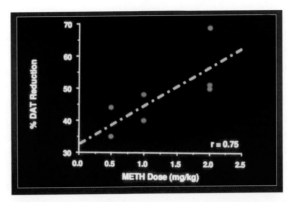

FIGURE 3. Reduction in DAT in the baboon brain after treatment with methamphetamine. (Reprinted, with permission, from Villemagne et al. 1998 [©Society for Neuroscience].)

Imaging Human Meth Users

A Meth study showed that the distribution volumes for WIN 35,428 (CFT), a marker for DAT, and dihydrotetrabenazine (DTBZ), a marker for the intracellular vesicular monoamine transporter 2 (VMAT-2), correlate quite well, pixel by pixel (see Table 3) (McCann et al. 1998b). It is believed that similar factors are probably being measured, but the entire series of studies must be completed to determine whether DTBZ binding is also reduced analogous to WIN binding.

DAT Regulatory Processes

When the first tropane-based probes for DAT were studied using PET by Wong et al. in 1990 (baboon) and in 1993 (human) (Wong et al. 1993), it was generally accepted that DAT was static, and not up- or down-regulated. Before the cloning of DAT and for a period subsequent to the cloning (Kilty et al. 1991; Shimada et al. 1991), there was no evidence for DAT internalization or other forms of regulation. Currently, it is widely accepted that DAT can be regulated by pharmacological administration and possibly by endogenous mechanisms (Pristupa et al. 1998; Daniels and Amara 1999; Melikian and Buckley 1999; Saunders et al. 2000; Zahniser and Zorkin 2004). Interestingly, DAT is primarily extrasynaptic (Nirenberg et al. 1997). The evidence for DAT regulation must be con-

TABLE 3. Reduction in DAT-binding Potential in the Human Striatum Compared with Methamphetamine Users (six) vs. Parkinson Subjects (three)

	n	Caudate nucleus (k_3/k_4)	Putamen (k_3/k_4)
Control	10	7.5 ± 1.7	7.3 ± 1.4
MA users	6	5.8 ± 1.0	5.5 ± 0.9
PD patients	3	4.0 ± 0.9	2.3 ± 0.4

Adapted, with permission, from McCann et al. 1998b (©Society for Neuroscience).

sidered when interpreting PET imaging data, with alternate indicators of DA neurons under consideration. For example, the VMAT-2 transporter, which is localized on synaptic vesicles (but is not specific for DA neurons), is also being assessed as a marker for DA neurons, using the radioligand [^{11}C]DTBZ (Fantegrossi et al. 2004). Imaging the VMAT-2 may not solve the "transporter regulation challenge" because VMAT-2 is also regulated (Holtje et al. 2003; Zucker et al. 2005).

An ongoing study to compare results from a DAT probe and a VMAT probe in controls and Meth users will help to clarify the issues stated above. An advantage of carbon-11 is that a participant can be scanned in the morning with one of these tracers, wait 1.5 hours, and then the other PET study can be conducted.

Imaging Human MDMA Users

Ecstasy (MDMA) is a very popular drug of abuse and has potent 5-HT neurotoxicity. Because MDMA has structural similarities to amphetamine, it is not surprising that the two compounds may share similar toxic mechanisms. The evidence for MDMA neurotoxic effects on serotonin neurons is based on long-term decreases in serotonin and its major metabolite (5-HIAA) and the serotonin transporter in rodents or primates (Clemens et al. 2004). Accordingly, imaging the serotonin transporter can offer a view of the status of serotonin neurons, with the same caveats described for the DAT (transporter trafficking and other forms of regulation).

The use of brain imaging to discern the neurotoxic effects of amphetamines (MDMA, amphetamine, and Meth) is a powerful tool for determining the extent of a lesion and its duration. When developing animal models for MDMA neurotoxicity that can be addressed by brain imaging, it is critical to establish appropriate experimental parameters (dose, route of administration, frequency of administration, and species differences) and to select a PET probe with properties in vivo suitable for detecting the 5-HT transporter. In our human studies, participants had taken MDMA no fewer than 25 times, with 400 times being the highest exposure. Figure 4 shows images illustrating PET imaging of the 5-HT transporter with McN5652.

Our results indicate a measurable reduction in the 5-HT transporter in MDMA abusers throughout multiple brain regions such as the midbrain, using distribution volume as the outcome measurement. In summary, the effects of MDMA on brain 5-HT neurons can be examined from the rodent to baboon to human, using appropriate PET imaging agents.

Imaging Human Cocaine Abusers

A unique collaborative study investigated the effects of cocaine on DAT in human cocaine abusers. Cocaine abusers were given 96 mg of intranasal cocaine for 5 days. After a 3-day washout to reduce direct brain cocaine levels, PET scanning of DAT was performed, and subjects were rescanned after a 30-day stay in the residential ward where they were unable to obtain cocaine. Obviously, one of the advantages of imaging is that both analog scales and behavioral scales can be used.

Using the test/retest method, a significant elevation of the binding potential was found only in the caudate on day 3. However, preliminary calculations showed that DAT den-

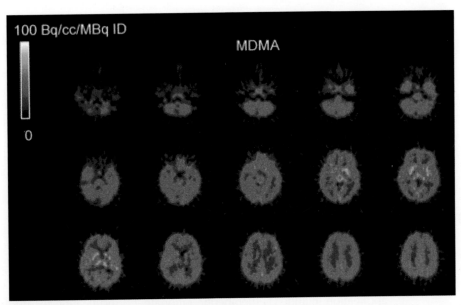

FIGURE 4. Effect of MDMA on [¹¹C](+)McN5652. (Adapted, with permission, from McCann et al. 1998a [©Elsevier].) Additional applications with [¹¹C]WIN 35,428.

sities returned to baseline levels on day 33, after abstinence (Wong et al. 1999a). This example illustrates the power of PET imaging to determine adaptive responses to cocaine and the duration of these changes in human subjects.

Cue-induced Cocaine Craving Increases DA Release

A study was conducted to examine the effect of cue-induced craving on human cerebral DA release (DAR). Preliminary studies showed that DAR during cue-induced craving is altered in cocaine users compared with control subjects. Furthermore, DAT-binding potential and cue-induced cravings were correlated with measures of craving and other behavioral parameters (Wong et al. 2003).

Cocaine Craving and Amphetamine-induced DAR

We tested the hypothesis that craving cues are associated with PET measures of DAT and amphetamine (AMP)-induced DAR. Subjects were presented with cues (visual and auditory) while viewing a neutral tape for 30 minutes, followed by a cocaine-craving tape for 30 minutes. Every 10 minutes, the cocaine-craving questionnaire (CCQ) (Tiffany et al. 1993) was administered. A neuropsychological battery then followed. The next day, subjects underwent two PET scans. The first PET consisted of an intravenous bolus of [¹¹C]methylphenidate (MP) over 90 minutes. The second PET followed with an [¹¹C]raclopride bolus and infusion over 90 minutes, and at 45 minutes, 0.3 mg/kg intravenous amphetamine was administered. DAT was modeled using a reference region technique for [¹¹C]methylphenidate. Ratio and reference tissue methods were used to estimate DAR.

Twelve cocaine users (average age 42.8) with the Diagnostic and Statistical Manual (DSM) of Mental Disorders diagnosis of cocaine use and eight controls (average age 37.1) were studied. Using intravenous MP, as previously reported (Volkow et al. 1997), we found a significant reduction in DAR in users compared with controls following intravenous AMP. Most interestingly, users ($n = 4$) who craved during the cocaine cues had significantly higher DAR than noncraving users or controls ($p < 0.05$) for most caudate putamen regions. Users ($n = 4$) who craved during the AMP-PET had significantly higher right anterior putamen DAR than controls ($p < 0.05$). There were no significant changes with DAT. This suggests that subjects who crave cocaine have a different response to amphetamine-induced DAR compared with controls (Wong et al. 2004).

Applications to Drug Development

PET and SPECT imaging has an important role in drug development (Wong et al. 2002). Many of these applications include imaging the biodistribution of candidate drug radioligands labeled with [^{11}C] and [^{18}F] for PET, and [^{123}I] and [^{99}Tc] for SPECT. The areas of application include (1) therapeutic rationale, (2) justification for study of specific biological systems, (3) mechanism of drug action, and (4) rational drug dosing (Wong et al. 2002).

Neuroreceptor Imaging Aids Drug Development for Substance Abuse

It is well known in schizophrenia research that D_2-receptor occupancy by antagonists correlates with their plasma concentrations. When a certain plasma concentration is reached, the dose may be adequate because receptor occupancy has plateaued; rather, it asymptotes. Past peak occupancy, therapeutic benefit does not increase, whereas side effects may increase. The same principle may be applicable to cocaine antagonists (Yokoi et al. 2002).

Human DAT Occupancy by GBR-12909

GBR-12909, a monoamine transporter inhibitor, is a potential cocaine treatment that has been developed with the use of imaging. GBR-12909 is currently being considered by the NIDA drug development group as a candidate drug for cocaine treatment, with the underlying rationale being an extension of work by Baumann et al. (1994).

GBR-12909 has favorable properties: slow dissociation, slow onset of action, and production of a modest elevation in intrasynaptic DA. It antagonizes increases induced by cocaine and suppresses cocaine administration in nonhuman primates (Glowa 1996). Figure 5 shows the stepwise change (the dose-dependent change) in GBR-12909 given at 1, 3, and 10 mg, and the DAT occupancy.

An important consideration when using GBR-12909 in humans is to obtain sufficiently high levels balanced against potential side effects. GBR-12909 was previously used in Europe, but it had been associated with some cardiotoxicity. It was important to establish, using a surrogate marker such as DAT occupancy, whether GBR-12909 could be given at

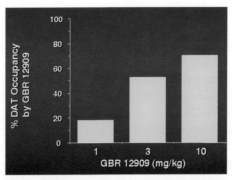

FIGURE 5. DAT percent occupancy by GBR-12909.

sufficiently high levels to produce a therapeutic benefit without exceeding cardiotoxic levels. An example is given in Figure 6 (Wong et al. 1999b).

Twelve volunteers in a safety dosing study (one individual received all doses) were used in PET studies to relate DAT occupancy to dose, a standard method for evaluation of drugs in development. In our view, PET or SPECT imaging should be a necessary component of a Phase I trial to assess receptor-mediated drugs. It was hypothesized that GBR-12909 might reach a plateau of <50%. Healthy humans received PET scans with [^{11}C]WIN 35,428 to monitor DAT occupancy. Multiple escalating doses of the DAT inhibitor GBR-12909 to inhibit DAT occupancy by [^{11}C]WIN 35,428 were then administered. Multiple mathematical methods for calculating DAT occupancy with and without plasma input function were used; the occupancy was low (<50%). Wong et al., in collaboration with Albert Gjedde of Aarhus University, modeled this with a new procedure for obtaining occupancy. The measure of occupancy can be confounded by a number of factors, including CFT binding to more than one affinity site at the DAT (Madras et al. 1989). Nevertheless, these studies offered guidelines for dosing of GBR-12909 within the parameters of safety. PET had a pivotal role in establishing guidelines, foremost because it indicated that GBR-12909 occupied DAT with a significant occupancy, permitting continuation of the study.

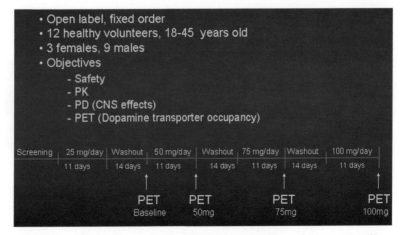

FIGURE 6. GBR-12909 multiple dose study in healthy volunteers: Timeline.

GBR-12909 Inhibits Intravenous Amphetamine-induced DAR

A common problem in human research is the balance between safety and scientific rationale. It is well known that a classic amphetamine challenge will cause an increase in extracellular DA in the rat that might be helpful for studying craving. It was important to determine whether GBR-12909 acts in a cocaine-like manner, i.e., promote accumulation of extracellular DA levels or dampen DA levels, or whether it attenuates accumulation of extracellular DA levels. To this end, we used an indirect method of monitoring DA levels, displacement of the D_2-receptor antagonists [^{11}C]raclopride with DA. To augment the DA signal, amphetamine was administered. A continuous-infusion technique of the PET ligand, [^{11}C]raclopride, was used with amphetamine challenge, and saline was administered for comparative purposes. The difference between amphetamine and saline, presumably, is the extracellular DAR. But, if GBR-12909 was given beforehand at an appropriate dose, the amphetamine challenge was less capable of demonstrating DA accumulation. Figure 7 shows that the binding was high with saline. With the amphetamine challenge, there is a significant difference between these two, indicative of robust DA release.

When GBR-12909 was administered before amphetamine, it also promoted an increase in the baseline DA. This result can be viewed as a positive outcome, because some increase in extracellular DA levels may reduce cocaine craving. When the amphetamine challenge is given, the concomitant proportional reduction in binding does not occur; thus, presumably, if this were given to a Meth or cocaine abuser, the same "high" would not be achieved.

Imaging Nicotine Occupancy of the α4β2 Nicotinic Cholinergic Receptor

The Wong laboratory collaborated with the London laboratory for many years to develop epibatidine derivatives to image the α4β2 nicotinic cholinergic receptor. The apparently high toxicity of this class of compounds shifted the focus to other selective probes developed by the industry for analgesia, but with apparently reduced toxicity. The SPECT ligand, [^{123}I]5-I-A85380 (Fig. 8) displayed very high binding for the receptor in baboons, as assessed by nicotine or cytosine blockade of radioligand binding (Musachio et al. 1997).

In a collaboration with Johns Hopkins University and the nuclear medicine group in Australia (Kassiou et al. 2001), a baboon, at baseline, was given 2 weeks of nicotine at 2

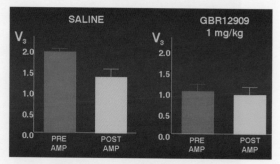

FIGURE 7. Saline vs. GBR-12909. (Reprinted, with permission, from Villemagne et al. 1999.)

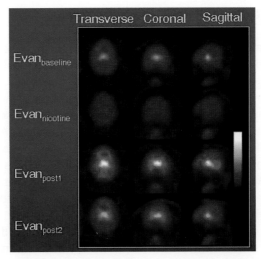

FIGURE 8. [^{123}I]5-I-A85380 can detect nAchR up-regulation in the baboon. (Reprinted, with permission, from Kassiou et al. 2001 [©Elsevier].)

mg/kg for 24 hours to achieve plasma levels of 27 mg/ml, approximately the values for an average human smoker using 20 cigarettes a day. Immediately after the nicotine infusion ended, nicotine competed with the radioligand for the receptor competition. One week later, the receptor sites were up-regulated compared with the baseline. Two weeks later, baseline levels began to return to normal. In a saline-treated baboon, there was no difference, indicating receptor up-regulation and confirmation of in vitro data (Kassiou et al. 2001; Darsow et al. 2005). Thus, this study illustrated the feasibility of demonstrating agonist-mediated up-regulation of the α4β2 nicotinic receptor. In smoking addiction, it will be possible to measure the effects of nicotine and other elements in cigarette smoke on various receptors and transporters, as more probes become available.

SUMMARY AND CONCLUSIONS

This chapter provided examples of imaging components of dopamine and serotonin neurotransmission of relevance to drug abuse. Historically, neuroreceptor imaging has emphasized the pathophysiology of major neuropsychiatric disorders such as schizophrenia, Parkinson's disease, and depression, and has also been used for drug development. The era of imaging neurochemistry in drug abuse is emerging as an effective means for clarifying the mechanisms of and neuroadaptive responses elicited by drugs of abuse. Currently, a wealth of radiotracers, for example, those for DA, 5-HT, cholinergic, and opiate systems, can be used to understand drug abuse.

Furthermore, the principles that have been developed in areas such as schizophrenia, i.e., receptor occupancy to help provide rational drug dosing and to examine mechanisms of action of drugs, are now applicable for substance abuse research. This will improve objectivity in research and clarify drug mechanisms and treatment pathways in the future.

GAPS AND FUTURE DIRECTIONS

A current and major challenge to imaging research in the field of substance abuse is a full understanding of the potential confounds of comorbidity. For example, it is difficult to find volunteers who have only one primary addiction. Hence, multiple comorbid addictions, such as nicotine, alcohol, cocaine, and heroin, may have a selective and specific neurochemical signature, but even when applying these sophisticated methods to such populations, the comorbidity issue remains a problem. Additional issues include comorbid mental illnesses such as depression, schizophrenia, and attention deficit disorder, which also may have unique neurochemical signatures that confound interpretation of the data. Animal models can circumvent these limitations, but this requires animal models that specifically and faithfully mimic the human disorders. This still remains an area of great debate.

Studies of knockout mice or nonhuman primate models of substance abuse are very promising in that they allow a selective set of experiments to be studied with these ever-increasingly sophisticated neurochemical markers, both before and after initiation of the model lesion. In addition, animal models can allow disease progression and effects of potential treatments to be studied.

The other challenge is to better understand the roles of receptor and enzyme interactions in improving the treatments for substance abuse. For example, in schizophrenia, it is well established, at least for some subtypes or for positive symptoms, that dopaminergic antagonists are important components of treatment. More recently DA, 5-HT, and perhaps DA partial agonists occupying a specific proportion of receptors (e.g., at least 80% in many cases and greater for partial agonists [Grunder et al. 2002]) are target occupancies for Phase I studies. This represents the initial dose used for Phase II clinical trials. However, as we found in our GBR-12909 studies, maximum DAT occupancy achieved was less than 50%. Is this the therapeutic threshold? PET imaging can help to establish the therapeutic threshold or minimum occupancy that must be achieved before costly drug development proceeds. For drug development in substance abuse to be effective, we must better understand the minimum occupancy threshold effect and the role of multiple receptors before the promise of neuroreceptor imaging can be brought to bear in dramatically changing therapeutic outcomes in substance abuse.

ACKNOWLEDGMENTS

This work was supported by Grant Nos. DA-00412, AA-12839, NS-38927, and HD-24448. We thank Kimberly Bell-Warren, M.A., Ayor Nandi, and Judy Buchanan, B.S., Department of Radiology, Johns Hopkins University for editing assistance. We are also grateful for the use of slides and reprinted published material from the laboratories of George Ricaurte and Una McCann, Johns Hopkins University Bayview Campus.

Appendix to Chapter 7

TECHNICAL ISSUES

Limitations in Resolution

Part of the reason that an increase in reduction with methylphenidate seen in vitro was not seen in vivo (Table 3) is, to some extent, because of the finite resolution of PET, a phenomenon known as "partial volume." In imaging instrumentation, there is a finite limit to the resolution of small structures (typically >2–3 mm). There is loss of resolution when objects are smaller or less than 2 or 3 times the full-width at half-maximum of the point source for any imaging agent. For example, the human cerebellum usually loses about 5% or 10% of its true activity when these studies are performed. Much smaller structures—certainly the accumbens and the caudate putamen—lose even more as a result of the partial volume effect.

This is also true with SPECT, and even more so, because it is a resolution issue. It has to do with the manner in which a three-dimensional image is obtained. With SPECT, perfect attenuation correction does not occur. With PET, the methods for attenuation correction are much more sophisticated and accurate. New techniques now exist to correct for such resolution losses (Rousset et al. 1993 and in prep.).

The resolution of our original PET scanner was about 8 mm. Our current scanner (GE Advance) is closer to 5 mm, and now PET scanners are approaching 2–3 mm for the high-resolution research tomograph (HRRT), which has the full-width-at-half-maximum ability to separate two points. Resolution is defined in many ways. A simpler way to think about it is that biological structures that are 0.5 cm to several centimeters in size show some loss of apparent radioactivity because of finite resolution. It is more difficult to resolve points that are smaller than 0.5 cm, especially if they have lower contrast (i.e., target activity similar to background).

The diagram of a compartmental model in Figure 9 shows the forward binding rate k_1 from the plasma to the brain; k_2 is the rate from the brain back to the plasma. The distri-

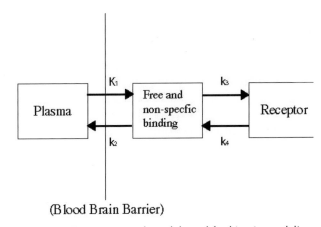

(Blood Brain Barrier)

FIGURE 9. Compartmental model used for kinetic modeling.

bution volume is simply the ratio of the rates of forward flux and backward flux. Various procedures are used to determine these numbers.

The standard three-compartment model is typically used for describing the kinetics. When first proposed by a number of groups in about 1984, it was quite novel because kinetic modeling into pharmacology was rather radical (the Michaelis-Menten equations assume that everything is at equilibrium, and all the derivatives go to zero). This is not always the case with PET. With PET, we cannot make assumptions that anything goes to zero.

With k_3/k_4, the ligand bound to receptors and the free ligand, the goal is to calculate B_{max}, where k_3 is k_{on} times the available number of receptors (B_{max}-bound receptors). This bound term is everything that would include endogenous neurotransmitter and exogenous administration of drug, such as haloperidol in the case of D_2. This k_{on} is the molecular association rate that would be obtained in an in vivo analogy to what would be done with in vitro studies. Then, k_4 is k_{off}. There are different ways of performing these calculations.

Limitations in Density Measures

For many procedures, especially in the DA system, B_{max} data collection has always been attempted, but it is not often practical, so most groups do not do it routinely. In the early 1980s, and still continuing, in vivo Scatchard plots of data obtained with humans have been generated. Our group at Johns Hopkins University and the Karolinska and Finland groups are among the few in the world that routinely obtain human B_{max} measures in vivo. This procedure consists of giving either an antagonist (e.g., haloperidol vs. [^{11}C]NMSP) carrier-added unlabeled raclopride or [^{11}C]raclopride in one of the PET scans and performing a Scatchard plot. The Karolinska group used up to five saturation points in each of four human subjects, measuring B_{max} and receptor affinity (K_D) (Farde et al. 1986). Measuring receptor subtypes is a function of having the appropriate ligand.

B_{max} and K_D can be calculated in a saturation study. But, if a single study is done without performing a saturation (as in most PET studies), a binding potential (B_{max}/K_D) is estimated from a three-compartment model (two tissue/one plasma model). Cystine can be given to baboons for PET studies for nAchR imaging, but it is not given to human beings.

Often, instead of two tissue compartments, there is one. When there is just a blood compartment and a brain compartment, the number of parameters is reduced. When there is just one compartment, it is assumed that the number of parameters rapidly equilibrate. This is called distribution volume, which is related to B_{max}/K_D, except that instead of doing a model fit where the free ligand is separated from the ligand bound to the receptors, the entire process is considered to be either plasma or brain.

Dynamic Studies for Estimating Parameters

How many timepoints are needed for a typical dynamic PET study? As increasingly smaller time intervals are used for sampling, there is less and less radioactivity. The variance is Poisson-noise determined, and so the variance and the mean counts—if increasingly shorter time intervals are counted, there is more and more variance. It is actually multiplied by a factor of many times when the investigator goes from Poisson statistics to the uncertainty in tomography, because that adds uncertainty in and of itself (in contrast to Poisson statistics where mean = variance).

If multiple, small, timepoints are taken, there is even more uncertainty in the curve. There is a trade-off between obtaining enough points to fit the curve and not having so many that huge confidence intervals do not occur on the curves. The grid size can be made as small as desired. Every parameter could be fitted to reveal a beautiful curve. Although a spline regression could be done, it does not mean anything. It is an issue of parameter "identifiability." There is a procedure called information theory, and there are ways of calculating this formally. The information content can be shown to decrease as a function of the number of parameters and as the variability increases. So, there is a trade-off mathematically.

There is a trade-off between having too many and not enough parameters. Many believe that the maximum that can probably be obtained from PET and SPECT scans, on a routine basis, is maybe four parameters, possibly five if a blood-volume term is estimated. Many investigators are just doing three parameters: blood volume, $k1'$, and $k2'$. That is the limitation for biological parameter estimation.

In any event, it turns out that many of these curves are now generated for all PET ligands and SPECT ligands. Some believe that fitting two compartments is the best that can be done. Thus, distribution volumes are being described in the literature by multiple compartmental methods. This is one of the limitations, but for all intents and purposes, it is a pretty good marker, although it is not precisely B_{max} and K_D, and it is not precisely k_{on} and k_{off}. It is more of a surrogate for these measurements.

Choosing a Radioligand for Some of the Described Studies

Regarding choices of radioligands, the following are some of the ligands used for assessing neurotoxicity: WIN, methylphenidate, and VMAT-2, the vesicular transporter. A number of new 5-HT transporter ligands are now available. Zsolt Szabo (Szabo et al. 1995) from the Johns Hopkins group and others developed the [^{11}C]McN5652 compound, which has both positive and negative isomers and binds to the 5-HT transporter.

REFERENCES

Armstrong B.D. and Noguchi K.K. 2004. The neurotoxic effects of 3,4-methylenedioxymethamphetamine (MDMA) and methamphetamine on serotonin, dopamine, and GABA-ergic terminals: An in-vitro autoradiographic study in rats. *Neurotoxicology* **25:** 905–914.

Baumann M.H., Char G.U., de Costa B.R., Rice K.C., and Rothman R.B. 1994. GBR12909 attenuates cocaine-induced activation of mesolimbic dopamine neurons in the rat. *J. Pharmacol. Exp. Therapeut.* **271:** 1216–1222.

Boja J.W., Cline E.J., Carroll F.I., Lewin A.H., Philip A., Dannals R., Wong D.F., Scheffel U., and Kuhar M.J. 1992. High potency cocaine analogs—Neurochemical, imaging, and behavioral-studies. *Ann. N.Y. Acad. Sci.* **654:** 282–291.

Brady L. 2005. PET and SPECT Consortium: Safety assessment of novel, high specificity, low mass radiotracers for use in research and drug discovery. National Institute of Mental Health, Bethesda, Maryland.

Breese G.R., Criswell H.E., Duncan G.E., Moy S.S., Johnson K.B., Wong D.F., and Mueller R.A. 1995. Model for reduced brain dopamine in Lesch-Nyhan syndrome and the mentally retarded: Neurobiology of neonatal-6-hydroxydopamine-lesioned rats. *Ment. Retard. Dev. Disabil. Res. Rev.* **1:** 111–119.

Canfield D.R., Spealman R.D., Kaufman M.J., and Madras B.K. 1990. Autoradiographic localization of cocaine binding sites by [3H]CFT ([3H]WIN 35,428) in the monkey brain. *Synapse* **6:** 189–195.

Clemens K.J., Van Niewenhuyzen P.S., Li K.M., Cornish J.L., Hunt G.E., and McGregor I.S. 2004. MDMA ("extasy"), methamphetamine and their combination long-term changes in social interaction and neurochemistry in the rat. *Psychopharmacology* **173:** 318–325.

Daniels G.M. and Amara S.G. 1999. Regulated trafficking of the human dopamine transporter. Clathrin-mediated internalization and lysosomal degradation in response to phorbol esters. *J. Biol. Chem.* **274:** 35794–35801.

Darsow T., Booker T.K., Pina-Crespo J.C., and Heinemann S.F. 2005. Exocytic trafficking is required for nicotine-induced up-regulation of alpha4beta2 nicotinic acetylcholine receptors. *J. Biol. Chem.* **280:** 18311–18320.

Di Chiara G. and Imperato A. 1988. Drugs abused by humans preferentially increase synaptic dopamine concentrations in the mesolimbic system of freely moving rats. *Proc. Natl. Acad. Sci.* **85:** 5274–5278.

Eckelman W.C., Reba R.C., Rzesotarski W.J., Gibson R.E., Hill T., Holman B.L. Budinger T., Conklin J.J., Eng R., and Grissom M.P. 1984. External imaging of cerebral muscarinic acetylcholine-receptors. *Science* **223:** 291–293.

Ernst M., Zametkin A.J., Jons P.H., Matochik J.A., Pascualvaca D., and Cohen R.M. 1999. High presynaptic dopaminergic activity in children with Tourette's disorder. *J. Am. Acad. Child Adolesc. Psychiatry* **38:** 86–94.

Ernst M., Zametkin A.J., Matochik J.A., Pascualvaca D., Jons P.H., Hardy K., Hankerson J.G., Doudet D.J., and Cohen R.M. 1996. Presynaptic dopaminergic deficits in Lesch-Nyhan disease. *N. Engl. J. Med.* **334:** 1562–1572.

Fantegrossi W.E., Woods J.H., Winger G., Kilbourn M., Sherman P., Woolverton W.L., Yuan J., Hatzidimitriou G., and Ricaurte G.A. 2004. Reply: MDMA and the loss of reinforcement in Fantegrossi et al. (2004). *Neuropsychopharmacology* **29:** 1942.

Farde L., Gustavsson J.P., and Jonsson E. 1997. D2 dopamine receptors and personality traits. *Nature* **385:** 590.

Farde L., Hall H., Ehrin E., and Sedvall G. 1986. Quantitative analysis of D_2 dopamine receptor binding in the living human brain by PET. *Science* **231:** 258–261.

Fowler J.S., Volkow N.D., Logan J., Wang G.J., Wolf A., Schlyer D., Macgregor R., Pappas N., Shea C., Alexoff D., Gatley S.J., Dorflinger E., Morawsky L., Yoo K., and Fazinni E. 1993. PET studies of reversible and irreversible mao-B inhibitors. *J. Nucl. Med.* **34:** 131–132.

Fowler J.S., Volkow N.D., Logan J., Franceschi D., Wang G.J., Macgregor R., Shea C., Garza V., Pappas N., Carter P., Netusil N., Bridge P., Liederman D., Elkashef A., Rotrosen J., and Hitzemann R. 2001. Evidence that L-deprenyl treatment for one week does not inhibit MAO A or the dopamine transporter in the human brain. *Life Sci.* **68:** 2759–2768.

Garnett E.S., Firnau G., and Nahmias C. 1983. Dopamine visualized in the basal ganglia of living man. *Nature* **305:** 137–138.

Glowa J.R. 1996. Dose-response analysis in risk assessment: Evaluation of behavioral specificity. *Environ. Health Perspect.* (suppl. 2) **104:** 391–396.

Grunder G., Schreckenberger M., Lochmann M., Lange-Asschenfeldt C., Siessmeier T., Hiemke C., Mann K., and Bartenstein P. 2002. Lorazepam modulates the thalamus as the generator of cortical EEG alpha rhythm. *Neuroimage* **16:** S52.

Guilarte T.R., Nihei M.K., McGlothan J.L., and Howard A.S. 2003. Methamphetamine-induced deficits of brain monoaminergic neuronal markers: Distal axotomy or neuronal plasticity. *Neuroscience* **122:** 499–513.

Holtje M., Winter S., Walther D., Pahner I., Hortnagl H., Ottersen O.P., Bader M., and Ahnert-Hilger G. 2003. The vesicular monoamine content regulates VMAT2 activity through Galphaq in mouse platelets. Evidence for autoregulation of vesicular transmitter uptake. *J. Biol. Chem.* **278:** 15850–15858.

Itzhak Y. and Achat-Mendes C. 2004. Methamphetamine and MDMA (Ecstasy) neurotoxicity: Neurotoxicity: Of Mice and Men. *IUBMB Life* **56:** 249–255.

Jiang X.-R., Wrona M.Z., Alguindigue S.S., and Dryhurst G. 2004. Reactions of the putative neurotoxin tryptamine-4,5-dione with L-cysteine and other thiols. *Chem. Res. Toxicol.* **17:** 357–369

Jönsson E.G., Abou Jamra R., Schumacher J., Flyckt L., Edman G., Forslund K., Mattila-Evenden M., Rylander G., Asberg M., Bjerkensted L., Wiesel F.A., Propping P., Cichon S., Nothen M.M., and Sedvall G.C. 2003. No association between a putative functional promoter variant in the dopamine beta-hydroxylase gene and schizophrenia. *Psychiatric Genet.* **13:** 175–178.

Kassiou M., Eberl S., Meikle S.R., Birrell A., Constable C., Fulham M.J., Wong D.F., and Musachio J.L. 2001. In vivo imaging of nicotinic receptor upregulation following chronic (–)-nicotine treatment in baboon using SPECT. *Nucl. Med. Biol.* **28:** 165–175.

Kilty J.E., Lorang D., and Amara S.G. 1991. Cloning and expression of a cocaine-sensitive rat dopamine transporter. *Science* **254:** 578–579.

Kraemer T. and Maurer H.H. 2002. Toxicokinetics of amphetamines: Metabolism and toxicokinetic data of designer drugs, amphetamine, methamphetamine, and their N-alkyl derivatives. *Therapeut. Drug Monitor.* **24:** 277–289.

Lloyd K.G., Hornykiewicz O., Davidson L., Shannak K., Farley I., Goldstein M., Shibuya M., Kelley W.N., and Fox I.H. 1981. Biochemical evidence of dysfunction of brain neurotransmitters in the Lesch-Nyhan syndrome. *N. Engl. J. Med.* **305:** 1106–1111.

Madras B.K., Meltzer P.C., Liang A.Y., Elmaleh D.R., Babich J., and Fischman A.J. 1998. Altropane, a SPECT or PET imaging probe for dopamine neurons: I. Dopamine transporter binding in primate brain. *Synapse* **29:** 93–104.

Madras B.K., Spealman R.D., Fahey M.A., Neumeyer J.L., Saha J.K., and Milius RA. 1989. Cocaine receptors labeled by [3H]2 beta-carbomethoxy-3 beta-(4-fluorophenyl)tropane. *Mol. Pharmacol.* **36:** 518–524.

McCann U.D., Szabo Z., Scheffel U., Dannals R.F., and Ricaurte G.A. 1998a. Positron emission tomographic evidence of toxic effect of MDMA ("Ecstasy") on brain serotonin neurons in human beings. *Lancet* **352:** 1433–1437.

McCann U.D., Wong D.F., Yokoi F., Villemagne V., Dannals R.F., and Ricaurte G.A. 1998b. Reduced striatal dopamine transporter density in abstinent methamphetamine and methcathinone users: Evidence from positron emission tomography studies with [^{11}C]WIN-35,428. *J. Neurosci.* **18:** 8417–8422.

Melikian H.E. and Buckley K.M. 1999. Membrane trafficking regulates the activity of the human dopamine transporter. *J. Neurosci.* **19:** 7699–7710.

Morgan D., Grant K.A., Gage H.D., Mach R.H., Kaplan J.R., Prioleau O., Nader S.H., Buchheimer N., Ehrenkaufer R.L., and Nader M.A. 2002. Social dominance in monkeys: Dopamine D2 receptors and cocaine self-administration. *Nat. Neurosci.* **5:** 169–174.

Musachio J.L., Villemagne V.L., Scheffel U., Stathis M., Finley P., Horti A., London E.D., and Dannals R.F. 1997. [125/123I]IPH: A radioiodinated analog of epibatidine for in vivo studies of nicotinic acetylcholine receptors. *Synapse* **26:** 392–399.

Nirenberg M.J., Chan J., Pohorille A., Vaughan R.A., Uhl G.R., Kuhar M.J., and Pickel V.M. 1997. The dopamine transporter: Comparative ultrastructure of dopaminergic axons in limbic and motor compartments of the nucleus accumbens. *J. Neurosci.* **17:** 6899–6907.

Pristupa Z.B., McConkey F., Liu F., Man H.Y., Lee F.J., Wang Y.T., and Niznik H.B. 1998. Protein kinase-mediated bidirectional trafficking and functional regulation of the human dopamine transporter. *Synapse* **30:** 79–87.

Rousset O., Ma Y., Kamber M., and Evans A.C. 1993. 3D simulations of radiotracer uptake in deep nuclei of human brain. *Comp. Med. Imaging Graph.* **17:** 373–379.

Sabol K.E., Lew R., Richards J.B., Vosmer G.L., and Seiden L.S. 1996. Methylenedioxymethamphetamine-induced serotonin deficits are followed by partial recovery over a 52-week period. Part I: Synaptosomal uptake and tissue concentrations. *J. Pharmacol. Exp. Therapeut.* **276:** 846–854.

Saunders C., Ferrer J.V., Shi L., Chen J., Merrill G., Lamb M.E., Leeb-Lundberg L.M., Carvelli L., Javitch J.A., and Galli A. 2000. Amphetamine-induced loss of human dopamine transporter activity: An internalization-dependent and cocaine-sensitive mechanism. *Proc. Natl. Acad. Sci.* **97:** 6850–6855.

Scheffel U., Pogun S., Stathis M., Boja J.W., and Kuhar M.J. 1991. In vivo labeling of cocaine binding sites on dopamine transporters with [3H]WIN 35,428. *J. Pharmacol. Exp. Therapeut.* **257:** 954–958.

Schneier F.R., Liebowitz M.R., Abi-Dargham A., Zea-Ponce Y., Lin S., and Laruelle M. 2000. Low dopamine D2 receptor binding potential in social phobia. *Am. J. Psychiatry* **157:** 457–459.

Seeman P., Guan H.C., and Niznik H.B. 1989. Endogenous dopamine lowers the domaine D2 receptor density as measured by [3H]raclopride: Implications for positron emission tomography of the human brain. *Synapse* **3:** 96–97.

Seiden L.S. and Sabol K.E. 1996. Methamphetamine and methylenedioxymethamphetamine neurotoxicity: Possible mechanisms of cell destruction. *NIDA Res. Monogr.* **163:** 251–276.

Shimada S., Kitayama S., Lin C.L., Patel A., Nanthakumar E., Gregor P., Kuhar M., and Uhl G. 1991. Cloning and expression of a cocaine-sensitive dopamine transporter complementary DNA. *Science* **254:** 576–578. (Erratum in *Science* [1992] **255:** 1195.)

Singer H.S., Szymanski S., Giuliano J., Yokoi F., Dogan A.S, Brasic J.R., Zhou Y., Grace A.A., and Wong D.F. 2002.

Elevated intrasynaptic dopamine release in Tourette's syndrome measured by PET. *Am. J. Psychiatry* **159:** 1329–1336.

Sonsalla P.K., Albers D.S., and Zeevalk G.D. 1998. Role of glutamate in neurodegeneration of dopamine neurons in several animal models of Parkinsonism. *Amino Acids* **14:** 69–74.

Szabo Z., Kao P.F., Scheffel U., Suehiro M., Mathews W.B., Ravert H.T., Musachio J.L., Marenco S., Kim S.E., Ricaurte G.A., Wong D.F., Wagner H.N., and Dannals R.F. 1995. Positron emission tomography imaging of serotonin transporters in the human brain using [C-11] (+)McN5652. *Synapse* **20:** 37–43.

Tiffany S., Singletin E., Haertzen C., and Heningfield J.E. 1993. The development of a cocaine craving questionnaire. *Drug Alcohol Depend.* **34:** 19–28.

Tsukada H., Nishiyama S., Fukumoto D., Ohba H., Sato K., and Kakiuchi T. 2004. Effects of acute acetylcholinesterase inhibition on the cerebral cholinergic neuronal system and cognitive function: Functional imaging of the conscious monkey brain using animal PET in combination with microdialysis. *Synapse* **52:** 1–10.

Villemagne V.L., Rothman R.B., Yokoi F., Clough D., Rice K.C., Matecka D., Stephane M., and Wong D.F. 1998a. Inhibition of amphetamine induced dopamine release by GBR 12909. *J. Nucl. Med.* **39:** 55.

Villemagne V., Yuan J., Wong D.F., Dannals R.F., Hatzidimitriou G., Mathews W.B., Ravert H.T., Musachio J., McCann U.D., and Ricaurte G.A. 1998b. Brain dopamine neurotoxicity in baboons treated with doses of methamphetamine comparable to those recreationally abused by humans: Evidence from [11C]WIN-35,428 positron emission tomography studies and direct in vitro determinations. *J. Neurosci.* **18:** 419–427.

Volkow N.D., Ding Y.S., Fowler J.S., Wang G.J., Logan J., Gatley S.J., Schlyer D.J., and Pappas N. 1995. A new PET ligand for the dopamine transporter: Studies in the human brain. *J. Nucl. Med.* **36:** 2162–2168.

Volkow N.D., Wang G.J., Fowler J.S., Logan J., Gatley S.J., Hitzemann R., Chen A.D., Dewey S.L., and Pappas N. 1997. Decreased striatal dopaminergic responsiveness in detoxified cocaine-dependent subjects. *Nature* **386:** 830–833.

Volkow N.D., Wang G.J., Fowler J.S., Franceschi D., Thanos P.K., Wong C., Gatley S.J., Ding Y.S., Molina P., Schyler D., Alexoff D., Hitzemann R., and Pappas N. 2000. Cocaine abusers show a blunted response to alcohol intoxication in limbic brain regions. *Life Sci.* **66:** PL161–PL167.

Wagner H.N., Jr., Burns H.D., Dannals R.F., Wong D.F., Langstrom B., Duelfer T., Frost J.J., Ravert H.T., Links J.M., Rosenbloom S.B., Lukas S.E., Kramer A.V., and Kuhar M.J. 1983. Imaging dopamine receptors in the human brain by positron tomography. *Science* **221:** 1264–1266.

Wilson A.A., Ginovart N., Hussey D., Meyer J., and Houle S. 2002. In vitro and in vivo characterization of [11C]-DASB: A probe for in vivo measurements of serotonin transporter by positron emission tomography. *Nucl. Med. Biol.* **29:** 509–515.

Wong D.F. and Pomper M.G. 2003. Predicting the success of a radiopharmaceutical for in vivo imaging of central nervous system neuroreceptor systems. *Mol. Imaging Biol.* **5:** 350–362.

Wong D.F., Gjedde A., and Wagner H.N., Jr. 1986a. Quantification of neuroreceptors in living human brain. Part I. Irreversible binding of ligands. *J. Cereb. Blood Flow Metab.* **6:** 137–146.

Wong D.F., Potter W.Z., and Brasic J.R. 2002. Proof of concept: Functional models for drug development in humans. In *Neuropsychopharmacology: The fifth generation of progress* (ed. K.L. Davis et al.), pp. 457–473. American College of Neuropharmacology, Nashville, Tennessee.

Wong D.F., Young D., Wilson P.D., Meltzer C.C., and Gjedde A. 1997a. Quantification of neuroreceptors in the living human brain: III. D2-like dopamine receptors: Theory, validation and results of normal aging. *J. Cereb. Blood Flow Metab.* **17:** 316–330.

Wong D.F., Rothman R.B., Contoreggi C., Yokoi F., Stephane M., Dogan S., Schretlen D., Kuhar M., and Gjedde A. 1999a. Dopamine transporter changes in cocaine users following one month abstinence. *J. Nucl. Med.* **40:** 109P.

Wong D.F., Cantilena L., Elkashef A., Yokoi F., Dogan A.S., Stephane M., Gjedde A., Rothman R.B., Mojsiak J., and Vocci F. 1999b. In vivo human dopamine transporter occupancy of a potential cocaine treatment agent, GBR12,909. *Soc. Neurosci.* **25:** 522–527.

Wong D.F., Pearlson G.D., Tune L.E., Young L.T., Meltzer C.C., Dannals R.F., Ravert H.T., Reith J., Kuhar M.J., and Gjedde A. 1997b. Quantification in living human brain: IV. Effects of aging and density elevations of D2-like receptors in schizophrenia and bipolar illness. *J. Cereb. Blood Flow Metab.* **17:** 331–342.

Wong D.F., Kuwabara H., Ye W., Kumar A., Zhou Y., Alexander M., Brasic J., Thomas M.L., Maris M., Schretlen D., London E.D., and Jasinski D.R. 2004. *Cocaine craving correlates with psychostimulant-induced dopamine release and dopamine transporters.* College on Problems of Drug Dependence Meeting, San Juan, Puerto Rico.

Wong D.F., Singer H.S., Brandt J., Shaya E., Chen C., Brown J., Kimball A.W., Gjedde A., Dannals R.F., Ravert H.T., Wilson P.D., and Wagner H.N., Jr. 1997c. D2-like dopamine receptor density in Tourette syndrome measured by PET. *J. Nucl. Med.* **38:** 1243–1247.

Wong D.F., Harris J.C., Naidu S., Yokoi F., Marenco S., Dannals R.F., Ravert H.T., Yaster M., Evans A., Rousset O., Bryan R.N., Gjedde A., and Kuhar M. 1996. Dopamine transporters are markedly reduced in Lesch-Nyhan disease in vivo. *Proc. Natl. Acad. Sci.* **83:** 5539–5543.

Wong D.F., Yung B., Dannals R.F., Shaya E.K., Ravert H.T., Chen C.A., Chan B., Folio T., Scheffel U., Ricuarte G.A., Neumeyer J., Wagner H.N., Jr., and Kuhar M.J. 1993. In vivo imaging of baboon and human dopamine transporters by positron emission tomography using [^{11}C]WIN 35,428. *Synapse* **15:** 130–142.

Wong D.F., Lee J.S., Zhou Y., Brasic J.R., Maini A., Kuwabara H., Kimes A.S., Contoreggi C., Ernst M., Schretlen D., Jasinski D., Zukin S.R., Bonson K., and London E.D. 2003. *Intrasynaptic dopamine release and cocaine craving induced by Video/audio cues.* College on Problems of Drug Dependence Meeting, Miami, Florida.

Wong D.F., Wagner H.N., Jr., Dannals R.F., Links J.M., Frost J.J., Ravert H.T., Wilson A.A., Rosenbaum A.E., Gjedde A., Douglass K.H., Petronis J.D., Folstein M.F., Toung J.K.T., Burns H.D., and Kuhar M.J. 1984. Effects of age on dopamine and serotonin receptors measured by positron tomography in the living human brain. *Science* **226:** 1393–1396.

Wong D.F., Wagner H.N., Jr., Tune L.E., Dannals R.F., Pearlson G.D., Links J.M., Tamminga C.A., Broussolle E.P., Ravert H.T., Wilson A.A., Toung J.K.T., Mala J.N., Williams J.A., O'uama L.A., Snyder S.H., Kuhar M.J., and Gjedde A. 1986b. Positron emission tomography reveals elevated D$_2$ dopamine receptors in drug naive schizophrenics. *Science* **234:** 1558–1563.

Yokoi F., Grunder G., Biziere K., Stephane M., Dogan A.S., Dannals R.F., Ravert H., Suri A., Bramer S., and Wong D.F. 2002. Dopamine D2 and D3 receptor occupancy in normal humans treated with the antipsychotic drug aripiprazole (OPC 14597): A study using positron emission tomography and [11C]raclopride. *Neuropsychopharmacology* **27:** 248–259.

Zahniser N.R. and Sorkin A. 2004. Rapid regulation of the dopamine transporter: Role in stimulant addiction? *Neuropharmacology* (suppl. 1) **47:** 80–91.

Zucker M., Weizman A., and Rehavi M. 2005. Repeated swim stress leads to down-regulation of vesicular monoamine transporter 2 in rat brain nucleus accumbens and striatum. *Eur. Neuropsychopharmacol.* **15:** 199–201.

PART 2
Development and Drug Abuse

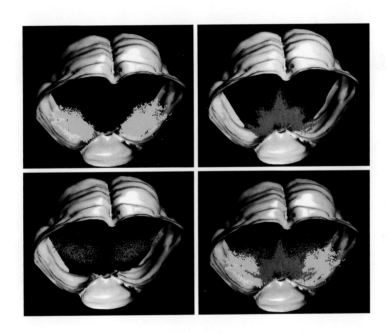

8 | Development of the Midbrain Dopaminergic Pathways

Margaret Cooper and Renping Zhou

Department of Chemical Biology, Rutgers College of Pharmacy and Department of Neuroscience and Cell Biology, Robert Wood Johnson Medical School, Piscataway, New Jersey 08854

ABSTRACT

The neurotransmitter dopamine regulates several vital functions including motor control and reward responses. These functions are mediated by a number of well-organized pathways emanating from the dopaminergic neurons located in the midbrain. The midbrain dopaminergic neurons are organized into three major groups: the substantia nigra, the ventral tegmental area, and the retrorubral field. These neurons from different groups make axon connections preferentially with different targets in the forebrain. Neurons in the substantia nigra project mainly to the dorsolateral striatum (the caudate/putamen), constituting the mesostriatal pathway, which functions to regulate motor behavior. In contrast, axons from neurons in the ventral tegmental area extend to the ventromedial striatum (nucleus accumbens and olfactory tubercle) and to the cerebral cortex, forming the mesolimbic and mesocortical pathways. These pathways are important for the reinforcement of natural reward and motivational functions. The ontogeny of the midbrain dopaminergic pathways consists of multiple developmental processes including neurogenesis, pathway differentiation, and innervation of the target tissues. Dopaminergic cell fate is determined through complex interactions of multiple extracellular and intracellular factors. Axons from the differentiated neurons are guided to the general target fields by multiple axon guidance molecules and organized into distinct pathways. Pathway differentiation occurs through the elimination of mistargeted axons and collaterals. The striatal targets are further organized into patch and matrix compartments, which have different neurochemical and connectional characteristics. Pathological changes in the midbrain dopaminergic pathways lead to many neurological disorders including drug addiction, schizophrenia, attention deficit hyperactivity disorder, depression, and Parkinson's disease. Understanding the molecular mechanisms underlying the construction of these pathways may provide insights into the treatment of neurological disorders involving the dopaminergic system.

INTRODUCTION

Dopaminergic System and Neurological Disorder

Dopamine is a neurotransmitter responsible for both the reinforcement of natural reward (Spanagel and Weiss 1999; Kelley and Berridge 2002; Salamone and Correa 2002) and the execution of learned motor behavior (Olanow and Tatton 1999; Groenewegen 2003).

137

Most dopaminergic neurons are located in the ventral mesencephalon, also referred to as the midbrain, just caudal to the hypothalamus and rostral to the cerebellum and pons. The complex array of vital roles played by these neurons is made possible through multiple pathways to the striatum and the cerebral cortex (Moore and Bloom 1978; Lindvall and Bjorklund 1983; Di Chiara et al. 1992). Disturbances in these pathways have been shown to contribute to schizophrenia, attention deficit hyperactivity disorder (ADHD), drug addiction, depression, and Parkinson's disease (Wise and Bozarth 1987; Hechtman 1994; Nieoullon and Coquerel 2003; Vernier et al. 2004).

Growing evidence suggests that the risk of suffering dopamine-related disorders may result from genetic and environmental insults affecting the integrity of the dopaminergic system (McGlashan and Hoffman 2000; Stanwood et al. 2001; Lewis and Levitt 2002; Carvey et al. 2003; Di Monte 2003). Many drugs, especially drugs of abuse, imitate the normal response to positive experience (Koob 1992; Spanagel and Weiss 1999). Cocaine, for example, blocks the reuptake of dopamine, increasing the availability to the postsynaptic neuron (Wise and Bozarth 1987). Schizophrenia has also been hypothesized to involve exaggerated dopamine stimulation (Snyder 1972; Angrist and Gershon 1974; Seeman and Van Tol 1994). Dopamine receptor agonists can induce psychotic reactions symptomatic of schizophrenia in normal individuals (Angrist et al. 1975). Consequently, the efficacy of many antipsychotic drugs is correlated with their capacity to block dopaminergic receptors (Seeman 1981; Kane 1993; Meltzer 1999). On the other hand, suppression of dopaminergic activity has been correlated with the reduction of sensory perception and learning capacity as well as ADHD (Hechtman 1994; Kostrzewa et al. 1994; Mangeot et al. 2001). Consistent with these observations, methylphenidate, a drug that stimulates dopaminergic activity, is used to treat children with ADHD (Malone et al. 1994).

First described in 1817 by the British physician, Dr. James Parkinson, Parkinson's disease (PD) is characterized by the loss of neurons in the substantia nigra. The dopaminergic population in the substantia nigra is particularly vulnerable to aging for reasons not yet understood. The primary symptoms of Parkinson's disease, including muscular rigidity, resting tremor, impairment of postural reflexes, and difficulty initiating (akinesia) and properly executing movement (bradykinesia), threaten the quality of life for millions of people (Olanow and Tatton 1999; Jenner 2003). L-Dopa, a precursor to dopamine, has been used to treat Parkinson's disease. However, L-Dopa treatment is often not sustainable, eventually leading to poor quality of life as patients suffer from involuntary repetitive movement (dyskinesia) when the drug loses its efficacy (Hadj Tahar et al. 2003; Jenner 2003). A promising alternative treatment aims at replacing lost dopaminergic neurons (Hermann et al. 2004). The success of cell replacement therapy lies in the restoration of the developmental environment to support new growth and establish proper dopaminergic connections. Implanted neurons often fail to form the appropriate neural circuits, resulting in undesirable side effects (Bjorklund et al. 2002; Ma et al. 2002).

GOALS OF THE CHAPTER

The midbrain dopaminergic neurons broadly project to the forebrain areas, forming several topographically organized pathways. These distinct pathways correlate with the differentiation of motivational and motor functions, suggesting that the topographic organization is critical. In this chapter, the anatomy of the midbrain dopaminergic pathways is described first, followed by a summary of the developmental mechanisms that underlie dopaminergic neurogenesis, differentiation, pathway formation, and target innervation. Understanding the developmental mechanisms that lead to the differentiation of the dopaminergic pathways may provide important insights into future interventions in the treatment of drug addiction, Parkinson's disease, and possibly other psychological disorders as well.

ANATOMIC FEATURES OF THE MIDBRAIN DOPAMINERGIC PATHWAYS

Cell Groups

Dahlstrom and Fuxe (1964) first reported three major groups of dopaminergic neurons in the midbrain. The nomenclature for the midbrain dopaminergic clusters laid down in the study persists to this day: A8 (retrorubral field [RRF]), A9 (substantia nigra [SN]), and A10 (ventral tegmental area [VTA]). The VTA flanks both sides of the midline, with the SN positioned further outward, extending almost to the lateral edge of the midbrain (Fig. 1). The dopaminergic neurons in these two regions together form an area with a characteristic "W" shape in coronal slices of the brain (German and Manaye 1993; Nelson et al. 1996). In the longitudinal dimension, the midbrain dopaminergic neurons begin rostrally at the level of the caudal hypothalamus and extend posteriorly about 2.5 mm to the rostral pons in the rat (German and Manaye 1993) and 2 mm in the mouse (Nelson et al. 1996). The substantia nigra is subdivided into the pars compacta (SNc), pars reticulata (SNr), and pars lateralis (SNl) (German and Manaye 1993; Nelson et al. 1996). Similarly, the VTA is also a montage of many smaller nuclei (Nelson et al. 1996). The third area, the retrorubral field, at the posterior end of substantia nigra, is considered to be an extension of this nucleus by some investigators (Fig. 1) (Di Carlo et al. 1973; Fallon and Moore 1976; Hokfelt et al. 1984; Nelson et al. 1996).

Species Differences in Numbers of Dopaminergic Neurons

The numbers of the midbrain dopaminergic neurons differ in different species. In addition, the distribution among different regions is also variable. Total tyrosine hydroxylase positive cells in the rat midbrain were found to number approximately 45,000 (German and Manaye 1993). The percentages of this total, shared among the RRF, SN, and VTA, were found to be 5%, 47%, and 48%, respectively. In contrast to the rat, mice have a higher ratio of neurons in the VTA than in the SN (Nelson et al. 1996). For example, divided among A8, A9, and A10, the percentages were 12%, 39%, and 49% of the total of approximately 30,000 neurons in the C57/B6 mouse midbrain (Nelson et al. 1996). Primates can have anywhere between three and seven times higher total midbrain DA neurons, with a significant increase in the proportion of neurons located in the SN (German and Manaye 1993). Humans, during their fourth decade of life, average about 590,000 cells, with 32% in VTA and 68% in the SN. By the sixth decade of life, this reduces to about 350,000 cells, with 21% in VTA and 89% in the SN, representing a natural loss of 40% overall over time (McGeer et al. 1977; German et al. 1989). It has been proposed that the reduced ratio of VTA to SN neurons in primates reflects the evolutionary increase in the volume of the caudate/putamen with respect to the nucleus accum-

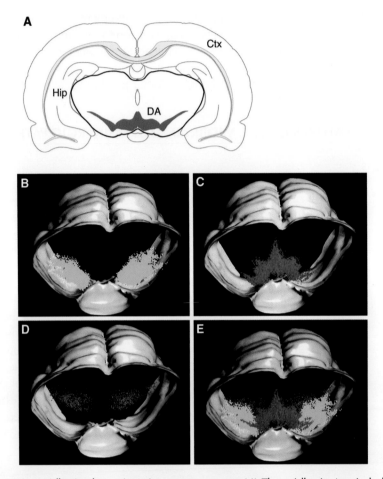

FIGURE 1. Major midbrain dopaminergic neuron groups. (*A*) The midbrain (encircled by a thick line) is located ventral and posterior to the cerebral cortex and hippocampus. The major dopaminergic neuron groups (DA) are positioned in the ventral midbrain. (Ctx) Cerebral cortex; (Hip) hippocampus. (*B–D*) The reconstruction of the rat midbrain and the position of dopaminergic neurons show three distinct dopaminergic groups, the substantia nigra (*B*), the ventral tegmental area (*C*), and the retrorubral field (*D*). *E* is a three-dimensional (3D) composite of the positions of all three groups of dopaminergic neurons. The rostral midbrain is in the foreground. (*B–E* reprinted, with permission, from German and Manaye 1993 [©Wiley-Liss].)

bens. Unlike the simple architecture of the SN in rodents, the SN is stratified in primates, including humans (Gerfen et al. 1987). A dorsal tier, most prominent rostrally, is calbindin- and calcium-binding protein (CaBP)-positive and therefore biochemically distinct from the ventral tiers, which do not express CaBP.

Dopaminergic Axon Pathways

Axon tracing studies over the last 40 years have shown that the midbrain dopaminergic neurons form extensive connections with several major brain regions including the caudate/putamen, nucleus accumbens, and the cerebral cortex. A number of techniques have been used over the years to define these axon pathways, and the following offers a brief description.

Techniques for Tracing Axons

Different techniques have been used in the studies of dopaminergic pathways, including anterograde and retrograde tracing methods and selective lesioning studies. In earlier studies, commonly used anterograde tracers, which diffuse toward axon terminals from the point of injection, were tritiated amino acids, such as proline or leucine. In these experiments, tritiated amino acids were injected into specific midbrain dopaminergic neuron groups. Approximately one week after injection, terminal regions could be visualized after exposure to X-ray emulsion. These experiments allowed for the mapping of terminal field topography of the midbrain dopaminergic cell efferents (Beckstead et al. 1979).

Retrograde tracing has been used to identify the origin of axon terminals (Vercelli et al. 2000). In retrograde tracing experiments, markers are injected into the target field with axon terminals, and allowed to transport retrogradely to the cell body, identifying the origin of axon terminals. A marker frequently used in a number of classical studies was horseradish peroxidase (HRP) (Van der Kooy 1979; Veening et al. 1980; Kalivas and Miller 1984). HRP, when injected into the nucleus accumbens, for instance, is taken up by axons and transported to the cell soma. The soma, located mainly in the VTA, can be reacted with 3,3´-diaminobenzidine (DAB), the substrate for horseradish peroxidase, which forms a brown precipitate. Another common method used in the early studies was lesioning of dopaminergic neurons (Anden et al. 1966; Lindvall et al. 1974; Simon et al. 1976). For example, injection of 6-hydroxydopamine (6-OHDA), a dopamine agonist, led to the death of dopaminergic cells in the vicinity of injection along with their axons and the terminals, revealing both axon pathways and targets of specific groups of dopaminergic neurons (Maler et al. 1973; Simon et al. 1976).

A variety of modern axon tracers are used today, each with different properties to be exploited (Lanciego et al. 2000; Perrin and Stoeckli 2000; Vercelli et al. 2000). Generally, tracers are chosen based on whether fluorescent or nonfluorescent markers are preferred and whether anterograde or retrograde tracing is performed. Many investigators prefer to use fluorescent markers to avoid the tedious handling of postlabeling procedures. Among the fluorescent tracers, the most commonly used are lipophilic carbocyanine dyes, which include the orange-red DiI and the greenish DiO (Perrin and Stoeckli 2000; Vercelli et al. 2000). Since they are lipophilic, their movement is restricted along the lipid membrane without diffusing trans-synaptically or to neighboring axons. An advantage of using these dyes is that they work in fixed tissue, allowing for pathway analysis at specific developmental stages. Another type of fluorescent tracer is fluorescent dextran, which includes compounds such as fluorogold and fluororuby (Lanciego et al. 2000; Vercelli et al. 2000). Dextrans are long-chain polysaccharide compounds that are hydrophilic. They can be covalently bound to amino acids, such as glutamine, that allow them to bind to other biomolecules. Finally, the use of fluorescent microspheres, which are fluorescent dyes coupled to polystyrene beads, is another option (Vercelli et al. 2000). Because they are not biodegradable, labeling with beads is very stable.

Nonfluorescent tracers are still very popular because they do not fade as fluorescent markers, but require secondary immunostaining. In addition, some tracers provide better resolution for fine terminals and synaptic boutons. Phaseolus vulgaris leucoagglutinin (PHA-L), a plant lectin, has been used for both retrograde and anterograde tracing (Lanciego et al. 2000; Vercelli et al. 2000). Because it marks axon terminals very well with high resolution, PHA-L is a popular anterograde tracer. The disadvantage is that it needs a very long survival time due to slow transport. Another anterograde option is biotinylated dextran (Lanciego et al. 2000; Vercelli et al. 2000). Similar to fluorescent dextran, biotinylated dextran is more stable, but requires an antibiotin antibody for visualization. Cholera toxin B (CTB), also used to trace anterogradely, may be the most sensitive and suitable for identifying different types of varicosities and synaptic boutons (Lanciego et al. 2000; Vercelli et al. 2000).

Tracers have been developed to label multiple segments of a neural circuit. These tracers must be able to move trans-synaptically into the target neurons. When coupled with wheat germ agglutinin (WGA-HRP), HRP is a useful anterograde tracer with the ability to trace both presynaptic and postsynaptic neurons (Vercelli et al. 2000). Today, viruses such as the herpes simplex virus, which can bind to specific neuronal receptors, are a promising, more sensitive way to trace the entire neural circuit because they can move trans-synaptically, to the next set of neurons. However, care must be taken to use an optimal concentration to ensure proper infection without rupturing the neuron (Vercelli et al. 2000).

Three Major Ascending Midbrain Dopaminergic Axon Pathways

Extensive studies using various tracing methods showed that the midbrain dopaminergic neurons project to form three major ascending pathways (Fig. 2) (Anden et al. 1966; Fallon and Moore 1976; Simon et al. 1976; Moore and Bloom 1978; Van der Kooy 1979; Veening et al. 1980; Lindvall and Bjorklund 1983; Kalivas and Miller 1984). Most dopaminergic neurons in the substantia nigra project to the caudate/putamen to form the nigrostriatal pathway. In general, medial SN neurons send axons to the medial caudate/putamen, whereas the lateral axons target the lateral caudate/putamen (Fallon and Moore 1976; Van der Kooy 1979; Veening et al. 1980). On the other hand, dopaminergic axons from the ventral tegmental area form connections with targets located in the ventral striatum preferentially, including the nucleus accumbens and the olfactory tubercle (Moore and Bloom 1978; Lindvall and Bjorklund 1983; Kalivas and Miller 1984), constituting the mesolimbic pathway. The nucleus accumbens is also targeted by dopaminergic axons from the medial substantia nigra (Lindvall and Bjorklund 1983). Thus, some overlap exists between the target fields of the substantia nigra and the ventral tegmental area. The third pathway, the mesocortical tract, consists of axon projections from the VTA neurons to the frontal cortex (Lindvall et al. 1974). Therefore, axons from the midbrain dopaminergic neurons are organized into three different pathways. Each pathway has been shown to have specific functions. For example, the nigrostriatal pathway is closely linked with motor controls, the degeneration of which leads to Parkinson's disease. In contrast, the mesolimbic pathway regulates reward signals and mediates addictive behavior (Spanagel and Weiss 1999; Nestler 2004). The correlation between functional differentiation and structural specification indicates that the topographic layout of the dopaminergic axons is crucial for function.

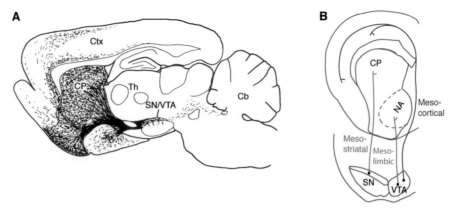

FIGURE 2. Midbrain dopaminergic pathways. Dopaminergic neurons in the ventral midbrain project massively to the striatum and to the frontal cortex. (*A*) Dopaminergic axons and their terminals are illustrated by stippling in a sagittal rat brain section. (Cb) Cerebellum; (Ctx) cerebral cortex; (Th) thalamus. (*B*) The pathways are divided into three major circuits. Axons from the substantia nigra (SN) project to the caudate/putamen (CP), constituting the mesostriatal pathway (*green*). DA neurons in the ventral tegmental area (VTA) send axons to the nucleus accumbens (NA) and the frontal cortex to form the mesolimbic (*red*) and mesocortical (*blue*) pathways. (*A* modified, with permission, from Fallon and Loughlin 1995 [©Elsevier]; *B* modified, with permission, from Yue et al. 1999 [©Wiley-Liss].)

DEVELOPMENT OF MIDBRAIN DOPAMINERGIC NEURAL PATHWAYS

Neurogenesis

Dopaminergic neural fate is determined by complex molecular interactions. Only a brief overview of the events leading to the differentiation of the dopaminergic neurons is provided here. For a more complete discussion, readers should consult an excellent recent review by Riddle and Pollock (2003). The events that set the stage for dopaminergic neuron induction are already in place by embryonic day 9 (e9) in the mouse (e11 in the rat) (Prakash and Wurst 2004). Progenitors are born at the ventral rim of the mesencephalic flexure and proliferate within the caudal region of the mesencephalon from e8.5–e10.5 (Burbach et al. 2003; Simon et al. 2003). These cells then exit the cell cycle and migrate to their appropriate positions. Birthdating studies in the mouse show that the midbrain dopaminergic neurons are born between e10 and e14 in a rostral to caudal gradient (Bayer et al. 1995). More laterally destined cells are born first and migrate before their more medial neighbors. In the rat, dopaminergic neurogenesis occurs between e12 and e16 (Altman and Bayer 1981; Bayer et al. 1995). However, the divisions between A9 and A10 do not become apparent until e18 in the rat (Golden 1972; Seiger and Olson 1973; Voorn et al. 1988). Soon after their final mitosis, dopaminergic neurons begin expressing tyrosine hydroxylase, the enzyme which converts tyrosine to L-Dopa, often before the end of their migration (Specht et al. 1981). Although neurons have been detected sprouting growth cones as early as e12.5 in the rat, cytological maturation is not seen until approximately e14.5 (Specht et al. 1981).

Patterning of the Midbrain

Patterning of the midbrain precedes the induction of the dopaminergic neurons. An organizing center at the midbrain-hindbrain boundary, the isthmus, dictates the expression coordinates of several transcription factors required for the proper development of the ventral midbrain. The homeodomain transcription factor, Otx-2, expressed at e7 in mice, gives rise to the anterior neuroectoderm (Simon et al. 2003; Prakash and Wurst 2004). Gbx-2, a second homeodomain transcription factor, is expressed more caudally. The interface of these two factors marks the midbrain-hindbrain boundary (Hynes and Rosenthal 1999; Simon et al. 2003; Prakash and Wurst 2004). Neither Otx-2 nor Gbx-2 are directly required for the activation of specific midbrain markers but both are important for the proper positioning of the midbrain (Martinez-Barbera et al. 2001; Acampora et al. 2003). Repression of Otx-2 expression results in the loss of both the forebrain and midbrain. Conversely, Gbx-2 ablation allows for the expansion of rostral phenotypes into the hindbrain (Prakash and Wurst 2004).

Following the establishment of the midbrain-hindbrain boundary, Pax-2, a paired box transcription factor, is expressed throughout the midbrain and in the rostral hindbrain, the first of several midbrain patterning signals including Pax-5, Wnt-1, En-1, and En-2 to specify neuronal fate (Rowitch and McMahon 1995; Simon et al. 2003; Prakash and Wurst 2004). Pax-2 has been shown to be necessary and sufficient to induce the expression of Fgf-8, a critical inducer of the midbrain dopaminergic neurons (Martinez et al. 1999; Ye et al. 2001). Inactivation of Pax-2 leads to a failure of neurotube closure at the midbrain (Schwarz et al. 1997). Pax-2 has also been shown to induce Wnt-1, a secreted protein, and En-1, a transcription factor (Rowitch and McMahon 1995). Wnt-1 is initially expressed broadly in the midbrain, but is later restricted to a rostral band adjacent to the organizer. In an experiment in which Wnt-1 was overexpressed, proliferation of DA neurons in the midbrain was greatly increased (Panhuysen et al. 2004). It has also been shown that Wnt-1 expression stabilizes the transcription factors En-1 and En-2, loss of which leads to the failure of the midbrain to develop (McMahon et al. 1992; Panhuysen et al. 2004). These observations indicate that the Wnt genes are required for continued proliferation of the anterior midbrain, possibly by supporting transcription factor Otx-2 expression (Martinez et al. 1999). Thus, the complex interactions among multiple extracellular and intracellular factors specify both the development of the midbrain in general and the fate of dopaminergic neurons.

Induction of Midbrain Dopaminergic Neurons

Midbrain dopaminergic precursors are born near the midbrain-hindbrain boundary driven by the influence of two inductive growth factor signals: Fgf-8, released from the isthmus, and sonic hedgehog, Shh, secreted from the ventral floor plate of the midline and notochord (Ye et al. 1998; Burbach 2000). Both factors are found to be necessary for the induction of dopaminergic neurons (Hynes et al. 1995b; Ye et al. 1998; Hynes and Rosenthal 1999). Early studies began with the discovery that the exogenous floor plate secreted a signal that could induce midbrain dopaminergic neurons both in vivo and in vitro (Hynes et al. 1995a). Later, Shh was confirmed to be the inductive signal expressed by the floor plate (Hynes et al. 1995b). Fgf-8, expressed just caudal to the midbrain-hindbrain boundary, can behave independently as an organizer (Crossley et al. 1996; Martinez et al. 1999).

Exogenously placed Fgf-8 coated beads or Fgf-8 expressing tissue can induce midbrain-specific markers and create a new midbrain boundary. However, Fgf-8 alone is not sufficient to induce dopaminergic cell fates (Ye et al. 1998; Martinez et al. 1999). Similarly, positional constraints were observed when ectopic Shh was used to induce dopaminergic phenotypes. Further studies showed that the intersecting expression of both Shh and Fgf-8 was required to confer the proper dorsoventral and anteroposterior positional information, respectively, to promote dopaminergic cell specificity (Ye et al. 1998).

Differentiation of Midbrain Dopaminergic Neurons

Once dopaminergic progenitors exit their last mitotic division, a new set of cellular factors is required to complete their differentiation, which takes place at approximately e11–e13 in the mouse. Tyrosine hydroxylase (TH), the rate-limiting enzyme for dopamine production, is under the control of an orphan nuclear receptor Nurr1 and possibly also the homeodomain transcription factor Ptx3 (Smidt et al. 1997; Perrone-Capano and Di Porzio 2000; Lebel et al. 2001). An unknown ligand is required for Nurr1 activation although there is evidence that it may function in a ligand-independent manner (Wang et al. 2003). Nurr1 is required for the development of dopaminergic neurons in the midbrain, but not for populations located elsewhere (Castillo et al. 1998). Nurr1$^{-/-}$ mice are generally characterized by the agenesis of midbrain dopaminergic neurons (Zetterstrom et al. 1997). However, in other studies, late dopaminergic precursor cells can position themselves correctly and express TH in Nurr1$^{-/-}$ mice, but they fail to complete their differentiation and form improper connections with the striatum. These cells consequently lose their phenotype and degenerate through apoptosis (Tornqvist et al. 2002). Recent data suggest that Nurr1 also has a role in regulating aspects of differentiation other than TH expression (Saucedo-Cardenas et al. 1998; Wallen et al. 1999; Witta et al. 2000). Several genes responsible for the final differentiation into dopaminergic neurons, including the dopamine transporter (DAT) (Sacchetti et al. 2001), vesicle-membrane-associated transporter (VMAT2), and Ret, the receptor of GDNF (an important survival factor for emerging dopaminergic neurons), are all regulated by Nurr1, noted by their absence in Nurr1 knockout animals (Simon et al. 2003). Mature dopaminergic neurons are absent in Nurr1$^{-/-}$ mice probably because the genes associated with dopamine synthesis, storage, and transport, as well as genes involved in cell survival, are compromised (Perrone-Capano et al. 2000).

The role of Ptx3 has not been fully uncoupled from Nurr1. Despite this, evidence suggests that Ptx3 has an important role in the survival and migration of dopaminergic neurons (Smidt et al. 2004). Ptx3 transcription is activated by the transcription factor Lmx1b, which is expressed as early as e7.5 and remains throughout development into the adult (Smidt et al. 1997, 2000). In the absence of Lmx1b, tyrosine-hydroxylase-positive midbrain neurons are generated, but fail to express Ptx3. Consequently, these cells die before maturation, possibly due to the lack of Ptx3 (Smidt et al. 2000; Nunes et al. 2003). Interestingly, Ptx3 can also activate the TH promoter. Three high-affinity binding sites for Ptx3 have been identified in the TH promoter (Lebel et al. 2001).

Maturation of dopaminergic cells occurs between e15 and e16 in the mouse (Perrone-Capano et al. 2000). When striatal connections are made, TH expression shifts from being localized mainly in the soluble fraction within the midbrain cell soma to the terminals at the same time as the appearance of striatal DAT expression (Coyle and Axelrod 1972;

Sacchetti et al. 2001). Between birth and one month of age, striatal tyrosine hydroxylase activity in the mouse increases about ten times (Coyle and Axelrod 1972).

Pathway Differentiation

Following the induction of dopaminergic neurons, the axons navigate toward the forebrain to form proper connections. The heterogeneous population of cells in the midbrain makes it all the more fascinating that each type of axon can be organized molecularly to form distinct pathways to their appropriate target fields, which include the caudate/putamen, nucleus accumbens, and cerebral cortex.

Timing

At e11 in the rat, short axons begin to extend dorsally, away from the midline (Nakamura et al. 2000). Between e12 and e13, axons turn sharply and begin extending rostrally toward the medial ganglionic eminence (Nakamura et al. 2000; Gates et al. 2004). The areas defined as A8, A9, and A10 have not differentiated apart from one another at this time. In the rat, these groups can be distinguished from one another at around e18 (Seiger and Olson 1973). These neurites extend into the medial ganglionic eminence, from which the basal ganglia is later formed, at e12.5 in the mouse and at e14 in the rat via the medial forebrain bundle (Voorn et al. 1988; Di Porzio et al. 1990). Axons from the substantia nigra reach the medial ganglionic eminence first, followed by neurites from the VTA (Seiger and Olson 1973). By e18 in the mouse (e19 in the rat), projection pathways from the midbrain dopaminergic neurons to the forebrain targets are basically established (Voorn et al. 1988; Hu et al. 2004).

Cellular Mechanisms

To project to the forebrain targets, midbrain DA axons make several pathfinding decisions. The axons first grow dorsally and then turn toward the anterior. When they finally reach the forebrain, they must decide in which target field to terminate: the caudate/putamen, nucleus accumbens, or frontal cortex. A variety of attractive and repulsive cues must guide the extending axons, although the nature of these molecules has not been well characterized.

To examine how the ascending dopaminergic pathways are guided during embryonic development, brain regions located in and around the axon tracts were tested for chemoattractive or repulsive activities (Gates et al. 2004). This analysis showed that the early medial forebrain bundle region, through which the midbrain dopaminergic axons travel, is highly attractive to these axons. In addition, strong attractive activity was also detected in embryonic striatal tissue, the target of the midbrain dopaminergic axons. The attractive activities appear to be temporally regulated to correlate with the arrival of the midbrain dopaminergic axons. In contrast, the brain stem tissue showed strong repulsive activity. The repulsive activity in the brain stem has been hypothesized to orient the axons rostrally. The rostral turning of these axons may also be regulated by polarized substrates over which these axons grow (Nakamura et al. 2000). Explant assays in which the polarity of the dorsal mesencephalon was reversed relative to the ventral mesencephalon led to neurite turning in the opposite direction, suggesting that a substrate-bound gradient was responsible for the rostral turning. Similar to the brain stem tissue, cerebral cortical explants repelled the midbrain dopaminergic axons, suggesting a function to restrict the axon terminals in the striatum (Gates et al. 2004).

The outgrowth into the lateral ganglionic eminence is initially widespread with no distinct specificity (Hu et al. 2004). Retrograde labeling of the midbrain neurons with DiI and DiA from the dorsolateral and ventromedial striatal areas in e15 mice demonstrates a lack of pathway specificity. The cells that innervate these areas originate from both the putative SN and VTA (Fig. 3). The early axons from the midbrain cover an extensive region of the striatum with many collaterals. Gradually, these connections become specialized. By e17

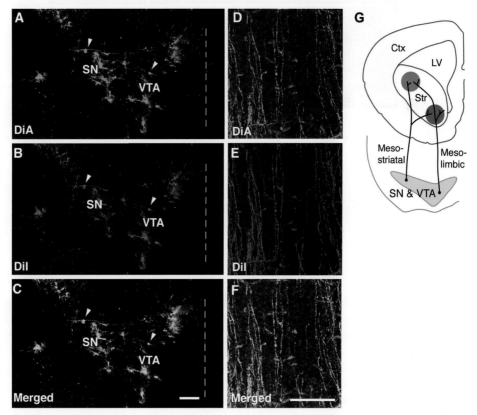

FIGURE 3. The midbrain to striatum projections exhibit no specificity in e15 mice. At e15, dopaminergic axons originating from both the presumptive SN and VTA project to the striatum with no topographic preference. Therefore, DiA placed in the dorsal striatum (*green*) is taken up by axons and carried back to both the presumptive SN and VTA (*A,D*). Similarly, DiI injected into the ventral striatum (*red*) also traces dopaminergic neurons and axons from the two midbrain regions (*B,E*). Merged images (*C,F*) indicate that the same group of neurons and axons is labeled by both DiA and DiI. Double labeling is indicated in yellow. Arrowheads in *A–C* denote examples of colocalization. These results reveal that the midbrain neurons send collaterals to both the dorsal and ventral striatum at e15, as illustrated in *G*. The substantia nigra and ventral tegmental area have not been fully differentiated at this age. Dashed lines in *A–C* indicate the midline. (Ctx) Cerebral cortex; (LV) lateral ventricle; (Str) striatum. Scale bar, 100 μm. (*A–F* modified, with permission, from Hu et al. [©Wiley-Liss].)

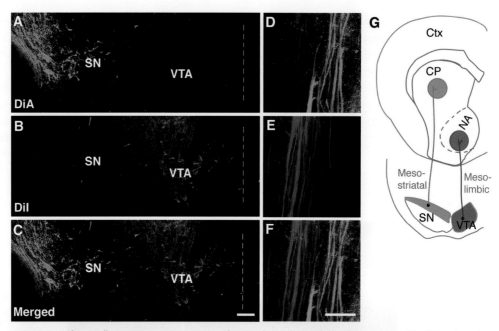

FIGURE 4. The midbrain to striatum DA pathways have been fully differentiated in P0 mice. Axon tracers now label only specific pathways. DiA placed in the caudate/putamen (CP) (*green*) retrogradely traces dopaminergic axons and neurons from the substantia nigra (SN) (*A,D*). In contrast, DiI injected in the nucleus accumbens (NA) (*red*) labels only dopaminergic neurons and axons from the ventral tegmental area (VTA) (*B,E*). Merged images show that DiA and DiI now each label a distinct set of midbrain neurons and axons (*C,F*). This study indicates that the midbrain to striatum DA pathways have been fully specified by P0, as illustrated in *G*. Dashed lines in *A–C* indicate the midline. Scale bar, 100 μm. (*A–E* modified, with permission, from Hu et al. 2004 [©Wiley-Liss.])

in the mouse, fibers extending from the midbrain have partially segregated into separate bundles to become the mesostriatal and mesolimbic tracts. At this stage, the midbrain dopaminergic neurons, retrogradely labeled by axon tracers injected into the dorsolateral and ventromedial striatum, no longer show significant overlap. By birth, mouse neonates show exclusive dye tracing from the dorsolateral striatum to the SN and ventromedial striatum into VTA (Fig. 4) (Hu et al. 2004). Some spatial overlap still exists in the axons forming the mesolimbic and mesostriatal tracts, but this disappears completely as pathways mature by postnatal day 7 (Hu et al. 2004). Pruning of mislocated collaterals, as opposed to apoptosis, is believed to confer specificity in the later stages of development (Fig. 5).

Molecular Basis

Pathfinding requires a coordinated effort by multiple guidance cues. Many axon guidance molecules participate in directing dopaminergic axon migration to the proper targets. Process outgrowth, navigation, and target selection may be mediated by ligand-receptor systems including Netrins, Ephrins, Semaphorins, and Slits. The understanding of the molecular biology of axon guidance is still quite fragmentary.

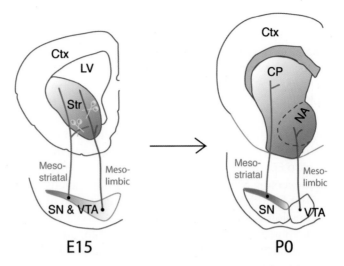

FIGURE 5. Schematic illustration of the development of the midbrain dopaminergic pathways. Dopaminergic axons and collaterals initially target both the dorsal and ventral striata. Mistargeted collaterals are then pruned (*scissors*) to provide specific pathways. The pathways are fully differentiated by P0. A gradient of Ephrin-B2 expression, highest in the ventral striatum (*blue*), may prevent EphB1-bearing SN axons (*green*) from targeting the ventromedial region. EphB1-poor VTA axons do not respond to Ephrin-B2 and can therefore innervate the nucleus accumbens. Whether these molecules contribute to the pruning during development is not known at present.

In the early embryo, the axons of dopaminergic axons migrate dorsally and then make an abrupt rostral turn (Nakamura et al. 2000). The role of Slit as a repulsive, soluble guidance cue is well established and thought to be a driving force in the initial exit of the midbrain neurites toward the ganglionic eminence (Hu 1999) and in the prevention of midline crossing. Both Slits1 and 2 are diffusible molecules expressed in the rhombic lip, caudal to the isthmus, while Slit1 is expressed early along the ventral midline of the brain including the mesencephalon (Yuan et al. 1999). Attempts at growing midbrain dopaminergic neurons in vitro over Slit-expressing cells showed that the molecule is repulsive to dopaminergic axons (R. Zhou, unpubl.). In addition, the Slit receptor Robo was expressed in the dopaminergic neurons. In Slit1 and 2 double knockout mice, DA axons are misrouted, in some instances crossing over the midline and traversing the hypothalamus instead of going straight to the forebrain targets (Bagri et al. 2002).

Another classic set of guidance molecules, the Eph/Ephrin family, has been implicated in directing the pathfinding of the dopaminergic system (Yue et al. 1999). The Eph family of the tyrosine kinase receptor has 14 members divided into two subgroups: EphA and EphB (Zhou 1998; Kullander and Klein 2002). The ligands to these receptors, the Ephrins, are also divided into two subgroups, Ephrin-A and Ephrin-B, and generally bind preferably to receptors within their own subfamily (Zhou 1998). All Ephrins and receptors are membrane bound and therefore function as contact-mediated guidance cues. Unlike most ligands, Ephrins are capable of reverse signaling (Kullander and Klein 2002; Murai and Pasquale 2003). These molecules have been shown to play key roles in the specification of topographical projections in the retinotectal, hippocamposeptal, and thalamocortical pathways (Zhou 1997, 1998; Knoll and Drescher 2002; Yue et al. 2002; Dufour et al. 2003). Ephrins appear to be capable of either repulsive or attractive regulation of migratory axons.

In situ hybridization of the dopaminergic midbrain revealed that two Eph receptors are expressed at significant levels: EphB1 and EphA5 (Yue et al. 1999). Elevated levels of EphB1 expression was found more specifically localized to the substantia nigra and at lower levels in the VTA. Complementary expression of the ligand Ephrin-B2 was found at high levels in the nucleus accumbens and at low levels in the dorsal striatum, suggesting an inhibitory interaction to restrict SN axons to their dorsolateral targets (Yue et al. 1999). In vitro Ephrin-B2 can inhibit the growth of EphB1-positive DA neurons, suggesting a role in preventing SN axons from targeting the nucleus accumbens (Fig. 5).

The Ephrin-A subclass may also have a role in dopaminergic axon targeting. EphA5 expression is found in both SN and VTA dopaminergic neurons (Yue et al. 1999). Its ligand, Ephrin-A5, is expressed in the striatum, and variation exists in the levels in different striatal areas (M.A. Cooper and R. Zhou, unpubl.). Seiber et al. (2004) showed that a soluble EphA receptor released in the brain of transgenic mice could reduce the number of substantia nigra dopaminergic neurons that could be traced from the striatum by 40–50%. Soluble forms of the Eph receptor act as antagonists because they can compete with functional, anchored forms of the receptor for a limited source of Ephrin ligand. Therefore, these data suggest that the A-Ephrins and EphA receptors participate in directing dopaminergic axons to their striatal targets.

Innervation of Dopaminergic Axon Terminals

After reaching their targets, the midbrain dopaminergic axon terminals go through a process of correction and reorganization. Up to 3 weeks postnatally, a decrease in dopaminergic synaptic density in the striatum is observed, probably corresponding to synaptic strengthening (Antonopoulos et al. 2002). This reduction is believed to be due to pruning of synapses formed between incoming dopaminergic axons and the soma or dendrites of the immature medium spiny neurons in the striatum (axosomatic and axodendritic synapses) to make room for dopaminergic synapses onto the mature neurons (Antonopoulos et al. 2002). Most dopaminergic neurons preferably form symmetric (inhibitory) synapses on the dendrites of medium spiny neurons. As the striatum matures, thin and diffuse fibers begin to replace the dopaminergic islands that first form, and varicosities begin to develop around postnatal day 2 (P2) (Antonopoulos et al. 2002). The nigrostriatal innervations may also be modulated through apoptotic cell death. A postnatal biphasic wave of apoptosis in the substantia nigra has been reported (Burke 2004). A five times higher rate of dopaminergic cell death is observed at P2 and P14, a period during which the dopaminergic synapses are actively remodeled.

The dopaminergic terminals throughout the striatum are initially organized as islands of dense innervation onto aggregated striatal neurons, the patches (or striosomes) (Seiger and Olson 1973). The surrounding regions are termed matrix (Seiger and Olson 1973; Gerfen 1989). The neurons in the patches and matrices exhibit different neural transmitter receptor phenotypes and express distinct protein markers. For example, patches are high in µ-opioid receptor-containing interneurons (Gerfen 1984; Kubota and Kawaguchi 1993). The matrix, on the other hand, contains high expression of acetylcholine esterase

(Martin et al. 1991; Gerfen 1992). In addition, the matrices are CaBP positive, whereas the patches do not express this protein (Gerfen 1992). The expression of the dopamine receptors D1 and D2 is also unevenly distributed in the patch-matrix compartments (Levey et al. 1993). Whereas D2 is found in both the patches and the matrices, D1 receptors are highly enriched only in patches. This differential distribution may be critical for dopamine functions, because D1 and D2 receptors have opposing biochemical and biological effects (Callier et al. 2003), suggesting that the patch-matrix arrangement is functionally important. After birth through the second week, projections to the medial striatum receive more diffuse dopaminergic innervation from the VTA, which obscures the patches, leaving only those in the dorsal striatum (Voorn et al. 1988). Regional differentiation also occurs in the nucleus accumbens (Meredith et al. 1996). Considerable variations in the expression of neuronal markers occur among different regions of the nucleus accumbens. The nucleus accumbens can be divided into the core and shell. The core region, which receives input from the VTA, exhibits more intense staining for calbindin than does the shell. In addition, within the shell, which receives input from the medial SNc (Lindvall and Bjorklund 1983), CaBP expression is more intense in the lateral than the medial shell regions (Meredith et al. 1996).

The patch-matrix differentiation is also correlated with different input-output patterns. The VTA, dorsal SNc, and RRF project to the matrices and are CaBP positive (Gerfen et al. 1987; Gerfen 1989), whereas the ventral SNc and SNr project to patches and are CaBP negative. In turn, neurons in the patches project back to the SNc to target the dopaminergic neurons, whereas neurons in matrices send axons to SNr, connecting with the GABAergic neurons in the ventral midbrain. Cortical inputs are also differentially distributed to the patches and the matrices. Deep layer V and VI neurons primarily project to the striatal patches. On the other hand, neurons from the superficial layer V, II, and III project to the matrix compartment. Thus, patch-matrix organization allows specific feedback and integration among multiple brain regions. The molecular mechanisms that regulate patch-matrix differentiation are not known. Interestingly, axon guidance molecules, Ephrin-A ligands, and EphA receptors may also be involved in defining patch-matrix compartmentalization (Janis et al. 1999). EphA receptors are expressed at high levels in the matrix, but not in the patches. Where the ligands are expressed in relation to the Eph receptors and whether they play a role in the patch-matrix differentiation are not known.

SUMMARY AND FUTURE DIRECTIONS

This chapter serves to briefly summarize what is known about dopaminergic system development; it is by no means comprehensive. The ontogeny of the midbrain dopaminergic neurons consists of multiple developmental processes that lead to the specification of the midbrain-hindbrain boundary and to the determination of dopaminergic neural fate. Axons from the differentiated dopaminergic neurons are guided to various targets by multiple axon guidance factors. Different dopaminergic pathways are differentiated possibly by selective pruning to remove mistargeted collaterals. Axon terminals in the synaptic targets are further organized into distinct cytoarchitectural components such as the patch and matrix in the striatum. Expression of axon guidance molecules has been shown to be induced by addictive drugs. Alterations in dendritic morphology and synaptic functions may underly addictive behavior. However, many questions on the development of the midbrain DA system remain unanswered. For example, the molecular mechanisms of specification of the rostrocaudal axis of the DA pathways are not known. Knowledge about the key regulators involved in dopaminergic axon pathfinding and target selection is also incomplete. The specific roles of Slit and Ephrin families of axon guidance molecules remain to be defined. Other known or unknown families of axon guidance molecules are also likely to contribute to the orientation of DA axons and to the selection of targets in the striatum. The exact nature of signals that prune dopaminergic collaterals needs to be elucidated, and molecules regulating synaptogenesis are also not known. Another unanswered question is whether changes in DA axon projections result in any behavioral abnormality. In addition, the intracellular signaling pathways underlying dopaminergic axon guidance are yet to be identified. A further challenge is to use this knowledge to help repair dopaminergic circuitry damaged by addiction or brain injury.

ACKNOWLEDGMENTS

We thank James Fallon and Dwight German for permission to use their figures, and Jonathan Pollock for critical review of the manuscript. Research was supported in part by grants (R.Z.) from the Michael J. Fox Foundation for Parkinson's Research and the National Institutes of Health.

REFERENCES

Acampora D., Annino A., Puelles E., Alfano I., Tuorto F., and Simeone A. 2003. OTX1 compensates for OTX2 requirement in regionalisation of anterior neuroectoderm. *Gene Expr. Patterns* **3**: 497–501.

Altman J. and Bayer S.A. 1981. Development of the brain stem in the rat. V. Thymidine-radiographic study of the time of origin of neurons in the midbrain tegmentum. *J. Comp. Neurol.* **198**: 677–716.

Anden N.-E., Dahlstom A., Fuxe K., Larsson L., Olson L., and Ungerstedt U. 1966. Ascending monoamine neurons to the telencephalon and diencephalon. *Acta Physiol. Scand.* **67**: 313–326.

Angrist B. and Gershon S. 1974. Dopamine and psychotic states: Preliminary remarks. *Adv. Biochem. Psychopharmacol.* **12**: 211–219.

Angrist B., Thompson H., Shopsin B., and Gershon S. 1975. Clinical studies with dopamine-receptor stimulants. *Psychopharmacologia* **44**: 273–280.

Antonopoulos J., Dori I., Dinopoulos A., Chiotelli M., and Parnavelas J.G. 2002. Postnatal development of the dopaminergic system of the striatum in the rat. *Neuroscience* **110**: 245–256.

Bagri A., Marin O., Plump A.S., Mak J., Pleasure S.J., Rubenstein J.L., and Tessier-Lavigne M. 2002. Slit proteins prevent midline crossing and determine the dorsoventral position of major axonal pathways in the mammalian forebrain [see comment]. *Neuron* **33**: 233–248.

Bayer S.A., Wills K.V., Triarhou L.C., and Ghetti B. 1995. Time of neuron origin and gradients of neurogenesis in midbrain dopaminergic neurons in the mouse. *Exp. Brain Res.* **105**: 191–199.

Beckstead R., Domesick V., and Nauta W. 1979. Efferent connections of the substantia nigra and ventral tegmental area in the rat. *Brain Res.* **175**: 191–217.

Bjorklund L.M., Sanchez-Pernaute R., Chung S., Andersson T., Chen I.Y.-C., McNaught K.S.P., Brownell A.-L., Jenkins B.G., Wahlestedt C., Kim K.-S., and Isacson O. 2002. Embryonic stem cells develop into functional dopaminergic neurons after transplantation in a Parkinson rat model. *Proc. Natl. Acad. Sci.* **99**: 2344–2349.

Burbach J.P.H. 2000. Genetic pathways in the developmental specification of hypothalamic neuropeptide and midbrain catecholamine systems. *Eur. J. Pharmacol.* **405**: 55–62.

Burbach J.P.H., Smits S., and Smidt M.P. 2003. Transcription factors in the development of midbrain dopamine neurons. *Ann. N.Y. Acad. Sci.* **991**: 61–68.

Burke R.E. 2004. Ontogenic cell death in the nigrostriatal system. *Cell Tissue Res.* **318**: 63–72.

Callier S., Snapyan M., Le Crom S., Prou D., Vincent J.D., and Vernier P. 2003. Evolution and cell biology of dopamine receptors in vertebrates. *Biol. Cell* **95**: 489–502.

Carvey P.M., Chang Q., Lipton J.W., and Ling Z. 2003. Prenatal exposure to the bacteriotoxin lipopolysaccharide leads to long-term losses of dopamine neurons in offspring: A potential, new model of Parkinson's disease. *Front. Biosci.* **8:** s826–s837.

Castillo S.O., Baffi J.S., Palkovits M., Goldstein D.S., Kopin I.J., Witta J., Magnuson M.A., and Nikodem V.M. 1998. Dopamine biosynthesis is selectively abolished in substantia nigra/ventral tegmental area but not in hypothalamic neurons in mice with targeted disruption of the Nurr1 gene. *Mol. Cell. Neurosci.* **11:** 36–46.

Coyle J.T. and Axelrod J. 1972. Tyrosine hydroxylase in rat brain: Developmental characteristics. *J. Neurochem.* **19:** 1117–1123.

Crossley P.H., Martinez S., and Martin G.R. 1996. Midbrain development induced by FGF8 in the chick embryo. *Nature* **380:** 66–68.

Dahlstom A. and Fuxe K. 1964. Evidence for the existence of monoamine-containing neurones in the central nervous system: I. Demonstration of monoamines in the cell bodies of brain stem neurones. *Acta Physiol. Scand.* **62:** 1–55.

Di Carlo V., Hubbard J.E., and Pate P. 1973. Fluorescence histochemistry of monoamine-containing cell bodies in the brain stem of the squirrel monkey (Saimiri sciureus). IV. An atlas. *J. Comp. Neurol.* **152:** 347–372.

Di Chiara G., Morelli M., Acquas E., and Carboni E. 1992. Functions of dopamine in the extrapyramidal and limbic systems: Clues for the mechanism of drug actions. *Arzneimittelforschung* **42:** 231–237.

Di Monte D.A. 2003. The environment and Parkinson's disease: Is the nigrostriatal system preferentially targeted by neurotoxins? *Lancet Neurol.* **2:** 531–538.

Di Porzio U., Zuddas A., Cosenza-Murphy D.B., and Barker J.L. 1990. Early appearance of tyrosine hydroxylase immunoreactive cells in the mesencephalon of mouse embryos. *Int. J. Dev. Neurosci.* **8:** 523–532.

Dufour A., Seibt J., Passante L., Depaepe V., Ciossek T., Frisen J., Kullander K., Flanagan J.G., Polleux F., and Vanderhaeghen P. 2003. Area specificity and topography of thalamocortical projections are controlled by ephrin/Eph genes. *Neuron* **29:** 453–465.

Fallon J. and Loughlin S.E. 1995. Substantia Nigra. In *The rat nervous system* (ed. G. Paxinos), pp. 215–237. Academic Press, San Diego.

Fallon J. and Moore R. 1976. Dopamine innervation of some basal forebrain areas in the rat. *Neurosci. Abstr.* **2:** 486.

Flanagan J.G. and Vanderhaeghen P. 1998. The ephrins and Eph receptors in neural development. *Annu. Rev. Neurosci.* **21:** 309–345.

Gates M.A., Coupe V.M., Torres E.M., Fricker-Gates R.A., and Dunnett S.B. 2004. Spatially and temporally restricted chemoattractive and chemorepulsive cues direct the formation of the nigro-striatal circuit. *Eur. J. Neurosci.* **19:** 831–844.

Gerfen C.R. 1984. The neostriatal mosaic: Compartmentalization of corticostriatal input and striatonigral output systems. *Nature* **311:** 461–464.

———. 1989. The neostriatal mosaic: Striatal patch-matrix organization is related to cortical lamination. *Science* **246:** 385–388.

———. 1992. The neostriatal mosaic: Multiple levels of compartmental organization. *Trends Neurosci.* **15:** 133–139.

Gerfen C.R., Baimbridge K.G., and Thibault J. 1987. The neostriatal mosaic: III. Biochemical and developmental dissociation of patch-matrix mesostriatal systems. *J. Neurosci.* **7:** 3935–3944.

German D.C. and Manaye K.F. 1993. Midbrain dopaminergic neurons (nuclei A8, A9, A10): Three-dimensional reconstruction in the rat. *J. Comp. Neurol.* **331:** 297–309.

German D.C., Manaye K., Smith W.K., Woodward D.J., and Saper C.B. 1989. Midbrain dopaminergic cell loss in Parkinson's disease: Computer visualization [see comment]. *Ann. Neurol.* **26:** 507–514.

Golden G.S. 1972. Embryologic demonstration of a nigro-striatal projection in the mouse. *Brain Res.* **44:** 278–282.

Groenewegen H.J. 2003. The basal ganglia and motor control. *Neural Plast.* **10:** 107–120.

Hadj Tahar A., Grondin R., Gregoire L., Calon F., Di Paolo T., and Bedard P.J. 2003. New insights in Parkinson's disease therapy: Can levodopa-induced dyskinesia ever be manageable? *Adv. Neurol.* **91:** 51–64.

Hechtman L. 1994. Genetic and neurobiological aspects of attention deficit hyperactive disorder: A review. *J. Psychiatry Neurosci.* **19:** 193–201.

Hermann A., Gerlach M., Schwarz J., and Storch A. 2004. Neurorestoration in Parkinson's disease by cell replacement and endogenous regeneration. *Expert Opin. Biol. Ther.* **4:** 131–143.

Hokfelt T., Martensson R., Bjorklund A., Kleinau S., and Goldstein M. 1984. Distribution maps of tyrosine-hydroxylase-immunoreactive neurons in the rat brain. In *Handbook of chemical neuroanatomy* (ed. A Bjorklund et al.), pp. 277–379. Elsevier Science, New York.

Hu H. 1999. Chemorepulsion of neuronal migration by Slit2 in the developing mammalian forebrain. *Neuron* **23:** 703–711.

Hu Z., Cooper M.A., Crockett D.P., and Zhou R. 2004. Differentiation of the midbrain dopaminergic pathways during mouse development. *J. Comp. Neurol.* **476:** 301–311.

Hynes M. and Rosenthal A. 1999. Specification of dopaminergic and serotonergic neurons in the vertebrate CNS. *Curr. Opin. Neurobiol.* **9:** 26–36.

Hynes M., Poulsen K., Tessier-Lavigne M., and Rosenthal A. 1995a. Control of neuronal diversity by the Floor Plate: Contact-mediated induction of midbrain dopaminergic neurons. *Cell* **80:** 95–101.

Hynes M., Porter J.A., Chiang C., Chang D., Tessier-Lavigne M., Beachy P.A., and Rosenthal A. 1995b. Induction of midbrain dopaminergic neurons by Sonic Hedgehog. *Neuron* **15:** 35–44.

Janis L.S., Cassidy R.M., and Kromer L.F. 1999. Ephrin-A binding and EphA receptor expression delineate the matrix compartment of the striatum. *J. Neurosci.* **19:** 4962–4971.

Jenner P. 2003. Dopamine agonists, receptor selectivity and dyskinesia induction in Parkinson's disease. *Curr. Opin. Neurol.* **16:** S3–S7.

Kalivas P. and Miller J. 1984. Neurotensin neurons in the ventral tegmental area project to the medial nucleus accumbens. *Brain Res.* **300:** 157–160.

Kane J. 1993. Newer antipsychotic drugs. A review of their pharmacology and therapeutic potential. *Drugs* **46:** 585–593.

Kelley A.E. and Berridge K.C. 2002. The neuroscience of natural rewards: Relevance to addictive drugs. *Neurosci.* **22:** 3303–3305.

Knoll B. and Drescher U. 2002. Ephrin-As as receptors in topographic projections. *Trends Neurosci.* **25:** 145–149.

Koob G.F. 1992. Drugs of abuse: Anatomy, pharmacology, and function of reward pathways. *Trends Pharmacol. Sci.* **13:** 177–184.

Kostrzewa R.M., Brus R., Kalbfleisch J.H., Perry K.W., and Fuller R.W. 1994. Proposed animal model of attention deficit hyperactivity disorder. *Brain Res. Bull.* **34:** 161–167.

Kubota Y. and Kawaguchi Y. 1993. Spatial distributions of chemically identified intrinsic neurons in relation to patch and matrix compartments of rat neostriatum. *J. Comp. Neurol.* **332:** 499–513.

Kullander K. and Klein R. 2002. Mechanisms and functions of Eph and ephrin signalling. *Nat. Rev. Mol. Cell Biol.* **3:** 475–486.

Lanciego J., Wouterlood F., Erro E., Arribas J., Gonzalo N., Urra X., Cervantes S., and Gimenez-Amaya J. 2000. Complex brain circuits studied via simultaneous and permanent detection of three transported neuroanatomical tracers in the same histological section. *J. Neurosci. Methods* **103:** 127–135.

Lebel M., Gauthier Y., Moreau A., and Drouin J. 2001. Pitx3 activates mouse tyrosine hydroxylase promoter via a high-affinity binding site. *J. Neurochem.* **77:** 558–567.

Levey A.I., Hersch S.M., Rye D.B., Sunahara R.K., Niznik H.B., Kitt C.A., Price D.L., Maggio R., Brann M.R., and Ciliax B.J. 1993. Localization of D1 and D2 dopamine receptors in brain with subtype-specific antibodies. *Proc. Natl. Acad. Sci.* **90:** 8861–8865.

Lewis D.A. and Levitt P. 2002. Schizophrenia as a disorder of neurodevelopment. *Annu. Rev. Neurosci.* **25:** 409–432.

Lindvall O. and Bjorklund A. 1983. Dopamine- and norepinephrin-containing neuron systems: Their anatomy in the rat brain. In *Chemical neuroanatomy* (ed. P.C. Emson), pp. 229–255. Raven Press, New York.

Lindvall O., Bjorklund A., Moore R.E., and Stenevi U. 1974. Mesencephalic dopamine neurons projecting to neocortex. *Brain Res.* **81:** 325–331.

Ma Y., Feigin A., Dhawan V., Fukuda M., Shi Q., Greene P., Breeze R., Fahn S., Freed C., and Eidelberg D. 2002. Dyskinesia after fetal cell transplantation for Parkinsonism: A PET study. *Ann. Neurol.* **52:** 628–634.

Maler L., Fibiger H.C., and McGeer P.L. 1973. Demonstration of the nigrostriatal projection by silver staining after nigral injections of 6-hydroxydopamine. *Exp. Neurol.* **40:** 505–515.

Malone M.A., Kershner J.R., and Swanson J.M. 1994. Hemispheric processing and methylphenidate effects in attention-deficit hyperactivity disorder. *J. Child Neurol.* **9:** 181–189.

Mangeot S.D., Miller L.J., McIntosh D.N., McGrath-Clarke J., Simon J., Hagerman R.J., and Goldson E. 2001. Sensory modulation dysfunction in children with attention-deficit-hyperactivity disorder. *Dev. Med. Child Neurol.* **43:** 399–406.

Martin L.J., Hadfield M.G., Dellovade T.L., and Price D.L. 1991. The striatal mosaic in primates: Patterns of neuropeptide immunoreactivity differentiate the ventral striatum from the dorsal striatum. *Neuroscience* **43:** 397–417.

Martinez S., Crossley P.H., Cobos I., Rubenstein J.L.R., and Martin G.R. 1999. FGF8 induces formation of an ectopic isthmic organizer and isthmocerebellar development via a repressive effect on Otx2 expression. *Development* **126:** 1189–1200.

Martinez-Barbera J.P., Signore M., Boyl P.P., Puelles E., Acampora D., Gogoi R., Schubert F., Lumsden A., and Simeone A. 2001. Regionalisation of anterior neuroectoderm and its competence in responding to forebrain and midbrain inducing activities depend on mutual antagonism between OTX2 and GBX2. *Development* **128:** 4789–4800.

McGeer P.L., McGeer E.G., and Suzuki J.S. 1977. Aging and extrapyramidal function. *Arch. Neurol.* **34:** 33–35.

McGlashan T.H. and Hoffman R.E. 2000. Schizophrenia as a disorder of developmentally reduced synaptic connectivity. *Arch. Gen. Psychiatry* **57:** 637–648.

McMahon A.P., Joyner A.L., Bradley A., and McMahon J.A. 1992. The midbrain-hindbrain phenotype of Wnt-1-/Wnt-1- mice results from stepwise deletion of engrailed-expressing cells by 9.5 days postcoitum. *Cell* **69:** 581–595.

Meltzer H.Y. 1999. Treatment of schizophrenia and spectrum disorders: Pharmacotherapy, psychosocial treatments, and neurotransmitter interactions. *Biol. Psychiatry* **46:** 1321–1327.

Meredith G.E., Pattiselanno A., Groenewegen H.J., and Haber S.N. 1996. Shell and core in monkey and human nucleus accumbens identified with antibodies to calbindin-D28k. *J. Comp. Neurol.* **365:** 628–639.

Moore R. and Bloom F. 1978. Central catecholamine neuron systems: Anatomy and physiology of the dopamine systems. *Annu. Rev. Neurosci.* **1:** 129–169.

Murai K.K. and Pasquale E.B. 2003. 'Eph'ective signaling: Forward, reverse and crosstalk. *J. Cell Sci.* **116:** 2823–2832.

Nakamura S.-i., Ito Y., Shirasaki R., and Murakami F. 2000. Local directional cues control growth polarity of dopaminergic axons along the rostrocaudal axis. *J. Neurosci.* **20:** 4112–4119.

Nelson E.L., Liang C.-L., Sinton C.M., and German D.C. 1996. Midbrain dopaminergic neurons in the mouse: Computer-assisted mapping. *J. Comp. Neurol.* **369:** 361–371.

Nestler E.J. 2004. Molecular mechanisms of drug addiction. *Neuropharmacology* (suppl. 1) **47:** 24–32.

Nieoullon A. and Coquerel A. 2003. Dopamine: A key regulator to adapt action, emotion, motivation and cognition. *Curr. Opin. Neurol.* (suppl. 2) **16:** S3–9.

Nunes I., Tovamasian L.T., Silva R.M., Burke R.E., and Goff S.P. 2003. Pitx3 is required for development of substantia nigra dopaminergic neurons. *Proc. Natl. Acad. Sci.* **100:** 4245–4250.

Olanow C.W. and Tatton W.G. 1999. Etiology and pathogenesis of Parkinson's disease. *Annu. Rev. Neurosci.* **22:** 123–144.

Panhuysen M., Vogt Weisenhorn D.M., Blanquet V., Brodski C., Heinzmann U., Beisker W., and Wurst W. 2004. Effects of Wnt1 signalling on proliferation in the developing mid-/hindbrain region. *Mol. Cell. Neurosci.* **26:** 101–111.

Perrin F.E. and Stoeckli E.T. 2000. Use of lipophilic dyes in studies of axonal pathfinding in vivo. *Microsc. Res. Technique* **48:** 25–31.

Perrone-Capano C. and Di Porzio U. 2000. Genetic and epigenetic control of midbrain dopaminergic neuron development. *Int. J. Dev. Biol.* **44:** 679–687.

Perrone-Capano C., Da Pozza P., and Di Porzio U. 2000. Epigenetic cues in midbrain dopaminergic neuron development. *Neurosci. Biobehav. Rev.* **24:** 119–124.

Prakash N. and Wurst W. 2004. Specification of midbrain territory. *Cell Tissue Res.* **318:** 5–14.

Riddle R. and Pollock J.D. 2003. Making connections: The development of mesencephalic dopaminergic neurons. *Dev. Brain Res.* **147:** 3–21.

Rowitch D.H. and McMahon A.P. 1995. Pax-2 expression in the murine neural plate precedes and encompasses the expression domains of Wnt-1 and En-1. *Mech. Dev.* **52:** 3–8.

Sacchetti P., Mitchell T.R., Granneman J.G., and Bannon M.J. 2001. Nurr1 enhances transcription of the human dopamine transporter gene through a novel mechanism. *J. Neurochem.* **76:** 1565–1572.

Salamone J.D. and Correa M. 2002. Motivational views of reinforcement: Implications for understanding the behavioral functions of nucleus accumbens dopamine. *Behav. Brain Res.* **137:** 3–25.

Saucedo-Cardenas O., Quintana-Hau J.D., Le W.-D., Smidt M.P., Cox J.J., De Mayo F., Burbach J.P.H., and Conneely O.M. 1998. Nurr1 is essential for the induction of the dopaminergic phenotype and the survival of ventral mesencephalic late dopaminergic precursor neurons. *Proc. Natl. Acad. Sci.* **95:** 4013–4018.

Schwarz M., Alvarez-Bolado G., Urbanek P., Busslinger M., and Gruss P. 1997. Conserved biological function between Pax-2 and Pax-5 in midbrain and cerebellum development: Evidence from targeted mutations. *Proc. Natl. Acad. Sci.* **94:** 14518–14523.

Seeman M. 1981. Pharmacologic features and effects of neuroleptics. *Can. Med. Assoc. J.* **125:** 821–826.

Seeman P. and Van Tol H. 1994. Dopamine receptor pharmacology. *Trends Pharmacol. Sci.* **15:** 264–270.

Seiber B.-A., Kuzmin A., Canals J.M., Danielsson A., Paratcha G., Arenas E., Alberch J., Ogren S.O., and Ibanez C.F. 2004. Disruption of EphA/ephrin-A signaling in the nigrostriatal system reduces dopaminergic innervation and dissociates behavioural responses to amphetamine and cocaine. *Mol. Cell. Neurosci.* **26:** 418–428.

Seiger A. and Olson L. 1973. Late prenatal ontogeny of central monoamine neurons in the rat: Fluorescence histochemical observations. *Z. Anat. Entwickl.-Gesch.* **140:** 281–318.

Simon H., Le Moal M., Galey D., and Cardo B. 1976. Silver impregnation of dopaminergic systems after radiofrequency and 6-OHDA lesions of the rat ventral tegmentum. *Brain Res.* **115:** 215–231.

Simon H.H., Bhatt L., Gherbassi D., Sgado P., and Alberi L. 2003. Midbrain dopaminergic neurons: Determination of their developmental fate by transcription factors. *Ann. N.Y. Acad. Sci.* **991:** 36–47.

Smidt M.P., Smits S., and Burbach J.P.H. 2004. Homeobox gene Pitx3 and its role in the development of dopamine neurons of the substantia nigra. *Cell Tissue Res.* **318:** 35–43.

Smidt M.P., Asbreuk C.H.J., Cox J.J., Chen H., Johnson R.L., and Burbach J.P.H. 2000. A second independent pathway for developement of mesencephalic dopaminergic neurons requires Lmx1b. *Nat. Neurosci.* **3:** 337–341.

Smidt M.P., van Schaick H.S.A., Lanctot C., Tremblay J.J., Cox J.J., van der Kleij A.A.M., Wolterink G., Drouin J., and Burbach J.P.H. 1997. A homeodomain gene Ptx3 has highly restricted brain expression in mesencephalic dopaminergic neurons. *Proc. Natl. Acad. Sci.* **94:** 13305–13310.

Snyder S.H. 1972. Catecholamines in the brain as mediators of amphetamine psychosis. *Arch. Gen. Psychiatry* **27:** 169–179.

Spanagel R. and Weiss F. 1999. The dopamine hypothesis of reward: Past and current status. *Trends Neurosci.* **22:** 521–527.

Specht L.A., Pickel V.M., Joh T.H., and Reis D.J. 1981. Light-microscopic immunocytochemical localization of tyrosine hydroxylase in prenatal rat brain. I. Early ontogeny. *J. Comp. Neurol.* **199:** 233–253.

Stanwood G.D., Washington R.A., Shumsky J.S., and Levitt P. 2001. Prenatal cocaine exposure produces consistent developmental alterations in dopamine-rich regions of the cerebral cortex. *Neuroscience* **106:** 5–14.

Tornqvist N., Hermanson E., Perlmann T., and Stromberg I. 2002. Generation of tyrosine hydroxylase-immunoreactive neurons in ventral mesencephalic tissue of Nurr1 deficient mice. *Dev. Brain Res.* **133:** 37–47.

Van der Kooy D. 1979. The organization of the thalamic, nigral, and raphe cells projection to the medial vs lateral caudate-putamen in rat. A fluorescent retrograde double labeling study. *Brain Res.* **169:** 381–387.

Veening J., Cornelissen F., and Lieven P. 1980. The topical organization of the afferents to the caudatoputamen of the rat. A horseradish peroxidase study. *Neuroscience* **5:** 1253–1268.

Vercelli A., Repici M., Garbossa D., and Grimaldi A. 2000. Recent techniques for tracing pathways in the central nervous system of developing and adult animals. *Brain Res. Bull.* **51:** 11–28.

Vernier P., Moret F., Callier S., Snapyan M., Wersinger C., and Sidhu A. 2004. The degeneration of dopamine neurons in Parkinson's disease: Insights from embryology and evolution of the mesostriatocortical system. *Ann. N.Y. Acad. Sci.* **1035:** 231–249.

Voorn P., Kalsbeek A., Jorritsma-Byham B., and Groenwegen H.J. 1988. The pre- and postnatal development of the dopaminergic cell groups in the ventral mesencephalon and the dopaminergic innervation of the striatum in the rat. *Neuroscience* **25:** 857–887.

Wallen A., Zetterstrom R.H., Solomin L., Arvidsson M., Olson L., and Perlmann T. 1999. Fate of mesencephalic AHD2-expressing dopamine progenitor cells in NURR1 mutant mice. *Exp. Cell Res.* **253:** 737–746.

Wang Z., Benoit G., Liu J., Prasad S., Aarnisalo P., Liu X., Xu H., Walker N.P., and Perlmann T. 2003. Structure and function of Nurr1 identifies a class of ligand-independent nuclear receptors. *Nature* **423:** 555–560.

Wise R.A. and Bozarth M.A. 1987. A psychomotor stimulant theory of addiction. *Psychol. Rev.* **94:** 469–492.

Witta J., Baffi J.S., Palkovits M., Mezey E., Castillo S.O., and Nikodem V.M. 2000. Nigrostriatal innervation is preserved in Nurr1-null mice, although dopaminergic neuron precursors are arrested from terminal differentiation. *Mol. Brain Res.* **84:** 67–78.

Ye W., Shimamura K., Rubenstein J.L.R., Hynes M., and Rosenthal A. 1998. FGF and Shh signals control dopaminergic and serotonergic cell fate in the anterior neural plate. *Cell* **93:** 755–766.

Ye W., Bouchard M., Stone D., Liu X., Vella F., Lee J., Nakamura H., Ang S.-L., Busslinger M., and Rosenthal A. 2001. Distinct regulators control the expression of the mid-hindbrain organizer signal Fgf8. *Nat. Neurosci.* **4:** 1175–1181.

Yuan W., Zhou L., Chen J.H., Wu J.Y., Rao Y., and Ornitz D.M. 1999. The mouse SLIT family: Secreted ligands for ROBO expressed in patterns that suggest a role in morphogenesis and axon guidance. *Dev. Biol.* **212:** 290–306.

Yue Y., Widmer D.A.J., Halladay A.K., Cerretti D.P., Wagner G.C., Dreyer J.-L., and Zhou R. 1999. Specification of distinct dopaminergic neural pathways: Roles of the Eph family receptor EphB1 and ligand Ephrin-B2. *J. Neurosci.* **19:** 2090–2101.

Yue Y., Chen Z.Y., Gale N.W., Blair-Flynn J., Hu T.J., Yue X., Cooper M., Crockett D.P., Yancopoulos G.D., Tessarollo L., and Zhou R. 2002. Mistargeting hippocampal axons by expression of a truncated Eph receptor. *Proc. Natl. Acad. Sci.* **99:** 10777–10782.

Zetterstrom R.H., Solomin L., Jansson L., Hoffer B.J., Olson L., and Perlmann T. 1997. Dopamine neuron agenesis in Nurr1-deficient mice. *Science* **276:** 248–249.

Zhou R. 1997. Regulation of topographic projection by the Eph family receptor Bsk (EphA5) and its ligands. *Cell Tissue Res.* **290:** 251–259.

———. 1998. The Eph family receptors and ligands. *Pharmacol. Ther.* **77:** 151–181.

PART 3

Cell Biology and Pharmacology

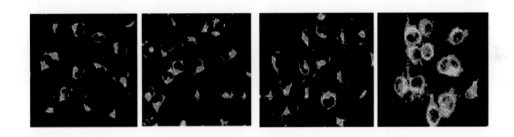

Transporter Structure and Function

Gary Rudnick

Department of Pharmacology, Yale University, New Haven, Connecticut 06520-8066

ABSTRACT

Many drugs of abuse target transporters for the neurotransmitters dopamine, serotonin, and norepinephrine. These proteins are located in the presynaptic plasma membrane of neurons that release biogenic amines, and they function to transport the transmitters back into the cells from which they were released. The transporters are likely to function by undergoing conformational changes that expose a binding site alternately to the extracellular and cytoplasmic faces of the plasma membrane. By allowing the conformational change to occur only when certain solutes are bound, the transport proteins can couple the energy in the transmembrane electrical potential and in ion concentration gradients to the influx of the neurotransmitter. The basic biochemistry of transporter structure and function described in this chapter provides a conceptual framework for understanding the process of neurotransmitter transport and how it is affected by drugs of abuse.

INTRODUCTION

Neurotransmitters released from nerve terminals act on presynaptic and postsynaptic receptors. The time course and extent of their actions are regulated by transport proteins in the plasma membrane of presynaptic cells, and sometimes in postsynaptic or glial plasma membranes. The recycling and storage of released neurotransmitters in synaptic vesicles rely on the transporters in both the plasma membrane and synaptic vesicle membrane. In the case of the biogenic amine neurotransmitters serotonin, dopamine, and norepinephrine (5-HT, DA, and NE, respectively) these proteins are targets for drugs, both therapeutics and drugs of abuse. To appreciate the mechanism by which these drugs act, it is essential to understand the basic structure and function of the transporters involved.

GOALS OF THE CHAPTER

This chapter lays the groundwork for the basic biochemistry of transport and the specific structural and functional characteristics of biogenic amine transporters in the plasma membrane, with particular focus on the 5-HT transporter (SERT) as a drug target and model for DA and NE transporters (DAT and NET, respectively). The many factors that drive transport and the importance of protein structure for transporter function are presented. The mechanisms by which amphetamine induces release of vesicular neurotransmitter through nonexocytotic means are discussed at the end of the chapter.

DRUGS THAT AFFECT NEUROTRANSMITTER TRANSPORT

Antidepressants Selectively Inhibit SERT and NET

The intense interest in biogenic amine transporters stems from their sensitivity to modulation by both therapeutic drugs and drugs of abuse. SERT is the target of many drugs used clinically in the treatment of depression and other psychiatric disorders (Schloss and Henn 2004). The prototype antidepressant drug, imipramine, was discovered to have antidepressant properties when tested as a potential antipsychotic medication. Many other compounds based on the tricyclic structure of imipramine were also found to have antidepressant properties as well, establishing a class of structurally related drugs known collectively as tricyclic antidepressants. Some of these, such as desipramine, were found to be more selective as NET inhibitors, leading to a divergent focus on both SERT and NET as potential therapeutic targets (Leonard 1999). Compounds designed specifically to inhibit SERT or NET were found to effectively treat depression, and the most selective compounds combined effective transport inhibition with a relative lack of side effects (Brosen 1993).

Cocaine Blocks Neurotransmitter Transporters

In wide use for centuries by indigenous peoples of South America, cocaine became problematic when introduced, as a purified substance, into the modern, industrialized world. As a stimulant, cocaine found its way into popular products such as Coca Cola, and it was also used medically as a local anesthetic. The local anesthetic property of cocaine resulted from its ability to block sodium channels, but this did not account for the stimulant properties responsible for its popular use (Byck and Van Dyke 1982). These properties are the result of avid cocaine binding to SERT, DAT, and NET—the biogenic amine neurotransmitter transporters of the plasma membrane (Gu et al. 1994). In fact, these transporters constitute the highest-affinity targets for cocaine in the body. On binding to these proteins, cocaine blocks their ability to transport neurotransmitters back into neurons after their release. Thus, in a cocaine-treated animal or human, transmitter levels are elevated, and remain so, long after a release event.

The stimulant properties of cocaine result from this elevated transmitter level, but in addition to acting as a stimulant, cocaine also acts as a potent activator of the brain's reward pathway by increasing extracellular transmitters in regions such as the ventral tegmental area (Koob 1992). This is believed to result primarily from elevated DA levels, but recent studies with knockout animals have shown that the rewarding properties of cocaine are observed even in DAT knockout mice, suggesting that other transporters are also involved (Sora et al. 1998). Mice lacking both DAT and SERT fail to respond to cocaine as a reward, although SERT knockout mice still respond, suggesting that inhibition of both transporters is involved in cocaine's rewarding properties (Sora et al. 2001).

Amphetamines Cause Release of Neurotransmitters

Similar increases in extracellular biogenic amine neurotransmitter levels result from amphetamine action (Sulzer et al. 1993). However, the ability of amphetamine and its derivatives to increase synaptic levels of biogenic amines is not caused by transporter inhibition but by a complicated interaction with both plasma membrane and vesicular transport systems. The result is a nonexocytotic release of transmitter that increases synaptic levels and depletes the neuron. The details of amphetamine action, as it is currently understood, are covered later in this chapter.

DRIVING FORCES FOR TRANSPORT OF NEUROTRANSMITTERS

Neurotransmitter transporters accumulate their substrate neurotransmitters to cytoplasmic and intravesicular concentrations hundreds of times higher than those present outside the cell. Therefore, the process requires energy. As we will see, the energy ultimately comes from ATP hydrolysis by ion pumps that generate transmembrane ion gradients and electrical potentials. It is these gradients and potentials that are utilized by the transporters as an energy source for accumulating neurotransmitters from the low concentration extracellularly to a higher intracellular level. As a framework for understanding how the transporters can accomplish this task, we discuss below the basic phenomena underlying membrane transport.

The Membrane Is a Barrier for the Entry and Exit of Neurotransmitters

The basic structure of the membrane bilayer causes it to be a barrier for hydrophilic solutes such as neurotransmitters. This is because the ability of neurotransmitters to dissolve in aqueous solutions requires them to form hydrogen bonds with water (Fig. 1A). Substances that are incapable of interacting with water in this way, such as oils, are insoluble (Fig. 1B). In the case of an oily, or hydrophobic, solute, the high energy required to separate water molecules to make room for a molecule of that solute is not compensated by formation of hydrogen bonds with water. For a polar solute, such as a neurotransmitter, the energy required to make room for the molecule (mostly the breaking of hydrogen bonds between water molecules) is recovered by the formation of hydrogen bonds between the neurotransmitter molecule and water. To move the solute from water to the interior of a lipid membrane requires breaking all those favorable bonds again. However, the energy is not recovered, because the interior of the bilayer does not have the hydrogen bonding ability of water. Therefore, it is extremely unfavorable energetically to move a hydrophilic molecule from aqueous solution to the membrane interior. This energy barrier effectively keeps neurotransmitters from crossing the membrane, unless their transport is mediated by a protein (Deamer and Bramhall 1986).

These same forces are responsible for the formation and maintenance of the membrane bilayer itself (Tanford 1973). Amphipathic molecules such as phospholipids and detergents contain a polar head group and a nonpolar hydrocarbon tail. To minimize the energy required to dissolve such compounds in aqueous solution, they organize themselves

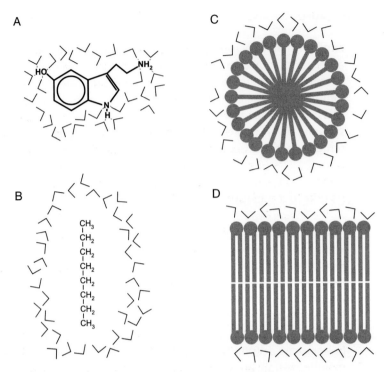

FIGURE 1. Interactions between water molecules and solutes. (*A*) Neurotransmitter molecules (5-HT is shown) are freely soluble because of the H-bond interactions between water and the polar hydroxyl, imino, and primary amine groups. (*B*) Hydrocarbons are not soluble because they do not form H bonds with water. (*C*) A detergent micelle interacts with water by forming H bonds at its polar surface without presenting exposed hydrocarbon regions to water. (*D*) Phospholipids, like detergents, form structures to present only their polar head groups to water, but the bulkier side chains stabilize a bilayer structure rather than a micelle.

into structures that contain all of the polar groups on the surface and all of the nonpolar tails in the interior, shielded from water. For a detergent, this structure is a micelle, which is roughly spherical (Fig. 1C). Detergent micelles have been useful for isolation of membrane proteins because they can associate with an integral membrane protein to satisfy the requirements of the protein to be in a lipid-like environment. At the same time, they expose only their polar head groups to keep the complex in solution. For a phospholipid, the two fatty acid tails per head group are too bulky to pack into a spherical micelle, and the optimal structure for keeping the tails shielded from water is the bilayer (Fig. 1D).

Forces Leading to Asymmetrical Distribution of Ions, Acids, and Bases

Even in a vesicle composed only of phospholipids, there can be driving forces for accumulation of solute inside the vesicle. The same forces that lead to accumulation under these conditions can be utilized by transport proteins embedded in cell membranes. However, as we will see, transporters have the ability to couple other driving forces to solute transport.

Transmembrane pH Differences Can Act as Driving Forces for Transport

An important mechanism for the distribution of organic acids and bases is nonionic diffusion. These compounds exist in protonated and nonprotonated forms. For a base, the protonated form has a positive charge and for an acid, the nonprotonated form has a negative charge. The charged forms are polar, interact strongly with water, and cross lipid bilayers very slowly. However, the neutral forms (the protonated acid and the unprotonated base) can cross the bilayer much faster, and in many cases, come to an equilibrium in which equal concentrations of the neutral form exist on both sides of the membrane. Because the distribution between neutral and charged forms of these compounds depends on pH, a pH difference across the membrane can lead to a concentration difference of the acid or base (Fig. 2).

For example, amphetamine is a weak base with a pKa of 9.8. The pKa is the pH at which half of the molecules are protonated. At pH 7.8, the cationic, protonated form of amphetamine will be 100 times greater than the neutral form. At pH 5.8, the protonated to neutral ratio rises to 10,000. For a membrane vesicle with an internal pH of 5.8, in a medium with a pH of 7.8, intravesicular amphetamine will be concentrated to 100 times the concentration in the medium. Assuming that the neutral form is in equilibrium across the membrane and at the same concentration inside and out, the total amphetamine concentration in the medium will be 101, assuming an arbitrary value of 1 for the neutral form. By the same reckoning, the concentration will be 10,001 on the inside, leading to a concentration gradient of 100-fold. Similarly, a weak acid would be concentrated within a vesicle whose internal pH was higher than that of the medium.

Transmembrane Electrical Potentials and Concentration Gradients Affect Ion Distribution

If a solute is permeant, two important forces will determine its distribution across the membrane. The first of these is the concentration difference of that solute between the

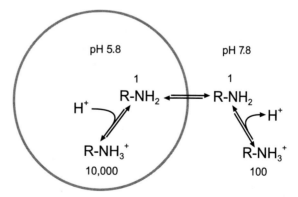

FIGURE 2. Accumulation of a weakly basic amine on the acidic side of the membrane. An amine (R-NH$_3^+$) such as amphetamine equilibrates rapidly across the membrane in its neutral form R-NH$_2$. The total concentration on each side is determined largely by the distribution between the neutral and charged forms. Because the vesicle interior in this example has a low pH, the concentration of internal amine is higher than the external concentration.

two sides of the membrane. The solute will spontaneously flow from the side with the higher concentration to the side with the lower concentration, because of entropy of mixing. A quantitative statement of the free energy available in the gradient is

$$\Delta G = RT \ln \frac{[A]_i}{[A]_o}$$

where the free energy, ΔG, depends on the natural log of the concentration ratio between the internal and external concentrations of solute A. R is the gas constant (1.987 cal deg^{-1} mol^{-1}) and T is the temperature in degrees Kelvin. The ΔG is favorable (negative) when the internal concentration is lower than the external one.

The other driving force is the transmembrane electrical potential difference, $\Delta\psi$, which is measured in millivolts (in minus out). The electrical driving force on a permeant ion is $\Delta G = Z_A F\Delta\psi$, where Z is the charge on the ion and F is Faraday's constant (23.061 cal mV^{-1} mol^{-1}). At equilibrium, where the two forces are balanced,

$$\Delta G = 0 \quad \text{and} \quad RT \ln \frac{[A]_i}{[A]_o} = -Z_A F\Delta\psi$$

Thus, a concentration difference can be opposed by a potential, leading to an equilibrium condition in which the solute is at unequal concentrations. This is defined as the Nernst equation, evaluated at approximately 29°C, where

$$\Delta\psi = \frac{60}{Z_A} \log \frac{[A]_o}{[A]_i}$$

Thus, for a monovalent ion, a membrane potential of 60 mV is sufficient to maintain a concentration gradient of tenfold. A practical example of this would be a diffusion potential. If a cell or vesicle contains ten times higher K$^+$ concentration than the medium, and the membrane is very permeable to K$^+$, then K$^+$ will begin to exit from the vesicle, and as it does, it will generate a potential, interior negative that will increase with the K$^+$ efflux until the potential is large enough that the electrical driving force pulling K$^+$ back into the vesicle is as great as the concentration gradient driving it out. That potential will be about 60 mV at normal ambient temperatures.

Ionophores Facilitate the Transport of Ions

One way to increase the permeability of a membrane to an ion, such as K$^+$, is with the use of ionophores (Pressman 1973). Some ionophores, such as gramicidin, form aqueous channels across the membrane, but these compounds are rather unselective. Many ionophores act by shuttling back and forth across the membrane, carrying an ion with them. Electrogenic ionophores such as valinomycin can cross the membrane either empty or with a K$^+$ ion. Others, such as nigericin, cannot cross without carrying an ion and consequently are electrically silent. The difference between pore-forming ionophores and carrier ionophores is similar to the difference between biological channels and the transporters we refer to as carriers and pumps.

Ion-conducting pores, whether they are in a biological channel or an ionophore, provide an aqueous or polar pathway through which ions can move across the membrane without any change in the channel itself. For carriers, either ionophores or transporter proteins, the carrier is an active participant in the transport process. This is represented by the movement of an ionophore such as valinomycin or nigericin from one side of the membrane to the other, or in the case of a transporter protein, by conformational changes that accompany solute transport (Jardetzky 1966). In both cases, however, a distinct kinetic step must occur for each ion or molecule transported, in addition to the binding and dissociation of the solute. This leads to a characteristic kinetic behavior that differentiates channels from carriers. With carriers, it is common that the presence of a substrate on one side of the membrane (e.g., the inside of a vesicle) facilitates the transport of a solute from the other side (inward flux in this example). The clearest case of this is for a strict exchanger, such as nigericin, which needs to transport an ion out of the vesicle to transport another ion to the inside. For a channel, the presence of internal ions that permeate the channel should not enhance, and may inhibit, influx of other ions.

Transporters Require Conformational Changes for Function

In contrast to channels, which can pass many ions without changing conformation, carriers require conformational changes as an integral step in the transport process. We envision these conformational changes as changing the accessibility of a substrate binding site from one side of the membrane to the other. In the simplest type of carrier, transport would involve a four-step cycle (Fig. 3): (1) binding of solute to the transporter from the external

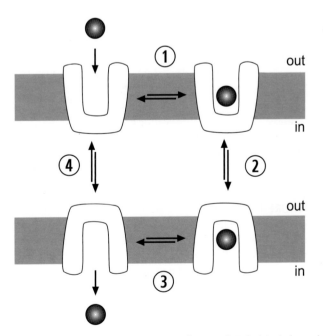

FIGURE 3. Simple four-step model for transport. (1) A solute molecule binds from the external medium. (2) A conformational change exposes the bound solute to the internal medium. (3) The solute dissociates internally. (4) The unoccupied transporter returns to its original state.

solution, (2) changing the accessibility of the bound solute from external to internal, (3) dissociation of the solute on the inside, and (4) return of the transporter to the initial conformation. One example of this simple type of transporter, or uniporter, is the glucose transporter of red blood cells, GLUT1 (Mueckler 1994). The process of uniport is frequently referred to as mediated or facilitated diffusion. The driving forces on such a system are the two that were discussed above—concentration gradients and membrane potentials—and the only influence the transporter can exert on these forces is to determine the form of the solute that is transported. For example, a solute that existed in different protonation states might be a substrate for transport in only one of those states.

Carriers Use Solute Gradients as Driving Forces

Although channels allow ion flux at rates enormously faster than those of carrier-mediated transport, carriers have the unique ability to couple the downhill movement of one solute to the uphill movement of another (see Fig. 4). For example, transporters for the neurotransmitters GABA, glycine, NE, and 5-HT have all been shown to drive the energetically uphill movement of a neurotransmitter into the cell by coupling it to the down-

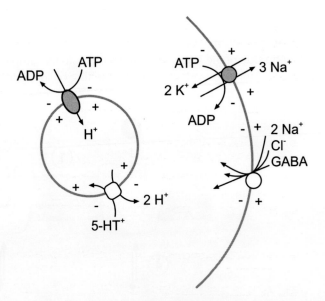

FIGURE 4. Coupling of ion gradients to neurotransmitter transport. The Na^+/K^+-ATPase of the plasma membrane uses the energy of hydrolysis of ATP to pump three Na^+ ions out of the cell and two K^+ ions in. The difference in charge movement creates a membrane potential (negative inside) that leads to passive Cl^- efflux and a Cl^- concentration gradient. The GABA transporter couples the downhill influx of two Na^+ ions and one Cl^- ion to the uphill transport of GABA—a process that also leads to positive charge entry. Thus, the Na^+ and Cl^- gradients and the membrane potential all act as driving forces for GABA uptake. At the synaptic vesicle membrane, the vacuolar ATPase hydrolyzes ATP and uses the energy to pump H^+ ions into the vesicle lumen, making it acidic and electrically positive relative to the cytoplasm. The vesicular monoamine transporter (VMAT) allows two H^+ ions to exit the vesicle in exchange for one molecule of biogenic amine, such as 5-HT. Both the membrane potential and the pH difference act as driving forces, concentrating 5-HT in the vesicle.

hill movement of Na^+ and Cl^- into the cell (Nelson and Rudnick 1979; Radian and Kanner 1983; Gu et al. 1996; Roux and Supplisson 2000). At the level of synaptic vesicles, the vesicular monoamine transporter (VMAT) couples the uphill influx of NE, DA, or 5-HT to the downhill movement of H^+ from the acidic vesicle interior to the cytoplasm (Johnson et al. 1979). Carriers are able to couple multiple solute fluxes through simple modifications of the four-step mechanism for glucose transport and through the imposition of rules that restrict the transitions between transporter states.

Symport Moves Two or More Solutes in the Same Direction

For the movement of two or more solutes in the same direction, or symport, we imagine that the transporter contains a binding site that simultaneously binds all substrates, for example, GABA, Na^+, and Cl^-. How these solutes bind, and the order of their binding, is not relevant to the overall energetics of the process. Once the substrates are bound, the carrier, or symporter, can undergo the conformational change that exposes the binding site, with all its bound solutes, to the other side of the membrane. After dissociation of the solutes, the symporter can return to its original conformation, ready to bind more substrates. Although this scheme provides a pathway for coordinated movement of GABA, Na^+, and Cl^-, it does not ensure coupling of the fluxes unless certain rules are imposed to restrict transitions between states. For a symporter such as the one in this example, the rules would restrict the conditions for interconversion of the inward- and outward-facing forms of the transporter. These transitions should occur only when the binding site is fully occupied with GABA, Na^+, and Cl^-, or when it is completely empty. If the symporter could move between states with only one, or even two, substrates bound, this would lead to uncoupled solute flux that would dissipate the transmembrane solute gradients without harnessing them for useful work.

An Antiport Couples Solute Fluxes in Opposite Directions and Follows Different Rules

In this case, the protein binds one solute on one side of the membrane and releases it on the other side, and then in a second step binds a second solute and releases it on the original side. For example, VMAT could bind an H^+ ion inside the vesicle, release it in the cytoplasm, then bind a molecule of DA, NE, or 5-HT from the cytoplasm, and release it into the vesicle. The rules required for coupling in an antiporter are that (1) the interconversion of internal- and external-facing forms should occur only when the binding site is occupied and (2) substrates moving in opposite directions should not be bound simultaneously.

We do not yet understand the mechanism used to implement a binding site that is alternately accessible from either side of the membrane. Two structural models give a hint as to how this might be accomplished. One is the lac permease/GlpT/OxlT structure (Hirai et al. 2002; Abramson et al. 2003; Huang et al. 2003), which suggests that the binding site lies in the well of a large cavity created by transmembrane α-helices. Transition between states would seem likely to involve large movements of these helices relative to one another to close the pathway to the cytoplasmic side and open it up to the periplasmic (external) side. At the other extreme, the putative structure of a glutamate transporter suggests that movements of small helical hairpins may open and close off the binding site from internal and external sides with little movement of the rest of the protein (Yernool et al. 2004). For those transporters that do not belong to the same family as lac permease or the glutamate transporter, we can only guess that the mechanism lies somewhere between those extremes.

Transport Stoichiometry Determines the Magnitude of Accumulation

For a transporter that couples Na$^+$ and substrate influx, as do all of the plasma membrane neurotransmitter transporters, the number of Na$^+$ ions transported is an important determinant of the amount of transmitter accumulation. The ultimate gradient achieved is related to the Na$^+$ gradient by the following general equation:

$$n_{NT}\ 60\ \log\ \frac{[NT]_i}{[NT]_o} = n_{Na}\ 60\ \log\ \frac{[Na^+]_o}{[Na^+]_i} - \Delta\psi\ (n_{Na} + n_{NT}Z_{NT})$$

where NT is the neurotransmitter, n_{NT} is the number of neurotransmitter molecules transported per cycle (usually 1), n_{Na} is the number of Na$^+$ ions per cycle, and Z_{NT} is the net charge on each neurotransmitter (Nicholls and Ferguson 2002). According to this equation, a tenfold Na$^+$ gradient will lead to a tenfold neurotransmitter gradient if n_{NT} and n_{Na^+} are both 1 and if there is no membrane potential. If n_{Na^+}, the number of Na$^+$ ions, is 2, then the gradient becomes 100-fold. This amplification is important for taking up the last traces of transmitter, especially in cases where the Na$^+$ gradient dissipates, as in ischemia. The EAAT3 transporter couples glutamate uptake to three Na$^+$ ions, generating a powerful driving force that can maintain low external levels even if the Na$^+$ gradient decreases. This is important because glutamate is an excitotoxin that can cause neuronal death if external levels rise for an extended period of time.

For an antiporter, such as VMAT, the stoichiometry equation is

$$n_{BA}\ 60\ \log\ \frac{[BA]_i}{[BA]_o} = n_H\ 60\ \log\ \frac{[H^+]_i}{[H^+]_o} + \Delta\psi(n_H - n_{BA}Z_{BA})$$

where BA represents the biogenic amine substrate. Some transporters, such as EAAT3 and SERT, couple neurotransmitter influx to both Na$^+$ influx and K$^+$ efflux. The equations for these transporters are more complex, but the same principles apply. Each additional gradient adds to substrate accumulation as an independent factor multiplied by the previous gradients to yield the final concentration ratio. The full reaction catalyzed by SERT is the symport of one 5-HT$^+$, one Na$^+$, and one Cl$^-$ with the antiport of one K$^+$. The overall equation for 5-HT transport by SERT is

$$60\ \log\ \frac{[\text{5-HT}^+]_i}{[\text{5-HT}^+]_o} = n_{Na}\ 60\ \log\ \frac{[Na^+]_o}{[Na^+]_i} + n_{Cl}\ 60\ \log\ \frac{[Cl^-]_o}{[Cl^-]_i}$$

$$+ n_K\ 60\ \log\ \frac{[K^+]_i}{[K^+]_o} - \Delta\psi(n_{Na} + n_{\text{5-HT}} - n_{Cl} - n_K)$$

This equation contains all these factors known to affect 5-HT$^+$ accumulation. The affinity of 5-HT to the transporter and the order in which 5-HT$^+$ and other ions bind are not relevant to the thermodynamics. These factors have an important role in the kinetics of transport, however, because it is optimal for the K_M of a substrate to be related to its typical concentration, and decreasing the internal affinity for 5-HT$^+$ and Na$^+$ would have the effect of decreasing the rate of the reverse transport reaction.

Pumps Drive Neurotransmitter Transporters

The Na$^+$ and K$^+$ gradients and membrane potentials that drive neurotransmitter transport are generated by the Na$^+$/K$^+$-ATPase, or Na$^+$ pump, of the plasma membrane. The Cl$^-$ gradients that are coupled to transport of many neurotransmitters are generated by passive diffusion of Cl$^-$ in response to the membrane potential. In the synaptic vesicle, the vacuolar ATPase is responsible for pumping H$^+$ ions into the vesicle at the expense of ATP. The mechanism of pump-mediated transport is, in a general sense, similar to that of carrier-mediated transport with the additional chemical coupling of one or more steps in the reaction cycle to a chemical reaction (in most cases, ATP hydrolysis) that pushes the transport cycle in one direction.

NEUROTRANSMITTER TRANSPORTER STRUCTURE AND FUNCTION

Transporter proteins fall into many families (for a review, see Chang et al. 2004). Plasma membrane neurotransmitter transporters are representatives of two families. The neurotransmitter sodium symporter (NSS) transporter family contains transporters for 5-HT, DA, NE, GABA, and glycine, and the dicarboxylate amino acid cation symporter (DAACS) family contains transporters for aspartate and glutamate. The vesicular transporters also fall into two families: The transporters for monoamines, glutamate, and acetylcholine are in the major facilitator superfamily (MFS) family and the vesicular GABA transporter is in the amino acid auxin permease (AAAP) family. For relevance to drug abuse and addiction, the relevant transporters are SERT, DAT, and NET, which form a subset of plasma membrane biogenic amine transporters within the NSS family and the vesicular transporter VMAT from the MFS family.

Closely Related Transporters Function with Different Stoichiometries

The close structural similarity among transport proteins does not guarantee that they catalyze the same reaction. Even within the NSS family, ion-coupling stoichiometry varies among family members (Rudnick 1998). As the name of the family implies, all of the NSS transporters couple substrate transport to Na$^+$ flux. Many, but not all, of the mammalian NSS transporters also couple transport to Cl$^-$, although this coupling can be less than obligatory (Nelson and Rudnick 1982; Radian and Kanner 1983; Zafra and Gimenez 1989; Keynan et al. 1992). At least one prokaryotic member of this family is insensitive to Cl$^-$ (Androutsellis-Theotokis et al. 2003), and some insect NSS transporters can utilize K$^+$ in place of Na$^+$ (Castagna et al. 1998). Among the NSS neurotransmitter transporters for 5-HT, NE, GABA, and glycine, all require Cl⁻ and seem to catalyze symport of neurotransmitter with both Na^+ and Cl^-, but the number of Na^+ ions transported and coupling to K^+ vary among the individual proteins. For example, the GABA transporter GAT-1 catalyzes symport of two Na^+ and one Cl^- ion per molecule of GABA (Radian and Kanner 1983). One of the glycine transporters, GlyT1, has the same stoichiometry but the other, GlyT2, transports three Na^+ ions (Roux and Supplisson 2000), and NET transports one Na^+ and one Cl^- with each NE molecule. SERT is unique among NSS transporters in that it couples 5-HT transport to symport of one Na^+ and one Cl^- but also to antiport of one K^+ (Nelson and Rudnick 1979). The stoichiometry of DAT has never been unequivocally determined, but kinetic experiments suggest involvement of two Na^+ ions (Gu et al. 1994). Thus, in spite of the high degree of sequence similarity, these transporters exhibit a surprising variety of coupling stoichiometries. Apparently, the structures of these transporters provide a framework within which similar conformational changes lead to coupling between ion and neurotransmitter flux, but the differences among transporters are responsible for which ions are coupled and how many are coupled to the transport of each neurotransmitter molecule.

Outside the NSS family is more variation. The glutamate transporter EAAT3 couples glutamate flux to symport of three Na^+ and one H^+ ion and antiport of one K^+ (Zerangue and Kavanaugh 1996). The vesicular transporters utilize a different driving force, the transmembrane pH difference generated by the vacuolar ATPase. Consequently, VMAT couples transport of each biogenic amine into the synaptic vesicle to the antiport of two H^+ ions (Kanner et al. 1980). Because the protonated amine transmitters are the true substrates for VMAT, there is one net charge and two protons transported per amine molecule. Therefore, amine transport is coupled more to the pH difference than to the membrane potential. As we will see, this fact has significant implications for the mechanism of amphetamine action.

Given the stoichiometry of SERT (for example), we can use the principles of alternating access to develop a possible reaction scheme (Rudnick 2002). A simplifying principle is that all substrates that are transported in the same direction are transported in the same step and that substrates transported in opposite directions are transported in different steps of the reaction cycle. Accordingly, we presume that the conformation of SERT with a binding site exposed to the extracellular medium binds 5-HT, Na^+, and Cl^-. Only after all of these solutes are bound is the transporter able to undergo its conformational transition, which closes off access to the binding site from the outside and opens it up from the inside. After Na^+, Cl^-, and 5-HT have dissociated on the inside, K^+ is able to bind, and that triggers the reverse conformational change, allowing K^+ to dissociate on the outside and regenerate the initial form of the transporter. For optimal coupling between the ion gradients and 5-HT accumulation, we would expect that the conformational transition would occur only from the forms with 5-HT, Na^+, and Cl^- bound or with K^+ bound.

Because 5-HT is transported in the cationic form, the net charge associated with SERT-mediated 5-HT transport is zero (Rudnick et al. 1989). Thus it was surprising that, when SERT was expressed in *Xenopus* oocytes, addition of 5-HT led to electrical current flow across the membrane (Mager et al. 1994). The SERT-mediated currents apparently represent ion flux uncoupled from transport (Lin et al. 1996) and are controlled by the association of SERT with syntaxin 1 (Quick 2003). Uncoupled currents were measured also for NET (Galli et al. 1996) and DAT (Sonders et al. 1997), suggesting that these proteins can also function as ion channels in addition to coupled transporters. The significance of the uncoupled current has not been established, but it may be a mechanism for signaling other components of the membrane when the transporters are functioning (Ingram et al. 2002).

The Structure of Neurotransmitter Transporters Is Important for Their Function

As of this writing, there are few high-resolution structures available for neurotransmitter transporters, and none for the NSS family that includes SERT, DAT, and NET. Consequently, all of the information about the structure of these proteins comes from biochemical experiments such as site-directed chemical modification. At present, we have only a rudimentary knowledge of NSS transporter structure. We have a relatively good idea about the general topological orientation of these proteins and a few clues about their overall fold. We know that most of them oligomerize in the membrane and we have identified parts that are likely to constitute the substrate-binding site. We also have some information about conformational changes involving the hydrophilic loops. Nevertheless, the structure, and how it relates to function, is still mostly a mystery that remains to be solved.

Hydropathy analysis of the cloned SERT, NET, and DAT sequences predicted 12 transmembrane domains with both NH_2 and COOH termini in the cytoplasm (Blakely et al. 1991; Hoffman et al. 1991). The predicted second external loop (EL2) between transmembrane (TM) segments 3 and 4 contains predicted sites of glycosylation in all NSS neurotransmitter transporters, and they were shown to be glycosylated in SERT, although the glycosylation was not essential for transport (Tate and Blakely 1994). By generating forms of SERT with a single lysine or cysteine in the predicted external loop structure, we were able to show that, in intact cells expressing SERT mutants, at least two positions in each external loop would react with reagents that could not cross the plasma membrane (Chen et al. 1998). Similar experiments were carried out using mutant forms of SERT that contained a single reactive cysteine residue in the predicted cytoplasmic loop domains. These mutants reacted with cysteine reagents only after the plasma membrane was disrupted but not in intact cells, demonstrating that each of the proposed cytoplasmic loops was exposed on the intracellular side of the plasma membrane (Androutsellis-Theotokis and Rudnick 2002). Thus, the originally proposed topology has been verified experimentally (Fig. 5).

Several approaches have been used to examine the possibility that neurotransmitter transporters associate to form homooligomers. We demonstrated by immunoprecipitation that two separate epitope-tagged forms of SERT were associated in detergent extracts of cells expressing both forms (Kilic and Rudnick 2000). Modification of a SERT mutant with a biotinylating reagent that blocked cocaine binding allowed precipitation of the protein from digitonin-solubilized membranes with immobilized streptavidin. However, no binding activity for a high-affinity cocaine analog was precipitated because modified transporters could no longer bind. Coexpression of this mutant with a SERT mutant resistant to modification led to precipitation of binding activity using the same assay, but expression of the resistant form alone did not lead to precipitation of activity because it could not be biotinylated. Experiments using coexpression of the sensitive and resistant forms of SERT in different molar ratios showed functional association of the two even in the intact cell. The extent of inactivation was consistent with a functional dimer that retained activity until both subunits were modified. Similar experiments with NET demonstrated oligomer formation by immunoprecipitation and inactivation of coexpressed sensitive and resistant forms (Kocabas et al. 2003).

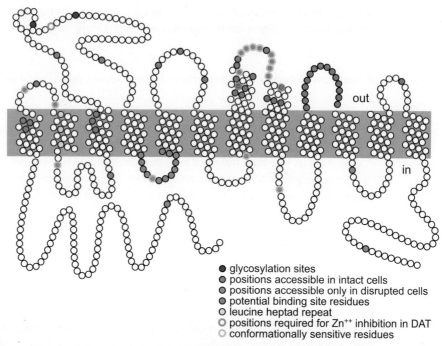

● glycosylation sites
● positions accessible in intact cells
● positions accessible only in disrupted cells
● potential binding site residues
○ leucine heptad repeat
○ positions required for Zn⁺⁺ inhibition in DAT
○ conformationally sensitive residues

FIGURE 5. Topological organization of SERT showing residues localized to extracellular and intracellular loop domains and possible binding sites. Each red or green residue was shown to be accessible in intact cells or only in disrupted cells, respectively. Unlabeled residues were not necessarily tested. Glycosylations sites, residues involved in Zn^{2+} inhibition in DAT, conformationally sensitive residues, and the leucine heptad repeat in TM2 are also shown.

Additional methods have also been applied to the question of transporter oligomerization. Cross-linking has demonstrated association of DAT monomers in intact cells (Hastrup et al. 2003), and DAT oligomerization also has been shown by expression of dominant negative constructs (Torres et al. 2003) and fluorescence resonance energy transfer (FRET) (Sorkina et al. 2003). The latter technique has been applied to SERT (Schmid et al. 2001) as well as GAT-1 (Scholze et al. 2002), and in GAT-1, one face of TM2 has been implicated as responsible for the association (Korkhov et al. 2004). Finally, among glutamate transporters, it has been shown by cross-linking that these proteins form trimers (Yernool et al. 2003; Gendreau et al. 2004). This trimer was also a feature of the high-resolution structure of a prokaryotic glutamate transporter homolog (Yernool et al. 2004). However, despite the abundant evidence for oligomeric transport proteins, no convincing evidence exists that the permeation pathway is made up of more than one subunit.

Within the transmembrane structure of the NSS transporters is little information to indicate which TM segments make contact with one another. However, there is some information about interaction between external loops. A high-affinity site for Zn^{2+} inhibition of DAT was found to require the presence of histidine residues in EL2 and EL4 (Fig. 5) (Norregaard et al. 1998). By histidine mutagenesis in EL4, Zn^{2+} binding sites were created between the proximal and distal ends of the loop in DAT, GAT-1, and SERT (Norregaard et al. 2000; MacAulay et al. 2001; Mitchell et al. 2004). In GAT-1, an interaction between IL4 and the NH_2 terminus was shown to be affected by the binding of syntaxin to the NH_2 terminus (Hansra et al. 2004). It is expected that intramolecular interactions between transmembrane domains and between loops have a key role in the mechanism of transport, and current studies are directed to understand these interactions.

Several studies have been directed toward identifying residues contributing to the substrate and inhibitor binding site and the permeation pathway. In SERT, the external loops were shown not to contain determinants for substrate and inhibitor selectivity (Smicun et al. 1999). In contrast, residues in TM3 were sensitive to modification by external reagents and were protected by substrate and inhibitor binding (Chen et al. 1997). Corresponding residues in other transporters were also shown to be critical for transport (Bismuth et al. 1997; Lee et al. 2000). Studies of TM1 in SERT and GAT-1 also identified residues with the properties expected of binding site residues (Henry et al. 2003; Zhou et al. 2004). Thus, TM1 and TM3 are likely to contribute to the substrate binding site for transporters in this family (Fig. 5). Similar studies with TM2 have not identified residues likely to participate in the binding site or the translocation pathway, although this region is known to contribute to transporter expression, oligomerization, and substrate selectivity (possibly an indirect effect) (Scholze et al. 2002; Wu and Gu 2003; Sato et al. 2004).

A critical element in the transport process is the conformational change that alternates access to the binding site between the two sides of the membrane. Evidence for conformational changes comes from studies of external and cytoplasmic loops (Fig. 5). The reactivity of residues in EL1, EL4, IL1, IL2, and IL3 of SERT is affected by the binding of 5-HT or cocaine (Androutsellis-Theotokis et al. 2001; Ni et al. 2001; Androutsellis-Theotokis and Rudnick 2002; Mitchell et al. 2004). In some cases, binding of 5-HT alone does not affect reactivity in the absence of Na^+ or, in some cases, the presence of both Na^+ and Cl^- (Chen and Rudnick 2000; Androutsellis-Theotokis and Rudnick 2002; Mitchell et al. 2004; Sato et al. 2004). These results suggest that a series of conformational changes occur as symported solutes bind to SERT. Other residues are affected by presence of K^+, which is required to restore the external 5-HT binding site, completing the transport cycle (Nelson and Rudnick 1979; Androutsellis-Theotokis and Rudnick 2002). Some conformational changes are observed only when all of the components of the transport cycle are present, suggesting that an intermediate state in the cycle is responsible for the change in reactivity (Mitchell et al. 2004; Sato et al. 2004).

METHODOLOGICAL CONSIDERATIONS

Many of the advances in understanding the mechanism of neurotransmitter transport resulted from the availability of model membrane vesicle systems in which the transport process could be studied in the absence of other cellular processes. The use of membrane vesicles for the study of membrane transport was pioneered by Kaback (1974) and also by Murer and Hopfer (1974). Further understanding of these systems, particularly of their electrical properties, came from the use of the *Xenopus laevis* oocyte system with cDNA clones encoding the transporters (Hediger et al. 1987; Kavanaugh et al. 1992). As more attention is focused on structure–function relationships, the use of mutagenesis and chemical modification has become widespread (Rudnick 2002). The challenges that remain are to relate a detailed structural model of the transporters to the catalytic mechanism responsible for transport. At this writing, a high-resolution structure is not available for the neurotransmitter transporter family, although a structure (Yernool et al. 2004) exists for a prokaryotic member of the glutamate transporter family, which is not structurally related to the transporters that are targets for drugs of abuse.

AMPHETAMINES AFFECT BOTH PLASMA MEMBRANE AND VESICULAR TRANSPORTERS

As described above, cocaine blocks the reuptake activity of DAT, NET, and SERT and therefore increases the amount and lifetime of synaptic DA, NE, and 5-HT released during normal activity (Shimizu et al. 1992; Saunders et al. 1994; Thomas et al. 1994). In contrast, amphetamine and its derivatives, including MDMA (3,4-methylenedioxymethamphetamine [ecstasy]), do not inhibit the transporters, but rather cause the release of biogenic amines stored in synaptic vesicles. This action results from two properties of the amphetamine molecule: They are highly membrane permeant and are substrates for the plasma membrane transporters.

Different amphetamine derivatives selectively release different amines. Amphetamine and methamphetamine are more effective at releasing DA and NE, but MDMA is more effective at releasing 5-HT (Wall et al. 1995). This selectivity results from their relative ability to be transported by DAT, NET, or SERT. The ability to serve as a substrate is one of the essential properties responsible for amphetamine action. Blocking a transporter, for example, with cocaine, prevents amphetamine from causing release (Fischer and Cho 1979). At the plasma membrane, amphetamine causes efflux of the biogenic amine neurotransmitter by exchanging with the cytoplasmic transmitter through the action of DAT, NET, or SERT (Wall et al. 1995). However, this plasma membrane exchange is not sufficient to cause much release, because the bulk of the neurotransmitter content in a neuron is intravesicular. It is because amphetamines are highly permeant that they are able to release intravesicular neurotransmitter into the cytoplasm, where the plasma membrane transporter can facilitate the exchange process that releases transmitter into the synapse (Fig. 6).

Amphetamines are weak bases, and as such, they equilibrate across membranes in the neutral, unprotonated form, and accumulate on the acidic side of the membrane. With the vacuolar ATPase pumping H^+ ions into their lumen, synaptic vesicles are acidic and amphetamines accumulate inside. Each molecule of amphetamine that enters the vesicle consumes one H^+ ion and raises the internal pH. As the internal pH rises, VMAT has less of a driving force available to accumulate neurotransmitter, and accumulated transmitter will begin to leak from the vesicle into the cytoplasm (Sulzer and Rayport 1990). In addition to this action, some amphetamine derivatives actually bind to VMAT and prevent the sequestration of cytoplasmic transmitter (Schuldiner et al. 1993). Furthermore, the ability of amphetamine derivatives to dissipate the pH difference across the vesicle membrane is greater than that of simple organic amine weak bases, suggesting that amphetamines may be permeant also in their protonated form (Rudnick and Wall 1992). This property would allow them to act as ionophores, diffusing into the vesicle in the unprotonated form, diffusing out with a bound H^+ ion, and thereby facilitating the efflux of H^+ from the vesicle (Fig. 6). Thus, it is the combination of the abilities of amphetamines to cross the plasma membrane, dissipate the pH difference at the vesicle membrane, and exchange with cytoplasmic transmitter through the plasma membrane transporter that is responsible for their unique ability to release vesicular transmitter by a nonexocytotic mechanism (Jones et al. 1998).

SUMMARY AND CONCLUSIONS

Neurotransmitter transporters function by binding their substrate together with cotransported ions (Na^+ and Cl^-) and changing the accessibility of the binding site so that substrate

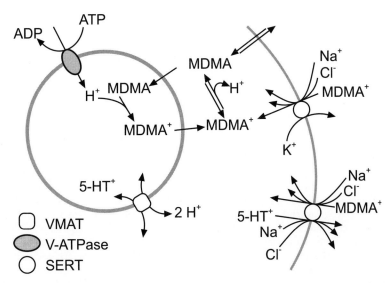

FIGURE 6. How the amphetamine derivative MDMA causes 5-HT release from synaptic vesicles. MDMA$^+$ is recognized as a substrate by SERT, which transports it into the cell along with symported Na$^+$ and Cl$^-$ and antiported K$^+$. By dissociating to the neutral form and passive efflux, MDMA engages a futile cycle that increases cellular Na$^+$, Cl$^-$, and H$^+$, and depletes K$^+$. This futile cycle may contribute to the toxicity of MDMA. In the cytoplasm, neutral MDMA can diffuse into the synaptic vesicle, bind an H$^+$, and exit as MDMA$^+$, thus dissipating the pH difference created by the vacuolar ATPase. The decreased H$^+$ gradient leads to 5-HT efflux into the cytoplasm, where, given the elevated Na$^+$ and Cl$^-$, it readily binds to SERT and is transported out of the cell in exchange for external MDMA$^+$.

bound on one side of the membrane can be released on the other side. Intracellular accumulation of the transmitters is driven by the energy in the transmembrane ion gradients and membrane potential. The transporters couple these driving forces to neurotransmitter accumulation by undergoing conformational changes. Inhibitors of transport, such as cocaine and antidepressant drugs, act as competitive blockers of substrate binding, while amphetamines are substrates for the biogenic amine transporters that also permeate the membrane passively. These basic properties of transporter function underlie the role that they have in regulating the synaptic concentration of neurotransmitters and the effect of that regulation in the function of the nervous system.

GAPS AND FUTURE DIRECTIONS

The mechanistic and structural analysis of neurotransmitter transporters is converging on a more fundamental understanding of how these proteins use transmembrane ion gradients to accumulate neurotransmitters. High-resolution structural information, such as that available for the glutamate transporter family, would represent a tremendous advance. However, accompanying functional studies will still be required to understand how the dynamic aspects of the structure lead to the accumulation of neurotransmitter from the synaptic cleft into the cell.

REFERENCES

Abramson J., Smirnova I., Kasho V., Verner G., Kaback H.R., and Iwata S. 2003. Structure and mechanism of the lactose permease of *Escherichia coli*. *Science* **301:** 610–615.

Androutsellis-Theotokis A. and Rudnick G. 2002. Accessibility and conformational coupling in serotonin transporter predicted internal domains. *J. Neurosci.* **22:** 8370–8378.

Androutsellis-Theotokis A., Ghassemi F., and Rudnick G. 2001. A conformationally sensitive residue on the cytoplasmic surface of serotonin transporter. *J. Biol. Chem.* **276:** 45933–45938.

Androutsellis-Theotokis A., Goldberg N.R., Ueda K., Beppu T., Beckman M.L., Das S., Javitch J.A., and Rudnick G. 2003. Characterization of a functional bacterial homologue of sodium-dependent neurotransmitter transporters. *J. Biol. Chem.* **278:** 12703–12709.

Bismuth Y., Kavanaugh M.P., and Kanner B.I. 1997. Tyrosine 140 of the gamma-aminobutyric acid transporter Gat-1 plays a critical role in neurotransmitter recognition. *J. Biol. Chem.* **272:** 16096–16102.

Blakely R., Berson H., Fremeau R., Caron M., Peek M., Prince H., and Bradely C. 1991. Cloning and expression of a functional serotonin transporter from rat brain. *Nature* **354:** 66–70.

Brosen K. 1993. The pharmacogenetics of the selective serotonin reuptake inhibitors. *Clin. Invest.* **71:** 1002–1009.

Byck R. and Van Dyke C. 1982. Cocaine. *Sci. Am.* **246:** 128–141.

Castagna M., Shayakul C., Trotti D., Sacchi V.F., Harvey W.R., and Hediger M.A. 1998. Cloning and characterization of a potassium-coupled amino acid transporter. *Proc. Natl. Acad. Sci.* **95:** 5395–5400.

Chang A.B., Lin R., Studley W.K., Tran C.V., and Saier M.H. 2004. Phylogeny as a guide to structure and function of membrane transport proteins (Review). *Mol. Membr. Biol.* **21:** 171–181.

Chen J.G. and Rudnick G. 2000. Permeation and gating residues in serotonin transporter. *Proc. Natl. Acad. Sci.* **97:** 1044–1049.

Chen J.G., Liu-Chen S., and Rudnick G. 1998. Determination of external loop topology in the serotonin transporter by site-directed chemical labeling. *J. Biol. Chem.* **273:** 12675–12681.

Chen J.G., Sachpatzidis A., and Rudnick G. 1997. The third transmembrane domain of the serotonin transporter contains residues associated with substrate and cocaine binding. *J. Biol. Chem.* **272:** 28321–28327.

Deamer D.W. and Bramhall J. 1986. Permeability of lipid bilayers to water and ionic solutes. *Chem. Phys. Lipids* **40:** 167–188.

Fischer J.F. and Cho A.K. 1979. Chemical release of dopamine from striatal homogenates: Evidence for an exchange diffusion model. *J. Pharm. Exp. Ther.* **208:** 203–209.

Galli A., Blakely R.D., and DeFelice L.J. 1996. Norepinephrine transporters have channel modes of conduction. *Proc. Natl. Acad. Sci.* **93:** 8671–8676.

Gendreau S., Voswinkel S., Torres-Salazar D., Lang N.,

Heidtmann H., Detro-Dassen S., Schmalzing G., Hidalgo P., and Fahlke C. 2004. A trimeric quaternary structure is conserved in bacterial and human glutamate transporters. *J. Biol. Chem.* **279:** 39505–39512.

Gu H., Wall S.C., and Rudnick G. 1994. Stable expression of biogenic amine transporters reveals differences in inhibitor sensitivity, kinetics, and ion dependence. *J. Biol. Chem.* **269:** 7124–7130.

———. 1996. Ion coupling stoichiometry for the norepinephrine transporter in membrane vesicles from stably transfected cells. *J. Biol. Chem.* **271:** 6911–6916.

Hansra N., Arya S., and Quick M.W. 2004. Intracellular domains of a rat brain GABA transporter that govern transport. *J. Neurosci.* **24:** 4082–4087.

Hastrup H., Sen N., and Javitch J.A. 2003. The human dopamine transporter forms a tetramer in the plasma membrane—Cross-linking of a cysteine in the fourth transmembrane segment is sensitive to cocaine analogs. *J. Biol. Chem.* **278:** 45045–45048.

Hediger M.A., Ikeda T., Coady M., Gundersen C.B., and Wright E.M. 1987. Expression of size-selected mRNA encoding the intestinal Na/glucose cotransporter in *Xenopus laevis* oocytes. *Proc. Natl. Acad. Sci.* **84:** 2634–2637.

Henry L.K., Adkins E.M., Han Q., and Blakely R.D. 2003. Serotonin and cocaine-sensitive inactivation of human serotonin transporters by methanethiosulfonates targeted to transmembrane domain I. *J. Biol. Chem.* **278:** 37052–37063.

Hirai T., Heymann J.A.W., Shi D., Sarker R., Maloney P.C., and Subramaniam S. 2002. Three-dimensional structure of a bacterial oxalate transporter. *Nat. Struct. Biol.* **9:** 597–600.

Hoffman B.J., Mezey E., and Brownstein M.J. 1991. Cloning of a serotonin transporter affected by antidepressants. *Science* **254:** 579–580.

Huang Y., Lemieux M.J., Song J., Auer M., and Wang D.-N. 2003. Structure and mechanism of the glycerol-3-phosphate transporter from *Escherichia coli*. *Science* **301:** 616–620.

Ingram S.L., Prasad B.M., and Amara S.G. 2002. Dopamine transporter-mediated conductances increase excitability of midbrain dopamine neurons. *Nat. Neurosci.* **5:** 971–978.

Jardetzky O. 1966. Simple allosteric model for membrane pumps. *Nature* **211:** 969–970.

Johnson R.G., Pfister D., Carty S.E., and Scarpa A. 1979. Biological amine transport in chromaffin ghosts: Coupling to the transmembrane proton and potential gradients. *J. Biol. Chem.* **254:** 10963–10972.

Jones S.R., Gainetdinov R.R., Wightman R.M., and Caron M.G. 1998. Mechanisms of amphetamine action revealed in mice lacking the dopamine transporter. *J. Neurosci.* **18:** 1979–1986.

Kaback H.R. 1974. Transport studies in bacterial membrane vesicles. *Science* **186:** 882–892.

Kanner B.I., Sharon I., Maron R., and Schuldiner S. 1980. Electrogenic transport of biogenic amines in chromaffin granule membrane vesicles. *FEBS Lett.* **111:** 83–86.

Kavanaugh M., Arriza J., North R., and Amara S. 1992. Electrogenic uptake of gamma-aminobutyric acid by a cloned transporter expressed in *Xenopus* oocytes. *J. Biol. Chem.* **267:** 22007–22009.

Keynan S., Suh Y.J., Kanner B.I., and Rudnick G. 1992. Expression of a cloned gamma-aminobutyric acid transporter in mammalian cells. *Biochemistry* **31:** 1974–1979.

Kilic F. and Rudnick G. 2000. Oligomerization of serotonin transporter and its functional consequences. *Proc. Natl. Acad. Sci.* **97:** 3106–3111.

Kocabas A.M., Rudnick G., and Kilic F. 2003. Functional consequences of homo- but not hetero-oligomerization between transporters for the biogenic amine neurotransmitters. *J. Neurochem.* **85:** 1513–1520.

Koob G.F. 1992. Drugs of abuse: Anatomy, pharmacology and function of reward pathways. *Trends Pharmacol. Sci.* **13:** 177–184.

Korkhov V.M., Farhan H., Freissmuth M., and Sitte H.H. 2004. Oligomerization of the {gamma}-aminobutyric acid transporter-1 is driven by an interplay of polar and hydrophobic interactions in transmembrane helix II. *J. Biol. Chem.* **279:** 55728–55736.

Lee S.H., Chang M., Lee K.H., Park B.S., Lee Y.S., and Chin H.R. 2000. Importance of valine at position 152 for the substrate transport and 2beta-carbomethoxy-3beta-(4-fluorophenyl) tropane binding of dopamine transporter. *Mol. Pharmacol.* **57:** 883–889.

Leonard B.E. 1999. Neuropharmacology of antidepressants that modify central noradrenergic and serotonergic function: A short review. *Hum. Psychopharmacol.* **14:** 75–81.

Lin F., Lester H.A., and Mager S. 1996. Single-channel currents produced by the serotonin transporter and analysis of a mutation affecting ion permeation. *Biophys. J.* **71:** 3126–3135.

MacAulay N., Bendahan A., Loland C.J., Zeuthen T., Kanner B.I., and Gether U. 2001. Engineered Zn^{2+} switches in the γ-aminobutyric acid (GABA) transporter-1. Differential effects on GABA uptake and currents. *J. Biol. Chem.* **276:** 40476–40485.

Mager S., Min C., Henry D.J., Chavkin C., Hoffman B.J., Davidson N., and Lester H.A. 1994. Conducting states of a mammalian serotonin transporter. *Neuron* **12:** 845–859.

Mitchell S.M., Lee E., Garcia M.L., and Stephan M.M. 2004. Structure and function of extracellular loop 4 of the serotonin transporter as revealed by cysteine-scanning mutagenesis. *J. Biol. Chem.* **279:** 24089–24099.

Mueckler M. 1994. Facilitative glucose transporters. *Eur. J. Biochem.* **219:** 713–725.

Murer H. and Hopfer U. 1974 Demonstration of electrogenic Na^+-dependent D-glucose transport in intestinal brush border membranes. *Proc. Natl. Acad. Sci.* **71:** 484–488.

Nelson P.J. and Rudnick G. 1979. Coupling between platelet 5-hydroxytryptamine and potassium transport. *J. Biol. Chem.* **254:** 10084–10089.

——. 1982. The role of chloride ion in platelet serotonin trans-port. *J. Biol. Chem.* **257:** 6151–6155.

Ni Y.G., Chen J.G., Androutsellis-Theotokis A., Huang C.J., Moczydlowski E., and Rudnick G. 2001. A lithium-induced conformational change in serotonin transporter alters cocaine binding, ion conductance, and reactivity of cys-109. *J. Biol. Chem.* **276:** 30942–30947.

Nicholls D.G. and Ferguson S.J. 2002. *Bioenergetics 3*, Academic Press, Boston, p. 297.

Norregaard L., Frederiksen D., Nielsen E.O., and Gether U. 1998. Delineation of an endogenous zinc-binding site in the human dopamine transporter. *EMBO J.* **17:** 4266–4273.

Norregaard L., Visiers I., Loland C.J., Ballesteros J., Weinstein H., and Gether U. 2000. Structural probing of a microdomain in the dopamine transporter by engineering of artificial Zn^{2+} binding sites. *Biochemistry* **39:** 15836–15846.

Pressman B.C. 1973. Properties of ionophores with broad range cation selectivity. *Fed. Proc.* **32:** 1698–1703.

Quick M.W. 2003. Regulating the conducting states of a mammalian serotonin transporter. *Neuron* **40:** 537–549.

Radian R. and Kanner B.I. 1983. Stoichiometry of sodium- and chloride-coupled gamma-aminobutyric acid transport by synaptic plasma membrane vesicles isolated from rat brain. *Biochemistry* **22:** 1236–1241.

Roux M.J. and Supplisson S. 2000. Neuronal and glial glycine transporters have different stoichiometries. *Neuron* **25:** 373–383.

Rudnick G. 1998. Bioenergetics of neurotransmitter transport. *J. Bioenerg. Biomembr.* **30:** 173–185.

——. 2002. Mechanisms of biogenic amine neurotransmitter transporters. In *Neurotransmitter transporters, structure, function, and regulation* (ed. M.E.A. Reith), pp. 25–52. Humana Press, Totowa, New Jersey.

Rudnick G. and Wall S.C. 1992. *p*-Chloroamphetamine induces serotonin release through serotonin transporters. *Biochemistry* **31:** 6710–6718.

Rudnick G., Kirk K.L., Fishkes H., and Schuldiner S. 1989. Zwitterionic and anionic forms of a serotonin analog as transport substrates. *J. Biol. Chem.* **264:** 14865–14868.

Sato Y., Zhang Y.W., Androutsellis-Theotokis A., and Rudnick G. 2004. Analysis of transmembrane domain 2 of rat serotonin transporter by cysteine scanning mutagenesis. *J. Biol. Chem.* **279:** 22926–22933.

Saunders R.C., Kolachana B.S., and Weinberger D.R. 1994. Local pharmacological manipulation of extracellular dopamine levels in the dorsolateral prefrontal cortex and caudate nucleus in the rhesus monkey: An in vivo microdialysis study. *Exp. Brain Res.* **98:** 44–52.

Schloss P. and Henn F.A. 2004. New insights into the mechanisms of antidepressant therapy. *Pharmacol. Therapeut.* **102:** 47–60.

Schmid J.A., Just H., and Sitte H.H. 2001. Impact of oligomerization on the function of the human serotonin transporter. *Biochem. Soc. Trans.* **29:** 732–736.

Scholze P., Freissmuth M., and Sitte H.H. 2002. Mutations within an intramembrane leucine heptad repeat disrupt oligomer formation of the rat GABA transporter 1. *J. Biol. Chem.* **277:**

43682–43690.

Schuldiner S., Steiner-Mordoch S., Yelin R., Wall S.C., and Rudnick G. 1993. Amphetamine derivatives interact with both plasma membrane and secretory vesicle biogenic amine transporters. *Mol. Pharmacol.* **44:** 1227–1231.

Shimizu N., Take S., Hori T., and Oomura Y. 1992. In vivo measurement of hypothalamic serotonin release by intracerebral microdialysis: Significant enhancement by immobilization stress in rats. *Brain Res. Bull.* **28:** 727–734.

Smicun Y., Campbell S.D., Chen M.A., Gu H., and Rudnick G. 1999. The role of external loop regions in serotonin transport. Loop scanning mutagenesis of the serotonin transporter external domain. *J. Biol. Chem.* **274:** 36058–36064.

Sonders M., Zhu S., Zahniser N., Kavanaugh M., and Amara S. 1997. Multiple ionic conductances of the human dopamine transporter: The actions of dopamine and psychostimulants. *J. Neurosci.* **17:** 960–974.

Sora I., Wichems C., Takahashi N., Li X.F., Zeng Z.Z., Revay R., Lesch K.P., Murphy D.L., and Uhl G.R. 1998. Cocaine reward models: Conditioned place preference can be established in dopamine- and in serotonin-transporter knockout mice. *Proc. Natl. Acad. Sci.* **95:** 7699–7704.

Sora I., Hall F.S., Andrews A.M., Itokawa M., Li X.F., Wei H.B., Wichems C., Lesch K.P., Murphy D.L., and Uhl G.R. 2001. Molecular mechanisms of cocaine reward: Combined dopamine and serotonin transporter knockouts eliminate cocaine place preference. *Proc. Natl. Acad. Sci.* **98:** 5300–5305.

Sorkina T., Doolen S., Galperin E., Zahniser N.R., and Sorkin A. 2003. Oligomerization of dopamine transporters visualized in living cells by fluorescence resonance energy transfer microscopy. *J. Biol. Chem.* **278:** 28274–28283.

Sulzer D. and Rayport S. 1990. Amphetamine and other psychostimulants reduce pH gradient in midbrain dopaminergic neurons and chromaffin granules: A mechanism of action. *Neuron* **5:** 797–808.

Sulzer D., Maidment N., and Rayport S. 1993. Amphetamine and other weak bases act to promote reverse transport of dopamine in ventral midbrain neurons. *J. Neurochem.* **60:** 527–535.

Tanford C. 1973. *The hydrophobic effect: Formation of micelles and biological membranes,* pp. viii, 200. Wiley, New York.

Tate C. and Blakely R. 1994. The effect of N-linked glycosylation on activity of the Na^+- and Cl^--dependent serotonin transporter expressed using recombinant baculovirus in insect cells. *J. Biol. Chem.* **269:** 26303–26310.

Thomas D.N., Post R.M., and Pert A. 1994. Focal and systemic cocaine differentially affect extracellular norepinephrine in the locus coeruleus, frontal cortex and hippocampus of the anaesthetized rat. *Brain Res.* **645:** 135–142.

Torres G.E., Carneiro A., Seamans K., Fiorentini C., Sweeney A., Yao W.D., and Caron M.G. 2003. Oligomerization and trafficking of the human dopamine transporter mutational analysis identifies critical domains important for the functional expression of the transporter. *J. Biol. Chem.* **278:** 2731–2739.

Wall S.C., Gu H., and Rudnick G. 1995. Biogenic amine flux mediated by cloned transporters stably expressed in cultured cell lines: Amphetamine specificity for inhibition and efflux. *Mol. Pharmacol.* **47:** 544–550.

Wu X. and Gu H.H. 2003. Cocaine affinity decreased by mutations of aromatic residue phenylalanine 105 in the transmembrane domain 2 of dopamine transporter. *Mol. Pharmacol.* **63:** 653–658.

Yernool D., Boudker O., Folta-Stogniew E., and Gouaux E. 2003. Trimeric subunit stoichiometry of the glutamate transporters from *Bacillus caldotenax* and *Bacillus stearothermophilus.* *Biochemistry* **42:** 12981–12988.

Yernool D., Boudker O., Jin Y., and Gouaux E. 2004. Structure of a glutamate transporter homologue from *Pyrococcus horikoshii.* *Nature* **431:** 811–818.

Zafra F. and Gimenez C. 1989. The role of chloride ions on the transport of glycine in plasma membrane vesicles from glial cells. *Biochim. Biophys. Acta* **979:** 147.

Zerangue N. and Kavanaugh M.P. 1996. Flux coupling in a neuronal glutamate transporter. *Nature* **383:** 634–637.

Zhou Y., Bennett E.R., and Kanner B.I. 2004. The aqueous accessibility in the external half of transmembrane domain I of the GABA transporter GAT-1 is modulated by its ligands. *J. Biol. Chem.* **279:** 13800–13808.

10 | Neuronal Nicotinic Acetylcholine Receptors and Nicotine Dependence

Andrew R. Tapper, Raad Nashmi, and Henry A. Lester

Division of Biology 156-29, California Institute of Technology, Pasadena, California 91125

ABSTRACT

Nicotine is the addictive component of tobacco smoke. It elicits its psychoactive effects by binding to and activating neuronal nicotinic acetylcholine receptors (nAChRs) in the brain. Nicotinic receptors are pentameric ligand-gated ion channels that are activated by the endogenous neurotransmitter, acetylcholine, as well as nicotine. They are expressed in presynaptic neuron terminals, where they modulate neurotransmitter release, and postsynaptically, where they contribute to neuronal excitability through the generation of excitatory postsynaptic potentials. Nicotinic receptors are expressed throughout the central nervous system, however receptors expressed in the mesocorticolimbic pathways are thought to be primarily responsible for initiating and maintaining a drug-dependent state. Acute nicotine exposure activates dopaminergic neurons in the midbrain, leading to the release of dopamine in the nucleus accumbens, a physiological response associated with pleasure. Chronic exposure to nicotine produces long-term physiological and behavioral changes associated with drug dependence including nAChR up-regulation, long-term potentiation at glutamatergic synapses, changes in gene expression, tolerance, sensitization, and withdrawal. Thus, physiological alterations induced by prolonged nicotine exposure and initiated by activation of nAChRs underlie nicotine addiction.

INTRODUCTION

Nicotine addiction elicited by smoking tobacco is responsible for over 3 million deaths annually, making it the largest cause of preventable mortality in the world. In addition, because of the increase of tobacco use in developing nations, it is predicted that this death toll will increase to more than 10 million per year over the next 30–40 years.

Nicotine is a naturally occurring alkaloid found in tobacco and is the primary addictive component of cigarette smoke. Vaporized nicotine is rapidly absorbed through the lungs where it enters the blood stream (Fig. 1). Within seconds of inhalation, nicotine base readily crosses the blood–brain barrier where it gains access to nAChRs expressed

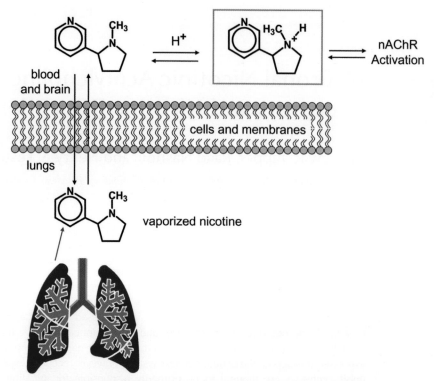

FIGURE 1. Nicotine: from the lungs to the blood and the brain.

throughout the central nervous system (CNS). In its protonated form, nicotine mimics the endogenous neurotransmitter, acetylcholine, and can activate nAChRs using the cholinergic system, which, under normal conditions, has an important role in reward, anxiety, cognition, attention, and many other physiological processes. This ability of nicotine to "hijack" nAChRs is thought to underlie the molecular basis of nicotine addiction.

GOALS OF THE CHAPTER

Because nicotine dependence is initiated through the activation of nAChRs, we begin with an introduction of the structure, function, and cell biology of neuronal nicotinic acetylcholine receptors. These receptors are expressed in the midbrain, where nicotine activates them and they produce the pleasurable effects associated with smoking. We focus on the circuitry of this brain region, as well as the effects of nicotine and the current work being done in mice to identify the specific nAChR subtypes critical for dependence. Finally, we emphasize the point that chronic nicotine exposure produces the long-lasting behavioral and physiological changes associated with addiction including increased synaptic strength, altered gene expression, and nAChR up-regulation.

BIOCHEMISTRY AND KINETICS OF NEURONAL NICOTINIC ACETYLCHOLINE RECEPTORS

Structure and Function

Nicotinic receptors are members of the ligand-gated ion channel superfamily that also includes GABA (A and C), glycine, and serotonin (5-HT3) receptors. Currently, 12 neuronal nAChR subunits have been identified ($\alpha2$–$\alpha10$ and $\beta2$–$\beta4$) and are generally grouped into α subunits that contain two adjacent cysteine residues essential for acetylcholine binding, and non-α (or -β) subunits, that lack these residues. Each nAChR gene encodes a protein subunit consisting of approximately 200 residue extracellular amino termini, four transmembrane segments (M1–M4), a variable intracellular loop (100–200 residues) between M3 and M4, and a 4–28-residue extracellular carboxyl terminus (Fig. 2A) (Corringer et al. 2000). The amino terminus contains the acetylcholine binding domain (Eisele et al. 1993). Five subunits coassemble to form ligand-gated channels (Cooper et al. 1991; Brejc et al. 2001). The M2 transmembrane segment of all five subunits forms the conducting pore of the channel, with regions in the M1–M2 intracellular loop contributing to ion selectivity (Fig. 2B,C). The subunit composition of each channel determines its electrophysiological properties, cation permeability, and agonist binding affinities (McGehee and Role 1995; Corringer et al. 2000). Thus, many nAChR subtypes exist because most subunits can form heteromeric channels, whereas a subset, $\alpha7$–$\alpha10$, may

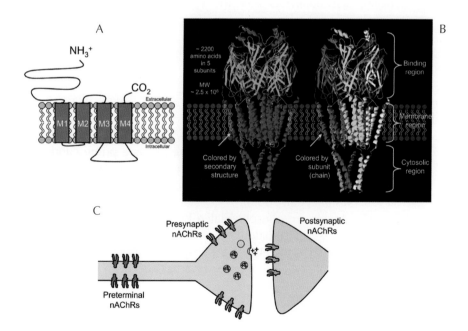

FIGURE 2. Neuronal nicotinic acetylcholine receptor: structure and localization. (*A*) Topology of a single subunit ($\alpha2$–$\alpha10$; $\beta2$–$\beta4$). (*B*) Probable structure of the entire receptor, based on the Torpedo structure postulated by Unwin (2005) (protein database structure 2BG9). (*C*) Various subcellular localizations of nAChRs.

form homomeric channels. The majority of neuronal nicotinic receptor subtypes falls into two categories: receptors that bind agonist with high affinity (nM concentrations) and those that bind with lower affinity (μM concentrations). Most high-affinity nAChRs in the CNS contain α4 and β2 subunits (denoted α4β2*), whereas most low-affinity receptors are presumably homomeric, α7 receptors that are also α-bungarotoxin sensitive.

Transition States

To understand the pharmacodynamics of nicotine, it is necessary to be familiar with the different transition states of nAChRs. Nicotinic receptors exist in three conformations: open, closed, and desensitized. When nicotine binds a closed receptor, it transitions into an active state, allowing cations to flux through the open channel and down their electrochemical gradient. However, the persistent presence of agonist drives the receptor into the desensitized, inactive state, which has a higher affinity for agonist (pM–nM) than either open or closed states of the receptor. The rate of entry into the desensitized state and recovery depends on the subunit composition of the receptor.

CELL BIOLOGY OF NICOTINIC RECEPTORS

The large intracellular loop between M3 and M4 for both mouse α4 and β2 nAChR subunits (272 and 150 amino acids in lengths, respectively) has an important role in the regulation of nicotinic receptors. The sequences contain many motifs that function as targeting signals, and they also contain sites for posttranslational modification such as phosphorylation. These signals ultimately control the level of expression, subcellular targeting, degradation, protein–protein interactions, and function of nicotinic receptors in cells.

Phosphorylation

The M3–M4 intracellular loop of α4 (275 amino acids in length) is the longest among the cys loop family of receptors. Therefore, α4 is a prime candidate for regulation. Many potential phosphorylation sites are in α4 including phosphorylation by PKC, PKA, casein kinase II, CAM kinase II, tyrosine kinase, and cyclin-dependent kinase 5 (Viseshakul et al. 1998; Nashmi et al. 2003).

α4 subunits are phosphorylated by PKA when heterologously expressed in oocytes (Hsu et al. 1997; Viseshakul et al. 1998). α4 subunits in human cloned epithelial cells stably transfected with α4β2 are phosphorylated by both PKA and PKC (Pacheco et al. 2003). Phosphorylation of α4 by PKA has also been found in rat brain tissue (Nakayama et al. 1993).

β2 has a significantly shorter cytoplasmic M3–M4 loop at 150 amino acids in length and contains far fewer putative phosphorylation sites, namely, PKC and casein kinase II (Viseshakul et al. 1998; Nashmi et al. 2003). Phosphorylation of the β2 subunit has been more challenging to detect (Viseshakul et al. 1998).

Does phosphorylation have a functional role in α4β2 nAChRs? Mutating the phosphorylation sites of γ and δ subunits of Torpedo nicotinic receptors results in nicotinic currents with significantly slower desensitization time constants (Hoffman et al. 1994). Activation of PKA and PKC both result in increased expression of $\alpha4\beta2$ receptors, as indicated by cytisine binding in HEK293 cells (Gopalakrishnan et al. 1997). Activation of PKA can stimulate more efficient subunit assembly and increase receptor expression for both Torpedo and mouse muscle nicotinic receptors (Ross et al. 1991). Furthermore, activation of PKC speeds the recovery from desensitization with prolonged nicotine application (Fenster et al. 1999) and increases surface expression of $\alpha4\beta2$ (Nashmi et al. 2003). Thus, phosphorylation can effect function, assembly, and subcellular expression of nicotinic receptors.

How can phosphorylation mediate these events? Overlap occurs between putative phosphorylation sites and subcellular targeting motifs in the M3–M4 cytoplasmic loop of $\alpha4$. This provides a mechanism for dynamic regulation of nicotinic receptor expression and targeting to and from the surface through phosphorylation. $\alpha4$ has a number of endoplasmic reticulum (ER) retention motifs (RXR or R/K-R/K) that overlap with potential phosphorylation sites. A similar mechanism of ER retention signal can be regulated by phosphorylation in N-methyl-D-aspartate (NMDA) receptors (Scott et al. 2001).

Palmitoylation

Palmitoylation is the thioester linkage of the long-chain fatty acid (palmitate) to cysteines in proteins (Smotrys and Linder 2004). Thioester linkage can include fatty acids other than palmitate. The main function of palmitoylation is to promote membrane association of soluble proteins and target proteins to lipid rafts, regions of plasma membrane rich in cholesterol and sphingolipids. For nicotinic receptors, palmitoylation of $\alpha7$ results in increased intracellular and surface expression of receptors, suggesting that palmitoylation enhances subunit assembly and targeting to the cell surface (Drisdel et al. 2004). Other ion channels that are palmitoylated include muscle nAChRs (Olson et al. 1984), $GABA_A$ receptors (Keller et al. 2004), and GluR6 kainate receptor subunits (Pickering et al. 1995).

Ubiquitination

Ubiquitination is the attachment of ubiquitin, a 76-amino-acid protein, to another protein by formation of an isopeptide bond between [76]Gly and the ε-amino group of a lysine residue on the target protein (Hicke and Dunn 2003). Polyubiquitin chains on proteins may serve as a signal for protein degradation via the 26S proteasome (Thrower et al. 2000). However, ubiquitination can also function as a trafficking signal. In muscle nicotinic receptors, ubiquitination inhibits Golgi-to-cell-surface trafficking (Keller et al. 2001). Neuronal nicotinic receptors contain lysine residues within the M3–M4 intracellular loop that are potential targets for ubiquitination. In addition, activity-dependent ubiquitination of postsynaptic proteins including NR1, NR2A, and NR2B NMDA receptor subunits is a mechanism linking receptor trafficking and degradation (Ehlers 2003).

Neuronal nAChR Localization and Scaffolding Proteins

Although most nicotinic receptor signaling appears to be presynaptic in the CNS (McGehee et al. 1995; Gray et al. 1996), there is some evidence for postsynaptic nicotinic signaling (Roerig et al. 1997; Alkondon et al. 1998). The effect of neuronal nAChR activation depends on the subcellular localization of the receptor. Nicotinic receptors expressed on dendrites and soma may mediate fast synaptic transmission and contribute to neuronal excitability through the generation of excitatory postsynaptic potentials (EPSPs). In addition, nAChRs are expressed at axon terminals where activation modulates neurotransmitter release through calcium influx and/or terminal depolarization (Fig. 2D). Nicotinic receptors modulate the release of norepinephrine, glutamate, GABA, and dopamine. Currently, the predominant role of nAChRs is thought to be in modulating neurotransmission presynaptically.

In chick ciliary ganglion, $\alpha 3$ and $\alpha 5$ target to the postsynaptic membrane, whereas $\alpha 7$ is perisynaptic. If one inserts the M3–M4 loop of $\alpha 3$ into the $\alpha 7$ nAChR subunit, $\alpha 7$ targets postsynaptically and reduces nAChR surface levels, whereas the intracellular loop of $\alpha 5$ inserted into $\alpha 7$ does neither (Williams et al. 1998). What is the mechanism of targeting to postsynaptic sites? Muscle nicotinic receptors must coassemble with MuSK and rapsyn to stabilize the receptor complex, as well as to ensure proper membrane localization (Mohamed et al. 2001). In the postsynaptic density of neurons, PSD-95 binds to the carboxyl terminus of NR2 NMDA subunits and induces clustering. PSD-95 acts as an organizing scaffold that tethers other constituents to NMDA receptors. Neuronal nicotinic receptors may be tethered by a similar mechanism through protein–protein interactions mediated by the M3–M4 intracellular loop. Yeast–two-hybrid screens and immunoprecipitation have shown that $\alpha 4$ nAChR subunits associate with the chaperone protein 14-3-3η. Coexpression of $\alpha 4$ nAChRs with 14-3-3η results in increased surface expression of $\alpha 4$ (Jeanclos et al. 2001). $\alpha 7$ colocalizes with lipid rafts in ciliary ganglion neurons. Lipid rafts may function in the clustering of $\alpha 7$ in somatic spines because when cholesterol is extracted from lipid rafts with methyl-β-cyclodextrin, $\alpha 7$ clusters disperse (Bruses et al. 2001). PSD-93a, a member of the PSD-95 family of proteins, shows strong association with $\alpha 3$ and $\alpha 5$ nAChR subunits, although it has virtually no association with $\alpha 4\beta 2$ and $\alpha 7$. SAP-102 preferentially associates with $\alpha 5$ (Conroy et al. 2003). Furthermore, adenomatous polyposis coli (APC) protein colocalizes with $\alpha 3$ nAChR subunits in postsynaptic sites of ciliary ganglion neurons and also binds to PSD-93, β-catenin, and microtubule end-binding protein EB1 (Temburni et al. 2004). So far, very little information has been garnered on the molecular scaffolding architecture of $\alpha 4\beta 2$ nAChRs.

MOLECULAR BASIS OF NICOTINE ADDICTION

The Mesocorticolimbic System

Although nAChRs are expressed throughout the CNS, nicotine's major addictive effects are thought to be mediated through the mesocorticolimbic dopamine (DA) pathways. DA neurons within this circuit originate in the ventral tegmental area (VTA) and project to the nucleus accumbens and prefrontal cortex (Fig. 3A). Nicotinic-receptor-expressing GABAergic neurons provide inhibitory control of DAergic neurons within the VTA and also project to the tegmental pedunculopontine (TPP) nucleus, which has a role in dopamine-independent reward signaling (Laviolette et al. 2002). In addition, glutamatergic neurons project down from the prefrontal cortex and can modulate both DAergic and

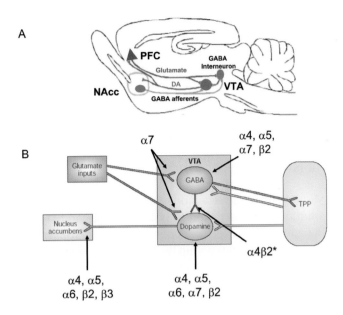

FIGURE 3. Nicotinic receptor subunits in the mesocorticolimbic pathways. (*A*) Mesocorticolimbic circuitry in the mouse brain. (*B*) Expression of α4–α7 and α2–β3 nAChR subunits in mesocortico-limbic pathways. (*B* modified, with permission, from Laviolette and van der Kooy 2004 [©Macmillan Magazines Ltd.].)

GABAergic pathways. The interplay among glutamate, dopamine, and GABA signaling is likely critical for the reinforcing effects of nicotine, and these interactions are orchestrated in the VTA. Rats readily self-administer nicotine through intravenous infusion (Corrigall and Coen 1989). However, this behavior is attenuated when nicotinic antagonists are directly infused into the VTA, indicating that the rewarding effects of nicotine are mediated through this brain region (Corrigall et al. 1994).

Nicotine concentrations that are self-administered and found in smokers' blood acutely activate dopaminergic neurons in the VTA, causing increased dopamine release in the nucleus accumbens (NAc) (Pidoplichko et al. 1997). Originally thought to directly mediate the rewarding effects of abused drugs, more recent evidence suggests that dopamine release may be a predictor of reward (Schultz 2004). Nevertheless, nicotine-induced release of dopamine is critical for the onset of addiction (Corrigall and Coen 1991; Picciotto et al. 1998). Pharmacological blockade of DA receptors, destruction of DA neurons, or lesioning of the nucleus accumbens reduces self-administration (Corrigall and Coen 1991; Corrigall et al. 1992).

Whereas nicotine may activate dopaminergic neurons directly via somatodendritic nAChR activation, a single exposure to nicotine can produce increased dopamine in the nucleus accumbens for more than 1 hour. Recently, the mechanism for this paradox has been explained. Nicotine, at doses found in smokers' blood, rapidly activates and then desensitizes the majority of high-affinity nAChRs expressed on dopaminergic VTA neurons. However, nicotine also desensitizes the high-affinity nAChRs expressed on the GABAergic interneurons that provide inhibitory control over DA VTA neurons, thereby decreasing inhibition. In addition, glutamatergic inputs from the prefrontal cortex remain active because they express presynaptic (most likely) $\alpha7$ nAChRs that desensitize less in response to low levels of nicotine. Thus, the DAergic VTA neurons receive an excitatory drive until the high-affinity nAChRs recover from desensitization, a process that can take up to 1 hour (Mansvelder et al. 2002).

Mouse Models of Nicotine Dependence

Multiple nicotinic receptor subtypes are expressed in the mesocorticolimbic system and are thought to be linked to both the acute and chronic effects of nicotine that underlie initiation and onset of dependence (Fig. 3B). For example, both dopaminergic and GABAergic neurons within the VTA express various nAChR subunits including $\alpha3$–$\alpha7$, $\beta2$, and $\beta3$ (Charpantier et al. 1998; Klink et al. 2001), whereas mainly $\alpha4$, $\beta2$, and $\alpha6$ nAChR subunits are located in dopaminergic terminals in the striatum (Champtiaux et al. 2003). Identification of these nAChR subtypes that are critical for nicotine dependence will provide insights into the pathophysiology of addiction and will also help to identify potential smoking cessation targets.

Genetically engineered mouse lines either missing individual nAChR subunits ("knock-out mice") or containing modified subunits ("knock-in mice") have been used to assess the role of specific nAChR subtypes in nicotine dependence. To date, $\alpha3$–$\alpha5$, $\alpha7$, $\beta2$, and $\beta4$ nAChR subunits have been targeted for deletion in mice. However, the majority of studies related to nicotine dependence have focused on the $\beta2$, $\beta4$, and $\alpha4$ knockouts. $\beta2$ knockout (KO) mice do not display an increase in dopamine release in the nucleus accumbens in response to acute intraperitoneal injections of nicotine, indicating that $\beta2^*$ receptors are necessary for dopamine release. In addition, $\beta2$ KO mice self-administer cocaine but fail to maintain self-administration when the cocaine is switched to nicotine, suggesting that $\beta2^*$ receptors are also necessary for the maintenance of self-administration (Picciotto et al. 1998) and reinforcement. $\alpha4$ KO mice express a higher basal level of striatal dopamine levels that does not increase in response to nicotine, supporting the idea that $\alpha4\beta2^*$ nAChRs are necessary for proper regulation of dopamine release. In addition, $\alpha4$ knockout animals exhibit a prolonged motor activity response to cocaine (Marubio et al. 2003). However, it is not known whether $\alpha4$ knockout mice have deficiencies in nicotine-induced reward or self-administration. Interestingly, $\beta4$ KO animals have milder withdrawal symptoms after chronic nicotine treatment is stopped, raising the interesting possibility that different nAChR subunits mediate reinforcement versus withdrawal (Salas et al. 2004).

More recently, knock-in mice expressing a single point mutation, Leu9´Ala, within the putative pore-forming M2 domain of the $\alpha4$ subunit (rendering $\alpha4^*$ receptors hypersensitive to nicotine), have been generated. In the absence of specific agonists, the Leu9´Ala mouse line allows for the selective activation of $\alpha4^*$ receptors with small doses of nicotine that do not activate other nAChR subtypes. Whereas KO animals provide answers to the question of necessity, the hypersensitive knock-in approach addresses the question of sufficiency. These animals are 50-fold more sensitive than wild type to nicotine-induced reward, tolerance, and sensitization, indicating that activation of $\alpha4^*$

receptors is sufficient for these phenomena (Tapper et al. 2004) and, perhaps, for nicotine dependence itself.

Chronic Nicotine Exposure, Gene Regulation, and Synaptic Plasticity

Acute nicotine exposure activates dopaminergic neurons within the VTA and elicits many physiological effects including reward, hypothermia, and, at high enough concentrations, seizures. However, smokers expose themselves to nicotine chronically. It is this chronic exposure that produces long-term physiological and behavioral changes associated with dependence. If laboratory animals, primates, and humans are persistently exposed to nicotine, they exhibit behaviors consistent with drug dependence including reinforcement, tolerance, sensitization ("reverse tolerance"), and withdrawal. At the molecular and neuronal levels, chronic nicotine produces nAChR desensitization and various long-term physiological alterations including nAChR up-regulation (Marks et al. 1983; Wonnacott 1990; Yates et al. 1995), modulation of gene expression (Konu et al. 2001; Li et al. 2002), and induction of long-term potentiation and depression at glutamatergic synapses (Mansvelder and McGehee 2000; Partridge et al. 2002).

How can chronic nicotine exposure produce these long-lasting physiological and behavioral changes? We have no definitive answers to this question. However, one possible mechanism may involve the ability of nAChRs to flux calcium or, alternatively, release internal calcium stores through depolarization and downstream activation of voltage-gated calcium channels. Calcium is a secondary messenger that can activate various signal transduction pathways and ultimately influence gene expression.

One potential molecule important for the adaptations associated with drug abuse is the cAMP responsive element binding (CREB) protein. CREB is regulated via phosphorylation by Ca^{2+}-/calmodulin-dependent kinases and by protein kinase A (PKA). Alternatively, CREB can be regulated by the mitogen-activated protein kinase (MAPK) pathway, a signaling cascade implicated in synaptic plasticity. When phosphorylated, CREB regulates expression of cAMP inducible genes (Nestler 2001). CREB participates in modulation of reward in response to cocaine, amphetamine, and morphine. For example, overexpression of CREB in the shell of the nucleus accumbens decreases cocaine- and morphine-induced reward. Decreasing CREB function through expression of a dominant negative form of CREB, on the other hand, increases cocaine reward and reinforcement (Carlezon et al. 1998). An analysis of nucleus accumbens CREB expression during chronic nicotine exposure followed by withdrawal in rats indicates that phosphorylated CREB (pCREB) and CREB expression are decreased during withdrawal but remain unchanged with acute nicotine exposure, indicating a potential role for CREB during nicotine reinforcement (Pluzarev and Pandey 2004). In a separate study, Brunzell et al. (2003) found a decrease in pCREB and CREB expression in the nucleus accumbens in mice chronically exposed to nicotine in their drinking water (Brunzell et al. 2003). These data suggest that CREB regulation may represent a common pathway associated with drug reinforcement.

Nicotine modulates synaptic strength at glutamatergic synapses of VTA dopaminergic neurons, potentially enhancing reward signaling and providing a neuronal correlate for addiction (Mansvelder and McGehee 2000). In whole-cell patch-clamp recordings of dopaminergic neurons in the VTA of brain slices, activation of both postsynaptic NMDA receptors and presynaptic nicotinic receptors was required for the production of long-term potentiation. Mansvelder and McGehee (2000) identified the nicotinic receptors as α7 because long-term potentiation was blocked by methyllycaconitine, an α7-specific antagonist (but see Klink et al. 2001).

The Potential Role of nAChR Up-regulation in Dependence

Steady-state concentrations of 0.1–0.5-μM nicotine are achieved in the bloodstream of smokers (Henningfield 1995). These nicotine concentrations have been shown to cause persistent desensitization of nAChRs in midbrain neurons. The cellular response to prolonged nAChRs desensitization is to increase expression of receptors at the cell surface, a process known as "up-regulation." Radioligand-binding studies have shown that smokers, as well as primates, mice, and rats continuously exposed to nicotine, contain more numerous [^{3}H]nicotine binding sites than normal controls. One possible molecular correlate of nicotine addiction may be the interplay between the initial, reinforcing feelings caused by the activation of nicotinic receptors, and tolerance and withdrawal caused by desensitization of nicotinic receptors. Up-regulation of nAChR would then normalize the pool of nondesensitized receptors in the presence of nicotine and increase the overall number of surface nAChRs that could undergo activation once nicotine has been completely cleared, thereby enhancing the positive reinforcement effects of nicotine. Supporting this idea, smokers generally report that the first cigarette of the day is the most enjoyable.

Alternatively, up-regulation of neuronal nicotinic receptors may result in hyperexcitability of dopaminergic neurons, perhaps causing anxiety or other withdrawal symptoms. Thus, obtaining a desensitizing dose of nicotine through smoking may be one possible strategy to achieve a normal level of neuronal excitability. On the basis of this hypothesis, smokers would self-administer nicotine not only as a reward but as a way to prevent withdrawal symptoms.

CONCLUSIONS

Nicotine is a naturally occurring alkaloid found in tobacco that is primarily responsible for addiction. At concentrations found in smokers' blood, chronic nicotine can activate and desensitize nicotinic acetylcholine receptors within the mesocorticolimbic pathways, producing increases in nucleus accumbens dopamine, reward signaling, up-regulation of nAChR, changes in synaptic plasticity, and modulation of gene expression. Together, this complicated combination of events that begins with activation of nAChRs can lead to behaviors associated with addiction including reinforcement, tolerance, sensitization, and withdrawal.

GAPS AND OPPORTUNITIES

Increasing evidence indicates that the first step toward nicotine dependence is activation of α4β2* neuronal nicotinic acetylcholine receptors by nicotine. However, little is known about the cascade of events that occurs from initial nicotine exposure to addiction.

As previously mentioned, one consequence of chronic nicotine exposure is nAChR up-regulation. Does up-regulation have a role in the onset or maintenance of addiction? To address this question, we must gain a deeper understanding of how nAChRs are regulated so that it may be possible to prevent or enhance up-regulation during nicotine exposure and determine how this affects dependence.

Although nAChR activation initiates dependence, likely hundreds of additional gene products downstream from receptor activation have a role in establishing a dependent state. Identifying these gene products and their role in nicotine dependence represents a major goal in addiction research. With the advent of microarray technology, differential expression of literally thousands of genes in a given neuronal population can be assayed during nicotine exposure. By combining this technique with various mouse models and nicotine treatments, many important genes critical for establishing and maintaining dependence may be characterized in the near future.

ACKNOWLEDGMENTS

Original research highlighted in this review was sponsored by grants from the National Institutes of Health (NS-11756, DA-17279), the California Tobacco-Related Disease Research Program (TRDRP), the Philip Morris External Research Program, and by fellowships from TRDRP (R.N.) and the National Institute of Neurological Disorders and Stroke (A.R.T.).

REFERENCES

Alkondon M., Pereira E.F., and Albuquerque E.X. 1998. α-Bungarotoxin- and methyllycaconitine-sensitive nicotinic receptors mediate fast synaptic transmission in interneurons of rat hippocampal slices. *Brain Res.* **810:** 257–263.

Brejc K., van Dijk W.J., Klaassen R.V., Schuurmans M., van Der Oost J., Smit A.B., and Sixma T.K. 2001. Crystal structure of an ACh-binding protein reveals the ligand-binding domain of nicotinic receptors. *Nature* **411:** 269–276.

Brunzell D.H., Russell D.S., and Picciotto M.R. 2003. In vivo nicotine treatment regulates mesocorticolimbic CREB and ERK signaling in C57Bl/6J mice. *J. Neurochem.* **84:** 1431–1441.

Bruses J.L., Chauvet N., and Rutishauser U. 2001. Membrane lipid rafts are necessary for the maintenance of the α7 nicotinic acetylcholine receptor in somatic spines of ciliary neurons. *J. Neurosci.* **21:** 504–512.

Carlezon W.A., Jr., Thome J., Olson V.G., Lane-Ladd S.B., Brodkin E.S., Hiroi N., Duman R.S., Neve R.L., and Nestler E.J. 1998. Regulation of cocaine reward by CREB. *Science* **282:** 2272–2275.

Champtiaux N., Gotti C., Cordero-Erausquin M., David D.J., Przybylski C., Lena C., Clementi F., Moretti M., Rossi F.M., Le Novere N., et al. 2003. Subunit composition of functional nicotinic receptors in dopaminergic neurons investigated with knock-out mice. *J. Neurosci.* **23:** 7820–7829.

Charpantier E., Barneoud P., Moser P., Besnard F., and Sgard F.

1998. Nicotinic acetylcholine subunit mRNA expression in dopaminergic neurons of the rat substantia nigra and ventral tegmental area. *Neuroreport* **9:** 3097–3101.

Conroy W.G., Liu Z., Nai Q., Coggan J.S., and Berg D.K. 2003. PDZ-containing proteins provide a functional postsynaptic scaffold for nicotinic receptors in neurons. *Neuron* **38:** 759–771.

Cooper E., Couturier S., and Ballivet M. 1991. Pentameric structure and subunit stoichiometry of a neuronal nicotinic acetylcholine receptor. *Nature* **350:** 235–238.

Corrigall W.A. and Coen K.M. 1989. Nicotine maintains robust self-administration in rats on a limited-access schedule. *Psychopharmacology* **99:** 473–478.

———. 1991. Selective dopamine antagonists reduce nicotine self-administration. *Psychopharmacology* **104:** 171–176.

Corrigall W.A., Coen K.M., and Adamson K.L. 1994. Self-administered nicotine activates the mesolimbic dopamine system through the ventral tegmental area. *Brain Res.* **653:** 278–284.

Corrigall W.A., Franklin K.B., Coen K.M., and Clarke P.B. 1992. The mesolimbic dopaminergic system is implicated in the reinforcing effects of nicotine. *Psychopharmacology* **107:** 285–289.

Corringer P.J., Le Novere N., and Changeux J.P. 2000. Nicotinic receptors at the amino acid level. *Annu. Rev. Pharmacol. Toxicol.* **40:** 431–458.

Drisdel R.C., Manzana E., and Green W.N. 2004. The role of palmitoylation in functional expression of nicotinic α7 receptors. *J. Neurosci.* **24:** 10502–10510.

Ehlers M.D. 2003. Activity level controls postsynaptic composition and signaling via the ubiquitin-proteasome system. *Nat. Neurosci.* **6:** 231–242.

Eisele J.L., Bertrand S., Galzi J.L., Devillers-Thiery A., Changeux J.P., and Bertrand D. 1993. Chimaeric nicotinic-serotonergic receptor combines distinct ligand binding and channel specificities. *Nature* **366:** 479–483.

Fenster C.P., Beckman M.L., Parker J.C., Sheffield E.B., Whitworth T.L., Quick M.W., and Lester R.A. 1999. Regulation of α4β2 nicotinic receptor desensitization by calcium and protein kinase C. *Mol. Pharmacol.* **55:** 432–443.

Gopalakrishnan M., Molinari E.J., and Sullivan J.P. 1997. Regulation of human α4β2 neuronal nicotinic acetylcholine receptors by cholinergic channel ligands and second messenger pathways. *Mol. Pharmacol.* **52:** 524–534.

Gray R., Rajan A.S., Radcliffe K.A., Yakehiro M., and Dani J.A. 1996. Hippocampal synaptic transmission enhanced by low concentrations of nicotine. *Nature* **383:** 713–716.

Henningfield J.E. 1995. Nicotine medications for smoking cessation. *N. Engl. J. Med.* **333:** 1196–1203.

Hicke L. and Dunn R. 2003. Regulation of membrane protein transport by ubiquitin and ubiquitin-binding proteins. *Annu. Rev. Cell. Dev. Biol.* **19:** 141–172.

Hoffman P.W., Ravindran A., and Huganir R.L. 1994. Role of phosphorylation in desensitization of acetylcholine receptors expressed in *Xenopus* oocytes. *J. Neurosci.* **14:** 4185–4195.

Hsu Y.N., Edwards S.C., and Wecker L. 1997. Nicotine enhances the cyclic AMP-dependent protein kinase-mediated phosphorylation of α4 subunits of neuronal nicotinic receptors. *J. Neurochem.* **69:** 2427–2431.

Jeanclos E.M., Lin L., Treuil M.W., Rao J., DeCoster M.A., and Anand R. 2001. The chaperone protein 14-3-3η interacts with the nicotinic acetylcholine receptor α4 subunit. Evidence for a dynamic role in subunit stabilization. *J. Biol. Chem.* **276:** 28281–28290.

Keller S.H., Lindstrom J., Ellisman M., and Taylor P. 2001. Adjacent basic amino acid residues recognized by the COP I complex and ubiquitination govern endoplasmic reticulum to cell surface trafficking of the nicotinic acetylcholine receptor α-subunit. *J. Biol. Chem.* **276:** 18384–18391.

Keller C.A., Yuan X., Panzanelli P., Martin M.L., Alldred M., Sassoe-Pognetto M., and Luscher B. 2004. The γ2 subunit of GABA_A receptors is a substrate for palmitoylation by GODZ. *J. Neurosci.* **24:** 5881–5891.

Klink R., de Kerchove d'Exaerde A., Zoli M., and Changeux J.P. 2001. Molecular and physiological diversity of nicotinic acetylcholine receptors in the midbrain dopaminergic nuclei. *J. Neurosci.* **21:** 1452–1463.

Konu O., Kane J.K., Barrett T., Vawter M.P., Chang R., Ma J.Z., Donovan D.M., Sharp B., Becker K.G., and Li M.D. 2001. Region-specific transcriptional response to chronic nicotine in rat brain. *Brain Res.* **909:** 194–203.

Laviolette S.R. and van der Kooy D. 2004. The neurobiology of nicotine addiction: Bridging the gap from molecules to behaviour. *Nat. Rev. Neurosci.* **5:** 55–65.

Laviolette S.R., Alexson T.O., and van der Kooy D. 2002. Lesions of the tegmental pedunculopontine nucleus block the rewarding effects and reveal the aversive effects of nicotine in the ventral tegmental area. *J. Neurosci.* **22:** 8653–8660.

Li M.D., Konu O., Kane J.K., and Becker K.G. 2002. Microarray technology and its application on nicotine research. *Mol. Neurobiol.* **25:** 265–285.

Mansvelder H.D. and McGehee D.S. 2000. Long-term potentiation of excitatory inputs to brain reward areas by nicotine. *Neuron* **27:** 349–357.

Mansvelder H.D., Keath J.R., and McGehee D.S. 2002. Synaptic mechanisms underlie nicotine-induced excitability of brain reward areas. *Neuron* **33:** 905–919.

Marks M.J., Burch J.B., and Collins A.C. 1983. Effects of chronic nicotine infusion on tolerance development and nicotinic receptors. *J. Pharmacol. Exp. Ther.* **226:** 817–825.

Marubio L.M., Gardier A.M., Durier S., David D., Klink R., Arroyo-Jimenez M.M., McIntosh J.M., Rossi F., Champtiaux N., Zoli M., and Changeux J.P. 2003. Effects of nicotine in the dopaminergic system of mice lacking the alpha4 subunit of neuronal nicotinic acetylcholine receptors. *Eur. J. Neurosci.* **17:** 1329–1337.

McGehee D.S. and Role L.W. 1995. Physiological diversity of nicotinic acetylcholine receptors expressed by vertebrate neurons. *Annu. Rev. Physiol.* **57:** 521–546.

McGehee D.S., Heath M.S.J., Gelber S., Devay P., and Role L.W. 1995. Nicotine enhancement of fast excitatory synaptic transmission in CNS by presynaptic receptors. *Science* **269:** 1692–1696.

Mohamed A.S., Rivas-Plata K.A., Kraas J.R., Saleh S.M., and Swope S.L. 2001. Src-class kinases act within the agrin/MuSK pathway to regulate acetylcholine receptor phosphorylation, cytoskeletal anchoring, and clustering. *J. Neurosci.* **21:** 3806–3818.

Nakayama H., Okuda H., and Nakashima T. 1993. Phosphorylation of rat brain nicotinic acetylcholine receptor by cAMP-dependent protein kinase in vitro. *Brain Res. Mol. Brain Res.* **20:** 171–177.

Nashmi R., Dickinson M.E., McKinney S., Jareb M., Labarca C., Fraser S.E., and Lester H.A. 2003. Assembly of α4β2 nicotinic acetylcholine receptors assessed with functional fluorescently labeled subunits: Effects of localization, trafficking, and nicotine-induced upregulation in clonal mammalian cells and in cultured midbrain neurons. *J. Neurosci.* **23:** 11554–11567.

Nestler E.J. 2001. Molecular basis of long-term plasticity underlying addiction. *Nat. Rev. Neurosci.* **2:** 119–128.

Olson E.N., Glaser L., and Merlie J.P. 1984. α and β subunits of the nicotinic acetylcholine receptor contain covalently bound lipid. *J. Biol. Chem.* **259:** 5364–5367.

Pacheco M.A., Pastoor T.E., and Wecker L. 2003. Phosphorylation of the α4 subunit of human α4β2 nicotinic receptors:

Role of cAMP-dependent protein kinase (PKA) and protein kinase C (PKC). *Brain Res. Mol. Brain Res.* **114:** 65–72.

Partridge J.G., Apparsundaram S., Gerhardt G.A., Ronesi J., and Lovinger D.M. 2002. Nicotinic acetylcholine receptors interact with dopamine in induction of striatal long-term depression. *J. Neurosci.* **22:** 2541–2549.

Picciotto M.R., Zoli M., Rimondini R., Lena C., Marubio L.M., Pich E.M., Fuxe K., and Changeux J.P. 1998. Acetylcholine receptors containing the beta2 subunit are involved in the reinforcing properties of nicotine. *Nature* **391:** 173–177.

Pickering D.S., Taverna F.A., Salter M.W., and Hampson D.R. 1995. Palmitoylation of the GluR6 kainate receptor. *Proc. Natl. Acad. Sci.* **92:** 12090–12094.

Pidoplichko V.I., DeBiasi M., Williams J.T., and Dani J.A. 1997. Nicotine activates and desensitizes midbrain dopamine neurons. *Nature* **390:** 401–404.

Pluzarev O. and Pandey S.C. 2004. Modulation of CREB expression and phosphorylation in the rat nucleus accumbens during nicotine exposure and withdrawal. *J. Neurosci. Res.* **77:** 884–891.

Roerig B., Nelson D.A., and Katz L.C. 1997. Fast synaptic signaling by nicotinic acetylcholine and serotonin 5-HT$_3$ receptors in developing visual cortex. *J. Neurosci.* **17:** 8353–8362.

Ross A.F., Green W.N., Hartman D.S., and Claudio T. 1991. Efficiency of acetylcholine receptor subunit assembly and its regulation by cAMP. *J. Cell Biol.* **113:** 623–636.

Salas R., Pieri F., and De Biasi M. 2004. Decreased signs of nicotine withdrawal in mice null for the beta4 nicotinic acetylcholine receptor subunit. *J. Neurosci.* **24:** 10035–10039.

Schultz W. 2004. Neural coding of basic reward terms of animal learning theory, game theory, microeconomics and behavioural ecology. *Curr. Opin. Neurobiol.* **14:** 139–147.

Scott D.B., Blanpied T.A., Swanson G.T., Zhang C., and Ehlers M.D. 2001. An NMDA receptor ER retention signal regulat-ed by phosphorylation and alternative splicing. *J. Neurosci.* **21:** 3063–3072.

Smotrys J.E. and Linder M.E. 2004. Palmitoylation of intracellular signaling proteins: Regulation and function. *Annu. Rev. Biochem.* **73:** 559–587.

Tapper A.R., McKinney S.L., Nashmi R., Schwarz J., Deshpande P., Labarca C., Whiteaker P., Marks M.J., Collins A.C., and Lester H.A. 2004. Nicotine activation of alpha4* receptors: Sufficient for reward, tolerance, and sensitization. *Science* **306:** 1029–1032.

Temburni M.K., Rosenberg M.M., Pathak N., McConnell R., and Jacob M.H. 2004. Neuronal nicotinic synapse assembly requires the adenomatous polyposis coli tumor suppressor protein. *J. Neurosci.* **24:** 6776–6784.

Thrower J.S., Hoffman L., Rechsteiner M., and Pickart C.M. 2000. Recognition of the polyubiquitin proteolytic signal. *EMBO J.* **19:** 94–102.

Unwin N. 2005. Refined structure of the nicotinic acetylcholine receptor at 4Å resolution. *J. Mol. Biol.* **346:** 967–989.

Viseshakul N., Figl A., Lytle C., and Cohen B.N. 1998. The α4 subunit of rat α4β2 nicotinic receptors is phosphorylated *in vivo*. *Brain Res. Mol. Brain Res.* **59:** 100–104.

Williams B.M., Temburni M.K., Levey M.S., Bertrand S., Bertrand D., and Jacob M.H. 1998. The long internal loop of the α3 subunit targets nAChRs to subdomains within individual synapses on neurons *in vivo*. *Nat. Neurosci.* **1:** 557–562.

Wonnacott S. 1990. The paradox of nicotinic acetylcholine receptor upregulation by nicotine. *Trends Pharmacol. Sci.* **11:** 216–219.

Yates S.L., Bencherif M., Fluhler E.N., and Lippiello P.M. 1995. Up-regulation of nicotinic acetylcholine receptors following chronic exposure of rats to mainstream cigarette smoke or alpha 4 beta 2 receptors to nicotine. *Biochem. Pharmacol.* **50:** 2001–2008.

11 | Opioids as a Model for Cell Biological Studies of Addictive Drug Action

Mark von Zastrow[1] and Christopher J. Evans[2]

[1]Departments of Psychiatry and Pharmacology, University of California at San Francisco, San Francisco, California 94143; [2]Neuropsychiatric Institute, University of California at Los Angeles, Los Angeles, California 90024

ABSTRACT

Opioids are among the most useful drugs in clinical medicine but are also widely abused. Although opioid drugs are highly effective analgesics when used acutely, their utility fades following prolonged or repeated administration. Chronic opioid exposure also produces complex physiological and neurobehavioral changes that underlie addiction. Opioid drugs bind to the same G-protein-coupled receptors as endogenously released opioid peptides, and opioid receptors also have a central role in modulating behavioral reward to natural stimuli and various non-opioid drugs of abuse. Early studies of opioid effects on cultured cells led the way to modern hypotheses about the cellular basis of drug action and adaptation, and established opioids as a model for mechanistic study of the cell biology of addiction. The past several years have seen considerable progress in elucidating specific mechanisms of opioid receptor regulation at the cellular level. Exciting current challenges are to determine how these mechanisms influence the effects of clinically relevant opioids in vivo and to rigorously test the role of specific cellular regulatory mechanisms in generating the complex neurobehavioral syndrome of addiction.

INTRODUCTION

Opioid abuse and addiction are societal issues of increasing significance. We typically think of opioid abuse in terms of illegal drugs such as heroin. However, commonly prescribed opiate analgesics such as oxycodone, morphine, and fentanyl are also widely abused. The 2003 National Survey on Drug Use and Health (NSDUH), which assessed patterns of drug usage within the United States during a 10-year period, indicates that both heroin use and the illicit (i.e., not medically directed) use of prescription opioids have increased dramatically in recent years. The Trends in Drug-related Emergency Department (ED) visits cite 93,519 ED visits to hospitals in the United States in 2002 as heroin related, compared with 63,158 in 1994. The lifetime prevalence for prescription

This chapter is written as a companion to lectures delivered by the authors for the Cold Spring Harbor Laboratory course entitled "Cellular Biology of Addiction."

opioid abuse in adults was estimated to be 6.8% in 1992, but 22.1% in 2002, with an approximately threefold increase in the estimated number of new users in 2001 compared to 1992. Illicit opioid injection is also associated with other important medical problems, most notably as a major route for transmission of HIV and hepatitis. Thus, opioid abuse clearly remains a major medical and societal problem, whose prevalence and deleterious consequences appear to be on the rise (see http://www.oas.samhsa.gov/ for compiled statistics).

Opioids have been used medicinally for many centuries and remain among the most effective and humane drugs available for providing relief from moderate to severe pain. Although societal standards regarding the appropriate use of opioids—and thus the definition of abuse—have differed among cultures and eras, the general problem of opioid dependence was recognized in Europe, the Middle East, and North Africa at least as far back as the fifth century B.C. Hippocrates (460–357 B.C.) recommended that physicians prescribe opium only in a sparing and strictly controlled manner because of its deleterious effects of excessive or prolonged use. During the 1800s, improved methods of opioid purification and synthetic modification, together with the application of hypodermic injection methods, improved the clinical utility of opioids but also enhanced problems of drug dependence and overdose. The modern concept of opioid addiction as a medical disorder emerged in the early 1900s. Recent progress in defining biological underpinnings of chronic opioid effects has contributed to a current (and still controversial) view that addiction represents a drug-induced brain disease (Booth 1998).

Opioid addiction has become an increasingly exciting area of fundamental neuroscience research because primary targets of opioid action are well defined, reliable animal models have been established for assessing complex physiological and behavioral effects of opioids, and a number of opioid actions have been rigorously described at the cellular level. It is increasingly apparent that opioid effects differ profoundly between acute and chronic (or repeated) administration, and environmental cues can have a major impact on opioid responses and addictive behaviors. It is also apparent that individuals differ quite substantially in their susceptibility to addiction following prolonged use of opioids, even though many neurobehavioral changes produced by chronic or repeated administration of opioids are similar across individuals. These considerations have led to the view that the nervous system undergoes fundamental and complex changes in response to chronic or repeated opioid exposure, and have motivated increased efforts to understand opioid regulatory effects at the level of individual signaling molecules, neurons, and neural circuits (Koob 2000; Williams et al. 2001; Berridge and Robinson 2003; Nestler and Malenka 2004).

GOALS OF THE CHAPTER

In this chapter, we review progress in understanding the effects of opioids by focusing on opioid receptors themselves and their regulation at the cellular level. We then use this information as a background for examining how cell biological hypotheses of opioid adaptation are being tested in animal models. Our overall goal is to provide one perspective on a specific approach to elucidating the biological basis of addiction. We hope the reader takes away an appreciation of some key concepts with regard to cellular mechanisms of opioid regulation, a context from which to consider the cellular basis of addiction, and a sense of some of the new challenges that lie ahead in this exciting field. Specific goals of the present chapter are as follows:

1. Sketch a brief overview of opioid drug effects and general concepts of physiological adaptation following chronic exposure.

2. Introduce the endogenous opioid peptide system and discuss its function in, and interdependence on, other endogenous signaling systems that modulate behavioral reward to various stimuli.

3. Provide insight into the historical basis for cell biological study of opioid effects and into some key hypotheses regarding the cellular underpinnings of opioid tolerance and dependence.

4. Summarize briefly a current understanding of opioid receptor regulation and evidence for distinct cellular regulatory effects of different opioid drugs.

5. Discuss current progress toward linking a specific cellular mechanism of opioid receptor regulation to clinically relevant effects on whole-animal physiology and behavior.

METHODOLOGICAL CONSIDERATIONS AND RELEVANCE TO ADDICTION

The present chapter is broad in scope and covers a wide range of experimental methods. The study of opium-derived and synthetic opioids has been an active area of research involving organic and natural product chemistry. The work leading to identification of endogenous opioid-binding sites and peptide ligands for these binding sites involved pharmacological methods, biochemical purification, and complementary DNA (cDNA) cloning techniques. Studies of opioid receptor regulation have involved cell culture methods, heterologous expression, and site-directed mutagenesis techniques, as well as various immunochemical and biochemical assays of receptor localization and modification. In vivo studies of opioid regulation have benefited from sophisticated behavioral assays of acute and chronic opioid effects, together with transgenic and gene knockout methodologies. Because opioid receptors are the major targets of addictive drugs such as morphine and heroin, the study of opioid biology is intrinsically relevant to addiction. In addition, as discussed in the following section, endogenous opioid signaling is involved in mediating the rewarding effects of a number of other important drugs that are widely abused and addictive but do not themselves bind to opioid receptors.

An Overview of Opioid Addiction and Differences between Acute and Chronic Drug Effects

A general view of how responses to acute and chronic opioids differ is schematized in Figure 1. A number of factors can motivate the initial administration of opioids. In the case of medicinal use, the acute motivation is often quite simple, involving a primary clinical indication such as pain or diarrhea. Factors motivating illicit opioid administration may be more complex and varied, and are thought to include genetic susceptibility, environmental history, beliefs, societal pressures, drug availability, mood, and prior drug experiences.

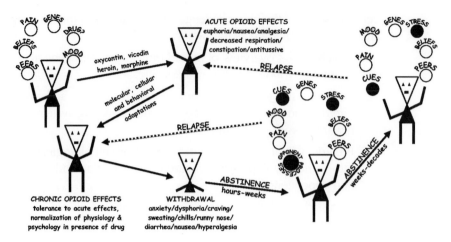

FIGURE 1. Differences between acute and chronic opioid effects.

Once the opioid is administered, most of the acute effects are physiologically robust. Acute morphine is an excellent analgesic, reliably induces respiratory depression, is antitussive (prevents coughing), and inhibits gut motility. In contrast, the psychological effects of acute opioids are remarkably variable. Some users find opiates irresistibly euphoric, whereas others experience them as uncomfortably mind altering. Following chronic or repeated administration, many opioid effects observed wane so that, with time, it becomes necessary to administer increasing doses of opioid to achieve similar effects. This phenomenon, called opioid tolerance, can develop to different rates and to different extents for various opioid effects. For example, tolerance to opioid-induced nausea often develops rapidly and is clinically useful because it allows opioids to be used as effective analgesics even in individuals who are initially susceptible to this aversive effect. Tolerance to the antinociceptive effects of opioids (often called analgesic tolerance) is clinically deleterious because it imposes a major limitation on the effectiveness of opioids for treating chronically painful conditions. Tolerance to the euphoric effects of opioids has been proposed to reduce the abuse liability of opioids in clinical populations, but this phenomenon may contribute to the tendency of some addicts to administer opioids at ever-increasing doses.

Chronic opioid administration, in addition to changing responsiveness to a subsequent opioid dose, can produce significant physiological and behavioral changes that are evident after opioid use is discontinued. If opioids are withdrawn after a period of prolonged or repeated administration, a variety of "withdrawal" effects are observed that are essentially opposite to the actions of acute opioids and include dysphoria, diarrhea, and enhanced sensitivity to painful stimuli (or hyperalgesia). The precipitation of withdrawal symptoms by cessation of opioid administration defines the phenomenon of opioid dependence. Most opioid withdrawal symptoms are aversive and can be quickly reversed by readministration of opioids. Thus, avoidance of the withdrawal syndrome is thought to be a powerful motivator of continued opioid administration in actively using addicts. The opioid withdrawal syndrome, like tolerance, is time limited, and withdrawal symptoms typically disappear within several weeks after discontinuation of drug use and clearance from the bloodstream. However, an additional consequence of chronic opioid administration is the development of a state in which the addict remains preoccupied with obtaining opioids (and perhaps other drugs of abuse) even after physiological withdrawal symptoms have passed. This more persistent drug-initiated state is generally called craving. Craving is often precipitated by exposure of the addict to environmental cues previously associated with drug procurement or use, and these cues can retain their salience for years after the last use of opioid drug. These persistent psychological processes are very important to the clinical management of opioid addiction, during which it is common for relapse to occur long after the abatement of overt withdrawal symptoms.

A basic hypothesis is that chronic opioids induce various changes in neural function that effectively normalize physiological responses in the face of persistent opioid stimulation. In principle, such homeostatic adaptation of the nervous system could be mediated by attenuation of opioid-dependent signaling processes themselves or by enhancement of other processes that effectively counteract opioid effects. Such adaptive responses, in addition to their likely contribution to physiological tolerance and dependence, may underlie opponent processes that have been described in motivational learning theory and have been invoked to explain some of the complex behavioral features of the drug-addicted state. In particular, great interest exists in understanding the role that opioids play in modulating neural circuitry mediating behavioral reward, because adaptive processes associated with these pathways are thought to have a major role in motivating opioid self-administration in addicts. Thus, a major goal of current opioid research is to define specific regulatory effects of opioid drugs on the physiology of neurons and neural circuits associated with reward (Koob 2000).

Opioids Act Via Specific Receptors that Modulate Diverse Neurobehavioral Processes

A series of seminal studies, beginning with the discovery of endogenous morphine-binding sites in brain membranes, led to the appreciation that opioid drugs bind to the same receptor sites that are normally activated by endogenously produced opioid peptide ligands (Evans 2004). Molecular cloning has identified three receptor types of opioid: MOP (μ-opioid peptide), DOP (Δ-opioid peptide), and KOP (κ-opioid peptide) receptors that are structurally homologous but encoded by distinct structural genes (Zaki et al. 1996; Kieffer 1999). These receptors, members of the large G-protein-coupled receptor (GPCR) superfamily, possess a conserved seven-transmembrane topology and are similar to rhodopsin. When expressed as recombinant proteins, the cloned opioid receptors show many (but not all; see Chapter 12, this volume) of the pharmacological properties of classical μ, Δ, and κ receptor types observed in native tissue preparations. A fourth gene product, known as the opioid receptor-like (ORL-1 or ORP) receptor, shares extensive sequence homology with classical opioid receptors (particularly, KOP receptors). However, the ORP receptor is not generally considered a "true" opioid receptor because it shows low affinity for most opioid drugs and endogenous opioid peptide ligands. Specific opioid peptides and receptors have been mapped to neuroanatomical pathways mediating a diverse array of important functions including pain perception, autonomic regulation, motivational processing, and learning.

A diverse array of ligands bind opioid receptors (see Fig. 2). Many of the alkaloid ligands are similar to the structure of morphine, including codeine (the N-methyl deriva-

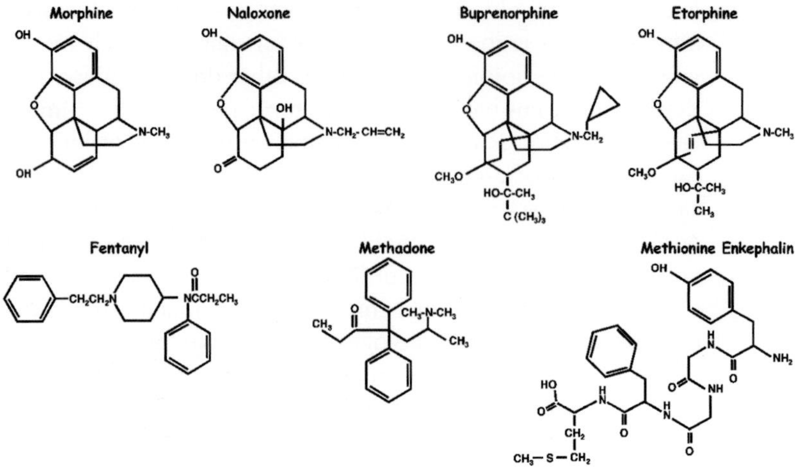

FIGURE 2. Examples of opioid ligand structures.

tive of morphine and also a natural alkaloid in the opium poppy); oxycodone (OxyContin/Percodan), hydrocodone (Vicodin), buprenorphine, and etorphine (derivatives of thebaine, the precursor of morphine in the opium poppy); heroin (the diacetyl ester of morphine); and the receptor antagonists naloxone and naltrexone. Fentanyl and methadone are other well-known opioids with very different structures from the morphine-like alkaloids and from opioid peptides such as met-enkephalin.

The endogenous opioid peptides are produced by proteolytic cleavage of three precursor polypeptides: proenkephalin, pro-opioimelanocortin, and prodynorphin, which are encoded by distinct genes. The various (more than 20) active opioid peptides produced from these precursors differ in relative potency for μ, Δ, and κ receptor activation, yet they share structural features required for receptor binding (Evans 2004). All known endogenous mammalian opioid peptides have an amino-terminal Tyr-Gly-Gly-Phe Met/Leu sequence and these peptides are clearly established to function as endogenous opioids. A large number of synthetic peptides have also been generated with opioid agonist activity but, to date, no additional peptides have been established as endogenous opioids. Some controversy surrounds two peptides, endomorphin 1 and 2, that were discovered by screening a synthetic peptide library and that have remarkably potent and specific μ agonist activity. However, despite the implication of their name, it remains unclear if these compounds are truly endogenous opioid ligands. No gene encoding the endomorphins or a putative precursor has been identified. Furthermore, knockout of the proenkephalin gene phenocopies many behavioral features of MOP receptor knockout mice, suggesting that proenkephalin-derived opioids are the major endogenous ligands for the μ-opioid receptor. Finally, although both enkephalins and endomorphins can promote rapid internalization of MOP receptors when infused into spinal cord slices, studies using peptidase inhibitors suggest that MOP receptor internalization produced by release of endogenous opioids is independent of endomorphins (Song and Marvizon 2003). Thus, although there is no doubt that endomorphins have potent μ agonist properties when administered exogenously, we cannot unequivocally classify them as endogenous opioids at this time.

The behavioral consequences of activating different members of the opioid receptor family can be opposing in some cases and similar in others. For example, both μ and κ agonists are analgesic. However, when tested in behavioral assays, μ agonists are rewarding and κ agonists are aversive. Opioid drugs have a complex in vivo pharmacology. Most opioid drugs modulate multiple members of the opioid receptor family and have receptor-active metabolites. Some opioid drugs can also affect non-opioid targets. Methadone, for example, modulates N-methyl-D-aspartate (NMDA)-type ionotropic glutamate receptors when administered at clinically relevant doses. Opioid drugs, in addition to differing in potency and pharmacokinetics, also differ widely in their intrinsic activity at opioid receptors. For example, methadone is a relatively efficacious agonist at μ-opioid receptors, whereas buprenorphine is a weak partial agonist. Finally, as discussed later in this review, the different opioid ligands can produce different regulatory effects on receptors and may also differ in signaling (Lecoq et al. 2004). It is also increasingly apparent that the precise regulatory effects of a given opioid are highly dependent on the cellular environment (Evans 2004).

Knockout mice have been very useful for defining the primary receptors mediating the effects of various opioid drugs (Kieffer and Gaveriaux-Ruff 2002; Contet et al. 2004). Morphine is a relatively straightforward drug in this regard. Morphine binds with highest affinity to μ-opioid receptors, but is also an efficacious (although less potent) agonist of Δ- and κ-opioid receptors. Remarkably, morphine-induced analgesia and dependence are absent in MOP receptor knockout mice. The principle biological effects of fentanyl are also absent in MOP knockout mice. However, other opioid receptors may have important roles for other drugs. For example, buprenorphine, a KOP receptor antagonist and weak partial agonist of MOP, DOP, and ORP receptors, retains its rewarding effects in MOP receptor knockout mice and these effects can be blocked by the opiate antagonist naloxone. These observations suggest that DOP activation, together with KOP antagonism, can produce significant rewarding activity independent of MOP receptor activation. Furthermore, buprenorphine's analgesic efficacy appears to be attenuated by coactivation of ORL-1 receptors (Eitan et al. 2003). Receptor knockout mice have also been important in teasing out the endogenous systems that mediate opioid drug reward. It is interesting that knockout of either the cannabinoid 1 (CB1) or the neurokinin 1 (NK1) receptor significantly impairs the ability of morphine to produce behavioral reward without affecting antinociceptive effects. Given that CB1 and neurokinin antagonists are clinically available (a cannabinoid antagonist is currently in phase 3 clinical trials for obesity), it is conceivable that novel combination therapies could be developed for maximizing the analgesic effects of opioids while attenuating their tendency to produce dependence.

The Endogenous Opioid System Modulates the Rewarding Effects of Diverse Stimuli

Beyond mediating the primary effects of opioid drugs, a long history documents the involvement of the endogenous opioid system in the rewarding activity of (1) other drugs of abuse and (2) natural rewards. Accumulating data from knockout mice suggest that the endogenous opioid system indeed has a critical role in the rewarding properties of many abused drugs. Animals lacking MOP receptors display behaviors indicating markedly diminished reward for alcohol, marijuana, and nicotine (Contet et al. 2004). Some behaviors that depend on natural rewards, such as maternal attachment, also appear to be disrupted in MOP receptor knockout mice (Moles et al. 2004). MOP receptor null pups appear unbothered on separation from the mother, showing no ultrasonic distress calls. Additionally, MOP receptor null pups show no preference for the mother's nest over the nest of another female. Interestingly, given the dependence of NK1 receptors for opioid reward, a similar phenotype has been observed with NK1 antagonists and in NK1 receptor knockout mice (Rupniak et al. 2000). Opioid receptor antagonists such as naloxone are highly aversive when administered to drug-naïve rodents, suggesting the importance of endogenous opioid signaling for natural hedonic homeostasis. Naloxone aversion is not observed in mice lacking the MOP receptor or in enkephalin knockout mice, leading to the hypothesis that naloxone produces aversion by antagonizing proenkephalin-derived activation of MOP receptor signaling that occurs under normal conditions in vivo. Taken together, these data suggest that the opioid system may be critical for producing many addictive behaviors as well as for modulating the rewarding effects of various natural stimuli.

Viewing Opioid Adaptations as a Cell Biological Problem

A cellular approach to the study of opioid adaptations had its genesis in the application of ex vivo tissue preparations, cultured cell models, and the study of opioid pharmacology. This is well illustrated by a series of seminal papers, published during the 1970s from the laboratory of Marshall Nirenberg, a remarkable investigator who in 1968 was awarded a Nobel Prize for his major contributions to defining the genetic code. Nirenberg and colleagues described an opioid-responsive cultured cell line (NG108-15 cells) in which morphine's ability to regulate cyclic AMP (cAMP) responses became dramatically attenuated following chronic exposure, and in which removal of morphine from the culture medium produced an opposing "overshoot" effect on cellular cAMP levels. These investigators proposed that these responses represent cellular correlates of opioid tolerance and dependence, respectively (Sharma et al. 1977). This led to a more general proposal that opioid addiction can be understood in terms of homeostatic (or possibly maladaptive) regulatory processes occurring at the level of individual opioid-responsive neurons. In the 30 years since this hypothesis was proposed, great progress has been made in defining opioid receptors and other components of opioid signaling networks, and in elucidating various cellular mechanisms of opioid regulation. However, specific relationships among cellular mechanisms of opioid regulation and their in vivo function in addiction—the core tenet of the Nirenberg hypothesis—remain poorly understood. A critical goal presently in the opioid field, and an exciting challenge in neuroscience in general, is to link cellular mechanisms of opioid drug action and regulation to a systems-level understanding of the complex neurobehavioral syndrome of addiction.

Regulation of Opioid Receptors Themselves

A basic finding from early studies, both of cultured cells and ex vivo tissue explants, was that prolonged exposure to opioids produces a reduced signaling response. Under some conditions, reduced signaling was observed without a net decrease in the number of receptors present in the cells or tissue explant. Such a reduction in opioid signaling was called functional desensitization to distinguish it from reduced responsiveness observed under other conditions that were associated with a net decrease, or down-regulation, of the total receptor complement. Desensitization and down-regulation of opioid receptors have been clearly shown to occur in native neural tissue and appear to reflect fundamental regulatory mechanisms that are now known to apply to ligand-mediated regulation of a wide variety of GPCRs (Lefkowitz 1996; Law et al. 2000).

Effects of Opioids on Receptor Regulation by Phosphorylation and Endocytosis

In recent years, and building on molecular mechanisms elucidated from the study of other GPCRs (such as rhodopsin and the β-adrenergic receptor [Lefkowitz et al. 1998]), considerable progress has been made in our understanding of the process by which opioid signaling can be rapidly and reversibly attenuated in response to receptor activation. The key features of this mechanism are that agonist-activated opioid receptors become functionally "uncoupled" from heterotrimeric G proteins and concentrate in clathrin-coated pits that subsequently undergo dynamin-dependent fission from the plasma membrane (Keith et al. 1996; Chu et al. 1997; Zhang et al. 1998). These events can occur within minutes after opioid receptor activation and are promoted by a highly conserved mechanism (Carman and Benovic 1998; Lefkowitz et al. 1998) involving phosphorylation of receptors by G-protein-coupled receptor kinases (GRKs) and association of receptors with "nonvisual" or "β" arrestins (Fig. 3A–C).

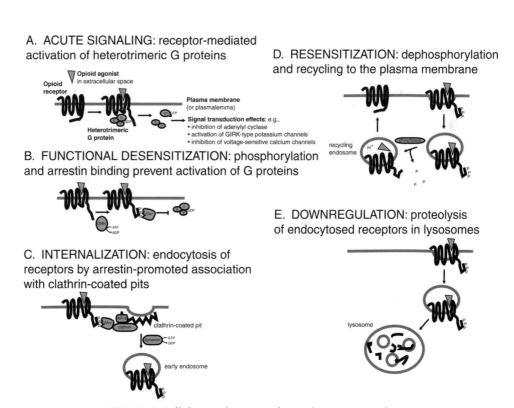

FIGURE 3. Cellular mechanisms of opioid receptor regulation.

Down-regulation (or not) of Opioid Receptors after Prolonged or Repeated Activation

The process of receptor down-regulation was originally defined by a net decrease in total opioid-binding sites detected in tissue lysates by a radioligand-binding assay. An extension of this radioligand-binding method to include subcellular fractionation suggested that opioid receptor down-regulation is mediated by proteolysis of receptors in lysosomes. It turns out that receptors can be down-regulated by other mechanisms as well, but the importance of lysosomal proteolysis in opioid receptor down-regulation has been supported in numerous subsequent studies using a variety of experimental approaches (Law et al. 2000). Activated opioid receptors are transported to lysosomes by membrane trafficking through the endocytic pathway. However, not all endocytosed receptors are down-regulated. For example, in some cases, internalized receptors can undergo efficient recycling to the plasma membrane, without any detectable proteolysis. It appears that the long-term functional consequences of ligand-induced endocytosis differ profoundly depending on the sorting of internalized receptors between divergent "downstream" membrane pathways (Tsao et al. 2001). Recycling of MOP receptors to the plasma membrane is thought to promote dephosphorylation of receptors and thereby enhance functional recovery of cellular signaling activity (Koch et al. 1998), similar to the paradigm of endocytosis-promoted "resensitization" of the β-adrenergic receptor (Lefkowitz et al. 1998). This signal-enhancing effect of endocytosis is essentially opposite from the signal-attenuating function of endocytosis in promoting down-regulation of DOP receptors by endocytic trafficking to lysosomes (Law et al. 1984; Tsao and von Zastrow 2000b). Nevertheless, both of these divergent "downstream" membrane pathways can be accessed from a shared early endocytic intermediate and can follow receptor endocytosis mediated via clathrin-coated pits (Tsao and von Zastrow 2000a). This observation has motivated considerable interest in understanding the specificity with which specific receptors are "sorted" among divergent downstream membrane pathways after endocytosis (Fig. 3D–E).

GPCRs can be sorted to lysosomes by a mechanism involving their covalent modification with ubiquitin. Ubiquitin-directed sorting of GPCRs, noted initially in studies of the yeast mating factor receptor, represents a highly conserved mechanism that is applicable to many membrane proteins in diverse organisms (Hicke 2001), including various mammalian GPCRs (e.g., see Marchese and Benovic 2001; Shenoy et al. 2001). However, additional machinery may contribute to the complex membrane trafficking of opioid receptors, because DOP receptors can undergo efficient lysosomal sorting in the apparent absence of ubiquitin modification in cultured mammalian cells (Tanowitz and von Zastrow 2002). This process uses similar membrane machinery as ubiquitination-directed sorting (Hislop et al. 2004) but can be influenced by a set of novel (and apparently ubiquitination-independent) protein interactions with receptors (Whistler et al. 2002; Simonin et al. 2004). Furthermore, MOP receptors contain a distinct cytoplasmic sequence that directs plasma membrane recycling rather than lysosomal sorting (Tanowitz and von Zastrow 2003). Taken together, these results suggest the existence of additional biochemical mechanisms controlling the endocytic trafficking (and hence down-regulation) of opioid receptors that could contribute significantly to functional plasticity in the endogenous opioid signaling system and may be important for distinguishing the acute and chronic regulatory effects of various opioid drugs.

Differences in Receptor Regulatory Effects among Various Opioid Peptides and Drugs

Early studies of radioligand binding to intact neuroblastoma cells suggested that morphine differs significantly from enkephalin in its ability to induce regulated endocytosis of DOP receptors (von Zastrow et al. 1993). This idea was extended to MOP receptors expressed in transfected fibroblast cells, under conditions of both acute (Keith et al. 1996, 1998) and chronic (Arden et al. 1995) receptor activation, and similar observations have been reported from the study of transfected hippocampal neurons (Bushell et al. 2002). Overexpression of GRK2 rendered morphine capable of promoting rapid phosphorylation and endocytosis of MOP receptors (Zhang et al. 1998). Overexpression of arrestin 2 or 3 (β-arrestin 1 or 2), in the absence of GRK overexpression, also promoted endocytosis of morphine-activated receptors (Whistler and von Zastrow 1998). Taken together, these results suggested that morphine-activated opioid receptors are relatively resistant to regulated endocytosis because they are not optimal substrates either for GRK-mediated phosphorylation or for binding to arrestins.

Although the existence of pronounced "agonist-selective" differences in the regulatory effects of various opiate drugs is now well accepted, the molecular pharmacology underlying these differences remains poorly understood (Fig. 4). Many opiate analgesics, including morphine, are partial agonists that have lower intrinsic efficacy than opioid peptides in assays of receptor-mediated activation of heterotrimeric G proteins. Agonist-specific differences in desensitization of cloned MOP receptors expressed in *Xenopus* oocytes correlated with differences in relative agonist efficacy, consistent with a simple two-state model of opioid receptor activation and arguing against the existence of signif-

A. Opioids differ only in the relative strength with which they promote or stabilize the **same** activated receptor conformation

B. Opioids differ also in their ability to promote or stabilize functionally **distinct** receptor conformations

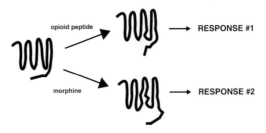

FIGURE 4. Hypotheses regarding the origin of ligand-specific opioid effects.

icant functional differences among opiate agonists (Kovoor et al. 1998). Similar studies conducted in transfected mammalian fibroblastic cells supported the importance of agonist efficacy but also suggested that certain opiates, such as morphine, induce regulatory phosphorylation (Yu et al. 1997) and endocytosis (Whistler et al. 1999) of receptors to a significantly smaller degree than would be predicted on the basis of a strict correlation with relative agonist efficacy. Furthermore, mutations of the cytoplasmic tail of the MOP receptor, which alter the degree to which various opiates promote regulatory endocytosis of receptors, do so without detectably changing the order of relative agonist efficacy (Whistler et al. 1999). Similar observations have been made in transfected PC12 neurosecretory cells, for which an additional distinction between MOP receptor desensitization (functional attenuation of signaling) and endocytosis has been suggested (Borgland et al. 2003), and in studies comparing the effects of a series of opioids on signaling and desensitization of endogenously expressed MOP receptors in acutely prepared slices from rat locus coeruleus (Alvarez et al. 2002).

It was also shown that a number of opiate partial agonists induce up-regulation of surface µ- and Δ-opioid receptors, in contrast to receptor down-regulation produced by several full (and some partial) agonists (Zaki et al. 2000). This effect, which may reflect ligand-dependent regulation of endocytic (Zaki et al. 2000) or biosynthetic (Petaja-Repo et al. 2002) pathways, further supports the idea that signaling and regulatory effects of opiates can be dissociated in some cases. Furthermore, evidence shows that distinct peptide and nonpeptide opioids can differ in the relative degree to which they promote internalization relative to down-regulation of DOP receptors, suggesting that opioids may differentially regulate the trafficking of opioid receptors after endocytosis as well (Okura et al. 2000; Marie et al. 2003). Pronounced ligand-specific effects have been observed in studies of the regulated membrane trafficking of some other GPCRs, such as serotonin receptors in which even certain antagonists induce rapid endocytosis (Willins et al. 1999). Thus, differential ligand effects on receptor regulation may not be unique to opioids. It is possible that these differential regulatory effects do indeed reflect the existence of distinct ligand-induced conformational states of the receptor, analogous to those inferred from biophysical studies of the β-2-adrenergic receptor (Seifert et al. 2001). However, it is not presently known if this hypothesis is correct in the case of opioid receptors and, if so, if distinct conformational effects of opioid drugs are relevant to understanding drug action in vivo.

Linking the GRK/Arrestin/Endocytic Mechanism to In Vivo Opioid Adaptation

With the elucidation of the GRK/arrestin/endocytic mechanism, it is now possible to explain a number of the cellular receptor regulatory effects that Nirenberg and others described in early studies. This advance has motivated efforts to undertake increasingly more definitive tests of the functional significance of the GRK/arrestin/endocytic mechanism in vivo, by examining acute and chronic opioid effects in rodent models. Considerable recent work has taken place on physiological tolerance, which was proposed in early studies to be a direct consequence of attenuated receptor signaling in target neurons. These recent efforts, although still at an early stage, are beginning to yield exciting results. The strongest data so far relate to an arguably simple type of antinociceptive tolerance induced by chronic opioids, called "intrinsic" tolerance because it is observed under conditions in which behavioral associations with (nondrug) environmental cues are minimized (Williams et al. 2001).

Relatively early studies in this series indicated that disrupting the gene encoding arrestin 3 (β-arrestin 2) both enhanced the potency of morphine for producing acute analgesia and reduced the development of intrinsic antinociceptive tolerance following chronic administration of morphine (Bohn et al. 1999, 2000). This observation led to the hypothesis that arrestin-dependent desensitization of opioid receptors may indeed contribute directly to the development of physiological tolerance in vivo. An interesting additional finding was that, although β-arrestin-2 knockout animals showed reduced antinociceptive tolerance to chronic morphine and heroin, antinociceptive tolerance induced by several other opioids (e.g., fentanyl and methadone) was not affected by the gene knockout (Bohn et al. 2002). However, disruption of the GRK3 gene reduced antinociceptive tolerance induced by fentanyl but did not affect tolerance induced by chronic morphine (Terman et al. 2004). A different experimental approach, in which the tendency of opioid receptors to exist in oligomeric complexes (see Chapters 12 and 15, this volume) was used to enhance endocytosis of morphine-activated MOP receptors in the spinal cord of wild-type rats, suggested that increasing receptor endocytosis reduced the development of antinociceptive tolerance to morphine (He et al. 2002). These various results, which seem inconsistent at first and could have multiple unforeseen pharmacological explanations, emphasize the importance of developing a more refined understanding of how distinct cellular mechanisms of opioid receptor regulation function in the setting of chronic drug exposure in vivo.

Using, as a starting point, the simple hypothesis for opioid tolerance proposed by Nirenberg and colleagues, one would expect that impairment of the GRK/arrestin-mediated receptor desensitization mechanism would attenuate the development of tolerance. This is consistent with the effects of deleting β-arrestin 2 on antinociceptive tolerance induced by morphine. However, it is not clear why opioids that engage the GRK/arrestin-mediated regulatory mechanism more strongly would not be affected to at least the same degree. Indeed, it is remarkable that opioids whose chronic effects are most affected by β-arrestin 2 knockout (such as morphine and heroin) are precisely those that engage the GRK/arrestin-dependent regulatory mechanism most weakly in heterologous cell models.

One possible explanation for this apparent discrepancy is that redundancy among arrestin isoforms renders opioids that strongly engage the GRK/arrestin-mediated regulatory mechanism relatively insensitive to knockout of any specific isoform. In contrast, regulation of the effects of opioids (such as morphine) that engage this regulatory machinery relatively weakly is limited by total GRK/arrestin activity and is thus highly sensitive to disruption—of even a single isoform (Bohn et al. 2004). Another possible explanation is suggested by the fundamentally nonredundant nature of GRK/arrestin-dependent receptor regulatory mechanisms when evaluated at the cellular level. Even though both functional uncoupling (desensitization) and endocytosis of opioid receptors are promoted by GRKs and arrestins, for receptors (such as MOP receptors) that recycle efficiently, these regulatory mechanisms mediate opposite net effects on opioid signaling (Fig. 3B,E). Thus one can envision desensitization and endocytosis of MOP receptors essentially as "opponent processes" driven by a shared biochemical mechanism. Opioids such as morphine, although they engage the GRK/arrestin mechanism relatively weakly, could produce a net attenuation of MOP receptor signaling with prolonged exposure because they promote endocytosis (and subsequent resensitization) even less strongly. In this case, reducing GRK/arrestin activities would be expected to produce a net *increase* in opioid signaling—attenuating the degree of tolerance induced by morphine. In contrast, opioids that engage the GRK/arrestin mechanism more strongly (such as methadone and fentanyl) also promote endocytosis of MOP receptors in a robust manner, driving both desensitization and endocytosis mechanisms strongly. In this case, partial reduction of GRK/arrestin activities would be expected to inhibit both opposing processes, producing relatively little *net* effect on opioid signaling; thus, tolerance to methadone or fentanyl would be unchanged.

These ideas, although they remain highly speculative (and probably overly simplistic), provide a conceptual framework for linking cellular mechanisms of opioid regulation to in vivo effects (Fig. 5). It will be important in future studies to examine regulatory effects more precisely in specific neurons or neural circuits mediating opioid-dependent behaviors.

A. Tolerance is a direct consequence of opioid receptor desensitization

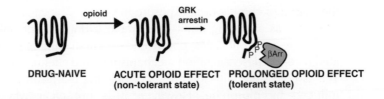

B. Tolerance reflects a balance between opposing effects of opioid receptor desensitization and endocytosis processes, together with adaptation of 'downstream' signaling machinery

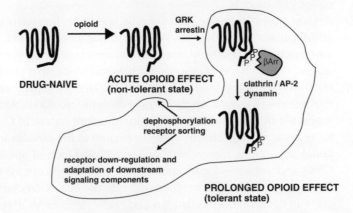

FIGURE 5. Models for the role of receptor desensitization and endocytosis mechanisms in physiological adaptations underlying opioid tolerance.

Downstream Adaptations Observed at the Cellular Level

Of course, regulation of receptors themselves is only part of the cellular adaptive response induced by opioids. A large number of other changes in cellular function have been described, and it is beyond the scope of the present chapter to describe them all. One adaptation of particular note is superactivation (also called up-regulation) of cellular cAMP signaling, which has been observed in many opioid-responsive cell types (and native tissue preparations) after chronic opioid exposure and can (again) be traced back to Nirenberg's early observations (Williams et al. 2001). This change in cAMP responsiveness can be understood most simply as a homeostatic adaptation to compensate for chronic inhibition of adenylyl cyclase activity produced by chronic opioid receptor activation. An interesting additional feature is that removal of the opioid agonist often produces a transient "overshoot" of cAMP signaling, above the normal levels observed in opioid-naïve cells. This "cellular withdrawal" response is what initially motivated the hypothesis that adenylyl cyclase superactivation represents a biochemical correlate of physiological dependence. Extensive work on second messenger signaling has established that the potential consequences of enhanced cAMP tone are varied and many. In particular, cAMP signaling has important functions in the regulation of neural gene expression that may be critical for long-lasting changes in neural function characteristic of the drug-addicted state (see Chapters 19 and 21, this volume). The overall idea is that adaptive responses to chronic opioids can be viewed as a network of regulatory processes, beginning with regulation of receptors themselves and extending to a variety of "downstream" steps of opioid signaling that tend to maintain homeostasis of neural signaling in the face of a pathologically prolonged activation of opioid receptors by exogenous opioid drugs. Although correlative data support this general hypothesis, a major challenge of future studies will be to perturb specific mechanisms of opioid receptor regulation in vivo and determine if the proposed homeostatic links to downstream adaptations, such as adenylyl cyclase superactivation, can be established causally.

SUMMARY, CONCLUSIONS, AND FUTURE DIRECTIONS

Opioid research is arguably where the cell biological approach to addiction originated. The observed physiological and behavioral effects of opioids are complex and involve multiple neurochemical systems. Nevertheless, the molecular targets of opioids are well defined, and considerable progress has been made in elucidating specific mechanisms of opioid regulation at the cellular level. Overall, support has increased for the basic hypothesis, proposed more than 30 years ago, that opioid drug action and addiction can be understood at the cellular level. However, it is clear that much remains to be learned before this goal can be truly achieved, in both elucidating mechanisms of regulation at the cellular level and testing the functional consequences of these mechanisms at the level of neural circuits and in the intact animal.

In describing the endogenous peptides that target opioid receptors, a number of questions remain. Why so many endogenous opioid peptides exist has historically been attributed to different opioid receptor selectivity and extracellular stability. But might more subtle interactions of endogenous opioids with individual receptors occur that result in selective regulation or signaling, paralleling differences observed among the opioid drugs?

Also important in the area of endogenous opioids is resolving the issue of endomorphins, with regard to both their existence and etiology. Understanding the endogenous opioid system becomes more pertinent to addiction as we learn that an intact endogenous opioid system appears to be critical for the rewarding effects of other drugs of abuse such as nicotine, cannabinoids, and alcohol, in addition to some natural rewarding stimuli.

When considering the cellular regulatory effects of opioids, it is important to recognize that most of our present mechanistic understanding is derived from studies of heterologous (non-neural) systems. One important area for future research is to test the applicability of this understanding to physiologically relevant neurons and search for potentially new features of opioid regulation that may not be evident in simpler cell models. Another very important challenge is to link specific cellular mechanisms of opioid regulation to the properties of functionally relevant neural circuits and ultimately to the complex effects of opioids on whole-animal physiology and behavior. Creative application of improved cell culture and electrophysiological and cellular imaging methods holds great promise for addressing the first question. Transgenic and gene knockout studies have already contributed greatly to our understanding of opioid neurochemistry and pharmacology, and are beginning to shed new light on the physiological and behavioral consequences of specific opioid regulatory mechanisms.

Most of the present work in this area has investigated links to acute opioid responsiveness and antinociceptive tolerance. It will be important to extend this analysis to understanding the cellular basis of opioid dependence and behavioral reward, which are of critical importance to addiction medicine. It is interesting that opioid receptors are expressed in neurons that modulate the rewarding effects of various non-opioid drugs and naturally rewarding stimuli, and that recent gene knockout studies support a rather general role of opioid signaling in reward processing. Thus, we anticipate that future studies in this area will be of fundamental interest to both the biology of addictive drug action and, more generally, will contribute to understanding mechanisms of neural plasticity underlying complex behavior.

ACKNOWLEDGMENTS

We thank many colleagues who have contributed to our ideas and experiments, and we regret that we were only able to cite a subset of the many important studies in this field. The authors are particularly indebted to the talented students and postdoctoral fellows who have contributed greatly to our progress and to the enjoyment of performing research. Critical financial support was provided by the National Institutes of Health and the National Institute on Drug Abuse.

REFERENCES

Alvarez V.A., Arttamangkul S., Dang V., Salem A., Whistler J.L., Von Zastrow M., Grandy D.K., and Williams J.T. 2002. mu-Opioid receptors: Ligand-dependent activation of potassium conductance, desensitization, and internalization. *J. Neurosci.* **22:** 5769–5776.

Arden J.R., Segredo V., Wang Z., Lameh J., and Sadee W. 1995. Phosphorylation and agonist-specific intracellular trafficking of an epitope-tagged mu-opioid receptor expressed in HEK 293 cells. *J. Neurochem.* **65:** 1636–1645.

Berridge K.C. and Robinson T.E. 2003. Parsing reward. *Trends*

Neurosci. **26:** 507–513.

Bohn L.M., Lefkowitz R.J., and Caron M.G. 2002. Differential mechanisms of morphine antinociceptive tolerance revealed in βarrestin-2 knock-out mice. *J. Neurosci.* **22:** 10494–10500.

Bohn L.M., Dykstra L.A., Lefkowitz R.J., Caron M.G., and Barak L.S. 2004. Relative opioid efficacy is determined by the complements of the G protein-coupled receptor desensitization machinery. *Mol. Pharmacol.* **66:** 106–112.

Bohn L.M., Gainetdinov R.R., Lin F.T., Lefkowitz R.J., and Caron M.G. 2000. Mu-opioid receptor desensitization by beta-arrestin-2 determines morphine tolerance but not dependence. *Nature* **408:** 720–723.

Bohn L.M., Lefkowitz R.J., Gainetdinov R.R., Peppel K., Caron M.G., and Lin F.T. 1999. Enhanced morphine analgesia in mice lacking beta-arrestin 2. *Science* **286:** 2495–2498.

Booth M. 1998. *Opium: A history,* 1st U.S. edition. St. Martin's Press, New York.

Borgland S., Conner M., Osborne P., Furness J., and Christie M. 2003. Opioid agonists have different efficacy profiles for G protein activation, rapid desensitization, and endocytosis of mu-opioid receptors. *J. Biol. Chem.* **278:** 18776–18784.

Bushell T., Endoh T., Simen A.A., Ren D., Bindokas V.P., and Miller R.J. 2002. Molecular components of tolerance to opiates in single hippocampal neurons. *Mol. Pharmacol.* **61:** 55–64.

Carman C.V. and Benovic J.L. 1998. G-protein-coupled receptors: Turn-ons and turn-offs. *Curr. Opin. Neurobiol.* **8:** 335–344.

Chu P., Murray S., Lissin D., and von Zastrow M. 1997. Delta and kappa opioid receptors are differentially regulated by dynamin-dependent endocytosis when activated by the same alkaloid agonist. *J. Biol. Chem.* **272:** 27124–27130.

Contet C., Kieffer B.L., and Befort K. 2004. Mu opioid receptor: A gateway to drug addiction. *Curr. Opin. Neurobiol.* **14:** 370–378.

Eitan S., Bryant C.D., Saliminejad N., Yang Y.C., Vojdani E., Keith D., Jr., Polakiewicz R., and Evans C.J. 2003. Brain region-specific mechanisms for acute morphine-induced mitogen-activated protein kinase modulation and distinct patterns of activation during analgesic tolerance and locomotor sensitization. *J. Neurosci.* **23:** 8360–8369.

Evans C.J. 2004. Secrets of the opium poppy revealed. *Neuropharmacology* (suppl. 1) **47:** 293–299.

He L., Fong J., von Zastrow M., and Whistler J.L. 2002. Regulation of opioid receptor trafficking and morphine tolerance by receptor oligomerization. *Cell* **108:** 271–282.

Hicke L. 2001. Protein regulation by monoubiquitin. *Nat. Rev. Mol. Cell. Biol.* **2:** 195–201.

Hislop J.N., Marley A., and von Zastrow M. 2004. Role of mammalian VPS proteins in endocytic trafficking of a non-ubiquitinated G protein-coupled receptor to lysosomes. *J. Biol. Chem.* **279:** 22522–22531.

Keith D.E., Murray S.R., Zaki P.A., Chu P.C., Lissin D.V., Kang L., Evans C.J., and von Zastrow M. 1996. Morphine activates opioid receptors without causing their rapid internalization.

J. Biol. Chem. **271:** 19021–19024.

Keith D.E., Anton B., Murray S.R., Zaki P.A., Chu P.C., Lissin D.V., Monteillet A.G., Stewart P.L., Evans C.J., and von Zastrow M. 1998. mu-Opioid receptor internalization: Opiate drugs have differential effects on a conserved endocytic mechanism in vitro and in the mammalian brain. *Mol. Pharmacol.* **53:** 377–384.

Kieffer B.L. 1999. Opioids: First lessons from knockout mice. *Trends Pharmacol. Sci.* **20:** 19–26.

Kieffer B.L. and Gaveriaux-Ruff C. 2002. Exploring the opioid system by gene knockout. *Prog. Neurobiol.* **66:** 285–306.

Koch T., Schulz S., Schröder H., Wolf R., Raulf E., and Hollt V. 1998. Carboxyl-terminal splicing of the rat mu opioid receptor modulates agonist-mediated internalization and receptor resensitization. *J. Biol. Chem.* **273:** 13652–13657.

Koob G. 2000. Drug addiction. *Neurobiol. Dis.* **7:** 543–545.

Kovoor A., Celver J.P., Wu A., and Chavkin C. 1998. Agonist induced homologous desensitization of mu-opioid receptors mediated by G protein-coupled receptor kinases is dependent on agonist efficacy. *Mol. Pharmacol.* **54:** 704–711.

Law P.-Y., Hom D.S., and Loh H.H. 1984. Down-regulation of opiate receptor in neuroblastoma x glioma NG108-15 hybrid cells: Chloroquine promotes accumulation of tritiated enkephalin in the lysosomes. *J. Biol. Chem.* **259:** 4096–4104.

Law P.-Y., Wong Y.H., and Loh H.H. 2000. Molecular mechanisms and regulation of opioid receptor signaling. *Annu. Rev. Pharmacol. Toxicol.* **40:** 389–430.

Lecoq I., Marie N., Jauzac P., and Allouche S. 2004. Different regulation of human delta-opioid receptors (hDOR) by SNC-80 and endogenous enkephalins. *J. Pharmacol. Exp. Ther.* **310:** 666–677.

Lefkowitz R.J. 1996. G protein-coupled receptors and receptor kinases: From molecular biology to potential therapeutic applications. *Nat. Biotechnol.* **14:** 283–286.

Lefkowitz R.J., Pitcher J., Krueger K., and Daaka Y. 1998. Mechanisms of beta-adrenergic receptor desensitization and resensitization. *Adv. Pharmacol.* **42:** 416–420.

Marchese A. and Benovic J.L. 2001. Agonist-promoted ubiquitination of the G protein-coupled receptor CXCR4 mediates lysosomal sorting. *J. Biol. Chem.* **276:** 45509–45512.

Marie N., Lecoq I., Jauzac P., and Allouche S. 2003. Differential sorting of human delta-opioid receptors after internalization by peptide and alkaloid agonists. *J. Biol. Chem.* **278:** 22795–22804.

Moles A., Kieffer B.L., and D'Amato F.R. 2004. Deficit in attachment behavior in mice lacking the mu-opioid receptor gene. *Science* **304:** 1983–1986.

Nestler E.J. and Malenka R.C. 2004. The addicted brain. *Sci. Am.* **290:** 78–85.

Okura T., Cowell S.M., Varga E., Burkey T.H., Roeske W.R., Hruby V.J., and Yamamura H.I. 2000. Differential down-regulation of the human delta-opioid receptor by SNC80 and [D-Pen(2),D-Pen(5)]enkephalin. *Eur. J. Pharmacol.* **387:** R11–R13.

Petaja-Repo U.E., Hogue M., Bhalla S., Laperriere A., Morello J.P., and Bouvier M. 2002. Ligands act as pharmacological chaperones and increase the efficiency of delta opioid receptor maturation. *EMBO J.* **21:** 1628–1637.

Rupniak N.M., Carlson E.C., Harrison T., Oates B., Seward E., Owen S., de Felipe C., Hunt S., and Wheeldon A. 2000. Pharmacological blockade or genetic deletion of substance P (NK(1)) receptors attenuates neonatal vocalisation in guinea-pigs and mice. *Neuropharmacology* **39:** 1413–1421.

Seifert R., Wenzel-Seifert K., Gether U., and Kobilka B.K. 2001. Functional differences between full and partial agonists: Evidence for ligand-specific receptor conformations. *J. Pharmacol. Exp. Ther.* **297:** 1218–1226.

Sharma S.K., Klee W.A., and Nirenberg M. 1977. Opiate-dependent modulation of adenylate cyclase. *Proc. Natl. Acad. Sci.* **74:** 3365–3369.

Shenoy S.K., McDonald P.H., Kohout T.A., and Lefkowitz R.J. 2001. Regulation of receptor fate by ubiquitination of activated beta 2-adrenergic receptor and beta-arrestin. *Science* **294:** 1307–1313.

Simonin F., Karcher P., Boeuf J.J., Matifas A., and Kieffer B.L. 2004. Identification of a novel family of G protein-coupled receptor associated sorting proteins. *J. Neurochem.* **89:** 766–775.

Song B. and Marvizon J.C. 2003. Peptidases prevent mu-opioid receptor internalization in dorsal horn neurons by endogenously released opioids. *J. Neurosci.* **23:** 1847–1858.

Tanowitz M. and von Zastrow M. 2002. Ubiquitination-independent trafficking of G protein-coupled receptors to lysosomes. *J. Biol. Chem.* **277:** 50219–50222.

——. 2003. A novel endocytic recycling signal that distinguishes the membrane trafficking of naturally occurring opioid receptors. *J. Biol. Chem.* **278:** 45978–45986.

Terman G.W., Jin W., Cheong Y.P., Lowe J., Caron M.G., Lefkowitz R.J., and Chavkin C. 2004. G-protein receptor kinase 3 (GRK3) influences opioid analgesic tolerance but not opioid withdrawal. *Br. J. Pharmacol.* **141:** 55–64.

Tsao P. and von Zastrow M. 2000a. Downregulation of G protein-coupled receptors. *Curr. Opin. Neurobiol.* **10:** 365–369.

——. 2000b. Type-specific sorting of G protein-coupled receptors after endocytosis. *J. Biol. Chem.* **275:** 11130–11140.

Tsao P., Cao T., and von Zastrow M. 2001. Role of endocytosis in mediating downregulation of G-protein-coupled recep-

tors. *Trends Pharmacol. Sci.* **22:** 91–96.

von Zastrow M., Keith D.E.J., and Evans C.J. 1993. Agonist-induced state of the delta-opioid receptor that discriminates between opioid peptides and opiate alkaloids. *Mol. Pharmacol.* **44:** 166–172.

Whistler J.L. and von Zastrow M. 1998. Morphine-activated opioid receptors elude desensitization by beta-arrestin. *Proc. Natl. Acad. Sci.* **95:** 9914–9919.

Whistler J.L., Chuang H.H., Chu P., Jan L.Y., and von Zastrow M. 1999. Functional dissociation of mu opioid receptor signaling and endocytosis: Implications for the biology of opiate tolerance and addiction. *Neuron* **23:** 737–746.

Whistler J.L., Enquist J., Marley A., Fong J., Gladher F., Tsuruda P., Murray S., and von Zastrow M. 2002. Modulation of post-endocytic sorting of G protein-coupled receptors. *Science* **297:** 615–620.

Williams J.T., Christie M.J., and Manzoni O. 2001. Cellular and synaptic adaptations mediating opioid dependence. *Physiol. Rev.* **81:** 299–343.

Willins D.L., Berry S.A., Alsayegh L., Backstrom J.R., Sanders-Bush E., Friedman L., and Roth B.L. 1999. Clozapine and other 5-hydroxytryptamine-2A receptor antagonists alter the subcellular distribution of 5-hydroxytryptamine-2A receptors in vitro and in vivo. *Neuroscience* **91:** 599–606.

Yu Y., Zhang L., Yin X., Sun H., Uhl G.R., and Wang J.B. 1997. Mu opioid receptor phosphorylation, desensitization, and ligand efficacy. *J. Biol. Chem.* **272:** 28869–28874.

Zaki P.A., Keith D.E., Brine G.A., Carroll F.I., and Evans C.J. 2000. Ligand-induced changes in surface mu-opioid receptor number: Relationship to G protein activation? *J. Pharmacol. Exp. Ther.* **292:** 1127–1134.

Zaki P.A., Bilsky E.J., Vanderah T.W., Lai J., Evans C.J., and Porreca F. 1996. Opioid receptor types and subtypes: The delta receptor as a model. *Annu. Rev. Pharmacol. Toxicol.* **36:** 379–401.

Zhang J., Ferguson S.S., Barak L.S., Bodduluri S.R., Laporte S.A., Law P.Y., and Caron M.G. 1998. Role for G protein-coupled receptor kinase in agonist-specific regulation of mu-opioid receptor responsiveness. *Proc. Natl. Acad. Sci.* **95:** 7157–7162.

12 Receptor–receptor Interactions Modulate Opioid Receptor Function

Ivone Gomes and Lakshmi A. Devi

Department of Pharmacology and Biological Chemistry,
Mount Sinai School of Medicine, New York, New York 10029

ABSTRACT

Opioid receptors are G-protein-coupled receptors (GPCRs) that are the primary clinical targets in the attenuation of pain. Most opiates used in pain management are also drugs of abuse because of their mood altering (i.e., euphoria) and rewarding properties. The traditional concept of a functional opioid receptor was that of monomeric proteins associated with heterotrimeric G proteins in a 1:1 stoichiometric ratio. However, recent evidence obtained through the use of biochemical techniques, such as coimmunoprecipitation of epitope-tagged receptors, and biophysical techniques, such as bioluminescence resonance energy transfer, show that these receptors exist as homo-dimers/oligomers and that they also form heterodimers with closely or distantly related GPCRs. These interactions have profound implications for opioid receptor function. In this chapter, we describe receptor–receptor interactions involving opioid receptors and how they affect the pharmacological, signaling, and trafficking properties of these receptors that could have ramifications for the study of drug addiction.

INTRODUCTION

Opioid receptors are members of the family A of G-protein-coupled receptors (GPCRs), characterized by the presence of seven α-helical transmembrane domains separated by three intracellular and extracellular loops. These receptors are classified as mu (μ), delta (δ), or kappa (κ), and their activation by agonists leads to the activation of associated inhibitory heterotrimeric G proteins (G_i). This is followed by the activation of multiple signal transduction pathways including the inhibition of adenylyl cyclase, modulation of inwardly rectifying K^+ channels or voltage-dependent Ca^{2+} channels, and regulation of mitogen-activated protein kinase (MAPK) activity (Herz 1993; Li and Chang 1996; Sarne et al. 1996; Jordan and Devi 1998). At the systemic level, opioid receptor activation induces a number of physiological responses including analgesia, euphoria, and decreased intestinal motility (Gutstein and Akil 2001). These receptors are distributed throughout the peripheral as well as central nervous system. In the latter case, they are present at presynaptic and postsynaptic sites in the dorsal horn of the spinal cord as well as in several brain regions including the brain stem, thalamus, and cortex (Inturrisi 2002).

Models describing the interaction between opioid receptors and their cognate G-protein targets have been based on the assumption that the receptors exist as monomers and couple to G proteins in a 1:1 stoichiometric ratio, despite indirect evidence to the contrary. For example, pharmacological studies show that dimeric enkephalin or morphine analogs show a higher affinity and potency for δ or μ receptors, respectively, than their monomeric counterparts (Hazum et al. 1982). Radiation inactivation as well as hydrodynamic studies suggest that the native opioid receptor exists as an oligomeric array of the receptor complexed with its G protein (McLawhon et al. 1983; Ott et al. 1986; Simon et al. 1986; Glasel 1987). Early pharmacological evidence shows that μ receptor-selective ligands inhibit the binding of δ receptor ligands in both a competitive and noncompetitive fashion (Rothman and Westfall 1981; Rothman et al. 1983). These studies suggest the presence of two types of δ receptors: those that are associated with μ receptors and those that are not. In addition, δ receptor-selective ligands can potentiate μ receptor-mediated analgesia, suggesting a functional interaction between these two subtypes of the opioid receptor (Russel et al. 1986; Heyman et al. 1989; Malmberg and Yaksh 1992).

Recent studies using a variety of techniques have shown that GPCRs can associate with one another (Gomes et al. 2001; Rios et al. 2001; Angers et al. 2002; Bai 2004; Kroeger et al. 2004; Abul-Husn et al. 2005). This association can be homomeric (between products of the same gene) or heteromeric (between products of distinct genes). The latter can involve associations between closely related proteins (such as members of the same receptor subfamily) or between distantly related proteins (such as members of distinct receptor subfamilies). In this chapter, we describe the various techniques that have been used in the detection and characterization of interactions among opioid receptors. These include biochemical detection using immunoprecipitation and Western blot analysis and in vivo detection using biophysical techniques such as bioluminescence resonance energy transfer (BRET). In addition, we describe the implications of receptor–receptor interactions on the pharmacological, signaling, and trafficking properties of opioid receptors.

GOALS OF THE CHAPTER

This chapter informs readers about recent advances in the field of opioid receptor regulation by receptor–receptor interactions. We describe how evidence obtained through the use of biochemical and biophysical techniques demonstrates that these receptors directly interact to form homodimers/oligomers; they also form heterodimers with closely or distantly related GPCRs. We also describe how receptor–receptor interactions involving opioid receptors modulate receptor pharmacology, as well as their signaling and trafficking properties. Finally, we discuss the ramifications of these interactions on pain treatment and drug addiction in relation to opiate abuse.

METHODOLOGICAL CONSIDERATIONS

Detection of Opioid Receptor Interactions by Coimmunoprecipitation of Differentially Epitope-tagged Receptors

The availability of opioid receptor complementary DNAs (cDNAs) has enabled the introduction of epitope tags for which selective and specific antibodies are commercially available. This has facilitated the isolation of interacting receptor complexes involving opioid receptors. Typically, the epitope tags (short peptide sequences from proteins that are not commonly expressed in eukaryotic proteins, such as Flag, *c-myc*, or hemagglutinin) are expressed at the amino or carboxyl termini of opioid receptors. To isolate interacting receptor complexes, cells are cotransfected with two differentially epitope-tagged opioid receptors, e.g., Flag- and *c-myc*-tagged receptors. Selective immunoprecipitation is achieved using antibodies to one of the epitopes (e.g., *c-myc*). The immunoprecipitate is then subjected to sodium dodecyl sulfate–polyacrylamide gel electrophoresis (SDS-PAGE) under nonreducing conditions followed by Western blotting with antibodies directed against the second epitope tag (e.g., Flag). A signal is detected in the blots only if physical interactions occur between the two cotransfected receptors. However, because of the inherent hydrophobic nature of GPCRs, it is possible to detect artifactual receptor complexes using this technique. This can be overcome through the use of stringent buffers for solubilization and immunoprecipitation. For example, the presence of a disulfide capping agent (such as iodoacetamide) in the buffers used for isolating receptor complexes can reduce the chance formation of spurious disulfide bonds. Treatment of cells with cross-linking agents followed by solubilization with a combination of detergents is also helpful in disrupting nonspecific receptor aggregation. A very important control, in this type of study, is to subject a mixture of cells individually expressing each differentially epitope-tagged receptor to identical solubilization and immunoprecipitation conditions as those for the cells coexpressing both receptors. An additional control that can be used is the coexpression of receptors that are known not to interact. Thus, with the use of appropriate controls, immunoprecipitation followed by Western blot analysis is a valid technique for the detection of opioid receptor interacting complexes (Gomes et al. 2002, 2003). This approach has been used to show the interaction of μ, δ, or κ receptors with one another or with other GPCRs (Table 1).

A number of studies have examined the effect of ligand treatment on the level of receptor interacting complexes detected by immunoprecipitation and Western blotting. Agonist treatment had no effect on the level of κ receptor dimers (Jordan et al. 2000), whereas several selective and nonselective agonists could decrease the level of δ receptor dimers (Cvejic and Devi 1997). In the case of μ-α_{2A} heteromeric complexes, treatment with the μ agonist, morphine, or with the α_{2A} agonist, clonidine, led to an increase in the level of α_{2A} receptors associated with μ receptors. However, cotreatment with morphine and clonidine led to a considerable decrease in the level of interacting complexes, suggesting that their combined presence disrupted the strength of μ-α_{2A} interactions (Jordan et al. 2003).

Detection of Opioid Receptor Interactions in Live Cells Using Biophysical Techniques

Proximity-based energy transfer assays such as BRET or fluorescence resonance energy transfer (FRET) have been used to study opioid receptor interactions in live cells. The BRET technique measures the energy transfer between a receptor tagged with a luminescent donor (luciferase) and a receptor tagged with a fluorescent acceptor, such as yellow fluorescent protein (YFP) or enhanced green fluorescent protein (EGFP). Luciferase acts on its substrate, coelenterazine *h*, to emit light at 470 nm that can excite the fluorescent acceptor; the resulting fluorescence emission at 530 nm is taken as a measure of the physical proximity between the two proteins. This energy transfer occurs only when the distance

TABLE 1. Homomeric and Heteromeric Interactions Involving Opioid Receptors

Receptors	Method	References
μ-μ opioid	Immunoprecipitation	Li-Wei et al. (2002)
	BRET	Gomes et al. (2002)
δ-δ opioid	Immunoprecipitation	Cvejic and Devi (1997); McVey et al. (2001); Ramsay et al. (2002)
	BRET	McVey et al. (2001); Gomes et al. (2002)
	FRET	McVey et al. (2001)
κ-κ opioid	Immunoprecipitation	Jordan et al. (2000)
	BRET	Ramsay et al. (2002)
μ-δ opioid	Immunoprecipitation	George et al. (2000); Gomes et al. (2000, 2004); Law et al. (2005)
	BRET	Gomes et al. (2004)
κ-δ opioid	Immunoprecipitation	Jordan et al. (1999); Ramsay et al. (2002)
	BRET	Gomes et al. (2002); Ramsay et al. (2002)
μ opioid-α_{2A} adrenergic	Immunoprecipitation	Jordan et al. (2003); Zhang and Limbird (2004)
	BRET	Jordan et al. (2003)
δ opioid-α_{2A} adrenergic	Immunoprecipitation	Rios et al. (2004)
	BRET	Rios et al. (2004)
δ opioid-β_2 adrenergic	Immunoprecipitation	Jordan et al. (2001); McVey et al. (2001)
κ opioid-β_2 adrenergic	Immunoprecipitation	Jordan et al. (2001); Ramsay et al. (2002)
	BRET	Ramsay et al. (2002)
μ opioid-CCR5 chemokine	Immunoprecipitation	Suzuki et al. (2002); Chen et al. (2004)
δ opioid-CCR5 chemokine	Immunoprecipitation	Suzuki et al. (2002)
κ opioid-CCR5 chemokine	Immunoprecipitation	Suzuki et al. (2002)
μ opioid-sst_{2A}	Immunoprecipitation	Pfeiffer et al. (2002)
μ opioid-substance P	Immunoprecipitation	Pfeiffer et al. (2003)
	BRET	Pfeiffer et al. (2003)

between the donor and acceptor is less than 100 Å (Angers et al. 2000). FRET is similar to BRET with the exception that the energy donor molecule, a variant of green fluorescent protein (GFP) (usually cyan fluorescent protein, CFP) is excited by an external light source (Overton and Blumer 2000). In addition to the inherent advantages and disadvantages in the use of either BRET or FRET (Gomes et al. 2002, 2003) limitations exist with regard to the information these techniques can provide. For example, observed agonist-induced increases or decreases in BRET could be caused by changes in receptor conformation that result in the movement of the receptor tags (luciferase and YFP) and not to an actual increase or decrease in the number of interacting receptors (Gomes et al. 2002, 2003).

BRET has been the technique of choice in investigating whether homomeric or heteromeric complexes involving opioid receptors are in close enough proximity to interact in live cells (Table 1). Studies examining the effect of ligand treatment on the level of receptor interacting complexes do not detect further modulation of the BRET signal, indicating that these receptors are present as constitutive homodimers or heterodimers in live cells (I. Gomes and L.A. Devi, pers. comm.; McVey et al. 2001; Ramsey et al. 2002; Gomes et al. 2004).

Modulation of Function by Interactions between Opioid Receptor Subtypes

Several approaches have been used to explore the effects of receptor–receptor interactions on GPCR function. These include pharmacological approaches such as examining ligand binding and signaling properties using selective ligands to one or both receptors, and internalization properties using differential epitope tags. Because heterodimeric receptors can be easily probed using receptor-selective ligands, we describe the functional changes observed on opioid receptor heterodimerization.

κ-δ *Receptor Interactions*

The modulation of receptor pharmacology was first examined with heterodimers of κ- and δ-opioid receptors. Cells coexpressing these receptors show altered ligand binding, signaling, and trafficking properties (Jordan and Devi 1999). The κ-δ heterodimers have reduced affinity for highly selective κ or δ receptor ligands, but have enhanced affinity for partially selective ligands (Jordan and Devi 1999). In the presence of a δ-selective agonist, a κ-selective agonist binds to the receptor with high affinity, and reciprocally, a κ-selective agonist increases the binding of a δ-selective agonist. Cells coexpressing κ and δ receptors also show synergistic effects on agonist-induced signaling, as measured by cyclic adenosine monophosphate (cAMP) and phosphorylated MAPK levels. In addition, the trafficking properties of these receptors are altered in cells coexpressing both receptors, because etorphine-induced trafficking of the δ receptor is significantly reduced. This suggests that δ receptors are retained at the cell surface as a result of dimerization with κ receptors (Jordan and Devi 1999).

Heterodimers between κ and δ receptors could occur, at the physiological level, if these receptors were present not only in the same cell but also in the same area of the cell. κ and δ receptors are present not only in the axons of the superficial dorsal horn of the spinal cord (one of the brain regions mediating the analgesic effects of opioids) but also in the same vesicle within the axons (Wessendorf and Dooyema 2001). In addition, studies have shown that the spinal coadministration of κ and δ receptor agonists leads to the potentiation of the analgesia compared with each individual agonist (Miaskowski et al. 1990). Therefore, the selective targeting of κ-δ heterodimers could lead to the development of potent analgesics with reduced side effects, compared to those currently available.

μ-δ *Receptor Interactions*

Studies with μ-δ heterodimers also report decreased binding affinity to selective synthetic agonists (George et al. 2000). The rank order of agonist affinities for the heterodimeric receptors is different from that of the individual receptors, suggesting modulation of the binding pocket on dimerization (George et al. 2000). In addition, treatment of cells coexpressing μ-δ receptors with very low doses of δ-selective ligands causes a significant increase in the binding of a μ-selective agonist (Gomes et al. 2000, 2004). It is possible that the heterodimeric μ-δ complex associates with pertussis toxin-insensitive G proteins because pertussis toxin treatment does not completely abolish the synergistic binding (Abul-Husn et al. 2005) or receptor signaling (George et al. 2000). Low doses of δ receptor ligands also enhance μ receptor-mediated signaling as assessed by GTPγS binding and measurement of cAMP and phosphorylated MAPK levels (Gomes et al. 2000, 2004). In addition, low doses of a δ antagonist can potentiate μ receptor-mediated analgesia by morphine (Gomes et al. 2004). The trafficking of μ and δ receptors was examined in a cell line that constitutively expresses μ receptors and in which δ receptor expression can be induced by an ecdysone-inducible mammalian expression system (Law et al. 2005). The results from these studies suggests that μ and δ receptors endocytose independently of each other (Law et al. 2005).

Colocalization of μ and δ receptors has been observed in the dorsal horn of the spinal cord (a region implicated in mediating the analgesic effects of clinically administered opiates) not only in the axonal terminals of single neurons but also in the plasmalemma (Arvidsson et al. 1995; Cheng et al. 1997). Studies with transgenic animals also suggest direct interactions between μ and δ receptors. In mice lacking μ receptors, the analgesic effects of morphine as well as morphine-induced place preference and physical dependence are abolished (Matthes et al. 1996). These animals also show dramatically reduced δ receptor-mediated analgesia (Sora et al. 1997), suggesting an interaction between the two receptors. Studies also show that animals treated with δ antagonists show diminished development of morphine tolerance and dependence (Abdelhamid et al. 1991; Zhu et al. 1999). Furthermore, the selective reduction of δ receptors by antisense nucleotides attenuates the development of morphine dependence (Sanchez-Blasquez et al. 1997). Therefore, the μ-δ heterodimer could serve as a model for developing potent analgesics with reduced side effects such as tolerance and dependence.

Studies examining the trafficking properties of the μ receptor have suggested that oligomerization influences its endocytic properties (He et al. 2002). Interestingly, as a consequence of this altered endocytosis, the development of tolerance to morphine is reduced, suggesting an approach for the development of opiate analogs with enhanced efficacy for the treatment of chronic pain (He et al. 2002).

Modulation of Function by Interactions between Opioid Receptors and Other G-protein-coupled Receptors

Opioid and α_{2A}-adrenergic Receptor Interactions

In the case of μ-α_{2A} receptors, the presence of α_{2A} receptors is sufficient to increase the efficacy of morphine (μ receptor agonist) signaling in cells coexpressing both receptors. However, coactivation of both μ and α_{2A} receptors leads to a decrease in receptor signaling in both heterologous cells and primary spinal cord neurons (Jordan et al. 2003). Interestingly, interactions between μ and α_{2A} receptors have no effects on receptor internalization, because treatment with α_{2A} receptor agonists leads to endocytosis of α_{2A} receptors but not μ receptors in cells coexpressing both receptors (Zhang and Limbird 2004). Similarly, treatment with a μ receptor agonist leads to endocytosis of μ receptors but not α_{2A} receptors in these cells (Zhang and Limbird 2004).

In the case of δ-α_{2A} receptor heterodimers, the functional outcome of receptor–receptor interactions was examined in Neuro-2A cells, which endogenously express δ-opioid receptors, using neurite outgrowth as a physiological readout. Deltorphin II, a δ receptor agonist, induces neurite outgrowth in these cells. This δ agonist-mediated neurite outgrowth is further enhanced on transfection of α_{2A} receptors into these cells (Rios et al. 2004).

Several studies have shown functional interactions between opioid and α_{2A}-adrenergic receptors in mediating pain attenuation in vivo (Ossipov et al. 1990; Grabow et al. 1999; Fairbanks et al. 2000). In addition, colocalization studies have shown the presence of α_{2A}-adrenergic receptors on μ receptor-containing neurons in the rat medial tractus solitarius (Glass and Pickel 2002), as well as extensive colocalization of δ receptors with adrenergic receptors in the spinal cord both at cellular and subcellular levels (Milner et al. 2002). Convincing evidence for interactions between the opioid and α_{2A}-adrenergic systems in pain attenuation has come from transgenic mouse models. In mice lacking μ-opioid receptors, synergistic interactions were observed between δ and α_{2A} receptor agonists in mediating analgesia (Guo et al. 2003). However, in mice lacking a functional α_{2A} receptor (Stone et al. 1997), a decrease in opioid (μ or δ)-mediated analgesia was observed compared to wild-type mice, suggesting functional interactions between opioid and adrenergic systems (Stone et al. 1997). Since then, studies with heterologous cell culture systems indicate that the mere presence of α_{2A} receptors increased the signaling of μ receptors (Jordan et al. 2003) or neurite outgrowth mediated by δ receptors (Rios et al. 2004). It is probable that α_{2A} receptors, via their long third intracellular loop, might sequester multiple signaling and scaffolding molecules for use by the "partner" opioid receptor and/or they may target associated opioid receptors to areas rich in signaling molecules, such as "lipid rafts." These interactions between opioid and α_{2A}-adrenergic receptors could have implications for pain attenuation.

κ- or δ-Opioid and β₂-adrenergic Receptor Interactions

Interactions between the β_2-adrenergic receptor (coupled to stimulatory G proteins) and either κ- or δ-opioid receptors (coupled to inhibitory G proteins) do not significantly alter the ligand-binding properties or signaling through adenylyl cyclase (Jordan et al. 2001). However, they affect receptor trafficking properties, because δ receptors associated with β_2-adrenergic receptors undergo isoproterenol-mediated endocytosis. Reciprocally, β_2-adrenergic receptors also undergo etorphine-induced endocytosis (Jordan et al. 2001). However, when coexpressed with κ-opioid receptors, β_2 receptors do not undergo either isoproterenol- or etorphine-induced endocytosis (Jordan et al. 2001). Because these receptors are coexpressed in the heart, it is likely that these interactions could have a role in modulating physiological responses to opiate administration.

μ-Opioid and Somatostatin 2A Receptor Interactions

Receptor–receptor interactions between μ-opioid and somatostatin (sst) 2A receptors do not have significant effects on receptor binding or signaling. However, treatment of cells coexpressing both receptors with a selective sst2A agonist causes increased phosphorylation, internalization, and desensitization of both the sst2A receptor and the μ receptor. Exposure of cells coexpressing these two receptors to a selective μ agonist induces phosphorylation and desensitization of both receptors, but it has no effect on the internalization of the sst2A receptor. Thus, it appears that in the case of μ-sst2A receptor interactions, heterodimerization provides a new regulatory mechanism for modulating the level of phosphorylation and/or desensitization of these receptors (Pfeiffer et al. 2002).

Extensive interactions between opioids and somatostatin in mediating analgesic responses have been shown (Betoin et al. 1994). In addition, a high degree of colocalization of μ and sst2A receptors is seen in the locus coeruleus, a brain region implicated in the expression of opioid withdrawal syndrome (Pfeiffer et al. 2002). In fact, patients administered with sst2A receptor agonists show attenuated opioid withdrawal symptoms (Bell et al. 1999). Therefore, interactions between μ and sst2A receptors could have profound implications for the management of opioid withdrawal.

μ-Opioid and Substance P Receptor Interactions

Interactions between μ-opioid and substance P receptors do not substantially alter the receptor ligand-binding and signaling properties in cells coexpressing both receptors. However, they dramatically alter the internalization and resensitization profile of these receptors, because a μ receptor agonist promoted the phosphorylation and internalization of the substance P receptor and vice versa. In cells expressing only μ receptors, following agonist treatment, β-arrestin directs the receptors to clathrin-coated pits but does not internalize with the receptor (Pfeiffer et al. 2003). However, in cells coexpressing μ and substance P receptors, agonist treatment induces the recruitment of β-arrestin to the plasma membrane and subsequent internalization of the heterodimer along with β-arrestin into the same endosomal compartment. This leads to the delayed resensitization of the μ receptor in these cells compared with cells expressing μ receptors alone (Pfeiffer et al. 2003).

Studies have shown the colocalization of μ-opioid and substance P receptors in the dendrites of some trigeminal dorsal horn neurons (Aicher et al. 2000), suggesting that both receptors can interact during nociceptive neurotransmission. Furthermore, both receptors are highly expressed not only in brain regions implicated in depression, anxiety, and stress but also in the nucleus accumbens, a region that mediates the motivational properties of drugs of abuse, including opioids (Pffeifer et al. 2003). Interestingly, mice lacking substance P receptors show an impairment in both the physical and motivational aspects of opiate withdrawal, i.e., loss of the rewarding effects of morphine, conditioned place preference, and the absence of many of the physical signs associated with opiate withdrawal (Murtra et al. 2000; Ripley et al. 2002). However, the mice do not show any impairment in morphine-mediated analgesia (De Filipe et al. 1998). This suggests that interactions between μ and substance P receptors might have an important role in mediating dependence and addiction to opiates.

Opioid and Chemokine CCR5 Receptor Interactions

Immunoprecipitation of differentially epitope-tagged receptors show that μ-, δ-, or κ-opioid receptors interact with CCR5 chemokine receptors in lymphocytes (Suzuki et al. 2002). Chemical cross-linking studies indicate that the receptors are less that 11.4 Å apart in cell membranes, and cotreatment of cells coexpressing these receptors with ligands to both receptors increases the stimulatory effect of μ agonists on CCR5 receptor expression (Suzuki et al. 2002). Another study, performed in Chinese hamster ovary cells (CHO) coexpressing μ and CCR5 receptors, shows that a μ receptor agonist enhances the phosphorylation of the CCR5 receptor and reduces chemokine-mediated GTPγS binding (Chen et al. 2004). Conversely, the CCR5 receptor agonist increases μ receptor phosphorylation and reduces μ receptor-mediated GTPγS binding in these cells (Chen et al. 2004).

It is to be noted that interactions between opioid and CCR5 receptors were not observed with the BRET assay (Gomes et al. 2003, 2004; Jordan et al. 2003; Rios et al. 2004). The BRET studies were performed in HEK-293 cells transiently expressing opioid and CCR5 receptors, and not in cell lines stably expressing these receptors. This suggests that either the luciferase- and YFP-tagged receptor constructs are not proximal enough for efficient energy transfer or that opioid and CCR5 receptors heterodimerize only in specific cell types. This will have to be taken into account when evaluating opioid-CCR5 receptor interactions.

Studies have shown that opioid and CCR5 receptors are present in cells of the immune system (Cairns and D'Souza 1998; McCarthy et al. 2001). It is well known that the CCR5 receptor is a coreceptor for the entry of the AIDS virus into the immune cells (Cairns and D'Souza 1998) and thereby facilitates HIV replication. Also, morphine treatment can induce the expression of the CCR5 receptor (Miyagi et al. 2000) and this may account for the rapid development of AIDS in patients addicted to opiates and other drugs of abuse (Holmberg 1996; Sorensen and Copeland 2000). Therefore, developing drugs that target μ-CCR5 heterodimers might improve the prognosis of drug addicts afflicted with AIDS.

RELEVANCE TO SUBSTANCE ABUSE AND OTHER PATHOLOGIES

A few examples in the literature suggest a role of GPCR homodimers/heterodimers under pathological conditions. For example, chronic pain causes an increase in $GABA_B$ receptor (an obligatory heterodimer) expression in the dorsal lumbar spinal cord (McCarson and Enna 1999). This suggests that the increase in the level of this heterodimeric receptor may have an important role in the mediation and perception of chronic pain. Also, an increase in the levels of angiotensin AT_1/bradykinin B_2 receptor heterodimers is observed in platelets and vessels of preeclamptic women, suggesting that these heterodimers contribute to the enhanced angiotensin II responsiveness observed in these patients (AbdAlla et al. 2001). In addition, increased AT_1 homodimer levels are detected in monocytes at the onset of this pathological condition (AbdAlla et al. 2004). Thus, it appears that GPCR homodimer or heterodimer levels might be elevated during certain disease states. Therefore, in the case of opioid receptors, it would be interesting to examine the levels of receptor homodimer/heterodimer under conditions of chronic pain as well as during chronic morphine administration. Such a scenario would open avenues for the development of dimer-selective drugs for the treatment of chronic pain and/or drug addiction.

GAPS AND FUTURE DIRECTIONS

To date, the majority of the studies examining receptor–receptor interactions involving opioid receptors have been performed in heterologous expression systems. Therefore, further studies are required to determine the physiological roles of opioid receptor homodimers/heterodimers in vivo under normal and pathological conditions such as drug addiction. This will require advances in the techniques used to investigate protein–protein interactions in vivo along with the development of reagents selective for the dimeric or oligomeric forms of GPCRs—such as selective antibodies and/or ligands.

Currently, available drugs used in the treatment of chronic pain are targeted to the orthosteric site (the site to which endogenous ligands bind) of μ-opioid receptors. Prolonged treatment with these drugs causes tolerance and addictive behavior. In addition, at high doses, these drugs can also bind and activate δ and κ receptors. The presence of heterodimers with one protomer being the μ receptor allows for a strategy to identify allosteric compounds that would modulate μ receptor function by binding the partner GPCR protomer. The development of such drugs would be of tremendous importance because, theoretically, they would be highly selective and should produce fewer side effects.

ACKNOWLEDGMENTS

We thank Noura Abul-Husn for critically reading this chapter. This work was supported by grants from the National Institutes of Health (Grant Nos. DA-088360 and DA-00458) to L.A.D.

REFERENCES

AbdAlla S., Lother H., Massiery A., and Quitterer U. 2001. Increased AT1 receptor heterodimers in preeclampsia mediate angiotensin II responsiveness. *Nat. Med.* **7:** 1003–1009.

AbdAlla S., Lother H., Langer A., Faramawy Y., and Quitterer U. 2004. Factor XIIIA transglutaminase crosslinks AT_1 receptor dimers on monocytes at the onset of atherosclerosis. *Cell* **119:** 343–354.

Abdelhamid E.E., Sultana M., Portoghese P.S., and Takemori A.E. 1991. Selective blockage of delta opioid receptors prevents the development of morphine tolerance and dependence in mice. *J. Pharmacol. Exp. Ther.* **258:** 299–303

Abul-Husn N.S., Gupta A., Devi L.A., and Gomes I. 2005. Modulation of receptor pharmacology by G-protein-coupled receptor dimerization. In *Contemporary clinical neuroscience: The G protein-coupled receptors handbook* (ed. L.A. Devi), pp. 323–346. Humana Press, New York.

Aicher S.A., Punnose A., and Goldberg A. 2000. mu-Opioid receptors often colocalize with the substance P receptor (NK1) in the trigeminal dorsal horn. *J. Neurosci.* **20:** 4345–4354.

Angers S., Salahpour A., and Bouvier M. 2002. Dimerization: An emerging concept for G-protein coupled receptor ontogeny and function. *Annu. Rev. Pharmacol. Toxicol.* **42:** 409–435.

Angers S., Salahpour A., Joly E., Hilairet S., Chelsky D., Dennis M., and Bouvier M. 2000. Detection of $β_2$-adrenergic receptor dimerization in living cells using bioluminescence resonance energy transfer (BRET). *Proc. Natl. Acad. Sci.* **97:** 3684–3689.

Arvidsson U., Riedl M., Chakrabarti S., Lee J., Nakano A., Dado R., Loh H., Law P., Wessendorf M., and Elde R. 1995. Distribution and targeting of a mu-opioid receptor (MOR1) in brain and spinal cord. *J. Neurosci.* **15:** 3328-3341.

Bai M. 2004. Dimerization of G-protein coupled receptors: Roles in signal transduction. *Cell. Signal.* **16:** 175–186.

Bell J.R., Young M.R., Masterman S.C., Morris A., Mattick R.P., and Bammer G. 1999. A pilot study of naltrexone-accelerated detoxification in opioid dependence. *Med. J. Aust.* **171:** 26-30.

Betoin F., Ardid D., Herbet A., Aumaitre O., Kemeny J.L., Duchene-Marullaz P., Lavarenne J., and Eschalier A. 1994. Evidence for a central long-lasting antinociceptive effect of vapreotide, an analog of somatostatin, involving an opioidergic mechanism. *J. Pharmacol. Exp. Ther.* **269:** 7–14.

Cairns J.S. and D'Souza M.P. 1998. Chemokines and HIV-1 second receptors: The therapeutic connection. *Nat. Med.* **4:** 563–568.

Chen C., Li J., Bot G., Szabo I., Rogers T.J., and Liu-Chen L.-Y. 2004. Heterodimerization and cross-desensitization between the μ-opioid receptor and the chemokine CCR5 receptor. *Eur. J. Pharmacol.* **483:** 175–185.

Cheng P., Liu-Chen L., and Pickel V. 1997. Dual ultrastructural immunocytochemical labeling of mu and delta opioid receptors in the superficial layers of the rat cervical spinal cord. *Brain Res.* **778:** 367–380.

Cvejic S. and Devi L.A. 1997. Dimerization of the δ opioid receptor: Implication for a role in receptor internalization. *J. Biol. Chem.* **272:** 26959–26964.

De Felipe C., Herrero J.F., O'Brien J.A., Palmer J.A., Doyle C.A., Smith A.J., Laird J.M., Belmonte C., Cervero F., and Hunt S.P. 1998. Altered nociception, analgesia and aggression in mice lacking the receptor for substance P. *Nature* **392:** 394–397.

Fairbanks C.A., Posthumus I.J., Kitto K.F., Stone L.S., and Wilcox G.L. 2000. Moxonidine, a selective imidazoline/a2 adrenergic receptor agonist, synergizes with morphine and deltrophin II to inhibit substance P-induced behavior in mice. *Pain* **84:** 13–20.

George S.R., Fan T., Xie Z., Tse R., Tam V., Varghese G., and

O'Dowd B.F. 2000. Oligomerization of mu and delta opioid receptors *J. Biol. Chem.* **275:** 26128–26135.

Glasel J.A. 1987. Physical characterization of native opiate receptors. Additional information from detailed analysis of a radiation-inactivated receptor. *Biochim. Biophys. Acta* **930:** 201–208.

Glass M.J. and Pickel V.M. 2002. α2A-Adrenergic receptors are present in μ-opioid receptor containing neurons in rat medial nucleus tractus solitarius. *Synapse* **43:** 208–218.

Gomes I., Filipovska J., and Devi L.A. 2003. Opioid receptor oligomerization. Detection and functional characterization of interacting receptors. *Methods Mol. Med.* **84:** 157–183.

Gomes I., Filipovska J., Jordan B.A., and Devi L.A. 2002. Oligomerization of opioid receptors. *Methods* **27:** 358–365.

Gomes I., Gupta A., Filipovska J., Szeto H.H., Pintar J.E., and Devi L.A. 2004. A role for heterodimerization of μ and δ opiate receptors in enhancing morphine analgesia. *Proc. Natl. Acad. Sci.* **101:** 5135–5139.

Gomes I., Jordan B.A., Gupta A., Rios C., Trapaidze N., and Devi L.A. 2001. G protein coupled receptor dimerization: Implications in modulating receptor function. *J. Mol. Med.* **79:** 226–242.

Gomes I., Jordan B.A., Gupta A., Trapaidze N., Nagy V., and Devi L.A. 2000. Heterodimerization of mu and delta opioid receptors: A role in opioid synergy. *J. Neurosci.* **20:** RC110 (1–5).

Grabow T.S., Hurley R.w., Banfor P.N., and Hammond D.L. 1999. Supraspinal and spinal delta(2) opioid receptor-mediated antinociceptive synergy is mediated by spinal alpha(2) adrenoceptors. *Pain* **83:** 47–55.

Guo X.H., Fairbanks C.A., Stone L.S., and Low H.H. 2003. DPDPE-UK14,304 synergy is retained in mu opioid receptor knock-out mice. *Pain* **104:** 209–217.

Gutstein H.B. and Akil H. 2001. Opioid analgesics. In *Goodman and Gilman's The pharmacological basis of therapeutics*, 10th edition (ed. J.G. Hardman and L.E. Limbird), pp. 569–619. McGraw-Hill Medical Publishing, New York.

Hazum E., Chang K-J., Leighton H.J., Lever O.W., and Cuatrecasas O.P. 1982. Increased biological activity of dimers of oxymorphone and enkephalin: Possible role of receptor cross-linking. *Biochem. Biophys. Res. Commun.* **104:** 347–353.

He L., Fong J., von Zastrow M., and Whistler J.L. 2002 Regulation of opioid receptor trafficking and morphine tolerance by receptor oligomerization. *Cell* **108:** 271–282.

Herz A. 1993. *Opioids*, Vol. 1. Springer-Verlag, Berlin.

Heyman J.S., Vaught J.L., Mosberg H.I., Haaseth R.C., and Porreca F. 1989. Modulation of μ-mediated antinociception by δ agonists in the mouse: Selective potentiation of morphine and normorphine by DPDPE. *Eur. J. Pharmacol.* **165:** 1–10.

Holmberg S.D. 1996. The estimated prevalence and incidence of HIV in 96 large US metropolitan areas. *Am. J. Public Health* **86:** 642–654.

Inturrisi C.E. 2002. Clinical pharmacology of opioids for pain. *Clin. J. Pain* **18:** S3–S13.

Jordan B.A. and Devi L.A. 1998. Molecular mechanisms of opioid receptor signal transduction. *Br. J. Anaesth.* **81:** 12–19.

——. 1999. G-protein coupled receptor heterodimerization modulates receptor function. *Nature* **399:** 697–700.

Jordan B.A., Cvejic S., and Devi L.A. 2000. Opioids and their complicated receptor complexes. *Neuropsychopharmacology* **23:** S5–S18.

Jordan B.A., Gomes I., Rios C., Filipovska J., and Devi L.A. 2003. Functional interactions between μ opioid and α2A-adrenergic receptors. *Mol. Pharmacol.* **64:** 1317–1324.

Jordan B.A., Trapaidze, N., Gomes I., Nivarthi R., and Devi L.A. 2001. Oligomerization of opioid receptors with β2-adrenergic receptors: A role in trafficking and mitogen-activated protein kinase activation. *Proc. Natl. Acad. Sci.* **98:** 343–348.

Kroeger K.M., Pfleger K.D.G., and Eidne K.A. 2004. G-protein coupled receptor oligomerization in neuroendocrine pathways. *Front. Neuroendocrinol.* **24:** 254–278.

Law P.Y., Erickson-Herbrandson L.J., Zha Q.Q., Solberg J., Chu J., Sarre A., and Loh H.H. 2005. Heterodimerization of μ- and δ-opioid receptors occurs at the cell surface only and requires receptor-G protein interactions *J. Biol. Chem.* **280:** 11152–11164.

Li L.Y. and Chang, K.J. 1996. The stimulatory effects of opioids on mitogen activated protein kinase in Chinese hamster ovary cells transfected to express mu-opioid receptors. *Mol. Pharmacol.* **50:** 599–602.

Li-Wei C., Can G., De-He Z., Qiang W., Xue-Jun X., Jie C., and Zhi-Qiang C. 2002. Homodimerization of human mu-opioid receptor overexpressed in Sf9 insect cells. *Protein Pept. Lett.* **9:** 145–152.

Malmberg A.B. and Yaksh T.L. 1992. Isobolographic and dose-response analyses of the interaction between mu and delta agonists: Effects of naltrindole and its benzofuran analog (NTB). *J. Pharmacol. Exp. Ther.* **263:** 264–275.

Matthes H.W., Maldonado R., Simonin F., Valverde O., Slowe S., Kitchen I., Befort K., Dierich A., Le Meur M., Dolle P., Tzavara E., Hanoune J., Roques B.P., and Kieffer B.L. 1996. Loss of morphine-induced analgesia, reward effect and withdrawal symptoms in mice lacking the mu-opioid-receptor gene. *Nature* **383:** 819–823.

McCarson K.E. and Enna S.J. 1999. Nociceptive regulation of GABAB receptor gene expression in rat spinal cord. *Neuropharmacology,* **38:** 1767–1773.

McCarthy L., Wetzel M., Sliker J.K., Eisenstein T.K., and Rogers T.J. 2001. Opioid, opioid receptors, and the immune response. *Drug Alcohol Depend.* **62:** 111–123.

McLawhon R.W., Ellory J.C., and Dawson G. 1983. Molecular size of opiate (enkephalin) receptors in neuroblastoma-glioma hybrid cells as determined by radiation inactivation analysis. *J. Biol. Chem.* **258:** 2102–2105.

McVey M., Ramsay D., Kellet E., Rees S., Wilson S., Pope A.J., and Milligan G. 2001. Monitoring receptor oligomerization using time-resolved fluorescence resonance energy transfer and bioluminescence resonance energy transfer. *J. Biol.*

Chem. **276:** 14092–14099.

Miaskowski C., Taiwo Y.O., and Levine J.D. 1990. Kappa- and delta-opioid agonists synergize to produce potent analgesia. *Brain Res.* **509:** 165–168.

Milner T.A., Drake T., and Aicher S.A. 2002. C1 adrenergic neurons are contacted by presynaptic profiles containing delta-opioid receptor immunoreactivity. *Neuroscience* **110:** 691–701.

Miyagi T., Chuang L.F., Doi R.H., Carlos M.P., Torres J.V., and Chuang R.Y. 2000. Morphine induces gene expression of CCR5 in human CEM x 174 lymphocytes. *J. Biol. Chem.* **275:** 31305–31310.

Murtra P., Sheasby A.M., Hunt S.P., and De Felipe C. 2000. Rewarding effects of opiates are absent in mice lacking the receptor for substance P. *Nature* **405:** 180–183.

Ossipov M.H., Lozito R., Messineo E., Green J., Harris S., and Lloyd P. 1990. Spinal antinociceptive synergy between clonidine and morphine, U69593, and DPDPE: Isobolographic analysis. *Life Sci.* **47:** PL71–PL76.

Ott S., Costa T., Wuster M., Hietel B., and Herz A. 1986. Target size analysis of opioid receptors. No difference between receptor types, but discrimination between two receptor states. *Eur. J. Biochem.* **155:** 621–630.

Overton M.C. and Blumer K.J. 2000. G-protein coupled receptors function as oligomers *in vivo*. *Curr. Biol.* **10:** 341–344.

Pfeiffer M., Koch T., Schroder H., Laugsch M., Hollt V., and Schulz S. 2002. Heterodimerization of somatostatin and opioid receptors cross-modulates phosphorylation, internalization and desensitization. *J. Biol. Chem.* **277:** 19762–19772.

Pfeiffer M., Kirscht S., Stumm R., Koch T., Wu D., Laugsch M., Schroder H., Hollt V., and Schulz S. 2003. Heterodimerization of substance P and μ-opioid receptors regulates receptor trafficking and resensitization. *J. Biol. Chem.* **278:** 51630–51637.

Ramsay D., Kellett E., McVey M., Rees S., and Milligan G. 2002. Homo- and hetero-oligomeric interactions between G-protein coupled receptors in living cells monitored by two variants of bioluminescence resonance energy transfer (BRET): Hetero-oligomers between receptor subtypes form more efficiently than between less closely related sequences. *Biochem. J.* **365:** 429–440.

Rios C., Gomes I., and Devi L.A. 2004. Interactions between δ opioid receptors and α_{2A}-adrenergic receptors. *Clin. Exp. Pharmacol. Physiol.* **31:** 833–836.

Rios, C.D., Jordan B.A., Gomes I., and Devi L.A. 2001. G-protein coupled receptor dimerization: Modulation of receptor function. *Pharmacol. Ther.* **92:** 71–87.

Ripley T.L., Gadd C.A., De Felipe C., Hunt S.P. and Stephens D.N. 2002. Lack of self-administration and behavioural sensitisation to morphine, but not cocaine, in mice lacking NK1

receptors. *Neuropharmacology* **43:** 1258–1268.

Rothman R. and Westfall T. 1981. Allosteric modulation by leucine-enkephalin of [3H]naloxone binding in rat brain. *Eur. J. Pharmacol.* **72:** 365–368.

Rothman R., Bowen W., Schumacher U., and Pert C. 1983. Effect of beta-FNA on opiate receptor binding: Preliminary evidence for two types of mu receptors. *Eur. J. Pharmacol.* **95:** 147–148.

Russell R.D., Leslie J.B., Su Y.F., Watkins W.D., and Chang K.J. 1986. Interaction between highly selective mu and delta opioids *in vivo* at the rat spinal cord. *NIDA Res. Monogr.* **75:** 97–100.

Sanchez-Blazquez P., Garcia-Espana A., and Garzon J. 1997. Antisense oligonucleotides to opioid mu and delta receptors reduced morphine dependence in mice: Role of delta-2 opioid receptors. *J. Pharmacol. Exp Ther.* **280:** 1423–1431.

Sarne Y., Fields A., Keren O., and Gafni M. 1996. Stimulatory effects of opioids on transmitter release and possible cellular mechanisms: Overview and original results. *Neurochem. Res.* **21:** 1353–1361.

Simon J., Benyhe S., Borsodi A., and Wollemann M. 1986. Hydrodynamic parameters of opioid receptors from frog brain. *Neuropeptides* **7:** 23–26.

Sora I., Funada M., and Uhl G.R. 1997. The mu-opioid receptor is necessary for [d-Pen2,D-Pen 5] enkephalin- induced analgesia. *Eur. J. Pharmacol.* **324:** R1–R2.

Sorensen J.L. and Copeland A.L. 2000. Drug abuse treatment as an HIV prevention strategy: A review. *Drug Alcohol Depend.* **59:** 17–31.

Stone L.S., MacMillan L.B., Kitto K.F., Limbird L.E., and Wilcox G.L. 1997. The α_{2A} adrenergic receptor subtype mediates spinal analgesia evoked by α_2 agonists and is necessary for spinal adrenergic-opioid synergy. *J. Neurosci.* **17:** 7157–7165.

Suzuki S., Chuang L.F., Yau P., Doi R.H., and Chuang R.Y. 2002. Interactions of opioid and chemokine receptors: Oligomerization of mu, kappa, and delta with CCR5 on immune cells. *Exp. Cell Res.* **280:** 192–200.

Wessendorf M.W. and Dooyema J. 2001. Coexistence of kappa- and delta-opioid receptors in rat spinal cord axons. *Neurosci. Lett.* **298:** 151–154.

Zhang Y.Q. and Limbird L.E. 2004. Hetero-oligomers of α_{2A}-adrenergic and μ-opioid receptors do not lead to transactivation of G-proteins or altered endocytosis profiles. *Biochem. Soc. Trans.* **32:** 856–860.

Zhu Y., King M.A., Schuller A.G., Nitsche J.F., Reidl M., Elde R.P., Unterwald E., Pasternak G.W., and Pintar J.E. 1999. Retention of supraspinal delta-like analgesia and loss of morphine tolerance in delta opioid receptor knockout mice. *Neuron* **24:** 243–252.

13 | The Endocannabinoid System: From Cell Biology to Therapy

Daniele Piomelli

Department of Pharmacology, Center for Drug Discovery, University of California, Irvine, California 92697-4625

ABSTRACT

The active principle of marijuana, Δ^9-tetrahydrocannabinol (Δ^9-THC), exerts its pharmacological effects by binding to selective receptors present on the membranes of neurons and other cells. These cannabinoid receptors are normally activated by a family of lipid mediators, called endocannabinoids, that are thought to participate in the regulation of a variety of physiological functions including pain, mood, appetite, and blood pressure. Drug abuse may lead to adaptive changes in endocannabinoid signaling, and these changes might contribute to the drugs' effects of marijuana as well as to the development of dependence and withdrawal. Here, I outline current views on how endocannabinoid substances are produced, released, and deactivated in the brain. In addition, I review recent progress on the development of agents that interfere with endocannabinoid deactivation and discuss their potential utility in the treatment of neuropsychiatric disorders and drug abuse.

INTRODUCTION

The active constituent of marijuana, Δ^9-THC, produces a complex palette of pharmacological effects by combining with selective receptors present on the membranes of cells in the brain, vasculature, and immune system. These G-protein-coupled receptors are engaged by a group of endogenous lipid substances called endocannabinoids that are released under various physiological conditions and regulate a diversity of biological processes that include pain, emotion, appetite, and blood pressure (Piomelli 2003). In recent years, our knowledge of the endocannabinoid signaling system has greatly expanded, allowing us to better understand the pharmacological properties of marijuana and to develop strategies to manage its dependence and exploit its therapeutic potential.

GOALS OF THE CHAPTER

This chapter describes the properties and distribution of CB_1 receptors—the predominant cannabinoid receptor subtype present in the central nervous system. I also review the synthesis, release, and deactivation of endogenous cannabinoids (also called endocannabinoids). Finally, I describe recent advances in the discovery of drugs that interfere with endocannabinoid deactivation and briefly discuss their potential utility in the treatment of neuropsychiatric disorders such as anxiety and drug abuse.

RECEPTORS AND LIGANDS OF THE ENDOCANNABINOID SYSTEM

CB_1 Receptors

CB_1 receptors are highly expressed in the brain of rodents and humans, and their activation accounts for most, if not all, behavioral actions attributed to Δ^9-THC. In fact, the four key effects exerted by this drug in rodents—hypothermia, catalepsy, analgesia, and decreased motor activity—are virtually absent in mutant CB_1-deficient mice (Ledent et al. 1999; Zimmer et al. 1999).

CB_1 Signaling Mechanisms

CB_1 is coupled to $G_{i/o}$ proteins and initiates signaling events typical of these transducing molecules. These include closure of Ca^{2+} channels, opening of K^+ channels, inhibition of adenylyl cyclase activity, and stimulation of tyrosine and serine/threonine kinases. Each of these events may have distinct roles in CB_1-mediated responses. For example, the ability of cannabinoid agonists to inhibit N- and P/Q-type voltage-activated Ca^{2+} channels (Caulfield and Brown 1992; Mackie and Hille 1992; Twitchell et al. 1997) may underlie CB_1-mediated depression of transmitter release at GABAergic synapses in the hippocampus or at glutamatergic synapses in the striatum (Hoffman and Lupica 2000; Gerdeman and Lovinger 2001; Huang et al. 2001). Similarly, cannabinoid regulation of voltage-gated K^+ currents may be involved in presynaptic inhibition at GABAergic and glutamatergic synapses in the cerebellum, amygdala, and nucleus accumbens (Daniel and Crepel 2001; Robbe et al. 2001; Azad et al. 2003). The sensitivity of these responses to pertussis toxin implies the participation of $G_{i/o}$, but it is still not known whether transduction occurs by direct modulation of channel activity or by affecting intracellular second messengers.

Although reduction in cAMP levels—a prototypical response to the G_i-coupled receptors such as CB_1—does not appear to be involved in ion-channel modulation, it may contribute to CB_1-mediated regulation of neuronal gene expression. This process is necessary to induce lasting modifications in synaptic strength and depends on the recruitment of interconnected networks of intracellular protein kinases. Two components of these networks (extracellular signal-regulated kinase [ERK] and focal adhesion kinase [FAK]) are activated by cannabinoid agonists in rat hippocampal slices (Derkinderen et al. 1996, 2003). Cell-permeant cAMP analogs block this activation, which suggests that it might result from a decrease in intracellular cAMP levels. The involvement of ERK and FAK in synaptic plasticity indirectly implicates these protein kinases in persistent neural adaptations that accompany cannabinoid administration and, possibly, long-term marijuana abuse (Hoffman et al. 2003).

Expression of CB_1 Receptors in the Brain

In the brain cortex, CB_1 receptors are primarily found on axon terminals of cholecystokinin (CCK-8)-releasing GABAergic interneurons (Tsou et al. 1998; Katona et al. 1999, 2001;

Marsicano and Lutz 1999; McDonald and Mascagni 2001). This expression pattern domi-nates the neocortex, hippocampal formation, and amygdala. There is also evidence that excitatory terminals in these regions contain CB_1: For example, cannabinoid agonists reduce glutamatergic transmission in the amygdala of normal mice, but fail to exert this effect in CB_1-deficient mutants (Freund et al. 2003). In addition, low levels of CB_1 mRNA have been found in many non-GABAergic neurons of the cortex (for a review, see Piomelli 2003).

CB_1 receptors are very densely expressed throughout the basal ganglia, where they outnumber even the highly abundant dopamine D_1 receptors. In the striatum, CB_1 recep-tors are localized to three distinct neuronal elements: glutamatergic terminals originating in the cortex (Gerdeman and Lovinger 2001; Huang et al. 2001), local-circuit GABAergic interneurons ("fast-spiking" interneurons that do not express CCK-8), and axon terminals of GABAergic projection neurons ("medium spiny neurons"). In the cerebellum, CB_1 is present on excitatory terminals of climbing and parallel fibers (but not on their postsynap-tic partners, the Purkinje neurons) as well as on GABAergic interneurons (Herkenham et al. 1990; Tsou et al. 1998). Although in lower numbers, CB_1 receptors are also found in the thalamus, hypothalamus, midbrain, medulla, and dorsal horn of the spinal cord (for a review, see Freund et al. 2003). Finally, CB_1 is expressed in peripheral sensory neurons, where it is localized to C and Aβ fibers (Freund et al. 2003).

Endocannabinoids

The endocannabinoids are lipid-derived mediators that share the key structural motif of an arachidonic-acid chain linked to ethanolamine or glycerol (Fig. 1). Because of this feature, the endocannabinoids superficially resemble the eicosanoids, a class of biologically active lipids produced through the enzymatic oxygenation of nonesterified arachidonic acid. However, the endocannabinoids are distinguished from the eicosanoids by their different biosynthetic routes, which do not primarily involve oxidative metabolism. The two best-characterized endocannabinoids—anandamide (arachidonoylethanolamide) (Devane et al. 1992) and 2-arachidonoylglycerol (2-AG) (Fig. 1) (Mechoulam et al. 1995; Sugiura et al. 1995)—are produced through cleavage of phospholipid precursors present in the mem-branes of neurons, glia, and other cells.

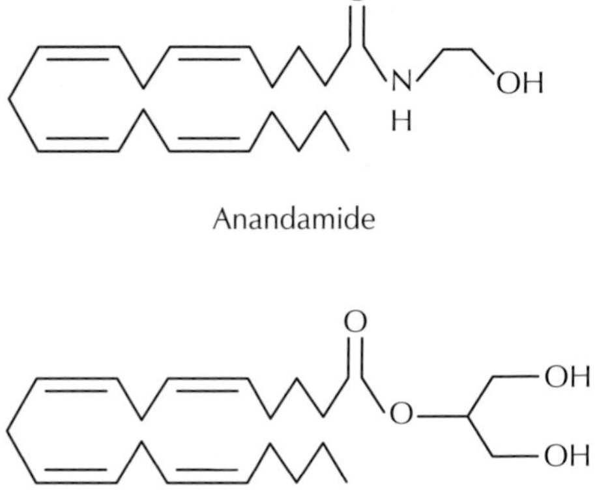

Anandamide

2-Arachidonoylglycerol (2-AG)

FIGURE 1. Chemical structures of endogenous compounds that bind to cannabinoid receptors.

Anandamide

Mechanism of Formation

Anandamide was the first endocannabinoid substance to be isolated and structurally characterized (Devane et al. 1992). Its formation in neural cells is thought to require two enzymatic steps that are illustrated in Figure 2. The first is activity-dependent cleavage of the phospholipid precursor N-arachidonoyl-PE (NAPE). This reaction, mediated by a unique D-type phospholipase (PLD) (Okamoto et al. 2004), produces anandamide and phosphatidic acid, which are recycled to produce other glycerol-containing phospholipids. The brain contains very small stores of NAPE—probably too small to sustain anandamide release for an extended period of time. These stores can be refilled, however, by an N-acyltransferase (NAT) activity that catalyzes the intermolecular transfer of an arachidonate moiety from the sn-1 position of phosphatidylcholine (PC) to the head group of phosphatidylethanolamine (PE) (Fig. 2). In brain neurons, NAT activity is regulated by Ca^{2+} ions, which are needed to activate the enzyme, and by cAMP, which stimulates protein kinase A-dependent protein phosphorylation to enhance NAT activity (Cadas et al. 1997). Although separate enzymes catalyze the syntheses of anandamide and NAPE, the two events are likely to occur simultaneously because Ca^{2+}-stimulated anandamide production is often accompanied by de novo formation of NAPE (Di Marzo et al. 1994; Cadas et al. 1997).

Cocaine and Dopamine Agonists Induce Anandamide Formation

Anandamide formation can be evoked both in vitro and in vivo by a variety of treatments that increase intracellular Ca^{2+} levels. For example, perfusion of a membrane-depolarizing

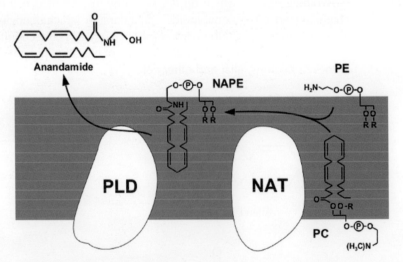

FIGURE 2. Mechanism of anandamide formation in neural cells. This hypothetical diagram illustrates two key steps of anandamide synthesis: (1) production of the anandamide precursor N-arachidonoyl-phosphatidylethanolamine (NAPE) from phosphatidylethanolamine (PE) and phosphatidylcholine (PC), catalyzed by the enzyme N-acyltransferase and (2) the hydrolysis of NAPE to yield anandamide, catalyzed by NAPE-specific phospholipase D (PLD).

concentration of K⁺ ions through a microdialysis probe evokes Ca^{2+}-dependent anandamide release in the dorsal striatum of awake rats (Giuffrida et al. 1999). Beside Ca^{2+} entry, occupation of certain G-protein-coupled receptors can also stimulate anandamide synthesis. Thus, administration of the dopamine D_2-receptor agonist quinpirole produces a marked increase in anandamide formation in rat basal ganglia, which is prevented by the D_2 antagonist raclopride (Giuffrida et al. 1999; Ferrer et al. 2003). Interestingly, cocaine elicits a similar response (Centonze et al. 2004), suggesting a role for anandamide in the actions of this psychostimulant drug. The ability of the anandamide transport inhibitor AM404 to reduce D_2 agonist-induced hyperactivity, discussed below, further supports this possibility (Beltramo et al. 2000).

Mechanisms of Anandamide Elimination

The biological deactivation of anandamide consists of a two-step process of high-affinity transport into cells and intracellular hydrolysis (Fig. 3).

Transport Into Cells

Primary cultures of brain neurons and astrocytes internalize [³H]anandamide with a mechanism that meets all defining criteria of carrier-mediated transport. Plots of the initial rates of [³H]anandamide internalization in rat brain neurons and astrocytes in culture

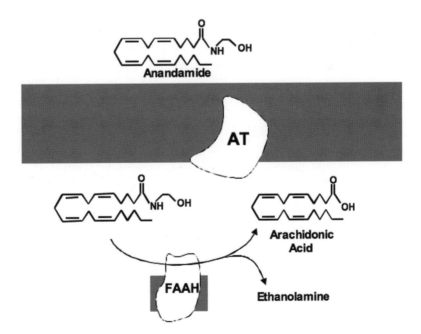

FIGURE 3. Mechanisms of anandamide deactivation in neural cells. Anandamide can be internalized by neurons and glia through a high-affinity transport (AT) system. After internalization, anandamide may be hydrolyzed by fatty-acid amide hydrolase (FAAH) present on mitochondria and endoplasmic reticulum membranes.

yield apparent Michaelis constants (K_M) that are consistent with a saturable reaction and comparable to the K_M values of other neurotransmitters for their respective transporters (Beltramo et al. 1997; Hillard et al. 1997; Piomelli et al. 1999). Furthermore, neurons and astrocytes in culture internalize [^{3}H]anandamide in a stereospecific manner (Piomelli et al. 1999). Finally, [^{3}H]anandamide uptake can be pharmacologically inhibited in a selective manner (Beltramo et al. 1997; Piomelli et al. 1999). Nevertheless, the fact that anandamide is internalized in a Na$^+$- and energy-independent manner (Beltramo et al. 1997; Hillard et al. 1997; Piomelli et al. 1999) differentiates its transport mechanism from those of other neurotransmitters.

Intracellular Hydrolysis

An amide hydrolase activity that catalyzes the hydrolysis of noncannabinoid fatty-acid ethanolamides such as palmitoylethanolamide (PEA) and oleoylethanolamide (OEA) (Calignano et al. 1998; Rodríguez de Fonseca et al. 2001; Fu et al. 2003; LoVerme et al. 2005) was first identified in 1984 (Schmid et al. 1985). Anandamide is a substrate for this activity, as indicated by both biochemical experiments (Di Marzo et al. 1994; Désarnaud et al. 1995; Ueda et al. 1995) and molecular cloning of the enzyme responsible (Cravatt et al. 1996, 2001). This enzyme, called fatty-acid amide hydrolase (FAAH), belongs to a group of proteins known as the "amidase signature family" (Cravatt et al. 1996; Giang and Cravatt 1997) and catalyzes the hydrolysis of not only anandamide and other fatty-acid ethanolamides, but also fatty-acid esters such as 2-AG (Goparaju et al. 1998; Lang et al. 1999). Site-directed mutagenesis and X-ray diffraction studies have shown that this broad substrate preference may be accounted for by a catalytic mechanism that involves the amino-acid residue lysine 142. This residue is thought to act as a general acid catalyst, favoring the protonation and consequent detachment of reaction products from the enzyme's active site (Patricelli et al. 1999; Bracey et al. 2002). Three additional serine residues, conserved in all amidase signature enzymes (S241, S217, and S218 in FAAH), may also be essential for enzymatic activity: Serine 241 is thought to serve as the enzyme's catalytic nucleophile, whereas serines 217 and 218 may modulate catalysis through an as-yet-unidentified mechanism (Patricelli et al. 1999). Electron microscopy studies have demonstrated that FAAH is predominantly found on mitochondria and endoplasmic reticulum membranes (Gulyas et al. 2004). Although FAAH appears to be the predominant route of anandamide hydrolysis in the brain, other enzymes are likely to participate in the breakdown of this endocannabinoid in peripheral tissues. For example, an acid amide hydrolase activity catalytically distinct from FAAH has been purified and molecularly cloned, although its role in anandamide deactivation has not yet been investigated (Tsuboi et al. 2005).

2-Arachidonoylglycerol (2-AG) Identified as Another Endocannabinoid

2-AG was identified as a second endocannabinoid substance in 1995 (Mechoulam et al. 1995; Sugiura et al. 1995). The multiple roles of this lipid compound in intermediate lipid metabolism have hindered our attempts to determine the biochemical route(s) involved in its physiological formation, but one pathway has emerged as a likely candidate (Fig. 4).

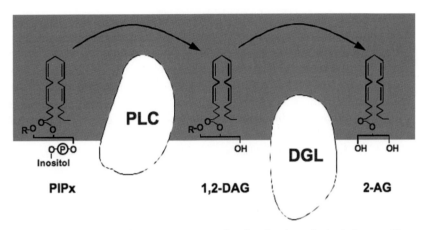

FIGURE 4. Mechanism of 2-AG formation in neural cells. This hypothetical diagram illustrates two steps of 2-AG synthesis: (1) production of 1,2-diacylglycerol (DAG) from phosphoinositides (PIPx), catalyzed by the enzyme phospholipase C (PLC) and (2) the hydrolysis of 1,2-DAG to yield 2-AG, catalyzed by diacylglycerol lipase (DGL).

This pathway begins with the phospholipase-mediated production of 1,2-diacylglycerol (DAG), which can serve as a substrate for two enzymes: DAG kinase, which catalyzes DAG phosphorylation to phosphatidic acid, and DAG lipase (DGL), which hydrolyzes DAG to monoacylglycerol (Bisogno et al. 2003). Pharmacological inhibition of phospholipase C (PLC) and DGL prevents the Ca^{2+}-dependent accumulation of 2-AG in rat cortical neurons, which suggests a key role of this pathway in 2-AG generation (Stella and Piomelli 2001). However, additional enzymes also may be involved in 2-AG synthesis including PLA_1, hormone-sensitive lipase, or lipid phosphatases (Piomelli 2003). In hippocampal slices, electrical stimulation of the Schaffer collaterals, a glutamatergic fiber tract that connects neurons in the CA3 and CA1 fields, causes a Ca^{2+}-dependent increase in 2-AG content (Stella et al. 1997). This stimulation has no effect on the levels of noncannabinoid monoacylglycerols, such as 1(3)-palmitoylglycerol, which indicates that 2-AG formation is unlikely to result from a nonspecific enhancement of lipid turnover. Furthermore, electrical stimulation of the Schaffer collaterals does not modify hippocampal anandamide levels, suggesting that the biochemical pathways leading to the production of 2-AG and anandamide may be independently controlled (Stella et al. 1997). Ca^{2+}-dependent 2-AG synthesis was also demonstrated in rat cortical neurons, where the Ca^{2+} ionophore ionomycin and the glutamate receptor agonist N-methyl-D-aspartate (NMDA) stimulate 2-AG production in a Ca^{2+}-dependent manner (Stella and Piomelli 2001).

Mechanisms of 2-AG Elimination

Transport into Cells

2-AG is internalized by neurons and glial cells through a mechanism formally similar to that involved in anandamide transport. Thus, human astrocytoma cells accumulate

[³H]anandamide and [³H]2-AG with similar kinetic properties, and accumulation of both compounds is prevented by the anandamide transport inhibitor AM404 (Piomelli et al. 1999; Beltramo and Piomelli 2000). Moreover, anandamide and 2-AG prevent each other's transport (Beltramo and Piomelli 2000). Nevertheless, there also appear to be differences between anandamide and 2-AG accumulation. For example, [³H]2-AG internalization in astrocytoma cells is reduced by exogenous arachidonic acid, whereas [³H]anandamide internalization is not. Arachidonic acid may produce this effect by preventing the enzymatic conversion of 2-AG to arachidonic acid through product inhibition of 2-AG hydrolysis, because a blocker of fatty acyl-coenzyme A synthetase, triacsin C, prevents [³H]2-AG uptake in astrocytoma cells (Beltramo and Piomelli 2000).

Intracellular Hydrolysis

Inside cells, 2-AG is hydrolyzed enzymatically to glycerol and arachidonic acid. Although FAAH can catalyze 2-AG hydrolysis in broken cell preparations, this enzyme is unlikely to contribute to 2-AG degradation in vivo for three reasons. First, the brain contains at least two 2-AG-hydrolase activities that are chromatographically distinct from FAAH (Goparaju et al. 1999). Second, inhibition of FAAH activity in intact neurons prevents the hydrolysis of anandamide, but not 2-AGs (Kathuria et al. 2003). Third, 2-AG breakdown is preserved in FAAH-deficient mice (Cravatt et al. 2001). These findings suggest that an enzyme different from FAAH is responsible for 2-AG degradation.

Monoacylglycerol Lipase May Be Responsible for 2-AG Degradation

One likely candidate for this role is monoacylglycerol lipase (MGL), a cytosolic serine hydrolase that cleaves 2- and 1-monoacylglycerols (Karlsson et al. 1997). The molecular cloning of rat brain MGL has recently allowed us to test this hypothesis (Dinh et al. 2002b). We observed that MGL is abundantly expressed in discrete areas of the rat brain—including the hippocampus, cortex, and cerebellum—where CB$_1$ receptors are also found. Furthermore, we have shown that adenovirus-induced overexpression of MGL enhances the hydrolysis of endogenously produced 2-AG in primary cultures of rat brain neurons (Dinh et al. 2002a). More recent experiments indicate that silencing the MGL gene through RNA interference impairs 2-AG degradation and elevates endogenous 2-AG levels in intact HeLa cells (Dinh et al. 2004). These findings suggest that MGL has an important role in 2-AG hydrolysis. Nevertheless, additional experiments with MGL-null mice and selective MGL inhibitors are needed to fully test this hypothesis.

Anandamide Transport and Its Inhibitors

Use of Anandamide Transport Inhibitors

Anandamide transport inhibitors such as the compound AM404 (Fig. 5) have provided important information about the properties of anandamide transport, aiding the in vitro characterization of this process and helping to reveal its possible physiological functions.

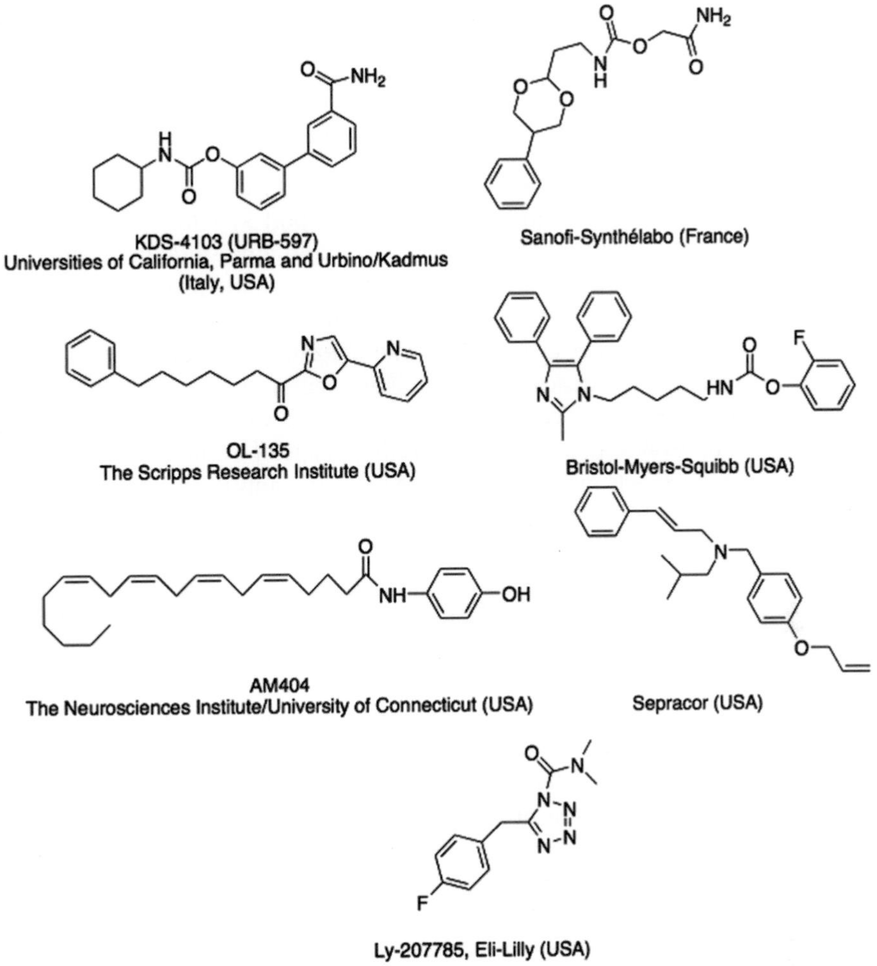

FIGURE 5. Chemical structures of various pharmacological inhibitors of anandamide transport and FAAH.

Moreover, the pharmacological profile exhibited by these agents suggests that anandamide transport could represent an innovative drug target in disease conditions in which the endocannabinoid system is hypofunctional (Beltramo and Piomelli 2000). One such condition may be opiate withdrawal, whose symptoms are markedly ameliorated by administration of AM404 (Del Arco et al. 2002). These ideas remain speculative, however, because the putative transport system responsible for anandamide internalization has not been molecularly characterized yet. In fact, its very existence has been questioned based on the observation that [^{3}H]anandamide uptake in certain cell lines is saturable at longer (> 5 min) but not at shorter (< 40 sec) incubation times (Glaser et al. 2003). This finding has been interpreted to suggest that FAAH may be responsible for the saturation of uptake noted at longer incubation times (Glaser et al. 2003). However, the result may also be explained on purely technical grounds, because the high concentration of serum albumin used in those experiments was previously shown to prevent [^{3}H]anandamide internalization (Di Marzo et al. 1994; Hillard and Jarrahian 2003).

Evidence for an Anandamide Transport System

Consistent with this interpretation, recent studies have provided additional evidence for the existence of an anandamide transport system independent of FAAH (Fegley et al. 2004b; Ligresti et al. 2004). For example, one study has demonstrated that cortical neurons deficient in FAAH internalize anandamide as efficiently as do neurons that express normal levels of the enzyme (Fegley et al. 2004b). The same study also showed that AM404 reduces anandamide internalization in neurons of FAAH-null and wild-type mice with equal efficacy. These findings suggest that FAAH neither drives anandamide uptake nor serves as a target for AM404. In vivo work further supports this possibility, showing that AM404 not only amplifies the effects of exogenous anandamide in FAAH-null mice, but also acts more effectively in these mutants than it does in wild-type animals (Fegley et al. 2004b). This suggests in turn that AM404 is not an FAAH inhibitor, as has been proposed (Glaser et al. 2003), but an FAAH substrate. Confirming this supposition, it was shown that brain membranes from normal mice rapidly hydrolyze AM404, whereas membranes from FAAH-deficient animals cannot carry out this reaction (Fegley et al. 2004b). If FAAH is not involved in anandamide internalization, what mechanism(s) provides the driving force for this process? One possibility is that an intracellular protein may sequester anandamide at the membrane, driving its internalization and facilitating its movement to the mitochondria and the endoplasmic reticulum, where FAAH is primarily localized (Gulyas et al. 2004). If selective for anandamide, such a protein might mediate the internalization process and act as a target for transport inhibitors. This hypothetical model is consistent with fatty-acid transport into cells, which is also thought to require the cooperation of membrane transporters and intracellular fatty-acid binding proteins (Black and DiRusso 2003).

Behavioral Effects of Anandamide Transport Inhibitors

In rodents, AM404 increases anandamide levels in the circulation and the brain (Giuffrida et al. 2000; Fegley et al. 2004b). This effect is associated with behavioral responses that are blocked by the CB$_1$ antagonist rimonabant (SR141716A), but are clearly different from those of direct-acting cannabinoid agonists. For example, administration of AM404 into the cerebral ventricles of rats decreases exploratory activity without producing catalepsy (rigid immobility) and analgesia, two typical signs of CB$_1$-receptor activation (Beltramo et al. 2000). Moreover, AM404 reduces two characteristic effects produced by dopamine D$_2$-receptor activation: apomorphine-induced yawning in mice and quinpirole-induced hyperlocomotion in rats (Beltramo et al. 2000). Importantly, these actions are caused by doses of AM404 that may selectively target anandamide transport (Beltramo et al. 2000). These observations differentiate the pharmacological profile of AM404 from that of direct-acting cannabinoid agonists, a distinction that may result from the ability of AM404 to magnify anandamide signaling in an activity-dependent manner, causing anandamide to accumulate in discrete regions of the brain and only when appropriate stimuli initiate its release. Pharmacological activation of D$_2$ receptors may represent one such stimulus, suggesting that blockade of anandamide transport might offer an innovative strategy to correct abnormalities associated with dysfunction in dopaminergic transmission. Initial tests of this hypothesis have shown that systemic administration of AM404 normalizes movement in spontaneously hypertensive rats, an inbred line in which hyperactivity and attention deficits have been linked to a defective regulation of mesocortico-limbic dopamine pathways (Beltramo et al. 2000).

FAAH INHIBITORS

A variety of potent and selective inhibitors of intracellular FAAH activity has been identified including substituted sulfonyl fluorides (Gifford et al. 1999), alpha-keto-oxazolopyridines (Boger et al. 2000), and carbamic-acid esters (Kathuria et al. 2003; Tarzia et al. 2003) (Fig. 5). The latter compounds were discovered during structure-activity relationship studies aimed at determining whether esters of carbamic acid such as the insecticide carbaryl inhibit FAAH activity. It was found that although carbaryl is ineffective in this regard, variations in its template display significant inhibitory potencies. Subsequent optimizations yielded a group of highly potent inhibitors, exemplified by the compound URB597 (Fig. 5). URB597 has no notable effect on cannabinoid receptor binding, anandamide transport, or rat brain MGL activity, but strongly inhibits brain FAAH activity in live animals. After injection of a maximal dose of compound (0.3 mg kg^{-1}, intraperitoneal), FAAH inhibition is rapid (< 15 min), persistent (> 16 hr), and accompanied by a time-dependent accumulation of anandamide in the brain. The inhibitor does not evoke overt cannabinoid-like responses when administered alone, but magnifies the effects exerted by exogenously administered anandamide (Kathuria et al. 2003; Fegley et al. 2004a). The latter effect is entirely due to blockade of FAAH activity, because it is absent in mutant FAAH-null mice (Fegley et al. 2004a).

Behavioral Effects of FAAH Inhibitors

The FAAH inhibitor URB597 exerts several pharmacological effects that might be therapeutically relevant. One such effect, the ability to reduce anxiety-like behaviors in rats, was demonstrated using the elevated "zero maze" test and the isolation-induced ultrasonic emission test (Kathuria et al. 2003). The zero maze consists of an elevated annular platform with two open and two closed quadrants and is based on the conflict between an animal's instinct to explore its environment and its fear of open spaces in which it may be attacked by predators (Shepherd et al. 1994). Benzodiazepines and other clinically used anxiolytic drugs increase the proportion of time spent in, and the number of entries made into, the open compartments. In a similar fashion, URB597 elicits anxiolytic-like responses at a dose (0.1 mg kg^{-1}, intraperitoneal) that corresponds to those required to inhibit brain FAAH activity. Moreover, these effects are prevented by the CB$_1$-selective antagonist rimonabant. Analogous results were obtained in the ultrasonic vocalization emission test, which measures the number of stress-induced vocalizations emitted by rat pups removed from their nest (Kathuria et al. 2003). If confirmed in other behavioral models, these findings would suggest that inhibition of intracellular FAAH activity might offer a novel target for the treatment of anxiety (for a review, see Gaetani et al. 2003), which is a prominent feature of marijuana withdrawal (Kouri et al. 1999; Kouri and Pope 2000; Budney et al. 2003).

SUMMARY

The last 10 years have produced an exciting series of discoveries in endocannabinoid biology. It has become clear that cells in the brain and peripheral tissues produce multiple endocannabinoid lipids using membrane constituents as starting material. It has been shown that these signaling lipids are quite different from traditional transmitters in that they are released in a nonsynaptic manner and activate cannabinoid receptors found near their sites of synthesis. Acting as local signals, the endocannabinoids contribute in important ways to the regulation of synaptic strength and, by doing so, modulate a diversity of brain functions including pain, emotion, and the rewarding properties of drugs of abuse.

GAPS AND OPPORTUNITIES

Despite these advances, many important questions remain. We must map the neuronal circuits that produce anandamide and 2-AG, which in turn requires the molecular characterization of all enzymes involved in these reactions. Important progress has recently been made in this area, but gaps remain in the identification of proteins involved in anandamide synthesis (e.g., NAT) and deactivation (the putative anandamide transporter). We must also understand how classical neurotransmitters and drugs of abuse interact with these circuits, and explore the functional consequences of such interactions. For example, although psychostimulant drugs such as cocaine have been shown to induce anandamide formation, the cellular mechanism of this effect and its impact on psychostimulant-induced behaviors remain undefined. Finally, we must continue to develop selective agents that target not only the different cannabinoid receptor subtypes, but also the mechanisms of endocannabinoid synthesis and deactivation. The availability of potent FAAH inhibitors is already helping to reveal new roles for endogenous anandamide. It is likely that the development of MGL inhibitors and a second generation of endocannabinoid transport inhibitors will have an even greater impact, particularly on our understanding of the biology of 2-AG (Hohmann et al. 2005). Addressing these issues may not only shed new light on fundamental mechanisms of drug addiction, but also help to develop innovative therapeutic strategies for drug abuse and other brain disorders.

ACKNOWLEDGMENTS

I thank all of the members of my lab for their extraordinary contributions, the National Institute on Drug Abuse for its financial support, and B. Cartwright for editorial assistance with this manuscript.

REFERENCES

Azad S.C., Eder M., Marsicano G., Lutz B., Zieglgansberger W., and Rammes G. 2003. Activation of the cannabinoid receptor type 1 decreases glutamatergic and GABAergic synaptic transmission in the lateral amygdala of the mouse. *Learn. Mem.* **10:** 116–128.

Beltramo M. and Piomelli D. 2000. Carrier-mediated transport and enzymatic hydrolysis of the endogenous cannabinoid 2-arachidonylglycerol. *Neuroreport* **11:** 1231–1235.

Beltramo M., Stella N., Calignano A., Lin S.Y., Makriyannis A., and Piomelli D. 1997. Functional role of high-affinity anandamide transport, as revealed by selective inhibition. *Science* **277:** 1094–1097.

Beltramo M., Rodríguez de Fonseca F., Navarro M., Calignano A., Gorriti M.A., Grammatikopoulos G., Sadile A.G., Giuffrida A., and Piomelli D. 2000. Reversal of dopamine D_2 receptor responses by an anandamide transport inhibitor. *J. Neurosci.* **20:** 3401–3407.

Bisogno T., Howell F., Williams G., Minassi A., Cascia M.G., Ligresti A., Matias I., Schiano-Moriello A., Paul P., Williams E.J., Gangadharan U., Hobbs C., Di Marzo V., and Doherty P. 2003. Cloning the first sn1-DAG lipases points to the spatial and temporal regulation of endocannabinoid signaling in the brain. *J. Cell. Biol.* **163:** 463–468.

Black P.N. and DiRusso C.C. 2003. Transmembrane movement of exogenous long-chain fatty acids: Proteins, enzymes, and vectorial esterification. *Microbiol. Mol. Biol. Rev.* **67:** 454–472.

Boger D.L., Sato H., Lerner A.E., Hedrick M.P., Fecik R.A., Miyauchi H., Wilkie G.D., Austin B.J., Patricelli M.P., and Cravatt B.F. 2000. Exceptionally potent inhibitors of fatty acid amide hydrolase: The enzyme responsible for degradation of endogenous oleamide and anandamide. *Proc. Natl. Acad. Sci.* **97:** 5044–5049.

Bracey M.H., Hanson M.A., Masuda K.R., Stevens R.C., and Cravatt B.F. 2002. Structural adaptations in a membrane enzyme that terminates endocannabinoid signaling. *Science* **298:** 1793–1796.

Budney A.J., Moore B.A., Vandrey R.G., and Hughes J.R. 2003. The time course and significance of cannabis withdrawal. *J. Abnorm. Psychol.* **112:** 393–402.

Cadas H., di Tomaso E., and Piomelli D. 1997. Occurrence and biosynthesis of endogenous cannabinoid precursor, *N*-arachidonoyl phosphatidylethanolamine, in rat brain. *J. Neurosci.* **17:** 1226–1242.

Calignano A., La Rana G., Giuffrida A., and Piomelli D. 1998. Control of pain initiation by endogenous cannabinoids. *Nature* **394:** 277–281.

Caulfield M.P. and Brown D.A. 1992. Cannabinoid receptor agonists inhibit Ca current in NG108-15 neuroblastoma cells via a pertussis toxin-sensitive mechanism. *Br. J. Pharmacol.* **106:** 231–232.

Centonze D., Battista N., Rossi S., Mercuri N.B., Finazzi-Agro A., Bernardi G., Calabresi P., and Maccarrone M. 2004. A critical interaction between dopamine D2 receptors and endocannabinoids mediates the effects of cocaine on striatal GABAergic transmission. *Neuropsychopharmacology* **29:** 1488–1497.

Cravatt B.F., Giang D.K., Mayfield S.P., Boger D.L., Lerner R.A., and Gilula N.B. 1996. Molecular characterization of an enzyme that degrades neuromodulatory fatty-acid amides. *Nature* **384:** 83–87.

Cravatt B.F., Demarest K., Patricelli M.P., Bracey M.H., Giang D.K., Martin B.R., and Lichtman A.H. 2001. Supersensitivity to anandamide and enhanced endogenous cannabinoid signaling in mice lacking fatty acid amide hydrolase. *Proc. Natl. Acad. Sci.* **98:** 9371–9376.

Daniel H. and Crepel F. 2001. Control of Ca(2+) influx by cannabinoid and metabotropic glutamate receptors in rat cerebellar cortex requires K(+) channels. *J. Physiol.* **537:** 793–800.

Del Arco I., Navarro M., Bilbao A., Ferrer B., Piomelli D., and Rodriguez De Fonseca F. 2002. Attenuation of spontaneous opiate withdrawal in mice by the anandamide transport inhibitor AM404. *Eur. J. Pharmacol.* **454:** 103–104.

Derkinderen P., Toutant M., Burgaya F., Le Bert M., Siciliano J.C., de Franciscis V., Gelman M., and Girault J.A. 1996. Regulation of a neuronal form of focal adhesion kinase by anandamide. *Science* **273:** 1719–1722.

Derkinderen P., Valjent E., Toutant M., Corvol J.C., Enslen H., Ledent C., Trzaskos J., Caboche J., and Girault J.A. 2003. Regulation of extracellular signal-regulated kinase by cannabinoids in hippocampus. *J. Neurosci.* **23:** 2371–2382.

Désarnaud F., Cadas H., and Piomelli D. 1995. Anandamide amidohydrolase activity in rat brain microsomes. Identification and partial characterization. *J. Biol. Chem.* **270:** 6030–6035.

Devane W.A., Hanus L., Breuer A., Pertwee R.G., Stevenson L.A., Griffin G., Gibson D., Mandelbaum A., Etinger A., and Mechoulam R. 1992. Isolation and structure of a brain constituent that binds to the cannabinoid receptor. *Science* **258:** 1946–1949.

Di Marzo V., Fontana A., Cadas H., Schinelli S., Cimino G., Schwartz J.C., and Piomelli D. 1994. Formation and inactivation of endogenous cannabinoid anandamide in central neurons. *Nature* **372:** 686–691.

Dinh T.P., Freund T.F., and Piomelli D. 2002a. A role for monoglyceride lipase in 2-arachidonoylglycerol inactivation. *Chem. Phys. Lipids.* **121:** 149–158.

Dinh T.P., Kathuria S., and Piomelli D. 2004. RNA interference suggests a primary role for monoacylglycerol lipase in the degradation of the endocannabinoid 2-arachidonoylglycerol. *Mol. Pharmacol.* **66:** 1260–1264.

Dinh T.P., Carpenter D., Leslie F.M., Freund T.F., Katona I., Sensi S.L., Kathuria S., and Piomelli D. 2002b. Brain monoglyceride lipase participating in endocannabinoid inactivation. *Proc. Natl. Acad. Sci.* **99:** 10819–10824.

Fegley D., Gaetani S., Duranti A., Tontini A., Mor M., Tarzia G., and Piomelli D. 2004a. Characterization of the fatty-acid amide hydrolase inhibitor URB597: Effects on anandamide and oleoylethanolamide deactivation. *J. Pharmacol. Exp. Ther. Online* jpet.104.078980V1.

Fegley D., Kathuria S., Mercier R., Li C., Goutopoulos A., Makriyannis A., and Piomelli D. 2004b. Anandamide transport is independent of fatty-acid amide hydrolase activity and is blocked by the hydrolysis-resistant inhibitor AM1172. *Proc. Natl. Acad. Sci.* **101:** 8512–8513.

Ferrer B., Asbrock N., Kathuria S., Piomelli D., and Giuffrida A. 2003. Effects of levodopa on endocannabinoid levels in rat basal ganglia: Implications for the treatment of levodopa-induced dyskinesias. *Eur. J. Neurosci.* **18:** 1607–1614.

Freund T.F., Katona I., and Piomelli D. 2003. Role of endogenous cannabinoids in synaptic signaling. *Physiol. Rev.* **83:** 1017–1066.

Fu J., Gaetani S., Oveisi F., Lo Verme J., Serrano A., Rodriguez de Fonseca F., Rosengarth A., Luecke H., Di Giacomo B., Tarzia G., and Piomelli D. 2003. Oleylethanolamide regulates feeding and body weight through activation of the nuclear receptor PPAR-alpha. *Nature* **425:** 90–93.

Gaetani S., Cuomo V., and Piomelli D. 2003. Anandamide hydrolysis: A new target for anti-anxiety drugs? *Trends Mol. Med.* **9:** 474–478.

Gerdeman G. and Lovinger D.M. 2001. CB1 cannabinoid receptor inhibits dynaptic release of glutamate in rat dorso-lateral striatum. *J. Neurophysiol.* **85:** 468–471.

Giang D.K. and Cravatt B.F. 1997. Molecular characterization of human and mouse fatty acid amide hydrolases. *Proc. Natl. Acad. Sci.* **94:** 2238–2242.

Gifford A.N., Bruneus M., Lin S., Goutopoulos A., Makriyannis A., Volkow N.D., and Gatley S.J. 1999. Potentiation of the action of anandamide on hippocampal slices by the fatty acid amide hydrolase inhibitor, palmitylsulphonyl fluoride (AM 374). *Eur. J. Pharmacol.* **383:** 9–14.

Giuffrida A., Rodríguez de Fonseca F., Nava F., Loubet-Lescoulié P., and Piomelli D. 2000. Elevated circulating levels of anandamide after administration of the transport inhibitor, AM404. *Eur. J. Pharmacol.* **408:** 161–168.

Giuffrida A., Parsons L.H., Kerr T.M., Rodríguez de Fonseca F., Navarro M., and Piomelli D. 1999. Dopamine activation of endogenous cannabinoid signaling in dorsal striatum. *Nat. Neurosci.* **2:** 358–363.

Glaser S.T., Abumrad N.A., Fatade F., Kaczocha M., Studholme K.M., and Deutsch D.G. 2003. Evidence against the presence of an anandamide transporter. *Proc. Natl. Acad. Sci.* **100:** 4269–4274.

Goparaju S.K., Ueda N., Taniguchi K., and Yamamoto S. 1999. Enzymes of porcine brain hydrolyzing 2-arachidonoylglycerol, an endogenous ligand of cannabinoid receptors. *Biochem. Pharmacol.* **57:** 417–423.

Goparaju S.K., Ueda N., Yamaguchi H., and Yamamoto S. 1998. Anandamide amidohydrolase reacting with 2-arachidonoyl-glycerol, another cannabinoid receptor ligand. *FEBS Lett.* **422:** 69–73.

Gulyas A.I., Cravatt, B.F., Bracey, M.H., Dinh, T.P., Piomelli, D., Boscia, F., and Freund, T.F. 2004. Segregation of two endo-cannabinoid-hydrolyzing enzymes into pre- and postsynaptic compartments in the hippocampus, cerebellum and amygdala. *Eur. J. Neurosci.* **20:** 441–458.

Herkenham M., Lynn A.B., Little M.D., Johnson M.R., Melvin L.S., de Costa B.R., and Rice K.C. 1990. Cannabinoid receptor localization in brain. *Proc. Natl. Acad. Sci.* **87:** 1932–1936.

Hillard C.J. and Jarrahian A. 2003. Cellular accumulation of anandamide: Consensus and controversy. *Br. J. Pharmacol.* **140:** 802–808.

Hillard C.J., Edgemond W.S., Jarrahian A., and Campbell W.B. 1997. Accumulation of N-arachidonoylethanolamine (anandamide) into cerebellar granule cells occurs via facilitated diffusion. *J. Neurochem.* **69:** 631–638.

Hoffman A.F. and Lupica C.R. 2000. Mechanisms of cannabinoid inhibition of GABA(A) synaptic transmission in the hippocampus. *J. Neurosci.* **20:** 2470–2479.

Hoffman A.F., Oz M., Caulder T., and Lupica C.R. 2003. Functional tolerance and blockade of long-term depression at synapses in the nucleus accumbens after chronic cannabi-noid exposure. *J. Neurosci.* **23:** 4815–4820.

Hohmann A.G., Suplita R.L., Bolton N.M., Neely M.H., Fegley D., Mangieri R., Krey J.F., Walker M.J., Holmes P.V., Crystal J.D., Duranti A., Tontini A., Mor M., Tarzia G., and Piomelli D. 2005. An endocannibinoid mechanical for stress-induced analgesia. *Nature* (in press).

Huang C.C., Lo S.W., and Hsu K.S. 2001. Presynaptic mechanisms underlying cannabinoid inhibition of excitatory synaptic transmission in rat striatal neurons. *J. Physiol.* **532:** 731–748.

Karlsson M., Contreras J.A., Hellman U., Tornqvist H., and Holm C. 1997. cDNA cloning, tissue distribution, and identification of the catalytic triad of monoglyceride lipase. Evolutionary relationship to esterases, lysophospholipases, and haloperoxidases. *J. Biol. Chem.* **272:** 27218–27223.

Kathuria S., Gaetani S., Fegley D., Valino F., Duranti A., Tontini A., Mor M., Tarzia G., La Rana G., Calignano A., Giustino A., Tattoli M., Palmery M., Cuomo V., and Piomelli D. 2003. Modulation of anxiety through blockade of anandamide hydrolysis. *Nat. Med.* **9:** 76–81.

Katona I., Rancz E.A., Acsády L., Ledent C., Mackie K., Hájos N., and Freund T.F. 2001. Distribution of CB1 cannabinoid receptors in the amygdala and their role in the control of GABAergic transmission. *J. Neurosci.* **21:** 9506–9518.

Katona I., Sperlagh B., Sik A., Kafalvi A., Vizi E.S., Mackie K., and Freund T.F. 1999. Presynaptically located CB1 cannabinoid receptors regulate GABA release from axon terminals of specific hippocampal interneurons. *J. Neurosci.* **19:** 4544–4558.

Kouri E.M. and Pope H.G.J. 2000. Abstinence symptoms during withdrawal from chronic marijuana use. *Exp. Clin. Psychopharmacol.* **8:** 483–492.

Kouri E.M., Pope H.G.J., and Lukas S.E. 1999. Changes in aggressive behavior during withdrawal from long-term marijuana use. *Psychopharmacology* **143:** 302–308.

Lang W., Qin C., Lin S., Khanolkar A.D., Goutopoulos A., Fan P., Abouzid K., Meng Z., Biegel D., and Makriyannis A. 1999. Substrate specificity and stereoselectivity of rat brain microsomal anandamide amidohydrolase. *J. Med. Chem.* **42:** 896–902.

Ledent C., Valverde O., Cossu G., Petitet F., Aubert J.F., Beslot F., Bohme G.A., Imperato A., Pedrazzini T., Roques B.P., Vassart G., Fratta W., and Parmentier M. 1999. Unresponsiveness to cannabinoids and reduced addictive effects of opiates in CB1 receptor knockout mice. *Science* **283:** 401–404.

Ligresti A., Morera E., Van Der Stelt M., Monory K., Lutz B., Ortar G., and Di Marzo V. 2004. Further evidence for the existence of a specific process for the membrane transport of anandamide. *Biochem. J.* **380:** 265–272.

LoVerme J.L., Fu J., Astarita G., La Rana G., Russo R., Calignano A., and Piomelli D. 2005. The nuclear receptor peroxisome proliferator-activated receptor-α mediates the anti-inflammatory actions of palmitoylethanolamide. *Mol. Pharmacol.* **67:** 15–19.

Mackie K. and Hille B. 1992. Cannabinoids inhibit N-type calcium channels in neuroblastoma-glioma cells. *Proc. Natl. Acad. Sci.* **89:** 3825–3829.

Marsicano G. and Lutz B. 1999. Expression of the canabinoid

receptor CB$_1$ in distinct neuronal subpopulations in the adult mouse forebrain. *Eur. J. Neurosci.* **11:** 4213–4225.

McDonald A.J. and Mascagni F. 2001. Localization of the CB1 type cannabinoid receptor in the rat basolateral amygdala: High concentrations in a subpopulation of cholecystokinin-containing interneurons. *Neuroscience* **107:** 641–652.

Mechoulam R., Ben-Shabat S., Hanus L., Ligumsky M., Kaminski N.E., Schatz A.R., Gopher A., Almog S., Martin B.R., Compton D.R., et al. 1995. Identification of an endogenous 2-monoglyceride, present in canine gut, that binds to cannabinoid receptors. *Biochem. Pharmacol.* **50:** 83–90.

Okamoto Y., Morishita J., Tsuboi K., Tonai T., and Ueda N. 2004. Molecular characterization of a phospholipase D generating anandamide and its congeners. *J. Biol. Chem.* **279:** 5298–5305.

Patricelli M.P., Lovato M.A., and Cravatt B.F. 1999. Chemical and mutagenic investigations of fatty acid amide hydrolase: Evidence for a family of serine hydrolases with distinct catalytic properties. *Biochemistry* **38:** 9804–9812.

Piomelli D. 2003. The molecular logic of endocannabinoid signalling. *Nat. Rev. Neurosci.* **4:** 873–884.

Piomelli D., Beltramo M., Glasnapp S., Lin S.Y., Goutopoulos A., Xie X.Q., and Makriyannis A. 1999. Structural determinants for recognition and translocation by the anandamide transporter. *Proc. Natl. Acad. Sci.* **96:** 5802–5807.

Robbe D., Alonso G., Duchamp F., Bockaert J., and Manzoni O.J. 2001. Localization and mechanisms of action of cannabinoid receptors at the glutamatergic synapses of the mouse nucleus accumbens. *J. Neurosci.* **21:** 109–116.

Rodríguez de Fonseca F., Navarro M., Gómez R., Escuredo L., Nava F., Fu J., Murillo-Rodríguez E., Giuffrida A., LoVerme J., Gaetani S., Kathuria S., Gall C., and Piomelli D. 2001. An anorexic lipid mediator regulated by feeding. *Nature* **414:** 209–212.

Schmid P.C., Zuzarte-Augustin M.L., and Schmid H.H. 1985. Properties of rat liver N-acylethanolamine amidohydrolase. *J. Biol. Chem.* **260:** 14145–14149.

Shepherd J.K., Grewal S.S., Fletcher A., Bill D.J., and Dourish C.T. 1994. Behavioural and pharmacological characterisation of the elevated "zero-maze" as an animal model of anxiety. *Psychopharmacology* **116:** 56–64.

Stella N. and Piomelli D. 2001. Receptor-dependent formation of endogenous cannabinoids in cortical neurons. *Eur. J. Pharmacol.* **425:** 189–196.

Stella N., Schweitzer P., and Piomelli D. 1997. A second endogenous cannabinoid that modulates long-term potentiation. *Nature* **388:** 773–778.

Sugiura T., Kondo S., Sukagawa A., Nakane S., Shinoda A., Itoh K., Yamashita A., and Waku K. 1995. 2-Arachidonoylglycerol: A possible endogenous cannabinoid receptor ligand in brain. *Biochem. Biophys. Res. Commun.* **215:** 89–97.

Tarzia G., Duranti A., Tontini A., Piersanti G., Mor M., Rivara S., Plazzi P.V., Park C., Kathuria S., and Piomelli D. 2003. Design, synthesis, and structure-activity relationships of alkylcarbamic acid aryl esters, a new class of fatty acid amide hydrolase inhibitors. *J. Med. Chem.* **46:** 2352–2360.

Tsou K., Brown S., Sañudo-Peña M.C., Mackie K., and Walker J.M. 1998. Immunohistochemical distribution of cannabinoid CB1 receptors in the rat central nervous system. *Neuroscience* **83:** 393–411.

Tsuboi K., Sun Y.X., Okamoto Y., Araki N., Tonai T., and Ueda N. 2005. Molecular characterization of N-acylethanolamine-hydrolyzing acid amidase, a novel member of the choloyl-glycine hydrolase family with structural and functional similarity to acid ceramidase. *J. Biol. Chem.* **280:** 11082–11092.

Twitchell W., Brown S., and Mackie K. 1997. Cannabinoids inhibit N- and P/Q-type calcium channels in cultured rat hippocampal neurons. *J. Neurophysiol.* **78:** 43–50.

Ueda N., Kurahashi Y., Yamamoto S., and Tokunaga T. 1995. Partial purification and characterization of the porcine brain enzyme hydrolyzing and synthesizing anandamide. *J. Biol. Chem.* **270:** 23823–23827.

Zimmer A., Zimmer A.M., Hohmann A.G., Herkenham M., and Bonner T.I. 1999. Increased mortality, hypoactivity, and hypoalgesia in cannabinoid CB1 receptor knockout mice. *Proc. Natl. Acad. Sci.* **96:** 5780–5785.

14 | Cocaine Neurobiology: From Targets to Treatment

Bertha K. Madras and Zhicheng Lin

Department of Psychiatry, Harvard Medical School, New England Primate Research Center, Southborough, Massachusetts 01772-9102

ABSTRACT

Dopamine, norepinephrine, and serotonin are widely implicated in precipitating the immediate pharmacological effects of psychostimulant drugs of abuse. Extracellular concentrations of these monoamines are partly regulated by monoamine transporters, which sequester the dopamine transporter (DAT), serotonin transporter (SERT), and norepinephrine transporter (NET) into neurons. Cocaine blocks transporter-mediated clearance of extracellular neurotransmitters, eliciting a surge of extracellular monoamines, activation of monoamine receptors, and recruitment of other neurotransmitter systems (e.g., glutamate and opioid). Following acute or repeated exposure, cocaine indirectly activates receptors to kindle adaptive responses. Cocaine proficiently transforms intercellular and intracellular signaling, metabolism, electrophysiological responses, morphology, and neural networks, processes implicated in cocaine addiction. Synchronous with this growing compendium of cocaine-induced cellular and molecular changes are a profusion of novel targets for medications to treat cocaine addiction.

INTRODUCTION

The alkaloid cocaine is synthesized by the plant *Erythroxylon coca*, native to the warm eastern slopes of the Andes (Fig. 1). Cultivated in Bolivia, Peru, Ecuador, and Columbia, the major varieties of coca plants produce cocaine at concentrations ranging from 0.35% to 0.72% (Rivier 1981). Compared with other tropical crops, coca plants are relatively insect-free, conceivably resulting from the insecticide action of cocaine. Cocaine sprayed on tomato leaves prevents herbivorous insects from consuming the leaves (Nathanson et al. 1993), a neurotoxic effect attributed to cocaine-induced blockade of transport of the insect-selective neurotransmitter octopamine.

The use of cocaine to enhance work capacity or resist fatigue by indigenous peoples living in the Andean mountains is a long-standing tradition. Cocaine penetrated western European scientific circles in 1860s, after Albert Neiman developed an effective isolation procedure from the coca leaf and described its anesthetic action. Anecdotal reports of its stimulant properties and other psychoactive effects rapidly led to incorporation of cocaine

FIGURE 1. Alkaloid cocaine is synthesized by the plant *Erythroxylon coca*, shown in this botanical print.

into "medicinal" wine and other beverages. By the late 1890s, the addictive properties of cocaine became apparent and the United States government responded in the early 1900s with two Acts to ban cocaine from consumables and eventually to make nonmedical use of cocaine illegal. Illicit use of cocaine reemerged in the 1970s and escalated rapidly in the 1980s.

Cocaine is self-administered via four routes: (1) Coca leaves can be sucked or chewed to deliver an oral dose; (2) cocaine · HCl is insufflated into nasal passages, with the high melting point of this salt preventing conversion to a stable, smokable form; (3) cocaine · HCl can be injected intravenously; and (4) a stable cocaine-bicarbonate salt ("crack cocaine") can be smoked. Intravenous, smoked, or insufflated cocaine produces euphoria, heightened energy, alertness, sensory perception, self-confidence, anxiety, and suspicion. Cocaine also decreases appetite, sleep, and fatigue, effects that are short-lived and lead to repeated, frequent use in a short span of time (Gawin and Ellinwood 1988).

Cocaine bears little structural similarity to dopamine, although both compounds contain a phenyl ring and an amine nitrogen (Fig. 2). The two ester links in the C2 and C3 positions of cocaine are readily hydrolyzed by cholinesterases, and resulting metabolites are without functional activity. Accordingly, the short duration of cocaine effects evokes a pattern of repeated frequent dosing. Interestingly, both esters are unnecessary for conferring potent blockade of monoamine transporters, and one or both have been omitted to yield potent cocaine analogs for probing the inhibitor-binding domain on transporters, in vivo imaging agents, in vitro probes, and medications development (see, e.g., Clarke et al. 1973; Madras et al. 1989b, 1990, 1996a,b, 1998a,b; Canfield et al. 1990; Carroll et al. 1991, 2005; Meltzer et al. 1993, 1994, 2002, 2004; Kozikowski et al. 1995; Davies et al. 2002; Dutta et al. 2003). These probes and a host of others have been extensively applied to investigate the complex biological targets of cocaine that can result in addiction. The paramount purpose of the research is to devise effective therapies to reverse addiction and prevent relapse.

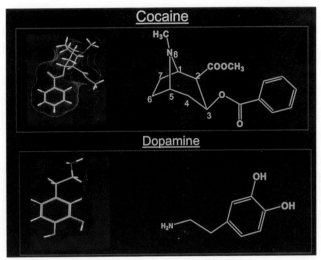

FIGURE 2. Structure of cocaine. (*Top*) Cocaine structure illustrated by two methods. (*Bottom*) Dopamine structure illustrated by two methods.

A foundation of objective research data is also needed to educate medical professionals and the public on the risks and consequences of cocaine. Equally significant, research on drug mechanisms, for example, psychostimulants, opioids, cannabinoids, and hallucinogens, has provided fundamental insights into brain function, neurotransmission, learning, and synaptic plasticity.

A number of factors may contribute to the subjective response elicited by the drug and progression to addiction. Polygenetic factors are considerable (Uhl and Grow 2004), and when combined with genetically based psychiatric disorders (e.g., schizophrenia, attention deficit hyperactivity disorder [ADHD], and bipolar disorder), they significantly increase the probability of cocaine use (Rounsaville et al. 1991). The pharmacokinetic and pharmacodynamic properties of cocaine also may uniquely contribute to abuse liability (Samaha and Robinson 2005). Finally, in developing a comprehensive view of the molecular and cellular substrates of cocaine addiction, it is important to recognize that various phases of cocaine use (Fig. 1 of Chapter 1; initiation, infrequent use, binge use, addiction, withdrawal, abstinence, craving, and relapse) can trigger unique or a continuum of neurobiological adaptations.

This overview capitalizes on the extensive literature of cocaine research, to serve as a paradigm for investigating drugs with addiction potential, and proffers a variety of caveats in the use of animals to model human addiction. Animal models and cell culture systems can be appropriate and are oftentimes essential for clarifying risk factors, consequences of reinforcing drugs, and the efficacy of candidate medications. Clinical research is optimal but frequently impractical or unethical. Cocaine users' self-reports of drug exposure histories are not necessarily reliable, drug dose and purity are uncontrollable in a nonlaboratory setting, and psychiatric comorbidity and polysubstance abuse are common in this population. Furthermore, genotype, environment, nutritional status, stress, medical problems, and other factors can confound interpretation of the data. Animal models can control for some of these factors. Nonetheless, for key evidence amassed from preclinical research, it is essential to attempt authentication noninvasively in human subjects or in postmortem human tissue. Noninvasive brain imaging approaches offer robust techniques to investigate and confirm in human subjects molecular and physiological changes uncovered in preclinical animal models (see Chapters 6 and 7, this volume). Unfortunately, the cellular, molecular, or network changes elicited by cocaine in animals that are confirmable with human brain imaging remain vanishingly low, with some notable exceptions (see, e.g., Porrino et al. 2004; Volkow and Li 2004). Intentionally, we introduce the topic by reviewing some risk factors in human patterns of cocaine use, to reflect on the validity of current animal models of cocaine addiction.

GOALS OF THE CHAPTER

This chapter provides an overview of the molecular, cellular, and neuronal responses to cocaine and the pertinence of these findings to developing cocaine medications. The content is designed to serve as a guide to molecular and cellular biologists who aspire to devise a preclinical, albeit translational, research program in drug abuse/addiction, with the ultimate objective of alleviating the personal and public health burden of drug addiction.

METHODOLOGICAL CONSIDERATIONS

In humans, addiction is a complex biobehavioral disorder that is influenced by multiple psychosocial factors. A number of these parameters are challenging to model with laboratory rodent or primate species. Equally challenging is the development of a model of cocaine withdrawal in humans, because withdrawal is primarily psychological, with irritability and dysphoria hallmark features. The not-inevitable continuum from use to addiction results from a complex interplay of environmental, individual, and genetic risk factors and patterns of drug use. According to the U.S. Substance Abuse and Mental Health Services Administration study (SAMHSA 2001), approximately 15% of the population in the United States report lifetime use of cocaine, but of these, it is estimated that 15–16% of cocaine users develop a profile of addictive behavior (Anthony and Petronis 1995; Wagner and Anthony 2002). Among recent users, a lower percentage (5–12%) report cocaine dependence in less than 24 months after initiation, with higher risk rates in females, young adults, and non-Hispanic Black/African-Americans (Chen and Anthony 2004; O'Brien and Anthony 2005). (Interestingly, excess risk for cannabis abuse within 24 months of first use is also associated with younger age of initiation [Chen et al. 2005].) Crack cocaine and injected cocaine are also associated with more rapid acquisition of dependence compared with cocaine. Most clinical features of addiction, including tolerance and narrowed behavioral repertoire, occur two to three times more often among crack smoking users, as compared to those using powder only. Accordingly, crack smoking may increase the risk of cocaine dependence after initiation of cocaine use, but individual susceptibility to using crack cocaine and becoming cocaine dependent cannot be ruled out (Chen and Anthony 2004). Interestingly, the rate at which cocaine is delivered via intravenous infusion to rodent brain affects the extent of DAT blockade, gene expression, and production of psychomotor sensitization, with more rapid infusion rates increasing all three parameters (Samaha et al. 2004). Conceivably, the smokable form of cocaine ("crack cocaine") can enter the brain more rapidly and at higher levels than insufflated cocaine (cocaine hydrochloride) to elicit contrasting adaptive responses, but corresponding human data are not available. These demographics signify that age, age of onset, gender, and ethnicity, as well as drug form (salt form) and route of administration are important variables to consider in the progression from cocaine use to addiction.

A host of other factors are associated with a propensity to use drugs or to avoid them, including stress, childhood environmental experiences (physical, psychological, or sexual abuse), academic achievement, parental and peer attitudes, psychiatric comorbidity, legal consequences, and polysubstance abuse (Hawkins et al. 1992; Smith et al. 1995; Lane et al. 1997). The clinical significance of these frequently capricious variables must be considered in experimental models. Nevertheless, it is feasible to quasimatch certain individual risk factors (gender, age, predisposition), some environmental circumstances (familiar or novel, stress, group dynamics, location-associated cues, other cues), drug-dosing regimens (passive or self-administration, dose, frequency, binging, route of administration, polysubstance exposure), withdrawal, stress- or drug-associated relapse, and sensation seeking (Porrino 1993; Giorgi et al. 1997, 2005; De Vries et al. 1999, 2002; Morgan et al. 2002, 2005a,b; Porcino et al. 2002, 2004; Lu et al. 2003, 2004a,b,c; Stansfield et al. 2004). Species differences in cocaine-responding or -seeking behavior also have been reported (see, e.g., Sabeti et al. 2003; Hanania et al. 2004), but the relevance of these rodent genotypes to human genotypes in cocaine response has not been systematically investigated. It is to be noted that sequences of genes encoding proteins of relevance to addiction are closely matched in nonhuman primates and humans (Miller and Madras 2005), suggesting that primates may serve as naturalistic models for assessing genetic contribution to cocaine-seeking behavior. Environmental influences, even within a laboratory setting, can produce biological changes of relevance to subjective responses to cocaine. In socially housed cynomolgus monkeys, cocaine was shown to be a reinforcing agent in subordinate monkeys, but not in dominant monkeys, and dopamine D_2-receptor density became elevated in dominant monkeys, but not in subordinate monkeys (Morgan et al. 2002).

Neuroadaptive changes that reflect and/or drive escalating use are critical occurrences, yet few animal studies model progressive escalation in drug-seeking or -using subjects (Ahmed and Koob 2004, 2005; Vanderschuren and Everitt 2004), a characteristic of susceptible human populations. Among the hypotheses proposed to decipher progression to addiction—establishment of an involuntary stimulus-response, loss of impulse control, sensitization of an incentive system, or disruption of a hedonic homeostatis (Jentsch and Taylor 1999; Everitt et al. 2001; Koob and Le Moal 2001; Robinson and Berridge 2001)—models for sensitization and incentive behavior have achieved a modicum of validity (Ahmed and Koob 2004). A pioneering measure of success in modeling pathological drug use in humans—compulsive and uncontrollable use despite adverse consequences—was achieved recently in rodents (Deroche-Gamonet et al. 2004; Vanderschuren and Everitt 2004). Three diagnostic criteria of human cocaine addiction were modeled in rodents: *Difficulty limiting cocaine intake* was shown by repeated drug-seeking behavior ("nose poking") during intervals at which the drug was not available; *high motivation to seek cocaine* was modeled by responding to a progressive ratio schedule; and compulsive *drug consumption despite adverse consequences* was measured in rats by their willingness to consume cocaine along with an aversive shock. Of the percentage of rats that fulfilled all three criteria, 17% was strikingly similar to the 15% of human cocaine users who progress to addiction (Deroche-Gamonet et al. 2004). Limited to modeling adverse consequences, Vanderschuren and Everitt (2004) showed that rodents with extended cocaine experience will self-administer cocaine paired to a foot shock, but will not do so with limited exposure or for a sucrose reward.

MOLECULAR TARGETS OF COCAINE

The immediate targets of cocaine can guide conceptual and experimental advances in clarifying the neuroadaptive cascades implicated in addiction and medications development. The biochemical targets and brain distribution of a drug frequently converge, but occasionally, they do not. Cocaine was recognized as an inhibitor of monoamine transport since the 1970s, and its binding properties in whole-brain homogenates and in select regions were described by Reith and colleagues (1980). Reith concluded that the majority of cocaine-binding sites in the brain were associated with the serotonin transporter. His subsequent discovery that drug potencies for blocking DAT correlated with their relative potencies for promoting locomotor activity in rodents was prescient (Reith et al. 1986). Expanding on this discovery, other groups showed a strong correlation between drug potencies at [^{3}H]cocaine- or [^{3}H]mazindol-binding sites on the dopamine transporter and their relative potencies for maintaining cocaine self-administration in primates (Ritz et al. 1987; Bergman et al. 1989; Madras et al. 1989a). Concurrently, cocaine was shown to raise extracellular dopamine levels in striatum and nucleus accumbens by microdialysis (Fig. 3), reflective of a physiological consequence to DAT blockade (Di Chiara and Imperato 1988; Bradberry and Roth 1989; Hurd and Ungerstedt 1989; Pettit and Justice 1989, 1991). During these early studies, it was apparent that the relatively low affinity of cocaine for DAT would compromise investigation of cocaine association with DAT. Higher-affinity cocaine congeners were developed to substitute for cocaine as a probe for DAT (Madras et al. 1989a,b, 1990). One of these, [^{3}H]WIN 35,428 bound virtually to the same sites as [^{3}H]cocaine in primate striatum, distributed predominately to dopamine regions of primate and rodent brain, and was metabolized relatively slowly (Madras et al. 1989a,b; Canfield et al. 1990; Kaufman and Madras 1991, 1992; Kaufman et al. 1991). The successful use of [^{3}H]WIN 35,428 rapidly expanded efforts to synthesize cocaine congeners based on WIN 35,428.

Brain Distribution of Cocaine

Are the pharmacological properties of cocaine in vitro, i.e., DAT blockade, reflected by its brain distribution? If not, can we identify other molecular targets of cocaine by surveying its brain distribution? Of four methods to map drug distribution in brain, the least sensitive is measuring drug levels in crude dissections of various brain regions (Misra et al. 1977). In vitro autoradiography can generate a high-resolution map in tissue sections, but it requires a radiolabeled form of the drug with a dissociation rate sufficiently slow to wash away nonspecifically bound radioligand. Technically, this is difficult to accomplish with cocaine because its dissociation rate from transporters occurs within seconds to minutes (Reith et al. 1980; Kaufman et al. 1992). It is feasible to develop a high-resolution map of cocaine disposition in primate brain following intravenous administration with [^{3}H]cocaine (Fig. 4) (Madras and Kaufman 1994), with blood flow clearing the radioli-

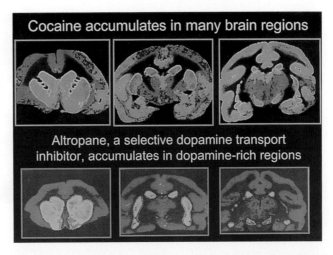

FIGURE 3. Ex vivo autoradiographic distribution of cocaine in squirrel monkey brain (*Saimiri sci-uerus*) and comparison with ex vivo distribution of altropane. (*Top panels*) Cocaine, a nonselective monoamine transport inhibitor, distributes widely in brain. The left panel is the most anterior plane and shows cocaine accumulation in the dopamine-rich caudate, putamen, and nucleus accum-bens. The middle panel is more posterior and illustrates cocaine accumulation in the caudate, puta-men, hypothalamus, and amygdala. The right panel is a more posterior plane and illustrates cocaine accumulation in various cortical regions, dorsal thalamus, substantia nigra, and hippocampus. (*Bottom panels*) Altropane, a relatively selective dopamine transport inhibitor, accumulates selec-tively in the dopamine-rich caudate, putamen, and nucleus accumbens (*left panel*), in the caudate, putamen, and substantia nigra (*middle panel*), and in the head and tail of the caudate and the sub-stantia nigra ventral tegmental area (VTA) (*right panel*). (Adapted from Madras and Kaufman 1994; Madras et al. 1998a.)

gand from low-affinity sites. Dopamine-rich brain regions preferentially bound pharma-cologically relevant high or low doses of cocaine (Fischman et al. 1976), but a number of other brain regions were highlighted as well. These results contrasted with more selec-tive dopamine transport inhibitors, CFT (WIN 35,428) or altropane, that distributed almost exclusively to dopamine-rich brain regions (Canfield et al. 1990; Kaufman et al. 1991; Madras et al. 1998b). The low dose of [³H]cocaine, within the dose range for inves-tigating cocaine disposition in living brain by positron emission tomography (PET) imag-ing with [¹¹C]cocaine, corresponded to data from this technique (Fowler et al. 1989; Volkow et al. 1992). Notwithstanding the limitations dictated by the resolution and sen-sitivity of the PET camera, these findings support the use of trace doses of [¹¹C]cocaine to monitor cocaine distribution by PET imaging in human brain (Fowler et al. 1989; Volkow et al. 1992). Conceivably, the widespread distribution of [³H]cocaine (caudate-putamen > nucleus accumbens, locus coeruleus, hippocampus, amygdala > medial septum, pre-frontal cortex, thalamus, pineal, substantia nigra, cortical regions, hypothalamus, and others) reflects a broad influence of cocaine on brain function, well beyond the dopamine-rich basal ganglia (London et al. 1990). Functional magnetic resonance imag-ing (fMRI) mapping of dynamic patterns of brain activation following cocaine infusion provided evidence of dynamically changing brain networks associated with cocaine-induced euphoria and craving (Breiter et al. 1997; Maas et al. 1998). Not surprisingly,

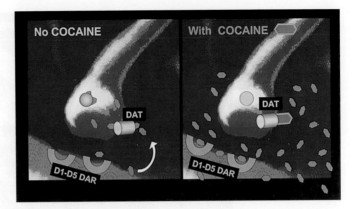

FIGURE 4. (*Left*) In a physiological state, DAT sequesters dopamine into presynaptic dopamine neurons, thereby circumscribing, temporally and spatially, dopamine activation of pre- or postsynaptic dopamine receptors (D_1–D_5). (*Right*) Cocaine blocks DAT and dopamine accumulates in the extracellular fluid. The inundation of dopamine is thought to contribute to the psychomotor stimulant and reinforcing effects of cocaine and to promote neuroadaptation.

cocaine use resulted in increased or decreased focal signals in a number of brain regions (Breiter et al. 1997), corresponding to regions identified by static autoradiography following cocaine administration (Madras and Kaufman 1994). Regions that showed early but sustained signal maxima were more correlated with craving than with rush ratings (nucleus accumbens, right parahippocampal gyrus, regions of lateral prefrontal cortex). It is important to point out that cocaine promotes vasoconstriction in brain, which conceivably can modify signal generation (Kaufman et al. 1998). Taken together, dopamine-rich brain regions are principal, albeit not exclusive, targets of pharmacologically relevant doses of cocaine in primate brain. On the basis of our unpublished data and receptor screening, SERT and NET are likely to be principal targets of cocaine in nondopaminergic brain regions.

The availability of transport inhibitors with varying selectivities for monoamine transporters created a unique opportunity to explore the hypothesis that blockade of DAT was a "necessary and sufficient" initial step in mediating the behavioral effects of cocaine (Reith et al. 1986; Ritz et al. 1987; Woolverton 1987, 1992; Bergman et al. 1989; Kleven et al. 1990). A clinical study reinforced this view: PET imaging revealed that cocaine occupancy of DAT was significantly correlated with self-reports of a cocaine-induced "high" (Volkow et al. 1997b). This categorical role of DAT was further sustained by the discovery that DAT-null mutant mice (DAT –/–) were unresponsive to cocaine (Giros et al. 1996), but this was subsequently tempered when DAT mutant mice (–/–) maintained the capacity to self-administer cocaine (Rocha et al. 1998; Rocha 2003). Intriguingly, double-mutant mice with DAT and SERT deletions (DAT –/– and SERT –/–), but not DAT and NET deletions, lost the capacity for cocaine-conditioned place preference and self-administration (Hall et al. 2004), indicative of a role for both DAT and SERT in producing the reinforcing effects of cocaine. Nonetheless, neurodevelopmental adaptations in null mutant mice may alter cocaine responses and yield an abnormal phenotype.

The Dopamine Transporter

DAT, a principal target of cocaine in the brain, is dynamically regulated by dopamine and cocaine. At the molecular level, the 12-membrane-spanning DAT contains a large extra-cellular loop with consensus sites for glycosylation (Kilty et al. 1991; Shimada et al. 1991) that regulate DAT stability and trafficking (Li et al. 2004), with potential phosphorylation sites (serine, threonine, and tyrosine) also serving as modulators (Mortensen and Amara 2003). Dopamine, released from dendrites and axons, may activate receptors locally or remotely. DAT limits the duration of dopamine activity or diffusion by sequestering dopamine (Cragg and Rice 2004). DAT is present on cell bodies, dendrites, and axons, but it apparently is not localized in the immediate active zone of the synapse (Lewis et al. 2001). Accordingly, DAT may reduce dopamine overflow into perisynaptic regions, but not within the synapse. DAT is expressed selectively in all dopamine neurons, including those originating in the substantia nigra and ventral tegmental area (VTA) (Ciliax et al. 1995) with neuronal projections to the striatum, nucleus accumbens, prefrontal cortex, and hypothal-amus, as well as the parietal cortex and dentate gyrus of the hippocampus. The relatively widespread distribution of cocaine, compared with the highly circumscribed distribution of DAT, most likely reflects cocaine association with NET and SERT, in addition to DAT. The relative concentrations of dopamine, DAT, and dopamine receptor densities are not con-sistent in all brain regions. In the caudate-putamen, nucleus accumbens, VTA, and sub-stantia nigra, the ratios of these three components of dopamine signaling are similar, and DAT is likely to be a major contributor to dopamine signaling strength and duration and to cocaine-induced effects. In other brain regions, the ratio of DAT to dopamine receptor expression levels (e.g., prefrontal cortex DAT and D_4 dopamine receptors) is lower, a mis-match that may permit dopamine clearance by metabolism, diffusion, or another trans-porter (De La Garza and Madras 2000). Accordingly, DAT control of dopamine signaling and cocaine control of this function are contingent on DAT density, activity regulation, and anatomical juxtaposition to dopamine release sites and dopamine receptors, as well as the capacity of other transporters to sequester dopamine.

The critical and brain-region-specific role of DAT in regulating dopamine neurotrans-mission and presynaptic homeostasis is strikingly borne out by targeted deletion of the DAT gene in mice. DAT-null mutant mice display significant phenotypic transformations, including hyperactivity, small size, skeletal abnormalities, pituitary hypoplasia, impaired care by females for their offspring, cognitive and sensorimotor gating deficits, and sleep dysregulation (Gainetdinov and Caron 2003). Dopamine signaling systems of the striatum are dramatically altered but not duplicated in the frontal cortex, which retains normal dopamine signaling. Beyond functioning as a dopamine carrier, DAT in the substantia nigra (and other brain regions) conceivably regulates dopamine release. DAT produces at least three types of ion-channel-like conductances (Sonders et al. 1997; Ingram et al. 2002; Sulzer and Galli 2003; Carvelli et al. 2004). In the substantia nigra, substrate transport by DAT augments somatodendritic dopamine release (Falkenburger et al. 2001; Ingram et al. 2002). DAT-mediated regulation of dopamine release differs from D_2 autoreceptor control of release. The D_2 autoreceptor attenuates dopamine release at relatively high dopamine concentrations, whereas DAT promotes dopamine release at relatively low dopamine lev-els (Ingram et al. 2002). In the striatum, dopamine clearance is a primary function of the DAT, whereas in the substantia nigra, DAT regulates extracellular dopamine levels by controlling both clearance and release. The complexity of region-specific DAT function is relevant to the pharmacological effects of cocaine. In the striatum, dopamine release following acute exposure to cocaine is enhanced and reduced in brains of abstinent cocaine users (Volkow et al. 1997a). For technical reasons, similar analyses have not been conducted in other regions of human brain.

Other Substrates for Monoamine Transporters

A largely unexplored function of monoamine transporters, of relevance to cocaine pharmacology, is their capacity to transport more than one substrate (for a review, see Madras et al. 2005). Two notable examples follow. NET is an effective dopamine carrier in frontal cortex, and it transports releasable dopamine into norepinephrine neurons (Carboni and Silvagni 2004), with amphetamine reportedly releasing dopamine from noradrenergic neurons in the prefrontal cortex (Shoblock et al. 2004). Furthermore, mice with null mutations of NET do not effectively clear dopamine in frontal cortex (Moron et al. 2002). Conceivably, in normal brain, cocaine-mediated blockade of NET may augment extracellular dopamine levels in the frontal cortex or other NET-rich brain regions, possibly accounting for retention of cocaine self-administration in DAT-null mutant mice (Rocha et al. 1998). In the frontal cortex, DAT may be marginally functional, because this brain region expresses much lower levels of DAT and dopamine autoreceptors, stores less dopamine, and relies more on dopamine synthesis than on vesicular recycling for dopamine release. Accordingly, the influence of cocaine on DAT in striatum cannot be liberally extrapolated to frontal cortex. Although highly speculative, cocaine clearance could result in dopamine sequestration by NET-expressing noradrenergic neurons, thereby engaging the locus coeruleus in regulation of dopamine release in the frontal cortex.

DAT also sequesters the trace amines phenylethylamine and tyramine into transporter-expressing cells. Cocaine is more potent in blocking phenylethylamine transport than dopamine transport (Madras et al. 2004a; Miller et al. 2005). These trace amines are agonists at the trace amine receptor 1 (TAR1) (Borowsky et al. 2001; Bunzow et al. 2001; Miller et al. 2005) and may function as indirect cocaine agonists at this receptor class. Intriguingly, coexpression of DAT with TAR1 in HEK cells facilitates phenylethylamine activation of TAR1 activity, suggesting a potential role for DAT in modulating this receptor (Miller et al. 2005), as well as another potential modulatory role of cocaine at a DAT-mediated function.

Addiction and Neuroadaptation

How does cocaine blockade of DAT promote adaptation and consequent addictive behaviors? The criteria for drug addiction, as defined by DSM IV criteria (American Psychiatric Association 2000) include increased drug intake over time, increased time and energy spent seeking and taking drugs, formidable internal conflict to remain abstinent, and use despite adverse consequences. Neuroadaptation, a hallmark of addiction, is implicitly thought to drive progression from casual use to compulsive drug seeking and taking (Koob et al. 2004). The underlying molecular and cellular events can inform reward mechanisms, experiential salience, learning and memory, motivation, volitional behavior, and neural plasticity. Enhanced drug seeking, use, and use despite adverse consequences can be modeled in rodents (Deroche-Gamonet et al. 2004; Vanderschuren and Everitt 2004; Morgan et al. 2005a,b). Behavioral sensitization to cocaine in animals has been considered a model of the progression to cocaine addiction. It can be elicited by repeated administration of a fixed dose of cocaine to produce a progressively greater behavioral response, following a challenge dose of identical magnitude (Kalivas and Duffy 1990, 1993a,b; Kalivas et al. 1998). In rodents, sensitization takes the form of enhanced locomotor activity, but in primates, locomotor and general activity decrease in response to a "sensitizing" regimen of cocaine (Farfel et al. 1992; Saka et al. 2004). Instead, behavioral sensitization is shown by increased head swinging and increased huddling (Saka et al. 2004) and reflected in enhanced dopamine release (Bradberry 2000). In humans, increased paranoia is the most prominent form of sensitization (Satel et al. 1991), although the entire spectrum of cocaine-induced drug-seeking and -craving behavior is considered a form of sensitization. In rodents, cocaine-induced stimulation can be augmented for months after discontinuing repeated cocaine injections, and the neuroadaptive changes mediating sensitization may contribute to the psychological and behavioral changes (paranoia, craving, relapse) that occur over extended periods of cocaine exposure in humans. Molecular changes that drive manifestations of sensitization may parallel or drive psychological and physiological adaptative mechanisms that contribute to compulsive drug craving (Kalivas and Duffy 1993a,b; Kalivas et al. 1998; Robinson and Berridge 2001; Steketee 2003). Certain molecular adaptations have been linked to behavioral sensitization in animals, but the association has not been made in humans.

We focus initially on the contribution of presynaptic dopamine/DAT to cocaine-mediated adaptive changes, bearing in mind that adaptation transpires beyond the purview of dopamine neurotransmission. Cocaine robustly increases dopamine dynamics in various brain regions, a process implicated in cocaine-induced behavioral sensitization, conditioned place preference, self-administration, and drug-induced reinstatement of cocaine self-administration. With repeated cocaine administration, presynaptic dopamine responses in the mesolimbic and nigrostriatal pathways are consistently enhanced (Vanderschuren and Kalivas 2000; Sabeti et al. 2003), along with postsynaptic changes in dopamine receptor activity, regulation, and postreceptor signal transduction (Anderson and Pierce 2005). Augmented dopamine response in brain regions that regulate locomotor activity and stereotypy is a logical explanation for behavioral sensitization, but the underlying mechanisms are incompletely understood.

DAT is not a passive recipient of substrates or cocaine and adapts acutely and after prolonged exposure to cocaine. In contrast to DAT substrates, which promote DAT internalization, cocaine increases DAT density, DAT surface expression, and dopamine transport, and attenuates DAT trafficking in various cell lines (Little et al. 1993, 2002; Saunders et al. 2000; Yatin et al. 2000; Daws et al. 2002; Kahlig and Galli 2003; Zahniser and Sorkin 2004). Although not uniformly replicated, higher DAT density has been corroborated in postmortem human brain striata (Little et al. 1993; Mash et al. 2002). Changing membrane expression of DAT may exquisitely regulate transport capacity in an adaptive effort to rectify dopamine levels, disrupted by cocaine-mediated attenuation of dopamine transport and DAT trafficking. Induction of the immediate-early c-*fos* gene is another presynaptic manifestation of cocaine-mediated adaptation in the substantia nigra, possibly via a DAT-mediated mechanism (Neisewander et al. 2000). In vitro DAT-expressing cells increase c-Fos protein following exposure to cocaine or dopamine (Yatin et al. 2002, 2005).

At peak concentrations of cocaine and as cocaine levels recede, initial excessive dopamine followed by augmented dopamine clearance conceivably promotes adaptive responses in postsynaptic neurons expressing dopamine receptors. Dopamine activates five dopamine receptor subtypes, D_1–D_5 (D_1 subtype: D_1 and D_5; D_2 subtype: D_2, D_3, and D_4), classified on the basis of pharmacological specificity, protein sequence homology, brain regional distribution, and signal transduction pathways (Sibley et al. 1993). D_1-subtype receptors activate adenylate cyclase via $G_{s\alpha}/G_{olf\alpha}$, whereas D_2-subtype receptors inhibit adenylate cyclase via $G_{i\alpha}/G_{o\alpha}$. Each receptor subtype also couples to other signal transduction pathways, with D_1 receptors activating phospholipase C, and D_2-like receptors increasing potassium conductance. The relevance of each receptor subtype to the pharmacology of cocaine has not achieved a consensus. Both D_1- and D_2-receptor subtypes are implicated in mediating the effects of self-administered cocaine, because representatives of each class of receptor agonists are self-administered, and antagonists of each receptor subtype can produce rightward shifts in the cocaine dose-response curves (Spealman et al. 1991; Caine et al. 1999). Notwithstanding these similarities, in a relapse model in rodents, D_1-receptor agonists were ineffective in promoting relapse to cocaine-seeking behavior, whereas D_2 agonists robustly triggered cocaine-seeking behavior (Self et al. 1996), and D_2 antagonists or partial D_3 agonists blocked cocaine-seeking behavior (DeVries et al. 1999, 2002; Pilla et al. 1999; Milivojevic et al. 2004).

Signal Transduction and Transcriptional Activity

Classical studies of neuroadaptation to cocaine have shown that a major consequence of dopamine receptor activation is up-regulation of the cAMP-CREB protein pathway. Repeated cocaine exposure bolsters adenylate cyclase levels, cAMP-dependent protein kinase A (PKA), and CDK5, while reducing levels of the inhibitory G_i protein, tyrosine hydroxylase, and producing an overall increase in signal transduction via the cAMP pathway (Trulson et al. 1987; Nestler et al. 1990; Beitner-Johnson and Nestler 1991; Terwilliger et al. 1991; Beitner-Johnson et al. 1992; Striplin and Kalivas 1993; Bibb et al. 2001; Nestler 2001 and Nestler, Chapter 21, this volume). These changes in enzymatic activity and protein levels have not been temporally correlated with sensitization or tolerance to cocaine (Hope et al. 2005). For example, PKA activity is increased by approximately 60% 1 day after cocaine treatment, but this increase diminishes 7 days later and does not correlate with locomotor response to cocaine.

Expression of transcription factors such as c-Fos and ΔFosB is also detectable and may, in turn, trigger a series of transcriptional events (Graybiel et al. 1990; Nestler et al. 1990, 1993; Cohen et al. 1991; Young et al. 1991; Bhat and Baraban 1993; Ang et al. 2001; Nestler 2001; Crombag et al. 2002). Acute cocaine administration may induce expression of other transcription factors, including Egr-1-3 in the shell of the nucleus accumbens or caudate-putamen, and Nac-1 in the nucleus acumbens, cerebellum, and caudate-putamen, but not in other examined areas (Cha et al. 1997; Humblot et al. 1998; Jouvert et al. 2002, 2004). Expression of specific immediate-early genes may be prolonged or of short duration (Nestler 2001; P.J. Wang et al. 2003; McClung et al. 2004; also see Nestler, Chapter 21, this volume). The temporal relationship between behavioral adaptation and region-specific transcriptional changes in immediate-early genes is of critical relevance for clarifying addictive processes.

Glutamate and various glutamate receptor subtypes also contribute to cocaine-induced locomotor activity, reinforcing effects, sensitization, and relapse (Baker et al. 2003; Kalivas et al. 2003; Kenny and Markou 2004), with a dopamine-glutamate conjunction partly driven by the neuroanatomical juxtaposition of mesocortical, mesolimbic, and mesoamygdaloid dopamine or glutamate projections (Vanderschuren and Kalivas 2000). Enhanced glutamate release, along with dopamine release, coincides with the development of behavioral sensitization to cocaine and is essential for expression of drug-seeking behavior (Pierce et al. 1996; Reid and Berger 1996; Cornish and Kalivas 2000), an association also observed with amphetamine- and μ-opioid-agonist-mediated sensitization (Vanderschuren and Kalivas 2000). Intriguingly, withdrawal from cocaine causes a reduction of extracellular glutamate in the nucleus accumbens that arises from a compromised cystine-glutamate exchange. This neuroadaptation conceivably is involved in cocaine addiction, because normalizing extracellular glutamate by stimulating cystine-glutamate exchange with N-acetylcysteine prevented cocaine-primed reinstatement (Baker et al. 2003). The mGluR5 metabotropic glutamate receptor contributes to the reinforcing properties of cocaine, because receptor blockade in normal mice attenuates cocaine self-administration and mice with null mutations of the mGluR5 do not acquire cocaine self-administration. Glutamate receptor antagonists also prevent relapse to cocaine-seeking behavior (Vanderschuren and Kalivas 2000).

Structural Plasticity

Repeated cocaine exposure may reorganize brain circuitry extensively by inducing changes in synaptic structure (Volkow and Li 2004). Within the striatum, multiple cocaine exposures shift c-Fos expression toward a higher striosome:matrix ratio and higher dorsal-caudate-putamen to ventral-caudate-putamen ratios (Fig. 5), indicative of neural circuit reorganization in this brain region (Graybiel et al. 1990; Saka et al. 2004). In both rodent and primate striatum, reorganization of neuronal activity is correlated with behavioral sensitization. Morphological changes are also induced following passive administration or self-administration of cocaine as spine densities and dendritic branching increase in the nucleus accumbens and medial prefrontal cortex, but not in orbital prefrontal cortex (for review, see Robinson and Kolb 2004). Intriguingly, these morphological changes may persist for months and dissipate when cocaine sensitization disappears (Kolb et al. 2003). Cocaine may also produce morphological changes in other cell types in the brain. For example, repeated exposure to cocaine transiently increases the expression of glial fibrillary acidic protein (a cytoskeletal intermediate filament protein exclusively expressed in astrocytes) and changes the density, cell size, and shape complexity of astrocytes in mouse dentate gyrus (Fattore et al. 2002). Because the degree of dendritic branching and density of spines directly influences the magnitude of neuronal input, cocaine-induced structural plasticity may contribute to cocaine reward.

These accumulating data reflect complex processes that require a comprehensive view of drug-induced molecular changes in the brain. Microarray analysis of cocaine-induced changes in gene expression has identified a spectrum of neuroadaptive changes, not feasible with single-target approaches (Freeman et al. 2001a,b, 2002; Toda et al. 2002; Lehrmann et al. 2003; Tang et al. 2003; Yuferov et al. 2003, 2005; Zhang et al. 2004; Bannon et al. 2005). Many genes up-regulated by cocaine in rat striata were immediate-

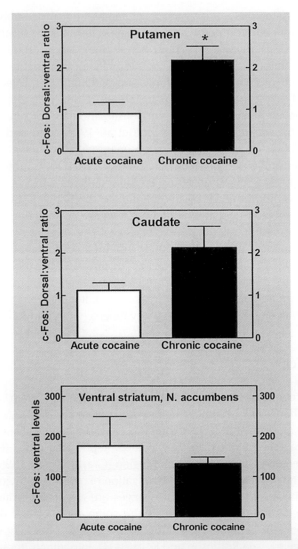

FIGURE 5. Daily repeated administration of cocaine to squirrel monkeys promotes reorganization in protein expression of the immediate-early c-*fos* gene, a measure of neuronal activity, in the caudate and putamen. (*Top panel*) c-Fos expression in the dorsal cap of the putamen is increased, relative to the ventral putamen in monkeys treated with 16 daily doses of cocaine, compared with monkeys treated with a single dose. (*Middle panel*) c-Fos expression in the dorsal cap of the caudate nucleus is increased, relative to the ventral caudate in monkeys treated with 16 daily doses of cocaine, compared with monkeys treated with a single dose. Behavioral stereotypies, which increase significantly after repeated cocaine compared with a single dose of cocaine, are correlated with this shift in c-Fos distribution. (*Bottom panel*) There are no significant differences in c-Fos distribution in the ventral striatum and nucleus accumbens with a single or repeated exposure to cocaine. With repeated cocaine exposure, expression levels in the striosome:matrix ratio increase compared with a single dose to drug-naïve animals, and this ratio correlates with enhanced behavioral stereotypies (data not shown). (Adapted from Saka et al. 2004.)

early genes for transcription factors and for "effector" proteins, and down-regulated genes included those associated with energy metabolism in mitochondria and regulation of G-protein signaling protein (Yuferov et al. 2003). In a microarray analysis of postmortem monkey brain, changes in gene expression were consistent with a signaling pathway induced by Fos, Jun, and cAMP response element binding proteins, previously shown to be associated with cocaine-induced behaviors in rodents (Freeman et al. 2001a,b). In postmortem nucleus accumbens of human cocaine abusers, interrogation of approximately 39,000 gene transcripts revealed changes in fewer than 1% of genes in cocaine abusers. Strikingly, myelin-related genes were substantially decreased in cocaine abusers, suggesting a possible dysfunction of myelin in cocaine abusers (Albertson et al. 2004; Bannon et al. 2005). Interspecies (human vs. rodent) inconsistencies are apparent in the microarray studies. Among the variables that could account for inconsistencies are subject-specific traits, dose, dosing regimen, route of administration, interspecies variation, anatomy, postmortem interval, microarray platforms, and methods for statistical analysis. To strengthen the validity of microarray data, it is necessary to confirm gene expression changes with quantitative reverse transcriptase–polymerase chain reaction (qRT-PCR), to measure whether the encoded proteins are also changed, to monitor changes in a brain-region- and neuron- or glia-specific manner, to conduct parallel studies in cohorts exposed to and withdrawn from cocaine for varying lengths of time, and to confirm findings in human postmortem tissues. Collectively, microarray analyses report robust cocaine-mediated changes in expression of genes involved in cytoskeletal structures, synaptogenesis, signal transduction, apoptosis, metabolism, and genes encoding proteins localized in almost every subcellular organelle, from nucleus to cytosol, mitochondria, and cell surfaces. Notwithstanding these impressive advances, a consistent and comprehensive portrait is yet to emerge of sequential cocaine-mediated changes in gene/protein expression or identification of changes most relevant to addiction.

Our knowledge of subcellular adaptive changes is more advanced than corresponding information of morphological and neural network plasticity. Deciphering which subcellular changes mediate morphological and neural network adaptations, and whether these are of core relevance to cocaine-induced behavioral and psychological transformations in humans, remains an ongoing challenge. Preliminary molecular mechanisms underlying cocaine-induced changes in morphology and neural networks are emerging. Axonal guidance molecules (semaphorins, ephrins, netrins, and receptors) are important contributors to neurodevelopment. Members of this class of proteins persist in the adult brain and are implicated in synaptic plasticity, particularly in hippocampal cells (Grunwald et al. 2004; Miyata et al. 2005). Eph family members are expressed in adult striatum (Xiao et al., in prep.), with acute cocaine administration enhancing EphB2 expression significantly (Yue et al. 1999). Cocaine can alter expression of a range of axonal guidance molecules, semaphorins, Ephs, ephrins, and neurophilins (Table 1) (Bahi and Dreyer 2005). A single dose of cocaine promoted major changes in gene expression in the striatum, a sensitization-dosing regimen resulted in the largest number of altered gene expression levels, and repeated exposure to cocaine uniquely affected gene expression most prominently in the hippocampus and nucleus accumbens. Noteworthy were consistent changes in semaphorin3E and neuropilin-1 gene expression across dosing regimens and brain regions. Collectively, these data indicate that genes encoding axonal guidance molecules, with their potential to alter neural networks, receptor activity, and synaptic morphology,

TABLE 1. Number of Axon Guidance Molecules whose Expression Levels Are Significantly Influenced by Seven Cocaine Treatment Paradigms in Four Brain Regions[a]

Cocaine paradigm[b]	Brain region				
	Hippocampus	NAc	Caudate putamen	VTA	Total
Acute	5	8	11	9	33
Binge	4	3	2	3	12
Chronic	9	10	6	7	32
S1	2	4	3	4	13
S2	2	4	5	2	13
S3	4	3	3	3	13
S4	9	9	9	14	41
Subtotal	35	41	39	42	157

Data from Bahi and Dreyer (2005).

[a]Out of 30 confirmed molecules, including 15 semaphorins, 9 Ephs, 5 ephrins, and 1 neuropilin.

[b](S1) Five-day sensitization with saline→3-day withdrawal→challenge with cocaine (saline as control); (S2) 5-day sensitization with saline→14-day withdrawal→challenge with cocaine (saline as control); (S3) 5-day sensitization with cocaine→3-day withdrawal→challenge with cocaine (saline as control); (S4) 5-day sensitization with cocaine→14-day withdrawal→challenge with cocaine (saline as control).

are significantly influenced by a specific cocaine-dosing paradigm. Confirmatory data in a different species and in human postmortem brain are needed to validate the clinical relevance of these findings.

In a recently completed study of EphA4 and EphB2 gene and protein distribution in primate brain, we discovered significant and highly circumscribed expression levels of EphA4 in the subgranular zone of the hippocampus, a region implicated in neurogenesis, and colocalization of EphA4 in tyrosine-hydroxylase-expressing cells in the substantia nigra (D. Xiao et al. 2004). On the basis of the known biochemical mechanisms of cocaine, we postulated that cocaine, acting via dopamine receptors and cAMP, could trigger changes in gene expression of axonal guidance molecules. To test this hypothesis and provide a link between cocaine's effects on dopamine neurons and expression of guidance molecules, we monitored expression of 14 axonal guidance molecules in the SK-N-MC human neuroepithelioma cell line that expresses the D_1 dopamine receptor, an indirect target of cocaine (Jassen et al. 2005). Forskolin (10 μM), which raises intracellular cAMP levels, increased EphA5, EphB2, and neuropilin-1 expression. The magnitude and direction of change for these three genes paralleled the findings in the rat hippocampus after cocaine treatment (Bahi and Dreyer 2005). The dopamine receptor agonist dihydrexidine (10 μM) promoted regulatory changes in a different set of genes. Conceivably, cocaine modifies axonal guidance molecule expression by blocking monoamine transporters to indirectly activate monoamine receptors and change intracellular cAMP production. On the basis of this in vitro evidence, axonal guidance molecule gene expression may be regulated, in part, by changes in cAMP and other signal transduction pathways, potentially linking monoamine receptor activation to signal transduction cascades, transcriptional regulation of axonal guidance molecules, and alterations in neural networks. From these preliminary data, we postulate that cocaine may elicit synaptic reconfiguration by indirectly altering combinations of axon guidance molecules, with potential consequences to behavior. Analysis of adaptation at this level may provide novel and creative targets for medications development.

Electrophysiological Effects of Cocaine

Daily cocaine followed by a 3-day withdrawal can enhance long-term potentiation (LTP) in rat hippocampus (Thompson et al. 2002, 2004). After 15 days of self-administration, a 100-day withdrawal period decreased LTP (Thompson et al. 2004). Because persistent changes in protein expression levels in amygdala after a 90-day withdrawal have been observed (Grimm et al. 2001, 2003; Hope et al. 2005), it is conceivable that expression levels of specific genes in the hippocampus are altered, as well. Microarray analysis of hippocampal tissues from long-term withdrawal in rats could address a possible correlation between protein synthesis and such LTP modulation.

Relapse and Craving

Craving and relapse to drug-seeking and -using behavior frequently occur during the abstinence phase and are significant obstacles to treatment. They can be provoked by a number of factors including stress, drug cues, or a single drug exposure. In animal models, relatively low doses of cocaine, stress, or cocaine-associated cues can reinstate cocaine self-administration (de Wit and Stewart 1981; Self et al. 1996; Self and Nestler 1998; Stewart 2000; Shaham et al. 2003; Lu et al. 2004b, 2005b). The central role of dopamine in mediating the reinforcing effects of cocaine would predict its role in relapse (Anderson and Pierce 2005). Dopamine transport inhibitors and D_2- but not D_1-receptor agonists precipitate cocaine-seeking behavior in rodents or primates (Wise et al. 1990; Self et al. 1996; DeVries et al. 1999, 2002; Khroyan et al. 2000; Schenk 2002). Although the contributions of dopamine in cocaine-seeking behavior in abstinence models are supported in vitro by changes in cAMP and associated signaling pathways, alterations in expression of other genes during a prolonged period of abstinence are not described. These patterns of change may provide important insights into molecular mechanisms underlying cocaine craving.

After repeated self-administration of cocaine and withdrawal, rats display enhanced rates of lever pressing for cocaine, as the abstinence period approaches 3 months (Grimm et al. 2001, 2003). Dopamine and glutamate are implicated in mediating enhanced cocaine-seeking behavior, a form of sensitization. During withdrawal, expression levels of glutamate receptors (GluR) are altered in a subregion-specific manner (Lu et al. 2005a). In the prefrontal cortex and nucleus accumbens, enhanced dopamine and glutamate release are required for reinstatement of cocaine-seeking behavior (Capriles et al. 2003; McFarland et al. 2004). Conceivably, adaptation in dopamine synapses of the prefrontal cortex promotes an increased incentive to seek drugs (over nondrug rewards), whereas drug-induced modulation of glutamate synapses in the nucleus accumbens is a prerequisite for reinstatement of cocaine-seeking behavior (Kalivas et al. 2005). These adaptive changes, combined with a modified cystine-glutamate exchanger (Moran et al. 2005), highlight glutamate receptor signaling as a potential target for medications development.

The escalation in drug seeking is correlated with increased BDNF (brain-derived neurotrophic factor) protein levels in various brain regions and increased phosphorylation of extracellular signal-regulated kinase (ERK) (Grimm et al. 2003; Lu et al. 2003, 2004a,b,c, 2005a,b). G-protein expression levels are also robustly altered during the window of relapse (Carrasco et al. 2004). In the paraventricular nucleus of the hypothalamus, $G\alpha_{11}$

expression levels in the membrane are gradually doubled after withdrawal with increasing days of cocaine exposure, whereas in amygdala, the increase is more rapid. For $G\alpha_q$, similar membrane expression patterns are observed for both brain regions. These neuroadaptive changes are observed for neither of the proteins in prefrontal cortex nor $G\alpha_z$ in any of these brain regions. In addition, withdrawal from repeated cocaine decreases extracellular glutamate concentration in nucleus accumbens but not in prefrontal cortex or striatum by a compromised cystine-glutamate exchanger, which seems to be a factor in cocaine relapse (Baker et al. 2003; Lu et al. 2004a) and a novel target for medications development. Collectively, these data indicate that chronic cocaine abuse can produce enduring molecular and morphological changes in brain, which can be viewed as a pathological state.

COCAINE MEDICATIONS

Coexisting psychiatric disorders, medical conditions, psychosocial environment, employment, family responsibilities, and a host of other factors must be addressed during treatment. Nevertheless, pharmacotherapy can be an effective, indispensable component of a comprehensive substance abuse treatment program. Pharmacotherapy (see Fig. 1 of Chapter 1) can intervene to promote abstinence during the addictive phase, alleviate withdrawal symptoms, and reduce craving and relapse. Despite significant efforts to develop drug therapies to treat cocaine addiction and the testing of more than 20 medications marketed for other indications, there are currently no pharmacotherapies approved for this purpose (for a review, see Gorelick et al. 2004). A number of candidate medications are in clinical trials to alleviate or reverse uncontrollable, compulsive use of the drug or drug craving. The development of innovative rational-based medications requires detailed knowledge of the immediate molecular targets of cocaine; of adaptive changes that occur at the molecular, cellular, and network level; and of toxic effects. Intended medications may be designed to (1) diminish or eliminate cocaine use by partially substituting for cocaine, (2) block the subjective effects of cocaine and extinguish cocaine seeking, (3) block cocaine craving to prevent relapse in the abstinent phase, (4) reduce the intense psychological withdrawal symptoms, (5) block cocaine overdose, and (6) attenuate the toxic effects of cocaine on cerebral vascular and cardiovascular systems. Medications may be designed as cocaine antagonists, full or partial replacements, and as modulators of neurotransmitters, receptors, or neuroadaptive processes. They furthermore should consider subjects who are highly motivated, ambivalent, or unwilling to stop drug use. Of the pharmacotherapies for heroin or nicotine addiction, replacement drugs are the most widely accepted, and antagonists are the least acceptable.

Appropriately, many candidate cocaine pharmacotherapies are modulators of dopamine, glutamate, GABA, and adrenergic systems, the latter implicated in withdrawal symptoms. Because cocaine users have reduced cerebral blood flow, cortical perfusion deficits, and vasoconstriction (Holman et al. 1993; Kaufman et al. 1998), cerebral vasodilators (e.g., amiloride or isradipine) are a potential target for cocaine pharmacotherapy. Collectively, these proposed drugs are active in the brain. An alternative approach is to attenuate or prevent cocaine entry into the brain, via immunotherapy (Haney and Kosten 2004). Biochemical restraint of cocaine and retention in the peripheral vasculature or tissues, via active or passive immunization or catabolic enzyme-linked antibodies, may restrict cocaine access to the brain.

In preclinical trials of candidate medications, animal models of psychomotor stimulation, drug discrimination, self-administration, relapse to self-administration, conditioned place preference, and sensitization have value for determining the potential of a candidate medication. Cross-species differences can confound interspecies replication (Leiderman et al. 2005), but unpredictable or uncontrollable variables encountered in human subjects, such as compliance, subject-specific drug side effects, psychological traits, genetic heterogeneity, polypharmacy, and accurate drug exposure records, can be minimized in animal models. With few exceptions, clinical trials have been conducted in experienced cocaine users, using drugs clinically approved for other purposes. The testing of candidate medications in animal models should consider that the cocaine-adapted brain may respond differently to a candidate medication than a drug-naïve brain.

The most intensive search for cocaine medications in the past decade focused on the direct target of cocaine, the presynaptic DAT, and on its indirect targets, dopamine receptors, with lesser efforts on compounds that modulate dopamine release, synthesis, or metabolism. Dopamine transport inhibitors with relatively high affinity, slow onset, and long duration of action were considered primary candidates as cocaine medications if they occupied the dopamine transporter and reduced cocaine intake. Indatraline (also known as Lu 19-005) was a prototypical compound for fulfilling these criteria (Rosenzweig-Lipson et al. 1992). The underlying assumption was that compounds with these characteristics would also block DAT in brain, but their slower onset and longer duration would significantly lower abuse liability compared with cocaine. The prolonged duration of action of indatraline led to undesirable side effects that attenuated enthusiasm for further development (Negus et al. 1999). Another objective was the discovery of a cocaine antagonist, a compound that would bind the dopamine transporter and permit dopamine transport, but block cocaine access to the transporter (Uhl et al. 1998; Uhl and Lin 2003). Underlying this strategy was mounting evidence that the cocaine-binding site on the dopamine transporter overlapped with, but was not identical to, the dopamine-binding domain (Madras et al. 1989a; Uhl and Lin 2003; W. Wang et al. 2003). A wide range of DAT inhibitors was developed for this purpose, including analogs of phenyltropane, benztropine, methylphenidate, piperazine, phenylindanamine, and nonamine analogs of these compounds (for a review, see Dutta et al. 2003). Multiple variants of each of these classes were explored, but no compound emerged from this range of dopamine transport inhibitors that bound significantly to a cocaine recognition site while permitting dopamine transport to proceed. During this effort, brain-imaging agents for clinical monitoring of DAT were serendipitously discovered and gave rise to a fascinating array of applications for DAT imaging in neuropsychiatric disorders, including Parkinson's disease (see, e.g., Madras et al. 1989b, 1996b, 1998b; Canfield et al. 1990; Kaufman and Madras 1991; Frost et al. 1993; Laruelle et al. 1993; Wong et al. 1993; Fischman et al. 1998). This period was also marked by the discovery of unanticipated pharmacological properties of several novel classes of compounds. For example, substituted benztropine analogs bound the dopamine transporter, with opposite enantioselectivity of cocaine (Meltzer et al. 1994). A different series of benztropine analogs displayed reduced cocaine-like effects in animal models (Hsin et al. 2002; Katz et al. 2004). Compounds in which the amine nitrogen was replaced by a carbon, oxygen, or sulfur atom gave rise to a novel concept that drugs need not contain an amine nitrogen in their structure to be effective transport inhibitors (Madras et al. 1996a, 2003; Meltzer et al. 2004). Figure 6 illustrates a working hypothesis that compounds associate with DAT by forming an ionic bond between the amine nitrogen of a drug and a counterion (an aspartate residue on transmembrane domain 1 [TM1] of DAT [or NET or SERT]), and by forming aromatic-aromatic interactions on another transmembrane domain of the transporter. Nonamines compromised this model, because they displayed potencies in the same range as their amine-based progenitors, even though oxa or carba analogs are unable to engage in ionic bonding and carba analogs cannot engage in hydrogen bonding (Fig. 7). Also introduced was the novel concept of a Trojan horse that would covalently bind to DAT in a locus that apparently is not essential for dopamine transport, cleave at a hydrolyzable bond, and leave a small residue bound to the transporter that interferes with cocaine binding (Meltzer et al. 2002). Several compounds in this class interfered with cocaine binding, permitted a higher proportion of dopamine transport to proceed, and significantly reduced the potency of cocaine to block DAT (Madras et al. 2004b). Whether one or more DAT-modulating agents will prove useful as a cocaine medication is unknown, but currently approved monoamine transport inhibitors, such as methylphenidate or antidepressants, have not succeeded in clinical trials. Other potential cocaine medications targeted to dopamine transmission systems include D_1 and D_3 dopamine receptor agonists, partial agonists or antagonists, dopamine-releasing agents (amantadine), and drugs that interfere with dopamine metabolism (disulfuram). Serotonergic agents, opioid receptor agonists and antagonists, compounds that increase or decrease glutaminergic and GABAergic neurotransmission, cannabinoids, neuropeptide receptor antagonists, and a host of others are also under consideration (for reviews, see De Vries et al. 1999, 2001; Childress and O'Brien 2000; Howell and Wilcox 2001; Dutta et al. 2003; Elkashef and Vocci 2003; Kalivas et al. 2003; Kleber 2003; Dackis 2004; Gorelick et al. 2004; Sofuoglu and Kosten 2004, 2005; Kampman et al. 2005). This extensive collection of candidate medications reflects the intensive commitment to develop an effective pharmacotherapy for cocaine addiction, based foremost on exploration of the molecular and cell biological substrates of cocaine and consequent adaptive responses.

Notwithstanding the rational approaches described above, it is feasible to serendipitously discover intriguing lead medications with indiscriminate, empirically based screening (Rothstein et al. 2005). To augment glutamate transporter function and reduce the neuro-

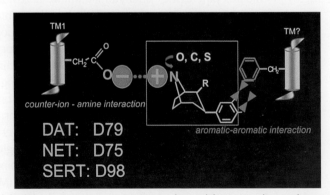

FIGURE 6. Model of drug-transporter association derived from site-directed mutagenesis of DAT. A highly conserved aspartate residue in transmembrane 1 of DAT was mutated with significant loss of DAT capacity and inhibitor binding (Kitayama et al. 1992). According to this model, nonamines, compounds in which the amine nitrogen is replaced by an oxa or carba, should not bind DAT or block DAT-mediated transport. Instead, as Fig. 7 illustrates, halogen-substituted nonamines retain high affinity for DAT.

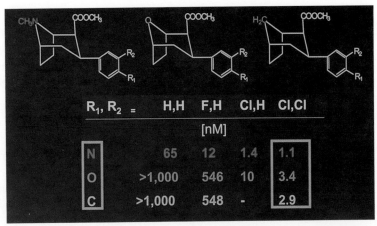

FIGURE 7. Nonamines retain high affinity for DAT, with halogenated substituents on the phenyl ring. Oxa, but not carba, analogs can engage in hydrogen bonding but not in ionic bonding.

toxic effects of glutamate in amyotrophic lateral sclerosis (ALS) and other neurodegenerative diseases, a random blind screening of 1040 FDA-approved drugs and nutritionals revealed that β-lactam antibiotics are potent stimulators of glutamate transporter 1 expression, an action mediated by increased transcription of the transporter gene.

SUMMARY

The past decade has been characterized by an explosive effort to clarify the molecular and cellular mechanisms underlying cocaine addiction and relapse. By blocking monoamine transporters and producing a surge in neurotransmitter levels, cocaine indirectly recruits monoamine and other receptors, and precipitates aberrant intracellular signaling cascades. These in turn activate transcription factors and induce adaptive changes in neurotransmitter availability, receptor function and regulation, electrophysiological activity, neuron morphology, and neural networks. The initial convergence on dopamine neurotransmission gradually extended to glutamate, opioid, and other neurotransmitter systems, and the limited vista of a single family of genes or proteins related to signal transduction and other intracellular events yielded to rapid expansion with microarray technology. The full spectrum of cocaine-mediated neuroadaptive responses now furnished a new level of clarity, but with added complexity and inconsistency. Our current challenge is to decipher which of the myriad of molecular changes are pivotal in veering human conduct toward obsessive drug-seeking and self-destructive behaviors. With the breathtaking level of information at hand, suitable medications that can reverse or alleviate the intense subjective effects of the cocaine-transformed brain are likely to emerge.

GAPS AND FUTURE DIRECTIONS

Molecular and cellular research are of fundamental value for clarifying the biological basis of cocaine-induced transformations of behavior. The information can also instruct the development of rationally based medications to treat the addiction. Major gaps and opportunities exist in the research. The field requires increased integration of molecular and behavioral data, improved approximation of animal models with clinical patterns of cocaine-using populations, and enhanced attention to the underlying molecular mechanisms of psychiatric comorbidity. Below is a brief summary of some of these issues.

1. Integration of preclinical and clinical investigation is needed to refine animal models to more closely approximate human patterns of use, environmental factors, and response to medications. The more rapid progression to addiction in younger humans must be clarified at the molecular level in preclinical models, as must psychiatric comorbidities. Although psychiatric phenotypes are a decidedly human condition, their corresponding genotypes may conceivably be identifiable in nonhuman primates, to yield more naturalistic models of the interplay between susceptibility to drug-seeking behavior and to psychiatric disorders.

2. Translational research is needed to verify in humans the pivotal molecular, cellular, and anatomical changes observed in animals. Compliance with this need will require significant advances in our ability to image specific events, proteins, and genes in the living human brain, or to discover peripheral surrogate makers of adaptive events in brain.

3. The newly discovered trace amine receptor family and its functional integration with monoamine neurotransmitters and transporters offers tantalizing leads for basic neuroscience and possibly medications development.

4. Evidence that cocaine can alter axonal guidance molecules requires intense scrutiny, because these proteins are capable of shaping glutamate receptor function, synaptic connectivity, and plasticity, and possibly contributing to neurogenesis. Can axonal guidance molecules provide the link between cocaine activation of receptors and restructuring of neuronal morphology, neural networks, and drug-induced mnemonic and behavioral changes?

5. Significant gaps persist in our understanding of the neurobiology of cocaine addiction. Conceptual models of neuroadaptation that consistently integrate drug-induced behaviors with adaptive changes in signal transduction cascades, transcription factors, morphology, electrophysiology, and neural networks are incomplete.

6. Exciting leads for cocaine medications have emerged in the past decade, and a number of promising candidates are in various phases of clinical testing. It is unlikely that a single compound or approach will emerge as a panacea. Genotype, personality, motivation, stage of use, and other factors are likely to require propitious pairing of medication with each addicted individual. Adroit selection of preclinical models and relevant molecular and cell biological changes engendered by cocaine will facilitate the development of effective treatments for this devastating addiction.

ACKNOWLEDGMENTS

The authors are grateful for the productive scientific and technical contributions of Jacob Bendor, Richard de la Garza, Ph.D., Michele Fahey, D.V.M., Alan Fishman, M.D., Ph.D., Claudia Goodrich, Martin Goulet, Ph.D., Ann Graybiel, Ph.D., Amy Jassen, Ph.D., Ryan

Johnson, Marc Kaufman, Ph.D., Peter Meltzer, Ph.D., Ava Meyerhoff, M.D., Gregory Miller, Ph.D., Helen Panas, Esen Saka, M.D., Chris Verrico, Ph.D., Danqing Xiao, Ph.D., Dina Yang, and Servet Yatin, Ph.D. We also thank Sandy Talbot and Jennifer Carter as well as the following National Institutes of Health agencies for support: NIDA (Grant Nos. DA-06303, DA-11558, and DA-15305) and NCRR (Grant No. RR-00168).

REFERENCES

Ahmed S.H. and Koob G.F. 2004. Changes in response to a dopamine receptor antagonist in rats with escalating cocaine intake. *Psychopharmacology* **172:** 450–454.

———. 2005. Transition to drug addiction: A negative reinforcement model based on an allostatic decrease in reward function. *Psychopharmacology* **180:** 473–490.

Albertson D.N., Pruetz B., Schmidt C.J., Kuhn D.M., Kapatos G., and Bannon M.J. 2004. Gene expression profile of the nucleus accumbens of human cocaine abusers: Evidence for dysregulation of myelin. *J. Neurochem.* **88:** 1211–1219.

American Psychiatric Association. 2000. *Diagnostic and statistical manual of mental disorders: DSM-IV-TR*, 4th edition. American Psychiatric Association, Washington, D.C.

Anderson S.M. and Pierce R.C. 2005. Cocaine-induced alterations in dopamine receptor signaling: Implications for reinforcement and reinstatement. *Pharmacol. Ther.* **106:** 389–403.

Ang E., Chen J., Zagouras P., Magna H., Holland J., Schaeffer E., and Nestler E.J. 2001. Induction of nuclear factor-kappaB in nucleus accumbens by chronic cocaine administration. *J. Neurochem.* **79:** 221–224.

Anthony J.C. and Petronis K.R. 1995. Early-onset drug use and risk of later drug problems. *Drug Alcohol Depend.* **40:** 9–15.

Bahi A. and Dreyer J.L. 2005. Cocaine-induced expression changes of axon guidance molecules in the adult rat brain. *Mol. Cell. Neurosci.* **28:** 275–291.

Baker D.A., McFarland K., Lake R.W., Shen H., Tang X.C., Toda S., and Kalivas P.W. 2003. Neuroadaptations in cystine-glutamate exchange underlie cocaine relapse. *Nat. Neurosci.* **6:** 743–749.

Bannon M., Kapatos G., and Albertson D. 2005. Gene expression profiling in the brains of human cocaine abusers. *Addict. Biol.* **10:** 119–126.

Beitner-Johnson D. and Nestler E.J. 1991. Morphine and cocaine exert common chronic actions on tyrosine hydroxylase in dopaminergic brain reward regions. *J. Neurochem.* **57:** 344–347.

Beitner-Johnson D., Guitart X., and Nestler E.J. 1992. Neurofilament proteins and the mesolimbic dopamine system: Common regulation by chronic morphine and chronic cocaine in the rat ventral tegmental area. *J. Neurosci.* **12:** 2165–2176.

Bergman J., Madras B.K., Johnson S.E., and Spealman R.D. 1989. Effects of cocaine and related drugs in nonhuman primates. III. Self-administration by squirrel monkeys. *J. Pharmacol. Exp. Ther.* **251:** 150–155.

Bhat R.V. and Baraban J.M. 1993. Activation of transcription factor genes in striatum by cocaine: Role of both serotonin and dopamine systems. *J. Pharmacol. Exp. Ther.* **67:** 496–505.

Bibb J.A., Chen J., Taylor J.R., Svenningsson P., Nishi A., Snyder G.L., Yan Z., Sagawa Z.K., Ouimet C.C., Nairn A.C., Nestler E.J., and Greengard P. 2001. Effects of chronic exposure to cocaine are regulated by the neuronal protein Cdk5. *Nature* **410:** 376–380.

Borowsky B., Adham N., Jones K.A., Raddatz R., Artymyshyn R., Ogozalek K.L, Durkin M.M., Lakhlani P.P., Bonini J.A., Pathirana S., Boyle N., Pu X., Kouranova E., Lichtblau H., Ochoa F.Y., Branchek T.A., and Gerald C. 2001. Trace amines: Identification of a family of mammalian G protein-coupled receptors. *Proc. Natl. Acad. Sci.* **98:** 8966–8971.

Bradberry C.W. 2000. Acute and chronic dopamine dynamics in a nonhuman primate model of recreational cocaine use. *J. Neurosci.* **20:** 7109–7115.

Bradberry C.W. and Roth R.H. 1989. Cocaine increases extracellular dopamine in rat nucleus accumbens and ventral tegmental area as shown by in vivo microdialysis. *Neurosci. Lett.* **103:** 97–102.

Breiter H.C., Gollub R.L., Weisskoff R.M., Kennedy D.N., Makris N., Berke J.D., Goodman J.M., Kantor H.L., Gastfriend D.R., Riorden J.P., Mathew R.T., Rosen B.R., and Hyman S.E. 1997. Acute effects of cocaine on human brain activity and emotion. *Neuron* **19:** 591–611.

Bunzow J.R., Sonders M.S., Arttamangkul S., Harrison L.M., Zhang G., Quigley D.I., Darland T., Suchland K.L., Pasumamula S., Kennedy J.L., Olson S.B., Magenis R.E., Amara S.G., and Grandy D.K. 2001. Amphetamine, 3,4-methylenedioxymethamphetamine, lysergic acid diethylamide, and metabolites of the catecholamine neurotransmitters are agonists of a rat trace amine receptor. *Mol. Pharmacol.* **60:** 1181–1188.

Caine S.B., Negus S.S., Mello N.K., and Bergman J. 1999. Effects of dopamine D(1-like) and D(2-like) agonists in rats that self-administer cocaine. *J. Pharmacol. Exp. Ther.* **291:** 353–360.

Canfield D.R., Spealman R.D., Kaufman M.J., and Madras B.K. 1990. Autoradiographic localization of cocaine receptors by [³H]CFT in monkey brain. *Synapse* **5:** 189–195.

Capriles N., Rodaros D., Sorge R.E., and Stewart J. 2003. A role for the prefrontal cortex in stress- and cocaine-induced reinstatement of cocaine seeking in rats. *Psychopharmacology* **168:** 66–74.

Carboni E. and Silvagni A. 2004. Dopamine reuptake by norepinephrine neurons: Exception or rule? *Crit. Rev. Neurobiol.* **16:** 121–128.

Carroll F.I., Blough B.E., Nie Z., Kuhar M.J., Howell L.L., and Navarro H.A. 2005. Synthesis and monoamine transporter binding properties of 3beta-(3′,4′-disubstituted phenyl) tropane-2beta-carboxylic acid methyl esters. *J. Med. Chem.* **48:** 2767–2771.

Carroll F.I., Gao Y., Rahman M.A., Abraham P., Parham K., Lewin A.H., Boja J.W., and Kuhar M.J. 1991. Synthesis, ligand binding, QSAR, and CoMFA study of 3β(p-substituted phenyl)tropane-2B-carboxylic acid methyl esters. *J. Med. Chem.* **4:** 2719–2725.

Carrasco G.A., Damjanoska K.J., D'Souza D.N., Zhang Y., Garcia F., Battaglia G., Muma N.A., and Van de Kar L.D. 2004. Short-term cocaine treatment causes neuroadaptive changes in $G\alpha_q$ and $G\alpha_{11}$ proteins in rats undergoing withdrawal. *J. Pharmacol. Exp. Ther.* **311:** 349–355.

Carvelli L., McDonald P.W., Blakely R.D., and Defelice L.J. 2004. Dopamine transporters depolarize neurons by a channel mechanism. *Proc. Natl. Acad. Sci.* **101:** 16046–16051.

Cha X.-Y., Pierce R.C., Kalivas P.W., and Mackler S.A. 1997. NAC1, a rat brain mRNA, is increased in the nucleus accumbens three weeks after chronic cocaine self-administration. *J. Neurosci.* **17:** 6864–6871.

Chen C.Y. and Anthony J.C. 2004. Epidemiological estimates of risk in the process of becoming dependent upon cocaine: Cocaine hydrochloride powder versus crack cocaine. *Psychopharmacology* **172:** 78–86.

Chen C.Y., O'Brien M.S., and Anthony J.C. 2005. Who becomes cannabis dependent soon after onset of use? Epidemiological evidence from the United States: 2000–2001. *Drug Alcohol Depend.* **79:** 11–22.

Childress A.R. and O'Brien C.P. 2000. Dopamine receptor partial agonists could address the duality of cocaine craving. *Trends Pharmacol. Sci.* **21:** 6–9.

Ciliax B.J., Heilman C., Demchyshyn L.L., Pristupa Z.B., Ince E., Hersch S.M., Niznik H.B., and Levey A.I. 1995. The dopamine transporter: Immunochemical characterization and localization in brain. *J. Neurosci.* **15:** 1714–1723.

Clarke R.L., Daum S.J., Gambino A.J., Aceto M.D., Pearl J., Levitt M., Cumiskey W.R., and Bogado E.F. 1973. Compounds affecting the central nervous system. 4. 3β-phenyltropane-2-carboxylic esters and analogs. *J. Med. Chem.* **16:** 1260–1267.

Cohen B.M., Nguyen T.V., and Hyman S.E. 1991. Cocaine-induced changes in gene expression in rat brain. *NIDA Res. Monogr.* **105:** 175–181.

Cornish J.L. and Kalivas P.W. 2000. Glutamate transmission in the nucleus accumbens mediates relapse in cocaine addiction. *J. Neurosci.* **20:** RC89 1–5.

Cragg S.J. and Rice M.E. 2004. DAncing past the DAT at a DA synapse. *Trends Neurosci.* **27:** 270–277.

Crombag H.S., Jedynak J.P., Redmond K., Robinson T.E., and Hope BT. 2002. Locomotor sensitization to cocaine is asso-

ciated with increased Fos expression in the accumbens, but not in the caudate. *Behav. Brain Res.* **136:** 455–462.

Dackis C.A. 2004. Recent advances in the pharmacotherapy of cocaine dependence. *Curr. Psychiatry Rep.* **6:** 323–331.

Davies H.M., Ren P., Kong N.X., Sexton T., and Childers S.R. 2002. Synthesis of iodinated 3β-aryltropanes with selective binding to either the dopamine or serotonin transporters. *Bioorg. Med. Chem. Lett.* **12:** 845–847.

Daws L.C., Callaghan P.D., Moron J.A., Kahlig K.M., Shippenberg T.S., Javitch J.A., and Galli A. 2002. Cocaine increases dopamine uptake and cell surface expression of dopamine transporters. *Biochem. Biophys. Res. Commun.* **290:** 1545–1550.

De La Garza R., II and Madras B,K. 2000. [3H]PNU-101958, a D4 dopamine receptor probe, accumulates in prefrontal cortex and hippocampus of non-human primate brain. *Synapse* **37:** 232–244.

Deroche-Gamonet V., Belin D., and Piazza P.V. 2004. Evidence for addiction-like behavior in the rat. *Science* **305:** 1014–1017.

De Vries T.J., Schoffelmeer A.N., Binnekade R., and Vanderschuren L.J. 1999. Dopaminergic mechanisms mediating the incentive to seek cocaine and heroin following long-term withdrawal of IV drug self-administration. *Psychopharmacology* **143:** 254–260.

De Vries T.J., Schoffelmeer A.N., Binnekade R., Raaso H., and Vanderschuren L.J. 2002. Relapse to cocaine- and heroin-seeking behavior mediated by dopamine D2 receptors is time-dependent and associated with behavioral sensitization. *Neuropsychopharmacology* **26:** 18–26.

De Vries T.J., Shaham Y., Homberg J.R., Crombag H., Schuurman K., Dieben J., Vanderschuren L.J., and Schoffelmeer A.N. 2001. A cannabinoid mechanism in relapse to cocaine seeking. *Nat. Med.* **7:** 1151–1154.

de Wit H. and Stewart J. 1981. Reinstatement of cocaine-reinforced responding in the rat. *Psychopharmacology* **75:** 134–143.

Di Chiara G. and Imperato A. 1988. Drugs abused by humans preferentially increase synaptic dopamine concentrations in the mesolimbic system of freely moving rats. *Proc. Natl. Acad. Sci.* **85:** 5274–5278.

Dutta A.K., Zhang S., Kolhatkar R., and Reith M.E.A. 2003. Dopamine transporter as target for drug development of cocaine dependence medications. *Eur. J. Pharmacol.* **479:** 93–106.

Elkashef A. and Vocci F. 2003. Biological markers of cocaine addiction: Implications for medications development. *Addict. Biol.* **8:** 123–139.

Everitt B.J., Dickinson A., and Robbins T.W. 2001. The neuropsychological basis of addictive behaviour. *Brain Res. Brain Res. Rev.* **36:** 129–138.

Falkenburger B.H., Barstow K.L., and Mintz I.M. 2001. Dendrodendritic inhibition through reversal of dopamine transport. *Science* **293:** 2465–2470.

Farfel G.M., Kleven M.S., Woolverton W.L., Seiden L.S., and Perry B.D. 1992. Effects of repeated injections of cocaine on catecholamine receptor binding sites, dopamine transporter binding sites and behavior in rhesus monkey. *Brain Res.* **578:** 235–243.

Fattore L., Puddu M.C., Picciau S., Cappai A., Fratta W., Serra

G.P., and Spiga S. 2002. Astroglial in vivo response to cocaine in mouse dentate gyrus: A quantitative and qualitative analysis by confocal microscopy. *Neuroscience* **110**: 1–6.

Fischman A.J., Bonab A.A., Babich J.W., Palmer E.P., Alpert N.M., Elmaleh D.R., Callahan R.J., Barrow S.A., Graham W., Meltzer P.C., Hanson R.N., and Madras B.K. 1998. Rapid detection of Parkinson's disease by SPECT with altropane: A selective ligand for dopamine transporters. *Synapse* **29**: 128–141.

Fischman M.W., Schuster C.R., Resnekov L., Shick J.F., Krasnegor N.A., Fennell W., and Freedman D.X. 1976. Cardiovascular and subjective effects of intravenous cocaine administration in humans. *Arch. Gen. Psychiatry* **33**: 983–989.

Fowler J.S., Volkow N.D., Wolf A.P., Dewey S.L., Schlyer D.J., Macgregor R.R., Hitzemann R., Logan J., Bendriem B., Gatley S.J., and Christman D. 1989. Mapping cocaine binding sites in human and baboon brain in vivo. *Synapse* **4**: 371–377.

Freeman W.M., Brebner K., Lynch W.J., Robertson D.J., Roberts D.C., and Vrana K.E. 2001a. Cocaine-responsive gene expression changes in rat hippocampus. *Neuroscience* **108**: 371–380.

Freeman W.M., Brebner K., Patel K.M., Lynch W.J., Roberts D.C., and Vrana K.E. 2002. Repeated cocaine self-administration causes multiple changes in rat frontal cortex gene expression. *Neurochem. Res.* **27**: 1181–1192.

Freeman W.M., Nader M.A., Nader S.H., Robertson D.J., Gioia L., Mitchell S.M., Daunais J.B., Porrino L.J., Friedman D.P., and Vrana K.E. 2001b. Chronic cocaine-mediated changes in non-human primate nucleus accumbens gene expression. *J. Neurochem.* **77**: 542–549.

Frost J.J., Rosier A.J., Reich S.G., Smith S.S., Ehlers M.D., Snyder S.H., Ravert H.T., and Dannals R.F. 1993. Positron emission tomographic imaging of the dopamine transporter with ^{11}C-WIN 35,428 reveals marked decline in mild Parkinson's disease. *Ann. Neurol.* **34**: 423–431.

Gainetdinov R.R. and Caron M.G. 2003. Monoamine transporters: From genes to behavior. *Annu. Rev. Pharmacol. Toxicol.* **43**: 261–284.

Gawin F.H. and Ellinwood E.H. 1988. Cocaine and other stimulants: Actions, abuse, treatment. *N. Engl. J. Med.* **318**: 1173–1182.

Giorgi O., Piras G., Lecca D., and Corda M.G. 2005. Behavioural effects of acute and repeated cocaine treatments: A comparative study in sensitisation-prone RHA rats and their sensitisation-resistant RLA counterparts. *Psychopharmacology* **180**: 530–538.

Giorgi O., Corda M.G., Carboni G., Frau V., Valentini V., and Di Chiara G. 1997. Effects of cocaine and morphine in rats from two psychogenetically selected lines: A behavioral and brain dialysis study. *Behav. Genet.* **27**: 537–546.

Giros B., Jaber M., Jones S.R., Wightman R.M., and Caron M.G. 1996. Hyperlocomotion and indifference to cocaine and amphetamine in mice lacking the dopamine transporter. *Nature* **379**: 606–612.

Gorelick D.A., Gardner E.L., and Xi Z.X. 2004. Agents in development for the management of cocaine abuse. *Drugs* **64**: 1547–1573.

Graybiel A.M., Moratalla R., and Robertson H.A. 1990. Amphetamine and cocaine induce drug-specific activation of the *c-fos* gene in striosome-matrix compartments and limbic subdivisions of the striatum. *Proc. Natl. Acad. Sci.* **87**: 6912–6916.

Grimm J.W., Hope B.T., Wise R.A., and Shaham Y. 2001. Neuroadaptation. Incubation of cocaine craving after withdrawal. *Nature* **412**: 141–142.

Grimm J.W., Lu L., Hayashi T., Hope B.T., Su T.P., and Shaham Y. 2003. Time-dependent increases in brain-derived neurotrophic factor protein levels within the mesolimbic dopamine system after withdrawal from cocaine: Implications for incubation of cocaine craving. *J. Neurosci.* **23**: 742–747.

Grunwald I.C., Korte M., Adelmann G., Plueck A., Kullander K., Adams R.H., Frotscher M., Bonhoeffer T., and Klein R. 2004. Hippocampal plasticity requires postsynaptic ephrinBs. *Nat. Neurosci.* **7**: 33–40.

Hall F.S., Sora I., Drgonova J., Li X.F., Goeb M., and Uhl G.R. 2004. Molecular mechanisms underlying the rewarding effects of cocaine. *Ann. N.Y. Acad. Sci.* **1025**: 47–56.

Hanania T., McCreary A.C., Salaz D.O., Lyons A.M., and Zahniser N.R. 2004. Differential regulation of cocaine-induced locomotor activity in inbred long-sleep and short-sleep mice by dopamine and serotonin systems. *Eur. J. Pharmacol.* **502**: 221–231.

Haney M. and Kosten T.R. 2004. Therapeutic vaccines for substance dependence. *Expert Rev. Vaccines* **3**: 11–18.

Hawkins J.D., Catalano R.F., and Miller J.Y. 1992. Risk and protective factors for alcohol and other drug problems in adolescence and early adulthood: Implications for substance abuse prevention. *Psychol. Bull.* **112**: 64–105.

Holman B.L., Mendelson J., Garada B., Teoh S.K., Hallgring E., Johnson K.A., and Mello N.K. 1993. Regional cerebral blood flow improves with treatment in chronic cocaine polydrug users. *J. Nucl. Med.* **34**: 723–727.

Hope B.T., Crombag H.S., Jedynak J.P., and Wise R.A. 2005. Neuroadaptations of total levels of adenylate cyclase, protein kinase A, tyrosine hydroxylase, cdk5 and neurofilaments in the nucleus accumbens and ventral tegmental area do not correlate with expression of sensitized or tolerant locomotor responses to cocaine. *J. Neurochem.* **92**: 536–545.

Howell L.L. and Wilcox K.M. 2001. The dopamine transporter and cocaine medication development: Drug self-administration in nonhuman primates. *J. Pharmacol. Exp. Ther.* **298**: 1–6.

Hsin L.W., Dersch C.M., Baumann M.H., Stafford D., Glowa J.R., Rothman R.B., Jacobson A.E., and Rice K.C. 2002. Development of long-acting dopamine transporter ligands as potential cocaine-abuse therapeutic agents: Chiral hydroxyl-containing derivatives of 1-[2-[bis(4-fluorophenyl)methoxy]ethyl]-4-(3-phenylpropyl)piperazine and 1-[2-(diphenylmethoxy)ethyl]-4-(3-phenylpropyl)piperazine. *J. Med. Chem.* **45**: 1321–1329.

Humblot N., Thiriet N., Gobaille S., Aunis D., and Zwiller J. 1998. The serotonergic system modulates the cocaine-induced expression of the immediate early genes egr-1 and *c-fos* in rat brain. *Ann. N.Y. Acad. Sci.* **844**: 7–20.

Hurd Y.L. and Ungerstedt U. 1989. Cocaine: An in vivo microdialysis evaluation of its acute action on dopamine transmission in rat striatum. *Synapse* **3**: 48–54.

Ingram S.L., Prasad B.M., and Amara S.G. 2002. Dopamine transporter-mediated conductances increase excitability of midbrain dopamine neurons. *Nat. Neurosci.* **5:** 971–978.

Jassen A.K., Yang H., Miller G.M., and Madras B.K. 2005. Gene expression of axon guidance molecules is regulated by forskolin and dopamine receptor activation: Implications for drug-induced neuroadaptation in brain. *Soc. Neurosci. Abstr.* (in press).

Jentsch J.D. and Taylor J.R. 1999. Impulsivity resulting from frontostriatal dysfunction in drug abuse: Implications for the control of behavior by reward-related stimuli. *Psychopharmacology* **146:** 373–390.

Jouvert P., Dietrich J.B., Aunis D., and Zwiller J. 2002. Differential rat brain expression of EGR proteins and of the transcriptional corepressor NAB in response to acute or chronic cocaine administration. *Neuromol. Med.* **1:** 137–151.

Jouvert P., Revel M.O., Lazaris A., Aunis D., Langley K., and Zwiller J. 2004. Activation of the cGMP pathway in dopaminergic structures reduces cocaine-induced EGR-1 expression and locomotor activity. *J. Neurosci.* **24:** 10716–10725.

Kahlig K.M. and Galli A. 2003. Regulation of dopamine transporter function and plasma membrane expression by dopamine, amphetamine, and cocaine. *Eur. J. Pharmacol.* **479:** 153–158.

Kalivas P.W. and Duffy P. 1990. Effect of acute and daily cocaine treatment on extracellular dopamine in the nucleus accumbens. *Synapse* **5:** 48–58.

——. 1993a. Time course of extracellular dopamine and behavioral sensitization to cocaine. I. Dopamine axon terminals. *J. Neurosci.* **13:** 266–275.

——. 1993b. Time course of extracellular dopamine and behavioral sensitization to cocaine. II. Dopamine perikarya. *J. Neurosci.* **13:** 276–284.

Kalivas P.W., Volkow N., and Seamans J. 2005. Unmanageable motivation in addiction: A pathology in prefrontal-accumbens glutamate transmission. *Neuron* **45:** 647–650.

Kalivas P.W., Pierce R.C., Cornish J., and Sorg B.A. 1998. A role for sensitization in craving and relapse in cocaine addiction. *J. Psychopharmacol.* **12:** 49–53.

Kalivas P.W., McFarland K., Bowers S., Szumlinski K., Xi Z.X., and Baker D. 2003. Glutamate transmission and addiction to cocaine. *Ann. N.Y. Acad. Sci.* **1003:** 169–175.

Kampman K.M., Leiderman D., Holmes T., LoCastro J., Bloch D.A., Reid M.S., Shoptaw S., Montgomery M.A., Winhusen T.M., Somoza E.C., Ciraulo D.A., Elkashef A., and Vocci F. 2005. Cocaine Rapid Efficacy Screening Trials (CREST): Lessons learned. *Addiction* (suppl. 1) **100:** 102–110.

Katz J.L., Kopajtic T.A., Agoston G.E., and Newman A.H. 2004. Effects of N-substituted analogs of benztropine: Diminished cocaine-like effects in dopamine transporter ligands. *J. Pharmacol. Exp. Ther.* **309:** 650–660.

Kaufman M.J. and Madras B.K. 1991. Severe depletion of cocaine recognition sites associated with the dopamine transporter in Parkinson's diseased striatum. *Synapse* **9:** 43–49.

——. 1992. Cocaine recognition sites labeled by [³H]CFT and [¹²⁵I]RTI-55 in monkey brain. II. Ex vivo autoradiographic distribution. *Synapse* **12:** 99–111.

Kaufman M.J., Spealman R.D., and Madras B.K. 1991. Distribution of cocaine recognition sites in monkey brain. I. In vitro autoradiography with [³H]CFT. *Synapse* **9:** 177–187.

Kaufman M.J., Levin J.M., Ross M.H., Lange N., Rose S.L., Kukes T.J., Mendelson J.H., Lukas S.E., Cohen B.M., and Renshaw P.F. 1998. Cocaine-induced cerebral vasoconstriction detected in humans with magnetic resonance angiography. *J. Am. Med. Assoc.* **279:** 376–380.

Kenny P.J. and Markou A. 2004. The ups and downs of addiction: Role of metabotropic glutamate receptors. *Trends Pharmacol. Sci.* **25:** 265–272.

Khroyan T.V., Barrett-Larimore R.L., Rowlett J.K., and Spealman R.D. 2000. Dopamine D1- and D2-like receptor mechanisms in relapse to cocaine-seeking behavior: Effects of selective antagonists and agonists. *J. Pharmacol. Exp. Ther.* **294:** 680–687.

Kilty J.E., Lorang D., and Amara S.G. 1991. Cloning and expression of a cocaine-sensitive rat dopamine transporter. *Science* **254:** 578–579.

Kitayama S., Shimada S., Xu H., Markham L., Donovan D.M., and Uhl G.R. 1992. Dopamine transporter site-directed mutations differentially alter substrate transport and cocaine binding. *Proc. Natl. Acad. Sci.* **89:** 7782–7785.

Kleber H.D. 2003. Pharmacologic treatments for heroin and cocaine dependence. *Am. J. Addict.* (suppl. 2) **12:** S5–S18.

Kleven M.S., Anthony E.W., and Woolverton W.L. 1990. Pharmacological characterization of the discriminative stimulus effects of cocaine in rhesus monkeys. *J. Pharmacol. Exp. Ther.* **254:** 312–317.

Kolb B., Gorny G., Li Y., Samaha A.N., and Robinson T.E. 2003. Amphetamine or cocaine limits the ability of later experience to promote structural plasticity in the neocortex and nucleus accumbens. *Proc. Natl. Acad. Sci.* **100:** 10523–10528.

Koob G.F. and Le Moal M. 2001. Drug addiction, dysregulation of reward, and allostasis. *Neuropsychopharmacology* **24:** 97–129.

Koob G.F., Ahmed S.H., Boutrel B., Chen S.A., Kenny P.J, Markou A., O'Dell L.E., Parsons L.H., and Sanna P.P. 2004. Neurobiological mechanisms in the transition from drug use to drug dependence. *Neurosci. Biobehav. Rev.* **27:** 739–749.

Kozikowski A.P., Eddine Saiah M.K., Johnson K.M., and Bergmann J.S. 1995. Chemistry and biology of the 2 beta-alkyl-3 beta-phenyl analogues of cocaine: Subnanomolar affinity ligands that suggest a new pharmacophore model at the C-2 position. *J. Med. Chem.* **38:** 3086–3093.

Lane J., Gerstein D., and Huang L. 1997. *Risk and protective factors for adolescent drug use: Findings from the 1997 National Household Survey on Drug Abuse.* Office of applied studies, National Opinion Research Center (NORC), Substance Abuse and Mental Health Services Administration (SAMSHA), U.S. Department of Health and Human Services, Rockville, Maryland.

Laruelle M., Baldwin R.M., Malison R.T., Zea-Ponce Y., Zoghbi S.S., Al-Tikriti M.S., Sybirska E.H., Zimmerman R.C., Wisniewski G., Neumeye J.L., Milius R.A., Wang S., Smith E.O., Roth R.H., Charney D.S., Hoffer P.B., and Innis R.B.

1993. SPECT imaging of dopamine and serotonin transporters with [123I]β-CIT: Pharmacological characterization of brain uptake in nonhuman primates. *Synapse* **13:** 295–309.

Lehrmann E., Oyler J., Vawter M.P., Hyde T.M., Kolachana B., Kleinman J.E., Huestis M.A., Becker K.G., and Freed W.J. 2003. Transcriptional profiling in the human prefrontal cortex: Evidence for two activational states associated with cocaine abuse. *Pharmacogenomics J.* **3:** 27–40.

Leiderman D.B., Shoptaw S., Montgomery A., Bloch D.A., Elkashef A., LoCastro J., and Vocci F. 2005. Cocaine Rapid Efficacy Screening Trial (CREST): A paradigm for the controlled evaluation of candidate medications for cocaine dependence. *Addiction* (suppl. 1) **100:** 1–11.

Lewis D.A., Melchitzky D.S., Sesack S.R., Whitehead R.E., Auh S., and Sampson A. 2001. Dopamine transporter immunoreactivity in monkey cerebral cortex: Regional, laminar, and ultrastructural localization. *J. Comp. Neurol.* **432:** 119–136.

Li L.B., Chen N., Ramamoorthy S., Chi L., Cui X.N., Wang L.C., and Reith M.E. 2004. The role of N-glycosylation in function and surface trafficking of the human dopamine transporter. *J. Biol. Chem.* **279:** 21012–21020.

Little K.Y., Elmer L.W., Zhong H., Scheys J.O., and Zhang L. 2002. Cocaine induction of dopamine transporter trafficking to the plasma membrane. *Mol. Pharmacol.* **61:** 436–445.

Little K.Y., Kirkman J.A., Carroll F.I., Clark T.B., and Duncan G.E. 1993. Cocaine use increases [3H]WIN 35428 binding sites in human striatum. *Brain Res.* **628:** 17–25.

London E.D., Cascella N.G., Wong D.F., Phillips R.L., Dannals R.F., Links J.M., Herning R., Grayson R., Jaffe J.H., and Wagner H.N., Jr. 1990. Cocaine-induced reduction of glucose utilization in human brain. A study using positron emission tomography and [fluorine 18]-fluorodeoxyglucose. *Arch. Gen. Psychiatry* **47:** 567–574.

Lu L., Hope B.T., and Shaham Y. 2004a. The cystine-glutamate transporter in the accumbens: A novel role in cocaine relapse. *Trends Neurosci.* **27:** 74–76.

Lu L., Dempsey J., Shaham Y., and Hope B.T. 2005a. Differential long-term neuroadaptations of glutamate receptors in the basolateral and central amygdala after withdrawal from cocaine self-administration in rats. *J. Neurochem.* **94:** 161–168.

Lu L., Grimm J.W., Hope B.T., and Shaham Y. 2004b. Incubation of cocaine craving after withdrawal: A review of preclinical data. *Neuropharmacology* (suppl. 1) **47:** 214–226.

Lu L., Grimm J.W., Shaham Y., and Hope B.T. 2003. Molecular neuroadaptations in the accumbens and ventral tegmental area during the first 90 days of forced abstinence from cocaine self-administration in rats. *J. Neurochem.* **85:** 1604–1613.

Lu L., Dempsey J., Liu S.Y., Bossert J.M., and Shaham Y. 2004c. A single infusion of brain-derived neurotrophic factor into the ventral tegmental area induces long-lasting potentiation of cocaine seeking after withdrawal. *J. Neurosci.* **24:** 1604–1611.

Lu L., Hope B.T., Dempsey J., Liu S.Y., Bossert J.M., and Shaham Y. 2005b. Central amygdala ERK signaling pathway is critical to incubation of cocaine craving. *Nat. Neurosci.* **8:** 212–219.

Maas L.C., Lukas S.E., Kaufman M.J., Weiss R.D., Daniels S.L., Rogers V.W., Kukes T.J., and Renshaw P.F. 1998. Functional

magnetic resonance imaging of human brain activation during cue-induced cocaine craving. *Am. J. Psychiatry* **155:** 124–126.

Madras B.K. and Kaufman M.J. 1994. Cocaine accumulates in dopamine-rich regions of primate brain after i.v. administration: Comparison with mazindol distribution. *Synapse* **18:** 261–275.

Madras B.K., Miller G.M., and Fischman A.J. 2005. The dopamine transporter and attention-deficit/hyperactivity disorder. *Biol. Psychiatry* **57:** 1397–1409.

Madras B.K., Verrico C., Jassen A., and Miller G.M. 2004a. Attention Deficit Hyperactivity Disorder (ADHD): New roles for old trace amines and monoamine transporters. *Neuropsychopharmacology* (suppl. 1) **29:** S137.

Madras B.K., Fahey M.A., Bergman J., Canfield D.R., and Spealman R.D. 1989a. Effects of cocaine and related drugs in nonhuman primates. I. [3H]cocaine binding sites in caudate-putamen. *J. Pharmacol. Exp. Ther.* **251:** 131–141.

Madras B.K., Jassen A., Meltzer P.C., Bonab A.A., Livni E., and Fischman A.J. 2004b. The Trojan-tropane horse and monoamine transporters: A cocaine antagonist strategy. Program No. 573.13, 2004, Abstract viewer/Itinerary planner, Society for Neuroscience, Washington D.C.

Madras B.K., Spealman R.D., Fahey M.A., Neumeyer J.L., Saha J.K., and Milius R.A. 1989b. Cocaine receptors labeled by [3H]2 β-carbomethoxy-3 β-(4-fluorophenyl)tropane. *Mol. Pharmacol.* **36:** 518–524.

Madras B.K., Gracz L.M., Meltzer P.C., Liang A.Y., Elmaleh D.R., Kaufman M.J., and Fischman A.J. 1998a. Altropane, a SPECT or PET imaging probe for dopamine neurons. II. Distribution to dopamine-rich regions of primate brain. *Synapse* **29:** 105–115.

Madras B.K., Pristupa Z.B., Niznik H.B., Liang A.Y., Blundell P., Gonzalez M.D., and Meltzer P.C. 1996a. Nitrogen-based drugs are not essential for blockade of monoamine transporters. *Synapse* **24:** 340–348.

Madras B.K., Kamien J.B., Fahey M.A., Canfield D.R., Milius R.A., Saha J.K., Neumeyer J.L., and Spealman R.D. 1990. N-modified fluorophenyltropane analogs of cocaine with high affinity for cocaine receptors. *Pharmacol. Biochem. Behav.* **35:** 949–953.

Madras B.K., Gracz L.M., Fahey M.A., Elmaleh D., Meltzer P.C., Liang A.Y., Stopa E.G., Babich J., and Fischman A.J. 1998b. Altropane, a SPECT or PET imaging probe for dopamine neurons. III. Human dopamine transporter in postmortem normal and Parkinson's diseased brain. *Synapse* **29:** 116–127.

Madras B.K., Jones A.G., Mahmood A., Zimmerman R.E., Garada B., Holman B.L., Davison A., Blundell P., and Meltzer P.C. 1996b. Technepine: A high-affinity 99m-technetium probe to label the dopamine transporter in brain by SPECT imaging. *Synapse* **22:** 239–246.

Madras B.K., Fahey M.A., Miller G.M., De La Garza R., Goulet M., Spealman R.D., Meltzer P.C., George S.R., O'Dowd B.F., Bonab A.A., Livni E., and Fischman A.J. 2003. Non-amine-based dopamine transporter (reuptake) inhibitors retain properties of amine-based progenitors. *Eur. J. Pharmacol.* **479:** 41–51.

Mash D.C., Pablo J., Ouyang Q., Hearn W.L., and Izenwasser S. 2002. Dopamine transport function is elevated in cocaine users. *J. Neurochem.* **81:** 292–300.

McClung C.A., Ulery P.G., Perrotti L.I., Zachariou V., Berton O., and Nestler E.J. 2004. DeltaFosB: A molecular switch for long-term adaptation in the brain. *Brain Res. Mol. Brain Res.* **132:** 146–154.

McFarland K., Davidge S.B., Lapish C.C., and Kalivas P.W. 2004. Limbic and motor circuitry underlying footshock-induced reinstatement of cocaine-seeking behavior. *J. Neurosci.* **24:** 1551–1560.

Meltzer P.C., Liang A.Y., and Madras B.K. 1994. The discovery of an unusually selective and novel cocaine analog: Difluoropine. Synthesis and inhibition of binding at cocaine recognition sites. *J. Med. Chem.* **37:** 2001–2010.

Meltzer P.C., Pham-Huu D.P., and Madras B.K. 2004. Synthesis of 8-thiabicyclo[3.2.1]oct-2-enes and their binding affinity for the dopamine and serotonin transporters. *Bioorg. Med. Chem. Lett.* **14:** 6007–6010.

Meltzer P.C., Liang A.Y., Brownell A.L., Elmaleh D.R., and Madras B.K. 1993. Substituted 3-phenyltropane analogs of cocaine: Synthesis, inhibition of binding at cocaine recognition sites, and positron emission tomography imaging. *J. Med. Chem.* **36:** 855–862.

Meltzer P.C., Liu S., Blanchette H., Blundell P., and Madras B.K. 2002. Design and synthesis of an irreversible dopamine-sparing cocaine antagonist. *Bioorg. Med. Chem.* **10:** 3583–3591.

Milivojevic N., Krisch I., Sket D., and Zivin M. 2004. The dopamine D1 receptor agonist and D2 receptor antagonist LEK-8829 attenuates reinstatement of cocaine-seeking in rats. *Naunyn-Schmiedebergs Arch. Pharmacol.* **369:** 576–582.

Miller G.M. and Madras B.K. 2005. Genetic variation and phenotype associations common to rhesus monkeys and humans. *Lab. Primate* (in press).

Miller G.M., Verrico C.D., Jassen A., Konar M., Yang H., Panas H., Bahn M., Johnson R., and Madras B.K. 2005. Primate trace amine receptor 1 modulation by the dopamine transporter. *J. Pharmacol. Exp. Ther.* **313:** 983–994.

Misra A.L., Giri V.V., Patel M.N., Alluri V.R., and Mule S.J. 1977. Disposition and metabolism of [3H] cocaine in acutely and chronically treated monkeys. *Drug Alcohol Depend.* **2:** 261–272.

Miyata S., Mori Y., Fujiwara T., Ikenaka K., Matsuzaki S., Oono K., Katayama T., and Tohyama M. 2005. Local protein synthesis by BDNF is potentiated in hippocampal neurons exposed to ephrins. *Brain Res. Mol. Brain Res.* **134:** 333–337.

Moran M.M., McFarland K., Melendez R.I., Kalivas P.W., and Seamans J.K. 2005. Cystine/glutamate exchange regulates metabotropic glutamate receptor presynaptic inhibition of excitatory transmission and vulnerability to cocaine seeking. *J. Neurosci.* **25:** 6389–6393.

Morgan D., Liu Y., and Roberts D.C. 2005a. Rapid and persistent sensitization to the reinforcing effects of cocaine. *Neuropsychopharmacology* (in press).

Morgan D., Smith M.A., and Roberts D.C. 2005b. Binge self-administration and deprivation produces sensitization to the reinforcing effects of cocaine in rats. *Psychopharmacology* **178:** 309–316.

Morgan D., Grant K.A., Gage H.D., Mach R.H., Kaplan J.R.,

Prioleau O., Nader S.H., Buchheimer N., Ehrenkaufer R.L., and Nader M.A. 2002. Social dominance in monkeys: Dopamine D2 receptors and cocaine self-administration. *Nat. Neurosci.* **5:** 169–174.

Moron J.A., Brockington A., Wise R.A., Rocha B.A., and Hope B.T. 2002. Dopamine uptake through the norepinephrine transporter in brain regions with low levels of the dopamine transporter: Evidence from knock-out mouse lines. *J. Neurosci.* **22:** 389–395.

Mortensen O.V. and Amara S.G. 2003. Dynamic regulation of the dopamine transporter. *Eur. J. Pharmacol.* **479:** 159–170.

Nathanson J.A., Hunnicutt E.J., Kantham L., and Scavone C. 1993. Cocaine as a naturally occurring insecticide. *Proc. Natl. Acad. Sci.* **90:** 9645–9648.

Negus S.S., Brandt M.R., and Mello N.K. 1999. Effects of the long-acting monoamine reuptake inhibitor indatraline on cocaine self-administration in rhesus monkeys. *J. Pharmacol. Exp. Ther.* **291:** 60–69.

Neisewander J.L., Baker D.A., Fuchs R.A., Tran-Nguyen L.T., Palmer A., and Marshall J.F. 2000. Fos protein expression and cocaine-seeking behavior in rats after exposure to a cocaine self-administration environment. *J. Neurosci.* **20:** 798–805.

Nestler E.J. 2001. Molecular basis of long-term plasticity underlying addiction. *Nat. Rev. Neurosci.* **2:** 119–128.

Nestler E.J., Hope B.T., and Widnell K.L. 1993. Drug addiction: A model for the molecular basis of neural plasticity. *Neuron* **11:** 995–1006.

Nestler E.J., Terwilliger R.Z., Walker J.R., Sevarino K.A., and Duman R.S. 1990. Chronic cocaine treatment decreases levels of the G protein subunits Gi alpha and Go alpha in discrete regions of rat brain. *J. Neurochem.* **55:** 1079–1082.

O'Brien M.S. and Anthony J.C. 2005. Risk of becoming cocaine dependent: Epidemiological estimates for the United States, 2000–2001. *Neuropsychopharmacology* **30:** 1006–1018.

Pettit H.O. and Justice J.B., Jr. 1989. Dopamine in the nucleus accumbens during cocaine self-administration as studied by in vivo microdialysis. *Pharmacol. Biochem. Behav.* **34:** 899–904.

———. 1991. Effect of dose on cocaine self-administration behavior and dopamine levels in the nucleus accumbens. *Brain Res.* **539:** 94–102.

Pierce R.C., Bell K., Duffy P., and Kalivas P.W. 1996. Repeated cocaine augments excitatory amino acid transmission in the nucleus accumbens only in rats having developed behavioral sensitization. *J. Neurosci.* **16:** 1550–1560.

Pilla M., Perachon S., Sautel F., Garrido F., Mann A., Wermuth C.G., Schwartz J.C., Everitt B.J., and Sokoloff P. 1999. Selective inhibition of cocaine-seeking behaviour by a partial dopamine D3 receptor agonist. *Nature* **400:** 371–375.

Porrino L.J. 1993. Functional consequences of acute cocaine treatment depend on route of administration. *Psychopharmacology* **112:** 343–351.

Porrino L.J., Dunais J.B., Smith H.R., and Nader M.A. 2004. The expanding effects of cocaine: Studies in a nonhuman primate model of cocaine self administration. *Neurosci. Biobehav. Rev.* **27:** 813–820.

Porrino L.J., Lyons D., Miller M.D., Smith H.R., Friedman D.P.,

Daunais J.B., and Nader M.A. 2002. Metabolic mapping of the effects of cocaine during the initial phases of self-administration in the nonhuman primate. *J. Neurosci.* **22**: 7687–7694.

Reid M.S. and Berger S.P. 1996. Evidence for sensitization of cocaine-induced nucleus accumbens glutamate release. *Neuroreport* **7**: 1325–1329.

Reith M.E., Sershen H., and Lajtha A. 1980. Saturable (3H)cocaine binding in central nervous system of mouse. *Life Sci.* **27**: 1055–1062.

Reith M.E.A., Meisler B.E., Sershen H., and Lajtha A. 1986. Structural requirements for cocaine congeners to interact with dopamine and serotonin uptake sites in mouse brain and to induce stereotyped behavior. *Biochem. Pharmacol.* **35**: 1123–1129.

Ritz M.C., Lamb R.J., Goldberg S.R., and Kuhar M.J. 1987. Cocaine receptors on dopamine transporters are related to self-administration of cocaine. *Science* **237**: 1219–1223.

Rivier L. 1981. Analysis of alkaloids in leaves of cultivated Erythroxylum and characterization of alkaline substances used during coca chewing. *J. Ethnopharmacol.* **3**: 313–335.

Robinson T.E. and Berridge K.C. 2001. Incentive-sensitization and addiction. *Addiction* **96**: 103–114.

Robinson T.E. and Kolb B. 2004. Structural plasticity associated with exposure to drugs of abuse. *Neuropharmacology* (suppl. 1) **47**: 33–46.

Rocha B.A. 2003. Stimulant and reinforcing effects of cocaine in monoamine transporter knockout mice. *Eur. J. Pharmacol.* **479**: 107–115.

Rocha B.A., Fumagalli F., Gainetdinov R.R., Jones S.R., Ator R., Giros B., Miller G.W., and Caron M.G. 1998. Cocaine self-administration in dopamine-transporter knockout mice. *Nat. Neurosci.* **1**: 132–137.

Rosenzweig-Lipson S., Bergman J., Spealman R.D., and Madras B.K. 1992. Stereoselective behavioral effects of Lu 19-005 in monkeys: Relation to binding at cocaine recognition sites. *Psychopharmacology* **107**: 186–194.

Rothstein J.D., Patel S., Regan M.R., Haenggeli C., Huang Y.H., Bergles D.E., Jin L., Dykes Hoberg M., Vidensky S., Chung D.S., Toan S.V., Bruijn L.I., Su Z.Z., Gupta P., and Fisher P.B. 2005. Beta-lactam antibiotics offer neuroprotection by increasing glutamate transporter expression. *Nature* **433**: 73–77.

Rounsaville B.J., Anton S.F., Carroll K., Budde D., Prusoff B.A., and Gawin F. 1991. Psychiatric diagnoses of treatment-seeking cocaine abusers. *Arch. Gen. Psychiatry* **48**: 43–51.

Sabeti J., Gerhardt G.A., and Zahniser N.R. 2003. Individual differences in cocaine-induced locomotor sensitization in low and high cocaine locomotor-responding rats are associated with differential inhibition of dopamine clearance in nucleus accumbens. *J. Pharmacol. Exp. Ther.* **305**: 180–190.

Saka E., Goodrich C., Harlan P., Madras B.K., and Graybiel A.M. 2004. Repetitive behaviors in monkeys are linked to specific striatal activation patterns. *J. Neurosci.* **24**: 7557–7565.

Samaha A.N. and Robinson T.E. 2005. Why does the rapid delivery of drugs to the brain promote addiction? *Trends Pharmacol. Sci.* **26**: 82–87.

Samaha A.N., Mallet N., Ferguson S.M., Gonon F., and

Robinson T.E. 2004. The rate of cocaine administration alters gene regulation and behavioral plasticity: Implications for addiction. *J. Neurosci.* **24**: 6362–6370.

SAMSHA (U.S. Substance Abuse and Mental Health Services Administration). 2001. *National survey on drug abuse and health.* Office of Applied Studies, U.S. Department of Health and Human Services, Rockville, Maryland.

Satel S.L., Southwick S.M., and Gawin F.H. 1991. Clinical features of cocaine-induced paranoia. *Am. J. Psychiatry* **148**: 495–498.

Saunders C., Ferrer J.V., Shi L., Chen J., Merrill G., Lamb M.E., Leeb-Lundberg L.M., Carvelli L., Javitch J.A., and Galli A. 2000. Amphetamine-induced loss of human dopamine transporter activity: An internalization-dependent and cocaine-sensitive mechanism. *Proc. Natl. Acad. Sci.* **97**: 6850–6855.

Schenk S. 2002. Effects of GBR 12909, WIN 35,428 and indatraline on cocaine self-administration and cocaine seeking in rats. *Psychopharmacology* **100**: 263–270.

Self D.W. and Nestler E.J. 1998. Relapse to drug-seeking: Neural and molecular mechanisms. *Drug Alcohol Depend.* **51**: 49–60.

Self D.W., Barnhart W.J., Lehman D.A., and Nestler E.J. 1996. Opposite modulation of cocaine-seeking behavior by D1- and D2-like dopamine receptor agonists. *Science* **271**: 1586–1589.

Shaham Y., Shalev U., Lu L., De Wit H., and Stewart J. 2003. The reinstatement model of drug relapse: History, methodology and major findings. *Psychopharmacology* **168**: 3–20.

Shimada S., Kitayama S., Lin C.L., Patel A., Nanthakumar E., Gregor P., Kuhar M., and Uhl G. 1991. Cloning and expression of a cocaine-sensitive dopamine transporter complementary DNA. *Science* **254**: 576–578.

Shoblock J.R., Maisonneuve I.M., and Glick S.D. 2004. Differential interactions of desipramine with amphetamine and methamphetamine: Evidence that amphetamine releases dopamine from noradrenergic neurons in the medial prefrontal cortex. *Neurochem. Res.* **29**: 1437–1442.

Sibley D.R., Monsma F.J.J., and Shen Y. 1993. Molecular neurobiology of dopaminergic receptors. *Int. Rev. Neurobiol.* **35**: 391–415.

Smith C., Lizotte A.J., Thornberry T.P., and Krohn M.D. 1995. Resilient youth: Identifying factors that prevent high-risk youth from engaging in delinquency and drug use. In *Delinquency and disrepute in the life course* (ed. J. Hagan), pp. 217–247. JAI Press, Greenwich, Connecticut.

Sofuoglu M. and Kosten T.R. 2004. Pharmacologic management of relapse prevention in addictive disorders. *Psychiatr. Clin. North Am.* **27**: 627–648.

———. 2005. Novel approaches to the treatment of cocaine addiction. *CNS Drugs* **19**: 13–25.

Sonders M.S., Zhu S.J., Zahniser N.R., Kavanaugh M.P., and Amara S.G. 1997. Multiple ionic conductances of the human dopamine transporter: The actions of dopamine and psychostimulants. *J. Neurosci.* **17**: 960–974.

Spealman R.D., Bergman J., Madras B.K., and Melia K.F. 1991. Discriminative stimulus effects of cocaine in squirrel monkeys: Involvement of dopamine receptor subtypes. *J. Pharmacol. Exp. Ther.* **258**: 945–953.

Stansfield K.H., Philpot R.M., and Kirstein C.L. 2004. An animal model of sensation seeking: The adolescent rat. *Ann. N.Y. Acad. Sci.* **1021:** 453–458.

Steketee J.D. 2003. Neurotransmitter systems of the medial prefrontal cortex: Potential role in sensitization to psychostimulants. *Brain Res. Brain Res. Rev.* **41:** 203–228.

Stewart J. 2000. Pathways to relapse: The neurobiology of drug- and stress-induced relapse to drug-taking. *J. Psychiatry Neurosci.* **25:** 125–136.

Striplin C.D. and Kalivas P.W. 1993. Robustness of G protein changes in cocaine sensitization shown with immunoblotting. *Synapse* **14:** 10–15.

Sulzer D. and Galli A. 2003. Dopamine transport currents are promoted from curiosity to physiology. *Trends Neurosci.* **26:** 173–176.

Tang W.X., Fasulo W.H., Mash D.C., and Hemby S.E. 2003. Molecular profiling of midbrain dopamine regions in cocaine overdose victims. *J. Neurochem.* **85:** 911–924.

Terwilliger R.Z., Beitner-Johnson D., Sevarino K.A., Crain S.M., and Nestler E.J. 1991. A general role for adaptations in G-proteins and the cyclic AMP system in mediating the chronic actions of morphine and cocaine on neuronal function. *Brain Res.* **548:** 100–110.

Thompson A.M., Gosnell B.A., and Wagner J.J. 2002. Enhancement of long-term potentiation in the rat hippocampus following cocaine exposure. *Neuropharmacology* **42:** 1039–1042.

Thompson A.M., Swant J., Gosnell B.A., and Wagner J.J. 2004. Modulation of long-term potentiation in the rat hippocampus following cocaine self-administration. *Neuroscience* **127:** 177–185.

Toda S., McGinty J.F., and Kalivas P.W. 2002. Repeated cocaine administration alters the expression of genes in corticolimbic circuitry after a 3-week withdrawal: A DNA macroarray study. *J. Neurochem.* **82:** 1290–1299.

Trulson M.E., Joe J.C., Babb S., and Raese J.D. 1987. Chronic cocaine administration depletes tyrosine hydroxylase immunoreactivity in the meso-limbic dopamine system in rat brain: Quantitative light microscopic studies. *Brain Res. Bull.* **19:** 39–45.

Uhl G.R. and Grow R.W. 2004. The burden of complex genetics in brain disorders. *Arch. Gen. Psychiatry* **61:** 223–229.

Uhl G.R. and Lin Z. 2003. The top 20 dopamine transporter mutants: Structure-function relationships and cocaine actions. *Eur. J. Pharmacol.* **479:** 71–82.

Uhl G., Lin Z., Metzger T., and Dar D.E. 1998. Dopamine transporter mutants, small molecules, and approaches to cocaine antagonist/dopamine transporter disinhibitor development. *Methods Enzymol.* **296:** 456–465.

Vanderschuren L.J. and Everitt B.J. 2004. Drug seeking becomes compulsive after prolonged cocaine self-administration. *Science* **305:** 1017–1019.

Vanderschuren L.J. and Kalivas P.W. 2000. Alterations in dopaminergic and glutamatergic transmission in the induction and expression of behavioral sensitization: A critical review of preclinical studies. *Psychopharmacology* **151:** 99–120.

Volkow N.D. and Li T.K. 2004. Drug addiction: The neurobiology of behaviour gone awry. *Nat. Rev. Neurosci.* **5:** 963–970.

Volkow N.D., Fowler J.S., Wolf A.P., Wang G.J., Logan J., Macgregor R., Dewy S.L., Schlyer D., and Hitzemann R. 1992. Distribution and kinetics of carbon-11-cocaine in the human body measured by PET. *J. Nucl. Med.* **33:** 521–525.

Volkow N.D., Wang G.J., Fowler J.S., Logan J., Gatley S.J., Hitzemann R., Chen A.D., Dewey S.L., and Pappas N. 1997a. Decreased striatal dopaminergic responsiveness in detoxified cocaine-dependent subjects. *Nature* **386:** 830–833.

Volkow N.D., Wang G.J., Fischman M.W., Foltin R.W., Fowler J.S., Abumrad N.N., Vitkun S., Logan J., Gatley S.J., Pappas N.R., Hitzemann R., and Shea C.E. 1997b. Relationship between subjective effects of cocaine and dopamine transporter occupancy. *Nature* **386:** 827–830.

Wagner F.A. and Anthony J.C. 2002. Into the world of illegal drug use: Exposure opportunity and other mechanisms linking the use of alcohol, tobacco, marijuana, and cocaine. *Am. J. Epidemiol.* **155:** 918–925.

Wang P.J., Stromberg M., Replenski S., Snyder-Mackler A., and Mackler S.A. 2003. The relationship between cocaine-induced increases in NAC1 and behavioral sensitization. *Pharmacol. Biochem. Behav.* **75:** 49–54.

Wang W., Sonders M.S., Ukairo O.T., Scott H., Kloetzel M.K., and Surratt C.K. 2003. Dissociation of high-affinity cocaine analog binding and dopamine uptake inhibition at the dopamine transporter. *Mol. Pharmacol.* **64:** 430–439.

Wise R.A., Murray A., and Bozarth M.A. 1990. Bromocriptine self-administration and bromocriptine-reinstatement of cocaine-trained and heroin-trained lever pressing in rats. *Psychopharmacology* **100:** 355–360

Wong D., Yung B., Dannals R.F., Shaya E.K., Ravert H.T., Chen C.A., Chan B., Folio T., Scheffel U., Ricaurte G.A., Neumeyer J.L., Wagner H.N., and Kuhar M.J. 1993. In vivo imaging of baboon and human dopamine transporters by positron emission tomography using [^{11}C]WIN 35,428. *Synapse* **15:** 130–142.

Woolverton W.L. 1987. Evaluation of the role of norepinephrine in the reinforcing effects of psychomotor stimulants in rhesus monkeys. *Pharmacol. Biochem. Behav.* **26:** 835–839.

———. 1992. Determinants of cocaine self-administration by laboratory animals. *Ciba Found. Symp.* **166:** 149–161.

Xiao D., Miller G.M., Westmoreland S.V., Pauley D., and Madras B.K. 2004. Eph/ephrin family members, implicated in cocaine-mediated neuroadaptation, are expressed in adult primate brain. In *43rd Annual Meeting of the American College of Neuropsychopharmacology*, December 2004, San Juan, Puerto Rico. Program No. 577.11, Society for Neuroscience, Washington D.C.

Yatin S.M., Miller G.M., and Madras B.K. 2005. Dopamine and norepinephrine transporter-dependent c-Fos production in vitro: Relevance to neuroadaptation. *J. Neurosci. Methods* **143:** 69–78.

Yatin S.M., Miller G.M., Norton C., and Madras B.K. 2002. Dopamine transporter-dependent induction of C-Fos in HEK cells. *Synapse* **45:** 52–65.

Yatin S.M., Miller G., Goulet M.. Alvarez X., and Madras B.K. 2000. Cocaine blocks dopamine-induced internalization of the dopamine transporter. *Drug Alcohol Depend.* **60:** S241.

Young S.T., Porrino L.J., and Iadarola M.J. 1991. Cocaine induces striatal c-fos-immunoreactive proteins via dopaminergic D1 receptors. *Proc. Natl. Acad. Sci.* **88:** 1291–1295.

Yue Y., Widmer D.A., Halladay A.K., Cerretti D.P., Wagner G.C., Dreyer J.L, and Zhou R. 1999. Specification of distinct dopaminergic neural pathways: Roles of the Eph family receptor EphB1 and ligand ephrin-B2. *J. Neurosci.* **19:** 2090–2101.

Yuferov V., Nielsen D., Butelman E., and Kreek M.J. 2005. Microarray studies of psychostimulant-induced changes in gene expression. *Addict. Biol.* **10:** 101–118.

Yuferov V., Kroslak T., Laforge K.S., Zhou Y., Ho A., and Kreek M.J. 2003. Differential gene expression in the rat caudate putamen after "binge" cocaine administration: Advantage of triplicate microarray analysis. *Synapse* **48:** 157–169.

Zahniser N.R. and Sorkin A. 2004. Rapid regulation of the dopamine transporter: Role in stimulant addiction? *Neuropharmacology* (suppl. 1) **47:** 80–91.

Zhang L., Lou D., Jiao H., Zhang D., Wang X., Xia Y., Zhang J., and Xu M. 2004. Cocaine-induced intracellular signaling and gene expression are oppositely regulated by the dopamine D1 and D3 receptors. *J. Neurosci.* **24:** 3344–3354.

15 | The Oligomerization of G-protein-coupled Receptors

Michael M.C. Kong,[1] Brian F. O'Dowd,[1,3] and Susan R. George[1-3]

Departments of [1]Pharmacology and [2]Medicine, University of Toronto; [3]The Centre for Addiction and Mental Health, Toronto, Ontario, Canada M5S 1A8

ABSTRACT

The molecular mechanisms underlying addiction involve an intricate network of proteins and signaling molecules that are modulated in response to acute and chronic exposure to substances of abuse. The superfamily of G-protein-coupled receptors (GPCRs) represents an important target of these drugs and is central in mediating their various behavioral effects. It is now well established that GPCRs form oligomers like many other cell-surface receptor families. A diversity of pharmacological, biochemical, and biophysical approaches have collectively shown that these GPCR oligomers have novel functional consequences and use different types of interactions to maintain their structural integrity. Although the physiological requirement for GPCR oligomerization is currently under investigation, evidence suggests that these oligomers have common cellular functions including potential roles in receptor trafficking, amplification of an agonist-induced signal resulting from receptor interactions within an oligomeric complex, and the generation of novel hetero-oligomeric signaling entities. These effects may have profound implications for understanding how GPCRs, including those involved in brain reward pathways, regulate biological processes such as addiction.

INTRODUCTION

Addiction is a biological process that occurs when substances of abuse cause complex molecular and cellular changes in neuronal circuits mediating brain reward and reinforcement. This leads to drug-seeking and -craving behavior that supersedes the normal physiological mechanisms that mediate reward, learning, and memory in the brain. Repeated use of these addictive substances can result in sustained and even permanent changes in brain circuitry, promoting continued abuse and relapse after abstinence (Nestler and Aghajanian 1997).

The initial direct substrate of action of abused drugs is the protein(s) with which it interacts to cause the acute rewarding effects of the drug. The primary site of cellular action for most drugs of abuse is cell-surface proteins, which are closely linked to other proteins that participate in their cascade of signal transduction and desensitization processes. The protein targets for many abused substances are known; for instance, opioid receptors for heroin and morphine, the dopamine transporter for cocaine, and cannabinoid (CB1)

271

receptors for marijuana. The various stimulatory and inhibitory pathways that are subsequently activated converge onto certain common mediators, such as activation of dopamine receptors and stimulation of the mesolimbic dopamine neuronal pathway. The dopamine receptor subtypes and the above-mentioned receptors belong to the superfamily of G-protein-coupled receptors (GPCRs) that represents an important direct or downstream mediator of the effects of these drugs. Many of these receptors, including the opioid, dopamine, and cannabinoid receptor subtypes, act through areas of the brain that are part of the limbic system and thus, are tightly connected to the reinforcing properties of abused drugs. The majority of GPCRs have been shown to form oligomers with various functional consequences. This may shed light on the receptor-mediated molecular mechanisms underlying addiction.

GOALS OF THE CHAPTER

This chapter describes our current understanding of GPCR oligomerization as it pertains to receptor structure and function. Various methodological approaches used to analyze GPCR oligomerization are evaluated, including some advantages and limitations associated with each method. The potential mechanisms of receptor interaction are discussed to provide insight into the structural nature of GPCR oligomers. Finally, the physiological implications of oligomeric complexes are addressed by establishing their existence in brain and their potential role in drug addiction.

GPCRs: AT A GLANCE

GPCRs represent the largest class of cell-surface receptors encoded by the largest and most conserved gene family (Pierce et al. 2002). Currently, we know of 367 GPCRs to endogenous ligands in the human genome (Vassilatis et al. 2003). GPCRs mediate signal transduction from a diverse array of molecules including peptides, hormones, neurotransmitters, ions, and drugs through a variety of intracellular effectors. They share a significant degree of structural topology and conserved architecture defined by seven transmembrane domains connected by alternating extracellular and intracellular loops. All GPCRs are coupled to a heterotrimeric G protein that dissociates on agonist binding and modulates effectors belonging to various downstream signaling cascades. These effectors trigger different responses including transcriptional regulation, calcium mobilization, neurotransmission, and changes in behavior. Much of what we know about GPCR structure and function was obtained from the crystal structure of bovine rhodopsin, a prototypical member of the class A family of GPCRs. This represented a significant advance in our understanding of GPCRs and will serve to provide additional clues about the structural and functional properties of this important family of receptors (Palczewski et al. 2000).

GPCR OLIGOMERIZATION

Traditionally, it has been thought that the mechanisms of GPCR ligand binding and signal transduction are modeled after a single receptor coupled to its ligand and G protein in a 1:1:1 stoichiometry. Despite the fact that many other classes of cell-surface receptors, such as the protein-tyrosine kinase receptors and cytokine receptors, function through either a ligand-dependent or constitutive dimerization mechanism (Heldin 1995), the monomeric model for GPCRs prevailed for quite some time. However, increasing evidence from diverse methodological approaches has successively shown that oligomerization also appears to be a fundamental characteristic of GPCR biology, as is discussed later in this chapter. Many of these approaches confirm earlier observations from radiation inactivation, chemical cross-linking, and radioligand-binding studies that GPCRs might exist as multimeric entities. The functional implications of receptor oligomerization are tantalizing because this phenomenon may implicate cross talk between directly associated receptors, resulting in alterations in effector signaling. Homo-oligomerization (oligomerization between the same receptor subtypes) may be a cell-surface mechanism of intracellular signal amplification. In the case of hetero-oligomers, the generation of novel binding pockets between different receptor subtypes presents exciting opportunities for identifying previously unrecognized drug targets. The following section describes the current evidence for GPCR oligomerization.

METHODOLOGICAL CONSIDERATIONS

Evidence for GPCR Oligomerization

Receptor Complementation

One of the first studies to bring the idea of receptor oligomerization into the spotlight was the work with α_{2C}-adrenergic/M3 muscarinic receptor chimeras by Maggio and colleagues (1993). In this report, they describe the generation of receptor chimeras in which the carboxy-terminal halves were exchanged between the α_{2C}-adrenergic receptor and the M3 muscarinic receptor, yielding α_{2C}/M3 and M3/α_{2C} fusion proteins. Individual expression of either of these chimeric constructs did not result in any detectable binding to either muscarinic or adrenergic ligands. However, coexpression of the chimeras resulted in binding to both muscarinic and adrenergic radioligands with binding properties comparable to the wild-type receptors. Furthermore, functional activity was restored on coexpression as phosphatidylinositol hydrolysis was detected following stimulation by a muscarinic agonist. These data infer a mechanism whereby receptors do not only simply interact but also engage in a reciprocal exchange of receptor domains.

This notion of functional complementation between receptor mutants, described as the domain swapping hypothesis of GPCR oligomerization (Gouldson et al. 2000), has also been reported in homo-oligomerization studies using mutants of the angiotensin II type-I receptor (Monnot et al. 1996) and the H1 histamine receptor (Bakker et al. 2004). Although conceptually, domain swapping is a plausible explanation for describing functional rescue of such receptor mutants, evidence calls into question the validity of this theory for other receptors (Schulz et al. 2000). In a study using vasopressin receptor mutants, investigators determined that reconstitution of a functional wild-type V2 vasopressin receptor could not be achieved by coexpressing an amino-terminal receptor fragment and a nonfunctional full-length vasopressin receptor, as would be predicted by the domain swapping model. Despite this, the two mutants could still interact, consistent with a model in which GPCRs simply form contact dimers. The contact dimer hypothesis proposes that two binding pockets are formed from regions donated within each monomer; it requires that receptors interact laterally to form dimers.

Coimmunoprecipitation

The use of differentially tagged proteins has been a popular biochemical tool for showing protein–protein interactions. It has also commonly been used to provide evidence for the existence of homo- and hetero-oligomers of GPCRs, including those that are involved in brain reward pathways (Jones et al. 1998; Kaupmann et al. 1998; White et al. 1998; Jordan and Devi 1999; Xie et al. 1999; George et al. 2000; Gomes et al. 2000; Scarselli et al. 2001; Lee et al. 2004). With this method, the two receptors of interest are usually modified by the addition of a distinct epitope or "tag," typically at the amino terminus of each receptor. When these receptors are coexpressed in cells, an antibody directed to one of the epitopes is used to immunoprecipitate the receptor. The receptor is then released from the antibody-antigen complex and separated on sodium-dodecyl-sulfate–polyacrylamide gel electrophoresis (SDS-PAGE) and then immunoblotted for the epitope on the second receptor. A direct interaction between the receptors is likely if both receptors can be coprecipitated with each other. This conclusion, however, must be interpreted with caution because adapter proteins could be involved in linking the two receptors together. In addition, artifactual aggregation between receptors is a potential problem that can occur when receptors are solubilized by detergents. When removed from the lipid environment of the plasma membrane, the GPCRs' hydrophobic nature endows them with the propensity to nonspecifically aggregate, possibly leading to erroneous conclusions. This can be carefully controlled by mixing cells that independently express each of the receptors.

In heterologous cell lines, nonspecific interactions may also occur in the cell as a result of overexpression of the two receptors of interest (Ramsay et al. 2002; Salim et al. 2002). As a result, it is more reliable, albeit more difficult, to coimmunoprecipitate receptors from native tissues where they are expressed endogenously at physiological levels.

Ultimately, the conclusions that can be drawn from coimmunoprecipitation experiments are limited. One can be confident that two receptors that coimmunoprecipitate with each other exist as a complex; however, to show a direct interaction would require further analysis using alternative techniques.

Cooperativity of Ligand Binding: Novel Pharmacology

Some of the earlier radioligand-binding studies performed on GPCRs showed that binding of a ligand to one site could increase or decrease the binding affinity of ligands to other sites, termed, respectively, positive and negative cooperativity. This phenomenon has often been used to suggest the presence of GPCR oligomers. Such work has recently been expanded to show that oligomers may not only modulate the pharmacological properties of individual receptors, but, in addition, alter their G-protein-coupling specificity (Levac et al. 2002). Some of the first evidence to show distinct pharmacological changes in hetero-oligomers came from studies performed with δ- and κ-opioid receptors (Jordan and Devi 1999). In these studies, coexpression of these two opioid receptor subtypes led to a reduction in binding affinity to individual δ and κ receptor-selective ligands; however, simultaneous treatment with δ- and κ-selective ligands resulted in a synergistic increase in high-affinity binding to the δ/κ hetero-oligomer. This observation was associated with enhanced effector activation as measured by cyclic AMP (cAMP) inhibition and mitogen-activated protein kinase (MAPK) phosphorylation (Jordan and Devi 1999), suggesting functional modulation by the hetero-oligomer. In a separate study, hetero-oligomers of the μ and δ receptor subtypes showed ligand affinity and potency properties distinct from each of the individually expressed receptor subtypes (George et al. 2000). Furthermore, the hetero-oligomer had novel G-protein-coupling features, indicative of the generation of a novel signaling unit; this is discussed in further detail later in this chapter. The inference from these pharmacological data that hetero-oligomers show novel binding properties unique from the receptor homo-oligomers has also been shown in other GPCRs (Maggio et al. 1999; Gines et al. 2000; Gomes et al. 2000; Rocheville et al. 2000; Armstrong and Strange 2001; Galvez et al. 2001).

Receptor Trafficking Studies

GPCR oligomerization has often been shown to be prerequisite to cell-surface trafficking of many receptors (Jones et al. 1998; Kaupmann et al. 1998; White et al. 1998; Nelson et al. 2001; Uberti et al. 2003, 2004; Balasubramanian et al. 2004; Hague et al. 2004a,b). When expressed alone in heterologous cell lines, these receptors are intracellularly retained; however, when coexpressed with another receptor subtype, cell-surface expression is restored. This effect can be mediated by a number of different mechanisms, including the masking of endoplasmic reticulum (ER) retention motifs on the intracellularly localized receptor or the dissociation of ER retention proteins (Hurtley et al. 1989; Zhang et al. 1997; Kapoor et al. 2003). In contrast, specific receptor variants have been shown to exert trafficking inhibition of the cognate wild-type receptor (Benkirane et al. 1997; Grosse et al. 1997; Zhu and Wess 1998; Le Gouill et al. 1999; Elmhurst et al. 2000; Karpa et al. 2000; Lee et al. 2000b). These mutants can be naturally occurring splice variants or genetically modified receptors that are typically nonfunctional. Both of these opposing phenomena support the concept that oligomerization occurs during transport through the distal secretory pathway and possibly even earlier, as in the ER where biosynthesis occurs (Fig. 1). It also suggests that significant functional consequences occur through these intermolecular interactions. For example, receptor variants of the D2 dopamine receptor, generated by mutagenesis of specific transmembrane residues, have been shown to inhibit cell-surface expression of the wild-type D2 receptor when coexpressed (Lee et al. 2000b). This suggests that certain oligomeric configurations are not compatible for transport to the plasma membrane.

The naturally occurring variant of the D3 dopamine receptor, D3nf, is an example of the physiological role that oligomerization may have in modulating receptor trafficking (Liu et al. 1994). In certain cell types, this truncated variant is retained intracellularly and shows very little cell-surface expression (Karpa et al. 2000). When coexpressed with the wild-type D3 receptor, it sequestered the wild-type receptor in the cell, thus attenuating its function (Elmhurst et al. 2000; Karpa et al. 2000). This has been suggested to occur by a direct receptor interaction and provides further evidence for constitutive GPCR homo-oligomerization.

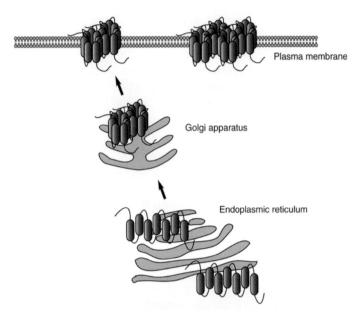

FIGURE 1. The majority of G-protein-coupled receptors have been shown to form oligomers constitutively before cell-surface expression, most probably in the endoplasmic reticulum. Maturation and transport of the oligomer occur through the distal secretory pathway where the oligomer is then targeted to the plasma membrane for functional activity.

Biophysical Techniques

One of the major shortcomings of studying GPCR oligomerization through the aforementioned biochemical and pharmacological approaches is the inability to detect oligomers physiologically in live cells. However, an increasing number of investigators are turning to biophysical methods to show that specific GPCRs are in close proximity to one another (Angers et al. 2000; Cheng and Miller 2001; Kroeger et al. 2001; McVey et al. 2001; Jensen et al. 2002; Overton and Blumer 2002; Canals et al. 2003). Two of these approaches, fluorescence resonance energy transfer (FRET) and bioluminescence resonance energy transfer (BRET), are based on the nonradiative transfer of energy between closely associated proteins (Bouvier 2001; Eidne et al. 2002). In FRET, the protein that transmits the signal and the protein that receives the signal are attached to donor and acceptor fluorophores, respectively. In BRET, the donor protein is fused to an enzyme called luciferase, derived from the sea pansy *Renilla reniformis*; the acceptor protein is attached to a variant of green fluorescent protein (GFP) from the jellyfish *Aequorea victoria*. Luciferase catalytically degrades the substrate, coelenterazine, which produces a bioluminescent signal that is transferred to the GFP acceptor. Energy transfer between donor and acceptor in FRET and BRET is critically dependent on the proximity between the proteins involved. Both methods require that proteins be within 100 Å (10 nm) of each other, a distance that would implicate a direct protein–protein interaction. In addition, both approaches rely on sufficient overlap between the emission and excitation spectra of the donor and acceptor fluorophores, respectively. Although these methods allow noninvasive detection of receptor oligomerization in live cells, their utility has two limitations. First, an increase in energy

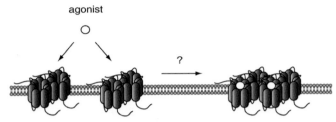

FIGURE 2. Ligand activation of constitutively formed GPCR oligomers may cause higher-order oligomers to form at the cell surface. Many studies that report this show an increase in energy transfer (using FRET or BRET) between receptors when exposed to the ligand. However, it is difficult to ascertain whether this is caused by ligand-induced oligomerization or ligand-induced conformational changes that bring the attached fluorophores in closer proximity to one another.

transfer efficiency, as may occur on ligand induction, can be interpreted either as enhanced oligomerization or simply a conformational change within the oligomer that may only bring the fluorophores in close proximity to one another (Fig. 2). The current challenge is differentiating between these two possibilities. Second, it is difficult to identify the precise cellular location at which oligomers are detected, particularly using BRET. Fortunately, some progress has been made in this regard using a variation of FRET (called time-resolved FRET), that substitutes the use of fluorescent fusion proteins with antibodies conjugated to fluorescent molecules (McVey et al. 2001; Gazi et al. 2003). Because antibodies are unable to permeate the plasma membrane, any detected energy transfer signal can be predicted to originate only from receptors at the cell surface. This advance can be used to detect the presence of GPCR oligomers or any ligand-induced changes in the conformation of oligomers, specifically at the cell surface.

In addition to measuring the proximity of receptors to one another, BRET has also been used to determine the affinity among receptors within a homo-oligomer or hetero-oligomer. Homo-oligomers of either the β_2-adrenergic receptor or κ-opioid receptor formed with high affinity because increased expression of receptor acceptor levels caused a linear increase in energy transfer (Ramsay et al. 2002). However, the β_2-adrenergic receptor and κ-opioid receptor were found to form hetero-oligomers less efficiently because overexpression of both receptors to nonphysiological levels was required to achieve energy transfer efficiency similar to that of homo-oligomers. This indicates that specificity and preferred interactions occur between receptors and that oligomerization is not a result of random collisions or nonspecific interactions in the cell.

Atomic Force Microscopy

Perhaps the most direct and convincing evidence to date for the existence of GPCR oligomers comes from their direct visualization by atomic force microscopy (AFM) studies by Liang et al. (2003). AFM is a technique used to study surface morphology and the properties of molecules, often at atomic resolution. The microscope is equipped with a cantilever attached to a tiny tip or probe that scans the surface of a specimen. In conjunction with a laser beam that reflects off the cantilever to a detector, the cantilever is raster-scanned to produce an image (by a linked computer via the detector) that represents the surface topography of the specimen. In this study, the investigators used AFM to reveal that rhodopsin molecules are organized as rows of dimers from native rod outer-segment disk membranes. A model of rhodopsin oligomers was produced that accounted for constraints imposed by the AFM data as well as the crystallographic data of rhodopsin. This model revealed potential contact points within and between dimers, and predicted the strength of these interactions on the basis of these empirical data. It will be interesting to determine whether this model applies to other receptors as well, including those involved in reward pathways; this will need to be proven using similar studies in brain.

Structural Features of GPCR Oligomers

Although the crystallization of bovine rhodopsin has provided detailed insight into the intramolecular intricacies of GPCR monomers (Palczewski et al. 2000), the configuration of GPCR oligomers and the intermolecular contacts involved remain to be fully elucidated. Although the above AFM study represents the first empirical evidence that class A GPCRs have a dimeric organization, it remains to be determined whether GPCRs in other families have a similar organization. Increasing evidence shows that receptors from different families use different mechanisms of oligomerization owing to inherent structural differences. However, specific fundamental mechanisms appear to be shared among families. Some of these are described below.

Disulphide Bonds

Disulfide covalent linkages among cysteine groups are important to the intramolecular stability of a receptor. Most rhodopsin-like GPCRs contain a conserved disulfide bond that links the first and second extracellular loops (Strader et al. 1994). However, it has been suggested that these cysteines may also participate in oligomer formation. Site-directed mutagenesis of these specific cysteines in the M3 muscarinic-acetylcholine receptor has been shown to yield only monomeric receptors (Zeng and Wess 1999). In addition, oligomers of the prostacyclin (Giguere et al. 2004), D1 dopamine (Lee et al. 2000a), 5-hydroxytryptamine 1B, and 1D serotonin (Lee et al. 2000a) receptors dissociate into monomers when exposed to reducing agents such as β-mercaptoethanol (2-ME) or dithiothreitol (DTT). It remains unclear whether this bridge participates in oligomer formation through a direct or indirect interaction. It is possible that this intramolecular disulfide bridge is not directly involved in maintaining interactions between receptors; instead, disruption may cause structural instability within the receptor that may hinder its ability to form oligomers.

A direct role of disulfide bonds in dimerization has been established in the metabotropic glutamate receptor (mGluR1) revealed by the crystal structure of its extracellular-binding domain (EBD) (Kunishima et al. 2000). This receptor was found to exist as a dimer in its resting and ligand-occupied state. Furthermore, a single disulfide bridge was shown to connect Cys-140 of each monomer within the dimeric complex. This residue cannot act as a structural scaffold because of its position within a disordered segment of the EBD, suggesting that other domains are critical for dimer assembly. Nevertheless, the role of disulfide bonds in the homodimerization of family 3 GPCRs within the EBD appears to be well conserved because mGluR5 (Romano et al. 1996) and the calcium-sensing receptor (Bai et al. 1998) also form disulfide linked dimers. These can be dissociated into their respective monomers by treatment with reducing agents.

Transmembrane Interactions

The oligomerization of many single-membrane-spanning proteins has been documented to involve their hydrophobic transmembrane domains with both structural and functional consequences (Furthmayr and Marchesi 1976; Manolios et al. 1990; Cosson et al. 1991). A notable example is the ligand-dependent dimerization of the epidermal growth factor receptor, which depends on these α-helical interactions for signal transduction (Ullrich and Schlessinger 1990). It has also been shown that single-membrane-spanning human glycophorin A (GpA) forms SDS-resistant dimers that appear to depend on a specific amino-acid motif (Lemmon et al. 1992). Evidence implicating the GpA motif in dimerization has been shown using the α-factor receptor from the yeast *Saccharomyces cerevisiae* (Overton and Blumer 2002; Overton et al. 2003). More recently, transmembrane domain 4 was found to form part of the dimerization interface of D2 dopamine receptors (Guo et al. 2003; Lee et al. 2003). However, disruption of this domain in the full-length D2 receptor did not affect the dimerization interface as extensively as in a similarly disrupted truncated version of the receptor (Lee et al. 2003). Thus, it is likely that oligomerization of the D2 receptor involves multiple sites of interaction. Packing models derived from the AFM studies (described earlier) on native rhodopsin predicted that the intradimeric interaction sites involved transmembranes 4 and 5, and the interdimeric interaction sites involved transmembranes 1 and 2 and the third cytoplasmic loop (Liang et al. 2003; Carrillo et al. 2004). This supports the model that multiple points of contact, particularly in the transmembrane regions, are involved in holding receptors in an oligomeric complex.

Intracellular and Extracellular Domain Interactions

Although the prevalent mode of oligomerization in some receptors belonging to family 3 GPCRs implicates disulfide bonding at the amino terminus (Tsuji et al. 2000), it is unknown whether this region is equally important in other GPCR families. The amino terminus has been implicated in agonist-induced oligomerization of the B2 bradykinin receptor, a family 1 GPCR, despite the absence of a large extracellular-binding domain (AbdAlla et al. 1999).

The carboxyl terminus has been reported as a site of intermolecular interaction in the $GABA_B$ receptor, a member of the family 3 GPCRs (Jones et al. 1998; Kaupmann et al. 1998; White et al. 1998). The mechanisms underlying $GABA_B$ receptor oligomerization are particularly interesting, because functional activity requires the interaction between two $GABA_B$ receptor subtypes, $GABA_BR1$ and $GABA_BR2$. Before the cloning of the $GABA_BR2$ subtype, it was known that $GABA_BR1$ failed to show any cell-surface expression in heterologous cell lines (Couve et al. 1998). The expression of the $GABA_BR2$ recombinant protein, however, resulted in normal cell-surface targeting with ligand-binding deficits, rendering it nonfunctional. Coexpression of both receptor subtypes, however, restored functional activity that was comparable to the natively expressed $GABA_B$ receptor. Subsequent studies have shown that specific regions of each receptor subtype contribute to the overall function of the hetero-oligomer, and that hetero-oligomerization is, in part, facilitated by a coiled-coil interaction involving the carboxyl tails of $GABA_BR2$ and $GABA_BR1$ (Margeta-Mitrovic et al. 2000, 2001; Calver et al. 2001; Galvez et al. 2001; Robbins et al. 2001; Carrillo et al. 2003). The coiled-coil interaction allows $GABA_BR2$ to mask an ER retention motif on $GABA_BR1$, thereby allowing the complex to traffic to the cell surface (Margeta-Mitrovic et al. 2000).

Agonist-induced Activation of GPCR Oligomers

GABA$_B$ receptors are largely found in the ventral tegmental area of the brain where they control firing of dopaminergic neurons. There is growing interest in the development of GABA$_B$ agonists to treat addictions to various drugs, particularly those that involve psycho-stimulants. The heterodimeric nature of the GABA$_B$ receptor is prerequisite for facilitating GABA-mediated activation. The initial observations that the GABA$_B$R1 subunit was required for ligand binding and the GABA$_B$R2 subunit was required for cell-surface expression suggested that each of these protomers had distinct nonredundant roles in the dimer. Indeed, it was shown that the GABAB$_R$2 subunit is sufficient for G-protein coupling (Galvez et al. 2001; Margeta-Mitrovic et al. 2001; Robbins et al. 2001; Duthey et al. 2002) and that activation of the G protein occurs in *trans* through binding of the GABA$_B$R1 subunit (Galvez et al. 2001). The effectiveness of each of these functional roles, however, is enhanced by its interaction with the other subunit. Hence, GABAB$_R$1 improves coupling efficiency of GABAB$_R$2 and GABAB$_R$2 improves the binding affinity of GABA$_B$R1, suggesting allosteric modulation between subunits within the dimer (Galvez et al. 2001). In agreement with these findings, it was also shown that strategically positioned disulfide bonds within the EBD of GABA$_B$R1 could generate a fully active GABA$_B$ receptor with a response magnitude comparable to the agonist-occupied receptor. Similarly positioned disulfide bonds in GABA$_B$R2 did not have an effect, indicating that the closed "active" conformation of the EBD of GABA$_B$R1 is sufficient to induce agonist-independent activation of the GABA$_B$ receptor. Thus, within a GABA$_B$ receptor, binding of the agonist to one protomer triggers closing of the large EBD; the resulting conformational change may alter the orientation of the G-protein-coupled protomer, resulting in activation and G-protein dissociation.

In contrast to this, ligand binding to one protomer of the homodimeric metabotropic glutamate receptor results in only partial activation of the receptor, whereas binding to both protomers results in full activation (Kniazeff et al. 2004). This coincides with the requirement of both EBDs to be in the closed "active" state for complete activity. In addition, whereas the GABA$_B$ receptor depends primarily on the GABA$_B$R2 protomer for activation, the mGluR5 receptor depends equally on both protomers for activation. This may imply that G-protein coupling to both protomers in the mGluR5 receptor, in contrast to binding to the single protomer in the GABA$_B$ receptor, is necessary for activation.

The scenario in which a single dimer couples to a single G protein would agree with the predicted stoichiometry of a rhodopsin dimer interacting with G$_t$ (transducin). In the resting state of rhodopsin, both G$_t$α and G$_t$γ interact with helix 8 of rhodopsin (Ernst et al. 2000; Marin et al. 2000). However, the crystal structure of transducin shows that these G-protein subunits are a relatively distant 42 Å apart from each other (Hamm 2001). These contact requirements would be satisfied by a dimeric model of rhodopsin that has been confirmed by rhodopsin AFM studies, as described previously. Furthermore, this 1:1 stoichiometry of the receptor dimer to the G-protein heterotrimer has been shown in mass spectrometry studies of the leukotriene B4 receptor, a class A rhodopsin-like GPCR (Baneres and Parello 2003). Activation of a single protomer within this functional pentameric complex has been shown to result in allosteric modulation of the partner protomer, similar to that observed with the family 3 GPCRs (Mesnier and Baneres 2004). It will be of interest to determine whether this receptor–G-protein arrangement and activation mechanism is conserved across the entire GPCR family.

GPCR Oligomers in Native Brain Tissue

One of the essential steps that must be taken when studying the oligomerization of GPCRs is establishing their physiological relevance in native tissues. It is important to realize that receptors that interact in heterologous cell lines may not do so in vivo because they normally do not colocalize in specific regions of the brain and thus have different expression profiles. In addition, mounting evidence suggests that a certain level of specificity exists within hetero-oligomers and that receptors, especially those from different families, do not interact in a promiscuous manner (Salim et al. 2002). For example, D1 and D2 dopamine receptors are capable of forming a mixed population of oligomers (Lee et al. 2004). The precise ratio of homo- to hetero-oligomers is governed by regulatory mechanisms in the neuron that may not exist in heterologous cell lines in which they would not normally be found.

The current methods of oligomer detection in brain are typically performed using immunological techniques such as Western blotting and immunoprecipitation. The detection of A1 adenosine receptors (A1R) in pig brain cortex was among the first of the reports showing GPCR dimers in brain. In this particular study, putative monomers and dimers were found to coexist (Ciruela et al. 1995). This study was subsequently expanded to show, through coimmunoprecipitation methods, that A1R and metabotropic glutamate 1α receptors formed hetero-oligomers in rat cerebellum (Ciruela et al. 2001). These data complement immunohistochemical studies in cerebellum that showed that both of these receptor subtypes colocalized in the dendritic tree of Purkinje and basket cells, as well as in large pyramidal cells of the human cerebral cortex (Ciruela et al. 2001). Several of the dopamine receptor subtypes have also been shown to exist as dimers and higher-order oligomers in brain. The D3 dopamine receptor and its naturally occurring truncation variant, D3nf, were shown to form homo- and hetero-oligomers in human brain (Nimchinsky et al. 1997). In a separate study, photoaffinity labeling of D2 dopamine receptors derived from human caudate nucleus revealed the presence of monomeric and dimeric D2 receptors. Through similar methods of detection, D2 dimers and higher-order oligomers were also found in rat striatum tissue (Zawarynski et al. 1998). Some of these receptor–receptor interactions are sensitive to reducing agents, suggesting the involvement of disulfide bonds (Zeng and Wess 1999; Gama et al. 2001), although noncovalent interactions among transmembrane domains may be the dominant interaction that holds the oligomer together. Dimers of the M3 muscarinic receptor (Zeng and Wess 1999) and heterodimers of the calcium-sensing receptor and metabotropic glutamate 1α receptor (Gama et al. 2001), from rat brain and bovine brain, respectively, have been shown to dissociate into monomeric subunits after DTT or β-mercaptoethanol treatment. Finally, the two subunits of the heteromeric $GABA_B$ receptor ($GABA_BR1$ and $GABA_BR2$) have been shown to coimmunoprecipitate in cortical membrane preparations (Kaupmann et al. 1998), implicating the physiological requirement of oligomeric assembly in presynaptic and postsynaptic function.

RELEVANCE TO SUBSTANCE ABUSE AND ADDICTION

Receptor cross talk has been traditionally thought to occur at the level of effector activation and second messenger production (Cordeaux and Hill 2002). For example, at the effector level, G_q-activated receptors can enhance G_s stimulation of adenylyl cyclase through G_q βγ subunits. At the second messenger level, G_q-activated phospholipase C triggers the release of calcium, which can act on specific G_s-associated adenylyl cyclase isoforms, causing cAMP accumulation (Defer et al. 2000). GPCR oligomerization presents new opportunities for refining the mechanism of various signal transduction pathways. Homo-oligomerization of receptors is suggestive of a signal amplification mechanism in which a single agonist molecule can have a physiological effect comparable to multiple agonist molecules. Hetero-oligomerization may be a means of facilitating cross talk among different receptor subtypes. This may result in the coupling of signaling pathways to increase the repertoire of ligands to which a receptor can respond or create novel signaling complexes to enhance pharmacological diversity. In any case, it appears that communication within a cell includes not only the network of intracellular molecules regulated by GPCRs but the interactions among the receptors themselves at the cell surface.

Receptor–receptor Cross Modulation and Signal Amplification

Behavioral sensitization is a dominant feature in drug addiction and is characterized by increases in a behavioral response to a given dose of drug after repeated treatments. Psychostimulants, such as cocaine and amphetamine, have often been implicated in the development of behavioral sensitization in which they increase synaptic levels of various neurotransmitters, including dopamine and serotonin, to trigger activation of reward pathways in the brain.

In search of the functional implications of GPCR oligomers, growing evidence now shows that interactions between GPCRs support cross talk at the level of the receptor itself, with behavioral consequences in sensitization. In functional studies of μ-opioid receptor/CCR5 chemokine receptor (Chen et al. 2004), μ-opioid receptor/somatostatin 2A receptor (Pfeiffer et al. 2002), and adenosine A_{2A}/D2 dopamine receptor (Hillion et al. 2002), hetero-oligomers have established that various cellular effects including G-protein uncoupling, phosphorylation, and desensitization can be conferred from one receptor subunit to another. These studies indicate that oligomerization can mediate *trans*-activation between two receptors in which selective activation of one receptor can activate neighboring receptors within an oligomeric complex (Fig. 3) (Pfeiffer et al. 2002; Carrillo et al. 2003; Chen et al. 2004). In the case of homo-oligomers, this mechanism may be a means of amplifying a response to a given agonist. In a physiological context, the response elicited by a given concentration of neurotransmitter in the synapse could be amplified through oligomeric receptors that are involved in reward, such as the dopamine or serotonin receptor subtypes. Such an efficient mechanism could enhance sensitization to a particular drug. In the case of hetero-oligomers, *trans*-activation may allow agonists to nondirectly activate receptors to which they normally would not bind. For example, cAMP inhibition by a somatostatin agonist is restored when a functionally deficient (but binding-competent) somatostatin-5 receptor (SSTR5) mutant is coexpressed with the D2 receptor (Rocheville et al. 2000). This suggests that D2-receptor-mediated cAMP inhibition can occur with somatostatin activation via the D2 interaction with SSTR5. Furthermore, if the wild-type hetero-oligomer is pretreated with a D2 agonist, the affinity to a SSTR5 agonist is substantially increased. Taken together, these observations support previous reports that somatostatinergic neurotransmission is modulated by dopamine in the brain (Izquierdo-Claros et al. 1997; Rodriguez-Sanchez et al. 1997) and suggests a mechanism of cross sensitization

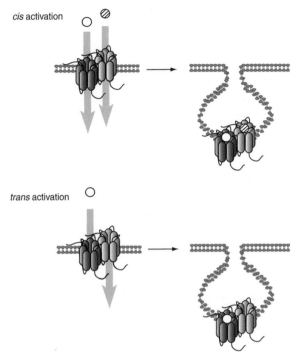

FIGURE 3. Activation of GPCR oligomers may occur in *cis* and/or in *trans* (as illustrated in the figure). *Cis*-activation occurs when a signal is transduced through the protomer to which the agonist was bound. *Trans*-activation occurs when the protomer that transduces the signal is distinct from the protomer that binds the agonist, thereby implicating direct receptor–receptor modulation. The receptors subsequently internalize into endosomes as an oligomeric complex.

in which behavioral sensitization through, for example, the D2 receptor can be initiated through nondopamine receptor agonists such as somatostatin.

Receptor Turnover and Tolerance

Acute activation of GPCRs by agonists is usually followed by functional uncoupling of the heterotrimeric G protein and termination of agonist action through desensitization of the receptor via receptor phosphorylation. The receptors are subsequently internalized and interacellularly sorted to be either recycled back to the cell surface or targeted for lysosomal degradation, resulting in down-regulation. This cellular control of receptor turnover functions to prevent overstimulation of receptors caused by prolonged agonist exposure. During chronic drug abuse, neuronal adaptations that involve rewiring of reward circuits in the brain, termed synaptic plasticity, can often modulate the response to these psychostimulants, frequently leading to symptoms such as sensitization and tolerance. These behavioral changes are important in the development of addiction. In contrast to traditional desensitization mechanisms, activation of the μ-opioid receptor (μOR) by morphine does not induce receptor internalization (Whistler et al. 1999). This is coupled with continued propagation of the cell-surface-activated signal without desensitization. Morphine's unusual inability to promote desensitization and endocytosis of μOR is directly linked to its ability to induce tolerance and dependence. Hence, endocytosis of the μOR is essential in alleviating these hallmarks of addiction. In a study by He et al. (2002), a traditional μOR agonist, [D-Ala^2-N-Me-Phe^4-Gly^5-ol]-enkephalin (DAMGO), was shown to facilitate morphine-induced internalization of μOR through receptor oligomerization. μORs exist as oligomers; when a subanalgesic dose of DAMGO was coadministered with an analgesic dose of morphine, DAMGO-occupied receptors could "drag" or cointernalize with morphine-occupied receptors into the cell. This produced the desired dual effect of morphine-induced analgesia and tolerance reduction of acute morphine treatment in mice. Furthermore, a low dose of DAMGO reduced cAMP superactivation following chronic morphine administration in rats. Superactivation of cAMP results from compensatory up-regulation of the cAMP pathway during sustained inhibition of cAMP production, as occurs during chronic opiate use (Sharma et al. 1975). This up-regulated system contributes to symptoms of opioid deficiency manifest as cellular withdrawal (Nestler and Aghajanian 1997); this, however, can be attenuated by inducing endocytosis of μOR (Finn and Whistler 2001). As a result, cointernalization of DAMGO and morphine-occupied μORs using this pharmacological regimen may also be used to alleviate the symptoms of opiate withdrawal. Collectively, this shows that oligomerization can induce changes in receptor turnover and that receptor oligomers can be pharmacologically manipulated to achieve a desired behavioral effect.

In contrast to the μOR response to morphine, desensitization and endocytosis processes directly contribute to the tolerance observed after chronic activation of other GPCRs. These include those that are involved in neurotransmission, such as the dopamine and adenosine receptor subtypes. A number of studies have shown that alterations in the sensitivity of D2 receptors modulate the locomotor-activating effects of the A_{2A} adenosine receptor antagonist, caffeine (Garrett and Holtzman 1994; Fenu et al. 2000). The oligomerization of adenosine A_{2A} ($A_{2A}R$) and dopamine D2 (Hillion et al. 2002) receptors has been suggested to mediate codesensitization and cointernalization of these receptors. Both of these effects may form a molecular mechanism to describe the cross tolerance observed between caffeine and different D2 agonists (Garrett and Holtzman 1994).

These data illustrate the different patterns of receptor turnover that can be achieved by manipulation of protomers within an oligomer. This may produce profound changes in the various neurochemical and behavioral responses to drugs of abuse.

Novel Signaling Complexes

One of the most exciting consequences of GPCR oligomerization is the prospect of generating novel G-protein interactions and thus novel signaling complexes (George et al. 2000; Lee et al. 2004; Kearn et al. 2005). These additional signal transduction pathways may reveal previously unappreciated modes of neurotransmission that may have a role in addiction.

The D1 and D2 dopamine receptors are fundamentally involved in the brain reward pathways that mediate substance abuse. Although they have an opposing ability to regulate cAMP through adenylyl cyclase, biochemical and electrophysiological evidence indicates that they act synergistically and are functionally linked. For example, coactivation of both receptors is required for potentiation of early gene expression (Gerfen et al. 1995; Svenningsson et al. 2000) and restoration of long-term synaptic depression (Calabresi et al. 1992). Furthermore, concurrent treatments with D1 and D2 agonists are required to elicit behavioral phenotypes of cocaine sensitization (Capper-Loup et al. 2002) and reward (Ikemoto et al. 1997) in rats. How can the activation of two functionally opposing receptors generate significant cellular and behavioral changes? This paradox was recently addressed by Lee et al. (2004) who showed that D1 and D2 receptors form hetero-oligomers in rat brain and that coactivation of both receptors results in the generation of a novel signaling unit characterized by increases in intracellular calcium. This change is shown to occur as a result of novel effector coupling of the hetero-oligomer to phospholipase C and implicates novel G-protein coupling that is unique from each of the individually functioning homo-oligomers (reviewed in Pollack 2004).

In support of this paradigm, μ- and δ-opioid receptor hetero-oligomers have shown pharmacological properties distinct from each receptor subtype, suggesting receptor cooperativity (George et al. 2000), as described earlier. Furthermore, in contrast to the individually expressed μ- and δ-opioid receptors, this complex was insensitive to pertussis toxin, a poison that prevents release of GDP from G_i, thereby keeping the receptor in an inactive state. This implies the formation of a unique hetero-oligomeric signaling unit and may explain the loss of some addiction phenotypes specific to the δ-opioid receptor that are observed in μ-opioid receptor knockout mice and vice versa (Sora et al. 1997; Matthes et al. 1998; Zhu et al. 1999; Hosohata et al. 2000; Hutcheson et al. 2001). Furthermore, such functionally and pharmacologically novel interactions may help to explain why only three opioid receptor genes have been cloned despite the existence of multiple pharmacologically defined opioid receptor subtypes (Jordan et al. 2000; Law et al. 2000).

In this context, hetero-oligomerization may be a resourceful means of using a common receptor for multiple cellular roles, thereby increasing its versatility and usefulness.

SUMMARY, CONCLUSION, AND FUTURE DIRECTIONS

The superfamily of GPCRs are important role players in the etiology of addiction. Many substances of abuse elicit their rewarding properties in the brain through activation of these receptors, either directly or indirectly. Much progress has been made in revealing the ubiquity of GPCR oligomerization and the various functional and pharmacological properties that are a consequence of this. However, many other functional and structural aspects of GPCR oligomers still remain to be elucidated. Because many of the current methods used to establish GPCR oligomerization are indirect, the conclusions that can be drawn from them are limited unless they are used collectively. Some fundamental questions remain, however, such as which regions of the receptor participate in the oligomerization interface and how the dimeric functional unit is activated, forms larger oligomeric complexes, and is functionally regulated. It will be important to refine and add to these current techniques to address these important questions.

An abundance of reports on hetero-oligomerization of distinct receptor subtypes have appeared, mostly within the same GPCR family. In determining whether a bona fide receptor interaction exists, it is important to establish if this occurs at endogenous expression levels in tissue. It should be emphasized that the physiological rationale of studying the interaction of two receptors is central to the study of GPCR hetero-oligomers. The expression of receptors in heterologous cell lines may be useful in dissecting out oligomerization contact points or other receptor domains involved in oligomer function. However, because they may not possess the full complement of proteins required for receptor function and often because of overexpression, they alone do not establish that a specific functional hetero-oligomer exists in nature.

The prevailing view that GPCRs structurally and functionally exist as oligomers may have profound implications regarding their responses to various drugs. The structural details of a GPCR oligomer in its basal and activated state will aid in the development of novel therapeutics that target these oligomers. Indeed, the use of bivalent ligands, which has already been shown to cause changes in receptor affinity and potency (Halazy et al. 1996; Dupuis et al. 1999), may be a useful tool in specifically coactivating receptors or modulating the properties of individual protomers within a dimer. The activation of GPCR oligomers may result in cross talk between receptors to regulate the magnitude of an intracellular signal or control cell-surface receptor turnover. This may have a number of effects, including exacerbating sensitization or modulating tolerance to drugs of abuse. Hetero-oligomerization may serve to couple receptors to new signaling pathways that may be involved in the rewarding properties of addiction. This potential for the generation of different signaling complexes through oligomerization presents novel substrates for drug action, including those that are involved in brain reward. Hence, GPCR oligomerization adds another dimension to the regulation of signal transduction and will provide further insight into the molecular mechanisms underlying many biological processes such as addiction.

ACKNOWLEDGMENTS

Research in our laboratory is supported by the National Institute on Drug Abuse, the National Institutes of Health, and the Canadian Institutes of Health Research. S.R.G. is the holder of a Canada Research Chair in Molecular Neuroscience.

REFERENCES

AbdAlla S., Zaki E., Lother H., and Quitterer U. 1999. Involvement of the amino terminus of the B(2) receptor in agonist-induced receptor dimerization. *J. Biol. Chem.* **274:** 26079–26084.

Angers S., Salahpour A., Joly E., Hilairet S., Chelsky D., Dennis M., and Bouvier M. 2000. Detection of beta 2-adrenergic receptor dimerization in living cells using bioluminescence resonance energy transfer (BRET). *Proc. Natl. Acad. Sci.* **97:** 3684–3689.

Armstrong D. and Strange P.G. 2001. Dopamine D2 receptor dimer formation: Evidence from ligand binding. *J. Biol. Chem.* **276:** 22621–22629.

Bai M., Trivedi S., and Brown E.M. 1998. Dimerization of the extracellular calcium-sensing receptor (CaR) on the cell surface of CaR-transfected HEK293 cells. *J. Biol. Chem.* **273:** 23605–23610.

Bakker R.A., Dees G., Carrillo J.J., Booth R.G., Lopez-Gimenez J.F., Milligan G., Strange P.G., and Leurs R. 2004. Domain swapping in the human histamine H1 receptor. *J. Pharmacol. Exp. Ther.* **311:** 131–138.

Balasubramanian S., Teissere J.A., Raju D.V., and Hall R.A. 2004. Hetero-oligomerization between GABA_A and GABA_B receptors regulates GABA_B receptor trafficking. *J. Biol.* *Chem.* **279:** 18840–18850.

Baneres J.L. and Parello J. 2003. Structure-based analysis of GPCR function: Evidence for a novel pentameric assembly between the dimeric leukotriene B4 receptor BLT1 and the G-protein. *J. Mol. Biol.* **329:** 815–829.

Benkirane M., Jin D.Y., Chun R.F., Koup R.A., and Jeang K.T. 1997. Mechanism of transdominant inhibition of CCR5-mediated HIV-1 infection by ccr5Δ32. *J. Biol. Chem.* **272:** 30603–30606.

Bouvier M. 2001. Oligomerization of G-protein-coupled transmitter receptors. *Nat. Rev. Neurosci.* **2:** 274–286.

Calabresi P., Maj R., Mercuri N.B., and Bernardi G. 1992. Coactivation of D1 and D2 dopamine receptors is required for long-term synaptic depression in the striatum. *Neurosci. Lett.* **142:** 95–99.

Calver A.R., Robbins M.J., Cosio C., Rice S.Q., Babbs A.J., Hirst W.D., Boyfield I., Wood M.D., Russell R.B., Price G.W., Couve A., Moss S.J., and Pangalos M.N. 2001. The C-terminal domains of the GABA(b) receptor subunits mediate intracellular trafficking but are not required for receptor signaling. *J. Neurosci.* **21:** 1203–1210.

Canals M., Marcellino D., Fanelli F., Ciruela F., de Benedetti P., Goldberg S.R., Neve K., Fuxe K., Agnati L.F., Woods A.S.,

Ferre S., Lluis C., Bouvier M., and Franco R. 2003. Adenosine A2A-dopamine D2 receptor-receptor heteromerization: Qualitative and quantitative assessment by fluorescence and bioluminescence energy transfer. *J. Biol. Chem.* **278:** 46741–46749.

Capper-Loup C., Canales J.J., Kadaba N., and Graybiel A.M. 2002. Concurrent activation of dopamine D1 and D2 receptors is required to evoke neural and behavioral phenotypes of cocaine sensitization. *J. Neurosci.* **22:** 6218–6227.

Carrillo J.J., Lopez-Gimenez J.F., and Milligan G. 2004. Multiple interactions between transmembrane helices generate the oligomeric α_{1b}-adrenoceptor. *Mol. Pharmacol.* **66:** 1123–1137.

Carrillo J.J., Pediani J., and Milligan G. 2003. Dimers of class A G protein-coupled receptors function via agonist-mediated *trans*-activation. *J. Biol. Chem.* **278:** 42578–42587.

Chen C., Li J., Bot G., Szabo I., Rogers T.J., and Liu-Chen L.Y. 2004. Heterodimerization and cross-desensitization between the mu-opioid receptor and the chemokine CCR5 receptor. *Eur. J. Pharmacol.* **483:** 175–186.

Cheng Z.J. and Miller L.J. 2001. Agonist-dependent dissociation of oligomeric complexes of G protein-coupled cholecystokinin receptors demonstrated in living cells using bioluminescence resonance energy transfer. *J. Biol. Chem.* **276:** 48040–48047.

Ciruela F., Casado V., Mallol J., Canela E.I., Lluis C., and Franco R. 1995. Immunological identification of A1 adenosine receptors in brain cortex. *J. Neurosci. Res.* **42:** 818–828.

Ciruela F., Escriche M., Burgueno J., Angulo E., Casado V., Soloviev M.M., Canda E.I., Mallol J., Chen W.Y., Lluis C., McIlhinney R.A., and Franco R. 2001. Metabotropic glutamate 1alpha and adenosine A1 receptors assemble into functionally interacting complexes. *J. Biol. Chem.* **276:** 18345–18351.

Cordeaux Y. and Hill S.J. 2002. Mechanisms of cross-talk between G-protein-coupled receptors. *Neurosignals* **11:** 45–57.

Cosson P., Lankford S.P., Bonifacino J.S., and Klausner R.D. 1991. Membrane protein association by potential intramembrane charge pairs. *Nature* **351:** 414–416.

Couve A., Filippov A.K., Connolly C.N., Bettler B., Brown D.A., and Moss S.J. 1998. Intracellular retention of recombinant GABA$_B$ receptors. *J. Biol. Chem.* **273:** 26361–26367.

Defer N., Best-Belpomme M., and Hanoune J. 2000. Tissue specificity and physiological relevance of various isoforms of adenylyl cyclase. *Am. J. Physiol. Renal Physiol.* **279:** F400–F416.

Dupuis D.S., Perez M., Halazy S., Colpaert F.C., and Pauwels P.J. 1999. Magnitude of 5-HT1B and 5-HT1A receptor activation in guinea-pig and rat brain: Evidence from sumatriptan dimer-mediated [35S]GTPgammaS binding responses. *Brain Res. Mol. Brain Res.* **67:** 107–123.

Duthey B., Caudron S., Perroy J., Bettler B., Fagni L., Pin J.P., and Prezeau L. 2002. A single subunit (GB2) is required for G-protein activation by the heterodimeric GABA(B) receptor. *J. Biol. Chem.* **277:** 3236–3241.

Eidne K.A., Kroeger K.M., and Hanyaloglu A.C. 2002. Applications of novel resonance energy transfer techniques to study dynamic hormone receptor interactions in living cells. *Trends Endocrinol. Metab.* **13:** 415–421.

Elmhurst J.L., Xie Z., O'Dowd B.F., and George S.R. 2000. The splice variant D3nf reduces ligand binding to the D3 dopamine receptor: Evidence for heterooligomerization. *Brain Res. Mol. Brain Res.* **80:** 63–74.

Ernst O.P., Meyer C.K., Marin E.P., Henklein P., Fu W.Y., Sakmar T.P., and Hofmann K.P. 2000. Mutation of the fourth cytoplasmic loop of rhodopsin affects binding of transducin and peptides derived from the carboxyl-terminal sequences of transducin alpha and gamma subunits. *J. Biol. Chem.* **275:** 1937–1943.

Fenu S., Cauli O., and Morelli M. 2000. Cross-sensitization between the motor activating effects of bromocriptine and caffeine: Role of adenosine A(2A) receptors. *Behav. Brain Res.* **114:** 97–105.

Finn A.K. and Whistler J.L. 2001. Endocytosis of the mu opioid receptor reduces tolerance and a cellular hallmark of opiate withdrawal. *Neuron* **32:** 829–839.

Furthmayr H. and Marchesi V.T. 1976. Subunit structure of human erythrocyte glycophorin A. *Biochemistry* **15:** 1137–1144.

Galvez T., Duthey B., Kniazeff J., Blahos J., Rovelli G., Bettler B., Prezeau L., and Pin J.P. 2001. Allosteric interactions between GB1 and GB2 subunits are required for optimal GABA(B) receptor function. *EMBO J.* **20:** 2152–2159.

Gama L., Wilt S.G., and Breitwieser G.E. 2001. Heterodimerization of calcium sensing receptors with metabotropic glutamate receptors in neurons. *J. Biol. Chem.* **276:** 39053–39059.

Garrett B.E. and Holtzman S.G. 1994. Caffeine cross-tolerance to selective dopamine D1 and D2 receptor agonists but not to their synergistic interaction. *Eur. J. Pharmacol.* **262:** 65–75.

Gazi L., Lopez-Gimenez J.F., Rudiger M.P., and Strange P.G. 2003. Constitutive oligomerization of human D2 dopamine receptors expressed in Spodoptera frugiperda 9 (Sf9) and in HEK293 cells. Analysis using co-immunoprecipitation and time-resolved fluorescence resonance energy transfer. *Eur. J. Biochem.* **270:** 3928–3938.

George S.R., Fan T., Xie Z., Tse R., Tam V., Varghese G., and O'Dowd B.F. 2000. Oligomerization of mu- and delta-opioid receptors. Generation of novel functional properties. *J. Biol. Chem.* **275:** 26128–26135.

Gerfen C.R., Keefe K.A., and Gauda E.B. 1995. D1 and D2 dopamine receptor function in the striatum: Coactivation of D1- and D2-dopamine receptors on separate populations of neurons results in potentiated immediate early gene response in D1-containing neurons. *J. Neurosci.* **15:** 8167–8176.

Giguere V., Gallant M.A., de Brum-Fernandes A.J., and Parent J.L. 2004. Role of extracellular cysteine residues in dimerization/oligomerization of the human prostacyclin receptor. *Eur. J. Pharmacol.* **494:** 11–22.

Gines S., Hillion J., Torvinen M., Le Crom S., Casado V., Canela E.I., et al. 2000. Dopamine D1 and adenosine A1 receptors form functionally interacting heteromeric complexes. *Proc. Natl. Acad. Sci.* **97:** 8606–8611.

Gomes I., Jordan B.A., Gupta A., Trapaidze N., Nagy V., and Devi L.A. 2000. Heterodimerization of mu and delta opioid receptors: A role in opiate synergy. *J. Neurosci.* **20:** RC110.

Gouldson P.R., Higgs C., Smith R.E., Dean M.K., Gkoutos G.V., and Reynolds C.A. 2000. Dimerization and domain swapping in G-protein-coupled receptors: A computational study. *Neuropsychopharmacology* **23:** S60–S77.

Grosse R., Schoneberg T., Schultz G., and Gudermann T. 1997. Inhibition of gonadotropin-releasing hormone receptor signaling by expression of a splice variant of the human receptor. *Mol. Endocrinol.* **11:** 1305–1318.

Guo W., Shi L., and Javitch J.A. 2003. The fourth transmembrane segment forms the interface of the dopamine D2 receptor homodimer. *J. Biol. Chem.* **278:** 4385–4388.

Hague C., Uberti M.A., Chen Z., Hall R.A., and Minneman K.P. 2004a. Cell surface expression of alpha1D-adrenergic receptors is controlled by heterodimerization with alpha1B-adrenergic receptors. *J. Biol. Chem.* **279:** 15541–15549.

Hague C., Uberti M.A., Chen Z., Bush C.F., Jones S.V., Ressler K.J., Hall R.A., and Minneman K.P. 2004b. Olfactory receptor surface expression is driven by association with the beta2-adrenergic receptor. *Proc. Natl. Acad. Sci.* **101:** 13672–13676.

Halazy S., Perez M., Fourrier C., Pallard I., Pauwels P.J., Palmier C., John G.W., Valentin J.P., Bonnafous P., and Martinez J. 1996. Serotonin dimers: Application of the bivalent ligand approach to the design of new potent and selective 5-HT(1B/1D) agonists. *J. Med. Chem.* **39:** 4920–4927.

Hamm H.E. 2001. How activated receptors couple to G proteins. *Proc. Natl. Acad. Sci.* **98:** 4819–4821.

He L., Fong J., von Zastrow M., and Whistler J.L. 2002. Regulation of opioid receptor trafficking and morphine tolerance by receptor oligomerization. *Cell* **108:** 271–282.

Heldin C.H. 1995. Dimerization of cell surface receptors in signal transduction. *Cell* **80:** 213–223.

Hillion J., Canals M., Torvinen M., Casado V., Scott R., Terasmaa A., et al. 2002. Coaggregation, cointernalization, and code-sensitization of adenosine A2A receptors and dopamine D2 receptors. *J. Biol. Chem.* **277:** 18091–18097.

Hosohata Y., Vanderah T.W., Burkey T.H., Ossipov M.H., Kovelowski C.J., Sora I., et al. 2000. delta-Opioid receptor agonists produce antinociception and [35S]GTPgammaS binding in mu receptor knockout mice. *Eur. J. Pharmacol.* **388:** 241–248.

Hurtley S.M., Bole D.G., Hoover-Litty H., Helenius A., and Copeland C.S. 1989. Interactions of misfolded influenza virus hemagglutinin with binding protein (BiP). *J. Cell. Biol.* **108:** 2117–2126.

Hutcheson D.M., Matthes H.W., Valjent E., Sanchez-Blazquez P., Rodriguez-Diaz M., Garzon J., Kieffer B.l., and Maldonado R. 2001. Lack of dependence and rewarding effects of deltorphin II in mu-opioid receptor-deficient mice. *Eur. J. Neurosci.* **13:** 153–161.

Ikemoto S,. Glazier B.S., Murphy J.M., and McBride W.J. 1997. Role of dopamine D1 and D2 receptors in the nucleus accumbens in mediating reward. *J. Neurosci.* **17:** 8580–8587.

Izquierdo-Claros R.M., Boyano-Adanez M.C., Larsson C., Gustavsson L., and Arilla E. 1997. Acute effects of D1- and D2-receptor agonist and antagonist drugs on somatostatin binding, inhibition of adenylyl cyclase activity and accumulation of inositol 1,4,5-trisphosphate in the rat striatum. *Brain Res. Mol. Brain Res.* **47:** 99–107.

Jensen A.A., Hansen J.L., Sheikh S.P., and Brauner-Osborne H. 2002. Probing intermolecular protein-protein interactions in the calcium-sensing receptor homodimer using bioluminescence resonance energy transfer (BRET). *Eur. J. Biochem.* **269:** 5076–5087.

Jones K.A., Borowsky B., Tamm J.A., Craig D.A., Durkin M.M., Dai M., et al. 1998. GABA(B) receptors function as a heteromeric assembly of the subunits GABA(B)R1 and GABA(B)R2. *Nature* **396:** 674–679.

Jordan B.A. and Devi L.A. 1999. G-protein-coupled receptor heterodimerization modulates receptor function. *Nature* **399:** 697–700.

Jordan B.A., Cvejic S., and Devi L.A. 2000. Opioids and their complicated receptor complexes. *Neuropsychopharmacology* **23:** S5–S18.

Kapoor M., Srinivas H., Kandiah E., Gemma E., Ellgaard L., Oscarson S., Helenius A., and Surolia A. 2003. Interactions of substrate with calreticulin, an endoplasmic reticulum chaperone. *J. Biol. Chem.* **278:** 6194–6200.

Karpa K.D., Lin R., Kabbani N., and Levenson R. 2000. The dopamine D3 receptor interacts with itself and the truncated D3 splice variant d3nf: D3-D3nf interaction causes mislocalization of D3 receptors. *Mol. Pharmacol.* **58:** 677–683.

Kaupmann K., Malitschek B., Schuler V., Heid J., Froestl W., Beck P., et al. 1998. GABA(B)-receptor subtypes assemble into functional heteromeric complexes. *Nature* **396:** 683–687.

Kearn C.S., Blake-Palmer K., Daniel E., Mackie K., and Glass M. 2005. Concurrent stimulation of cannabinoid CB1 and dopamine D2 receptors enhances heterodimer formation: A mechanism for receptor crosstalk? *Mol. Pharmacol.* **67:** 1697–1704.

Kniazeff J., Bessis A.S., Maurel D., Ansanay H., Prezeau L., and Pin J.P. 2004. Closed state of both binding domains of homodimeric mGlu receptors is required for full activity. *Nat. Struct. Mol. Biol.* **11:** 706–713.

Kroeger K.M., Hanyaloglu A.C., Seeber R.M., Miles L.E., and Eidne K.A. 2001. Constitutive and agonist-dependent homo-oligomerization of the thyrotropin-releasing hormone receptor. Detection in living cells using bioluminescence resonance energy transfer. *J. Biol. Chem.* **276:** 12736–12743.

Kunishima N., Shimada Y., Tsuji Y., Sato T., Yamamoto M., Kumasaka T., Nakanishi S., Jingami H., and Monkawa K. 2000. Structural basis of glutamate recognition by a dimeric metabotropic glutamate receptor. *Nature* **407:** 971–977.

Law P.Y., Wong Y.H., and Loh H.H. 2000. Molecular mechanisms and regulation of opioid receptor signaling. *Annu. Rev. Pharmacol. Toxicol.* **40:** 389–430.

Lee S.P., O'Dowd B.F., Rajaram R.D., Nguyen T., and George S.R. 2003. D2 dopamine receptor homodimerization is mediated by multiple sites of interaction, including an intermolecular interaction involving transmembrane domain 4. *Biochemistry* **42:** 11023–11031.

Lee S.P., Xie Z., Varghese G., Nguyen T., O'Dowd B.F., and George S.R. 2000a. Oligomerization of dopamine and sero-tonin receptors. *Neuropsychopharmacology* **23:** S32–S40.

Lee S.P., O'Dowd B.F., Ng G.Y., Varghese G., Akil H., Mansour A., Nguyen T., and George S.R. 2000b. Inhibition of cell sur-face expression by mutant receptors demonstrates that D2 dopamine receptors exist as oligomers in the cell. *Mol. Pharmacol.* **58:** 120–128.

Lee S.P., So C.H., Rashid A.J., Varghese G., Cheng R., Lanca A.J., O'Dowd B.F., and George S.R. 2004. Dopamine D1 and D2 receptor co-activation generates a novel phospholipase C-mediated calcium signal. *J. Biol. Chem.* **279:** 35671–35678.

Le Gouill C., Parent J.L., Caron C.A., Gaudreau R., Volkov L., Rola-Pleszczynski M., and Stankova J. 1999. Selective mod-ulation of wild type receptor functions by mutants of G-pro-tein-coupled receptors. *J. Biol. Chem.* **274:** 12548–12554.

Lemmon M.A., Flanagan J.M., Hunt J.F., Adair B.D., Bormann B.J., Dempsey C.E., and Engelman D.M. 1992. Glycophorin A dimerization is driven by specific interactions between trans-membrane alpha-helices. *J. Biol. Chem.* **267:** 7683–7689.

Levac B.A., O'Dowd B.F., and George S.R. 2002. Oligomerization of opioid receptors: Generation of novel signaling units. *Curr. Opin. Pharmacol.* **2:** 76–81.

Liang Y., Fotiadis D., Filipek S., Saperstein D.A., Palczewski K., and Engel A. 2003. Organization of the G protein-coupled receptors rhodopsin and opsin in native membranes. *J. Biol. Chem.* **278:** 21655–21662.

Liu K., Bergson C., Levenson R., and Schmauss C. 1994. On the origin of mRNA encoding the truncated dopamine D3-type receptor D3nf and detection of D3nf-like immunoreactivity in human brain. *J. Biol. Chem.* **269:** 29220–29226.

Maggio R., Vogel Z., and Wess J. 1993. Coexpression studies with mutant muscarinic/adrenergic receptors provide evi-dence for intermolecular "cross-talk" between G-protein-linked receptors. *Proc. Natl. Acad. Sci.* **90:** 3103–3107.

Maggio R., Barbier P., Colelli A., Salvadori F., Demontis G., and Corsini G.U. 1999. G protein-linked receptors: Pharma-cological evidence for the formation of heterodimers. *J. Pharmacol. Exp. Ther.* **291:** 251–257.

Manolios N., Bonifacino J.S., and Klausner R.D. 1990. Transmembrane helical interactions and the assembly of the T cell receptor complex. *Science* **249:** 274–277.

Margeta-Mitrovic M., Jan Y.N., and Jan L.Y. 2000. A trafficking checkpoint controls GABA(B) receptor heterodimerization. *Neuron* **27:** 97–106.

———. 2001. Function of GB1 and GB2 subunits in G protein coupling of GABA(B) receptors. *Proc. Natl. Acad. Sci.* **98:** 14649–14654.

Marin E.P., Krishna A.G., Zvyaga T.A., Isele J., Siebert F., and Sakmar T.P. 2000. The amino terminus of the fourth cytoplas-mic loop of rhodopsin modulates rhodopsin-transducin interaction. *J. Biol. Chem.* **275:** 1930–1936.

Matthes H.W.D., Smadja C., Valverde O., Vonesch J.L., Foutz A.S., Boudinot E., et al. 1998. Activity of the δ-opioid recep-tor is partially reduced, whereas activity of the κ-receptor is maintained in mice lacking the μ-receptor. *J. Neurosci.* **18:**

7285–7295.

McVey M., Ramsay D., Kellett E., Rees S., Wilson S., Pope A.J., and Milligan G. 2001. Monitoring receptor oligomerization using time-resolved fluorescence resonance energy transfer and bioluminescence resonance energy transfer. The human delta-opioid receptor displays constitutive oligomerization at the cell surface, which is not regulated by receptor occupan-cy. *J. Biol. Chem.* **276:** 14092–14099.

Mesnier D. and Baneres J.L. 2004. Cooperative conformational changes in a G-protein-coupled receptor dimer, the leukotriene B(4) receptor BLT1. *J. Biol. Chem.* **279:** 49664–49670.

Monnot C., Bihoreau C., Conchon S., Curnow K.M., Corvol P., and Clauser E. 1996. Polar residues in the transmembrane domains of the type 1 angiotensin II receptor are required for binding and coupling. Reconstitution of the binding site by co-expression of two deficient mutants. *J. Biol. Chem.* **271:** 1507–1513.

Nelson G., Hoon M.A., Chandrashekar J., Zhang Y., Ryba N.J., and Zuker C.S. 2001. Mammalian sweet taste receptors. *Cell* **106:** 381–390.

Nestler E.J. and Aghajanian G.K. 1997. Molecular and cellular basis of addiction. *Science* **278:** 58–63.

Nimchinsky E.A., Hof P.R., Janssen W.G., Morrison J.H., and Schmauss C. 1997. Expression of dopamine D3 receptor dimers and tetramers in brain and in transfected cells. *J. Biol. Chem.* **272:** 29229–29237.

Overton M.C. and Blumer K.J. 2002. The extracellular N-termi-nal domain and transmembrane domains 1 and 2 mediate oligomerization of a yeast G protein-coupled receptor. *J. Biol. Chem.* **277:** 41463–41472.

Overton M.C., Chinault S.L., and Blumer K.J. 2003. Oligomerization, biogenesis, and signaling is promoted by a glycophorin A-like dimerization motif in transmembrane domain 1 of a yeast G protein-coupled receptor. *J. Biol. Chem.* **278:** 49369–49377.

Palczewski K., Kumasaka T., Hori T., Behnke C.A., Motoshima H., Fox B.A., et al. 2000. Crystal structure of rhodopsin: A G protein-coupled receptor. *Science* **289:** 739–745.

Pfeiffer M., Koch T., Schroder H., Laugsch M., Hollt V., and Schulz S. 2002. Heterodimerization of somatostatin and opi-oid receptors cross-modulates phosphorylation, internaliza-tion, and desensitization. *J. Biol. Chem.* **277:** 19762–19772.

Pierce K.L., Premont R.T., and Lefkowitz R.J. 2002. Seven-trans-membrane receptors. *Nat. Rev. Mol. Cell. Biol.* **3:** 639–650.

Pollack A. 2004. Coactivation of D1 and D2 dopamine recep-tors: In marriage, a case of his, hers, and theirs. *Sci. STKE* **2004:** pe50.

Ramsay D., Kellett E., McVey M., Rees S., and Milligan G. 2002. Homo- and hetero-oligomeric interactions between G-pro-tein-coupled receptors in living cells monitored by two vari-ants of bioluminescence resonance energy transfer (BRET): Hetero-oligomers between receptor subtypes form more effi-ciently than between less closely related sequences. *Biochem. J.* **365:** 429–440.

Robbins M.J., Calver A.R., Filippov A.K., Hirst W.D., Russell

R.B., Wood M.D., et al. 2001. GABA(B2) is essential for G-protein coupling of the GABA(B) receptor heterodimer. *J. Neurosci.* **21:** 8043–8052.

Rocheville M., Lange D.C., Kumar U., Patel S.C., Patel R.C., and Patel Y.C. 2000. Receptors for dopamine and somatostatin: Formation of hetero-oligomers with enhanced functional activity. *Science* **288:** 154–157.

Rodriguez-Sanchez M.N., Puebla L., Lopez-Sanudo S., Rodriguez-Martin E., Martin-Espinosa A., Rodriguez-Pena M.S., Juarranz M.G., and Arilla E. 1997. Dopamine enhances somatostatin receptor-mediated inhibition of adenylate cyclase in rat striatum and hippocampus. *J. Neurosci. Res.* **48:** 238–248.

Romano C., Yang W.L., and O'Malley K.L. 1996. Metabotropic glutamate receptor 5 is a disulfide-linked dimer. *J. Biol. Chem.* **271:** 28612–28616.

Salim K., Fenton T., Bacha J., Urien-Rodriguez H., Bonnert T., Skynner H.A., et al. 2002. Oligomerization of G-protein-coupled receptors shown by selective co-immunoprecipitation. *J. Biol. Chem.* **277:** 15482–15485.

Scarselli M., Novi F., Schallmach E., Lin R., Baragli A., Colzi A., et al. 2001. D2/D3 dopamine receptor heterodimers exhibit unique functional properties. *J. Biol. Chem.* **276:** 30308–30314.

Schulz A., Grosse R., Schultz G., Gudermann T., and Schoneberg T. 2000. Structural implication for receptor oligomerization from functional reconstitution studies of mutant V2 vasopressin receptors. *J. Biol. Chem.* **275:** 2381–2389.

Sharma S.K., Klee W.A., and Nirenberg M. 1975. Dual regulation of adenylate cyclase accounts for narcotic dependence and tolerance. *Proc. Natl. Acad. Sci.* **72:** 3092–3096.

Sora I., Funada M., and Uhl G.R. 1997. The mu-opioid receptor is necessary for [D-Pen2,D-Pen5]enkephalin-induced analgesia. *Eur. J. Pharmacol.* **324:** R1–R2.

Strader C.D., Fong T.M., Tota M.R., Underwood D., and Dixon R.A. 1994. Structure and function of G protein-coupled receptors. *Annu. Rev. Biochem.* **63:** 101–132.

Svenningsson P., Fredholm B.B., Bloch B., and Le Moine C. 2000. Co-stimulation of D(1)/D(5) and D(2) dopamine receptors leads to an increase in c-fos messenger RNA in cholinergic interneurons and a redistribution of c-fos messenger RNA in striatal projection neurons. *Neuroscience* **98:** 749–757.

Tsuji Y., Shimada Y., Takeshita T., Kajimura N., Nomura S., Sekiyama N., Otomo J., Usukura J., Nakanishi S., and Jingami J. 2000. Cryptic dimer interface and domain organization of the extracellular region of metabotropic glutamate receptor subtype 1. *J. Biol. Chem.* **275:** 28144–28151.

Uberti M.A., Hall R.A., and Minneman K.P. 2003. Subtype-specific dimerization of alpha 1-adrenoceptors: Effects on receptor expression and pharmacological properties. *Mol. Pharmacol.* **64:** 1379–1390.

Uberti M.A., Hague C., Oller H., Minneman K.P., and Hall R.A. 2004. Heterodimerization with beta-2-adrenergic receptors promotes surface expression and functional activity of alpha-1D-adrenergic receptors. *J. Pharmacol. Exp. Ther.* **313:** 16–23.

Ullrich A. and Schlessinger J. 1990. Signal transduction by receptors with tyrosine kinase activity. *Cell* **61:** 203–212.

Vassilatis D.K., Hohmann J.G., Zeng H., Li F., Ranchalis J.E., Mortrud M.T., et al. 2003. The G protein-coupled receptor repertoires of human and mouse. *Proc. Natl. Acad. Sci.* **100:** 4903–4908.

Whistler J.L., Chuang H.H., Chu P., Jan L.Y., and von Zastrow M. 1999. Functional dissociation of mu opioid receptor signaling and endocytosis: Implications for the biology of opiate tolerance and addiction. *Neuron* **23:** 737–746.

White J.H., Wise A., Main M.J., Green A., Fraser N.J., Disney G.H., Barnes A.A., Emson P., Foord S.M., and Marshall F.H. 1998. Heterodimerization is required for the formation of a functional GABA(B) receptor. *Nature* **396:** 679–682.

Xie Z., Lee S.P., O'Dowd B.F., and George S.R. 1999. Serotonin 5-HT1B and 5-HT1D receptors form homodimers when expressed alone and heterodimers when co-expressed. *FEBS Lett.* **456:** 63–67.

Zawarynski P., Tallerico T., Seeman P., Lee S.P., O'Dowd B.F., and George S.R. 1998. Dopamine D2 receptor dimers in human and rat brain. *FEBS Lett.* **441:** 383–386.

Zeng F.Y. and Wess J. 1999. Identification and molecular characterization of m3 muscarinic receptor dimers. *J. Biol. Chem.* **274:** 19487–19497.

Zhang J.X., Braakman I., Matlack K.E., and Helenius A. 1997. Quality control in the secretory pathway: The role of calreticulin, calnexin and BiP in the retention of glycoproteins with C-terminal truncations. *Mol. Biol. Cell.* **8:** 1943–1954.

Zhu X. and Wess J. 1998. Truncated V2 vasopressin receptors as negative regulators of wild-type V2 receptor function. *Biochemistry* **37:** 15773–15784.

Zhu Y., King M.A., Schuller A.G., Nitsche J.F., Reidl M,. Elde R.P., Unterwald E., Pasternak G.W., and Pintar J.E. 1999. Retention of supraspinal delta-like analgesia and loss of morphine tolerance in delta opioid receptor knockout mice. *Neuron* **24:** 243–252.

16 | The Critical Role of Adenosine A2a Receptors and Gi βγ Subunits in Alcoholism and Addiction: From Cell Biology to Behavior

Ivan Diamond[1–5] and Lina Yao[1,2,5]

[1]Ernest Gallo Clinic and Research Center, Emeryville, California 94608; Departments of [2]Neurology, [3]Cellular and Molecular Pharmacology, and the [4]Neuroscience Graduate Program, University of California, San Francisco, California 94110; [5]CV Therapeutics, Department of Neuroscience, Palo Alto, California 94304

ABSTRACT

This chapter summarizes emerging information about the role of adenosine and $G_{i/o}$ βγ stimulated cAMP signal transduction in the pathophysiology of alcoholism and addiction. We describe a novel role for adenosine A2a receptors in mediating responses to ethanol and Gi βγ subunits in mediating responses to dopamine D2, opiate, and cannabinoid receptors. Each activates cyclic AMP (cAMP) production, protein kinase A (PKA) translocation, and cAMP responsive element (CRE)-mediated gene expression. We also find synergy between these receptors for activation of cAMP/PKA signaling. Synergy requires adenosine and A2a receptors and is mediated by $G_{i/o}$ βγ stimulation of adenylyl cyclase II and IV. Indeed, βγ subunits appear to be required for alcohol drinking and reinstatement of heroin-seeking behavior.

INTRODUCTION

Adenosine Mediates Neural Responses to Ethanol

Adenosine is a global inhibitory neuromodulator in the brain; several reviews provide extensive information about adenosine signaling in the nervous system (Dunwiddie and Masino 2001; Latini and Pedata 2001; Ribeiro et al. 2003). Extracellular concentration of adenosine in the brain increases with neural activity and the physiological effects of adenosine are terminated, in part, by reuptake into the cell. Ethanol inhibits adenosine reuptake (Nagy et al. 1990) and accumulating evidence from several laboratories, including our own, implicates adenosine in mediating many of the neuronal and behavioral responses to ethanol (Dar and Clark 1992; Malec et al. 1996; Diamond and Gordon 1997; Barwick and Dar 1998; Jarvis and Becker 1998; Gatch et al. 1999; Kaplan et al. 1999; Dar 2001; El Yacoubi et al. 2001; Naassila et al. 2002; Mailliard and Diamond 2004).

GOALS OF THE CHAPTER

In this chapter, we present relevant background information and recent experimental data from the investigators' laboratory with relevance to ethanol, opiates, and cannabinoids, and we provide evidence that adenosine A2a antagonists could be novel therapeutic agents in the treatment of alcoholism and addiction.

THE UNDERLYING BIOLOGY OF ADDICTIVE DRUGS

Ethanol Metabolism in the Liver Modifies Acetate and Adenosine Metabolism in the Brain

Several studies have shown that extracellular adenosine can be increased in tissues and organs in vivo as a consequence of ethanol metabolism in the liver (Carmichael et al. 1987, 1988, 1991; Orrego et al. 1988a). Hepatic ethanol metabolism generates acetate via the action of alcohol and acetaldehyde dehydrogenase (Orrego et al. 1988b). Acetate is further metabolized to acetyl CoA, consuming ATP in the process; this generates adenosine (Israel et al. 1994). Adenosine released into the circulation from the liver freely crosses the blood–brain barrier (Cornford and Oldendorf 1975). In addition, acetate generated by alcohol metabolism in the liver is also released into the circulation where it too crosses the blood–brain barrier. Once in the brain, acetate is readily converted to acetyl CoA, consuming ATP in the process (Berl and Frigyesi 1969). Thus, acetate metabolism in the brain can generate adenosine in situ. Acetate, like adenosine, is a central nervous system (CNS) depressant. The effect of acetate in the brain appears to be mediated by adenosine because it is prevented by adenosine receptor inhibitors (Israel et al. 1994; Campisi et al. 1997).

Ethanol Inhibits Adenosine Reuptake

Not only can ethanol metabolism in the liver lead to increases in adenosine, ethanol also increases extracellular adenosine levels directly by blocking adenosine reuptake into cells (Nagy et al. 1990). This raises the possibility that ethanol inhibition of adenosine uptake could potentiate ethanol-induced increases in brain adenosine in the context of alcohol consumption (Mailliard and Diamond 2004). Extracellular adenosine levels are tightly regulated by the activity of nucleoside transporters, which mediate adenosine reuptake. In mammalian cells, there are two major classes of nucleoside transporters: (1) sodium-dependent active transporters that drive adenosine uptake against a concentration gradient and (2) facilitative transporters that enable adenosine to equilibrate across the membrane (Thorn and Jarvis 1996; Cass et al. 1998; Latini and Pedata 2001). Our studies suggest that ethanol inhibition of adenosine uptake via the equilibrative nucleoside transporter ENT1 (Cass et al. 1998) accounts exclusively for the ethanol-induced increase in extracellular adenosine in cultured cells (Nagy et al. 1990; Krauss et al. 1993).

Moreover, it appears that PKA-mediated phosphorylation of ENT1 and/or an associated regulatory protein might regulate the sensitivity of ENT1 to ethanol inhibition (Nagy et al. 1991; Coe et al. 1996b). Protein phosphatase activity, probably regulated by protein kinase C (PKC), appears to eliminate ethanol inhibition of adenosine uptake (Coe et al. 1996a). We cloned and characterized ENT1, but did not identify a specific phosphorylation site that might confer sensitivity to ethanol (Choi et al. 2000; Handa et al. 2001).

These results suggest, therefore, that a protein associated with ENT1 is likely to be the phosphorylated PKA substrate that confers sensitivity to ethanol inhibition of adenosine uptake. Studies in ENT1 null mice suggest that ENT1 may modulate ethanol intoxication and preference in mice (Choi et al. 2004).

Addictive Drugs Activate Nucleus Accumbens Neurons

Drug abuse is a worldwide, biomedical, public health problem (McIntire and Diamond 2002). A fundamental question in addiction biology is why all drugs of abuse cause addiction, even though each addictive drug produces a unique clinical response. This suggests that a molecular mechanism shared by addicting drugs in a specific brain region could contribute to the development of addiction.

The nucleus accumbens (NAc) is involved in reinforcement and reward. One clue toward identifying a common molecular mechanism in addiction is that all addicting drugs activate NAc neurons (Robbins and Everitt 1999) and increase dopamine levels (Imperato and DiChiara 1986; Weiss et al. 1993). Dopamine receptors (D2) in the NAc are particularly important for mediating the locomotor-activating and rewarding effects of ethanol (Phillips et al. 1998; Cunningham et al. 2000; Risinger et al. 2000) and drugs of abuse (Koob and Nestler 1997; Self 1998; Wise 1998). The postsynaptic δ-opioid receptor (DOR) (Svingos et al. 1996; Wang and Pickel 2001), μ-opioid receptor (MOR), and cannabinoid (CB1) receptors are also expressed in the NAc (Rodriguez et al. 2001; Wang and Pickel 2001). In addition, these receptors are coupled to $G\alpha_{i/o}$, like D2 (Nestler 2001).

Opioid peptides, like opiate drugs, have rewarding properties. Opiates bind to three opioid receptors, DOR, MOR, and the κ-opioid receptor (KOR), mimicking the actions of the endogenous opioid peptides, endorphins, endomorphins, enkephalins, and dynorphins. MOR is critical for the rewarding effects of heroin and morphine. Blockade of MOR attenuates opiate self-administration, and constitutive deletion of MOR attenuates conditioned place preference (CPP) (Negus et al. 1993; Matthes et al. 1996). Work that has taken place during the past several years has suggested that DOR is also involved in mediating the physical dependence that develops to morphine (Moriguchi et al. 1993). Selective blockade of either DOR or MOR is sufficient to induce conditioned aversive effects in morphine-dependent animals (Funada et al. 1996).

Although marijuana is widely abused, the mechanisms of its euphoriant and addictive effects remain unclear. THC, a psychoactive and addictive constituent of marijuana and hashish, also induces CPP in mice (Ghozland et al. 2002) and rats (Lepore et al. 1995). Cannabinoids activate various cellular effectors via CB1 and CB2 receptors. Only CB1 is found extensively throughout the mammalian CNS, including the NAc (Howlett 1995; Pertwee 1997; Ameri 1999). CB1 also participates in the rewarding effects of opiates, because both morphine self-administration (Ledent et al. 1999) and place preference (Martin et al. 2000) are decreased in mice lacking CB1 receptors. These findings suggest the existence of cross interactions between opioid and cannabinoid systems in addictive behavioral responses.

cAMP/PKA Signaling Is a Major Pathway in the Cellular Response to Addictive Drugs

One of the best-established molecular mechanisms of addiction is modulation of the cAMP-dependent PKA second messenger pathway, which occurs in many neuronal cell types, including NAc neurons (Khasar et al. 1995; Buzas et al. 1997; Glass and Felder 1997; Nestler and Aghajanian 1997). Our findings suggest that activation of the cAMP/PKA pathway is a major signaling response to ethanol in neural preparations. Prevailing current opinion is that acute opiate exposure inhibits adenylyl cyclase (AC), whereas chronic opiate exposure leads to a compensatory up-regulation of the cAMP pathway (Childers 1991). However, during the past few years, several groups have found that brief activation of DOR, MOR, and CB1 receptors actually increases cAMP production (Chan et al. 1995; Bilecki et al. 2000; Rubovitch et al. 2003). These observations are consistent with our finding that activation of DOR, CB1, and D2 increases cAMP levels at 10 minutes and CRE-mediated gene expression 5 hours later (Yao et al. 2002, 2003, 2005). The studies described below suggest that adenosine A2a receptors and $G_{i/o}$ βγ subunits are critical molecular components that enable ethanol, opiates, and cannabinoids to synergistically activate cAMP/PKA signaling in the NAc.

ETHANOL: METHODOLOGICAL CONSIDERATIONS

Ethanol-induced increases in extracellular adenosine in the brain, as a consequence of ethanol inhibition of adenosine uptake, activate adenosine receptors and promote receptor-coupled signal transduction and gene expression. This section summarizes the consequences of ethanol-induced adenosine A2 receptor activation. An overview is presented in Figure 1.

Adenosine Receptors

Adenosine receptors are activated by increases in extracellular adenosine. Adenosine A1 and Adenosine A3 receptors are coupled to the inhibitory G protein $G_{i/o}$, whereas A2 is

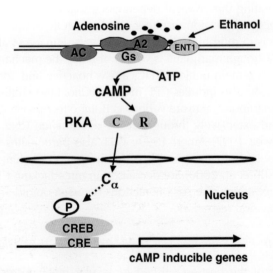

FIGURE 1. Ethanol activates PKA signaling via A2 receptors. Ethanol inhibits adenosine uptake, leading to activation of A2. This increases cAMP, causing PKA translocation and stimulation of cAMP-dependent gene expression.

coupled to the stimulatory G protein G_s (in most cells) or primarily G_{olf} in brain (Kull et al. 2000). Therefore, the complement of adenosine receptors on specific cells appears to determine the response of those cells to ethanol-induced increases in extracellular adenosine. For example, acute ethanol exposure increases cAMP levels in NG108-15 neural cells, which express only G_s-coupled A2 (Gordon et al. 1986; Nagy et al. 1989; Sapru et al. 1994). Under our conditions, these findings cannot be the result of reported direct ethanol activation of AC VII (Yoshimura and Tabakoff 1999), because ethanol stimulation of cAMP production is completely prevented by A2 receptor blockade. In contrast, hepatic cells that express $G_{i/o}$-coupled A1 exhibit a decrease in cAMP in response to acute ethanol (Nagy 1994).

The regional distribution of adenosine receptors in the brain appears to provide clues about the importance of certain adenosine receptors for some of the behavioral responses to ethanol. For example, A1 receptors are expressed abundantly in the cortex, cerebellum, hippocampus, and spinal cord (Ribeiro et al. 2003), and studies from Dar and colleagues (Barwick and Dar 1998; Dar 2001) suggest that A1 receptors mediate acute ethanol-induced incoordination. In contrast, the highest concentration of A2a is in the striatum and NAc (Jarvis et al. 1989; Parkinson and Fredholm 1990; Wan et al. 1990; Martinez-Mir et al. 1991; Nonaka et al. 1994; Svenningsson et al. 1997; Fredholm et al. 1998; Rosin et al. 1998; El Yacoubi et al. 2001). Mice lacking A2a exhibit increased alcohol consumption (Naassila et al. 2002). This may be related to their increased anxiety and aggressiveness (Ledent et al. 1997) as well as to pathophysiologic changes that occur as a result of developmental compensations in several neurotransmitter and hormone systems (Snell et al. 2000a,b; Dassesse et al. 2001; Berrendero et al. 2003; Wang et al. 2003). Clearly, further work is necessary to dissect out the direct role of the A2a receptor in regulating alcohol consumption from compensatory mechanisms as a consequence of gene deletion.

Ethanol Activates the cAMP/PKA Pathway

Ethanol inhibition of adenosine uptake leads to increases in extracellular adenosine with subsequent activation of A2 receptors in NG108-15 cells and primary neurons in culture. Activation of A2 receptors leads, in turn, to $G\alpha_s$-mediated stimulation of AC activity, cAMP production, and stimulation of cAMP-dependent PKA activity. PKA is a tetrameric holoenzyme consisting of two regulatory (R) and two catalytic (C) subunits. PKA is activated by cAMP binding to R, causing the release of C, which translocates to different sites in the cell and phosphorylates nearby substrates. Substrate specificity is largely determined by the subcellular localization of PKA achieved through interactions with the A-kinase-anchoring protein (AKAP) family that binds PKA (Lester and Scott 1997).

Several in vivo studies have demonstrated a robust link between alterations in the cAMP/PKA signaling pathway and sensitivity to and consumption of ethanol (Moore et al. 1998; Park et al. 2000; Thiele et al. 2000; Pandey et al. 2001b; Wand et al. 2001). Moore et al. (1998) identified a *Drosophila* mutant whose increased ethanol sensitivity appears to be attributed to decreased function of the cAMP/PKA signaling pathway. Further studies indicate that a PKA inhibitor only effects ethanol sensitivity when expressed in selected regions of fly brain (Rodan et al. 2002). Mice with a targeted disruption of the stimulatory G-protein allele or with reduced neuronal PKA activity are more sensitive to the sedative effects of ethanol and consume less alcohol. Additional studies with these mice suggest that the development of tolerance to ethanol involves cAMP/PKA signaling (Abel et al. 1997; Wand et al. 2001; Yang et al. 2003). Two major subtypes of PKA in neurons are RIIβ-Cα and RIβ-Cβ. Thiele et al. (2000) reported that mice lacking the RIIβ subunit of PKA consume more ethanol and are less sensitive to the sedative effects of ethanol. A *Drosophila* mutant deficient in RIIβ is also less sensitive to ethanol (Park et al. 2000); mice lacking the PKA RIβ regulatory subunit showed no change in ethanol consumption or sensitivity (Thiele et al. 2000).

The reason for discrepancies relating regulatory subunit deletions with alcohol drinking behavior is unclear, but may reflect differential compensation in genetically engineered rodents and flies. Nevertheless, it does seem, in general, that PKA modulates voluntary alcohol consumption, and that alcohol consumption is often but not always inversely related to sensitivity to alcohol intoxication. More recent findings suggest that activation of cAMP/PKA signaling is required for ethanol-induced increases in CRE-mediated cAMP-dependent gene expression (Thibault et al. 2000; Asher et al. 2002; Constantinescu et al. 2002; Hassan et al. 2003), whereas decreased cAMP signaling may induce alcohol withdrawal anxiety (Pandey et al. 2003). Our studies implicate cAMP/PKA signaling in a model of voluntary alcohol consumption (Yao et al. 2002), but this remains to be established via mechanistic intervention studies.

Ethanol Induces PKA Translocation

Important consequences of ethanol-induced increases in cAMP production in NG108-15 cells and primary neurons in culture are activation and translocation of PKA. Several years ago, we first reported sustained translocation of PKA Cα into the nucleus after chronic exposure to ethanol (Dohrman et al. 1996). It is now apparent that ethanol-induced PKA Cα translocation occurs in two distinct phases: an early and a late phase. These two phases are temporally distinct and independently regulated (Dohrman et al. 2002). The early phase of ethanol-induced PKA Cα translocation reaches its peak at 10 minutes and requires an increase in extracellular adenosine and activation of A2 receptors (Dohrman et al. 2002; Yao et al. 2002). In the continued presence of ethanol, however, PKA Cα exits the nucleus within 1 hour (Dohrman et al. 2002) and reenters the nucleus during the late phase several hours later. Exposing NG108-15 cells to ethanol for 2 days caused PKA Cα to remain in the nucleus as long as ethanol was present (Dohrman et al. 1996). Ethanol-induced nuclear Cα is functionally active, resulting in cAMP responsive element binding (CREB) protein phosphorylation. Ordinarily, translocation of PKA Cα to the nucleus is transient, and Cα exits the nucleus within minutes (Nigg et al. 1985). Export out of the nucleus is mediated by PKA Cα binding to the heat-stable PKA inhibitor PKI, which contains a nuclear export signal (Meinkoth et al. 1993; Fantozzi et al. 1994; Constantinescu et al. 1999). Persistent activation and localization of PKA Cα in the nucleus may be due, in part, to ethanol inhibition of Cα reassociation with R subunits, as well as direct ethanol interference with PKI (Constantinescu et al. 1999).

Ethanol, CREB Phosphorylation, and cAMP-dependent Gene Expression

Ethanol-induced PKA Cα localization in the nucleus appears to have pathophysiologic significance. PKA can increase gene expression by phosphorylating Ser-133 on CREB, an important molecular mechanism in memory and addiction (Berke and Hyman 2000; Nestler 2001). Phosphorylation of CREB initiates transcription at genes containing CRE. Using a luciferase reporter construct under CRE control, we found that ethanol promotes CREB-mediated cAMP-dependent gene expression in NG108-15 cells (Asher et al. 2002). Acute exposure to ethanol can increase CREB phosphorylation (Yang et al. 1996; Li et al. 2003b), whereas chronic exposure can decrease CREB phosphorylation (Yang et al. 1998; Pandey et al. 2001a, 2003; Li et al. 2003a,b) in brain. Microarray studies indicate that a high percentage of ethanol-responsive genes requires PKA (Thibault et al. 2000; Hassan et al. 2003). Ethanol-induced activation of A2 causing increases in cAMP and PKA Cα translocation in 10 minutes is followed by activation of cAMP-dependent gene expression 5 hours after ethanol is removed (Yao et al. 2002). This observation is consistent with current concepts that suggest that significant cAMP-dependent gene expression persists long after cAMP produced by receptor activation has been degraded (Schwartz 2001). Indeed, these observations may help to explain the recent finding that a single dose of ethanol potentiates GABAergic synaptic function for more than 7 days (Melis et al. 2002). Ethanol-induced potentiation in that study was PKA dependent and appeared to contribute to increased ethanol consumption.

We have presented evidence that ethanol-induced CREB phosphorylation and cAMP-dependent gene expression are differentially mediated by two PKA isoforms (Constantinescu et al. 2002). NG108-15 cells express type-II PKA consisting of RIIβ-Cα in the Golgi and type-I PKA consisting of RIβ-Cβ in the cytoplasm. On ethanol exposure, the type-II PKA subunits, RIIβ and Cα, translocate to the nucleus where Cα phosphorylates CREB on Ser-133, inducing cAMP-dependent gene expression (Asher et al. 2002; Yao et al. 2002). In contrast, the type-I PKA subunits, RIβ and Cβ, remain in the cytoplasm on ethanol exposure. Nevertheless, inhibition of type-I PKA prevents ethanol-induced cAMP-dependent gene expression, even though type-I PKA does not enter the nucleus and does not phosphorylate CREB. This indicates that in NG108-15 cells, ethanol-induced activation of CRE-mediated gene transcription requires the activation of both type-I and -II isoforms of PKA (Constantinescu et al. 2002). This is the first report of differential regulation of cAMP-dependent gene expression by these two isoforms of PKA. It appears that type-I PKA, activated by ethanol/A2, affects a downstream pathway in the cytoplasm that ultimately activates other transcription factors that promote CREB-mediated increases in cAMP-dependent gene expression (Constantinescu et al. 2004).

The Mesolimbic Dopamine System Is Activated during Addictive Behaviors

The mesolimbic dopamine system, which includes the ventral tegmental area (VTA) projecting to the NAc, is activated during addictive behaviors (Hyman and Malenka 2001) and is implicated in alcohol self-administration. Extensive evidence shows that dopaminergic mechanisms contribute to the reward and reinforcement of ethanol consumption (Phillips et al. 1998; Cunningham et al. 2000; Risinger et al. 2000). Indeed, alcohol-preferring rats self-administer ethanol directly into the VTA cell body region of mesolimbic dopamine neurons (Gatto et al. 1994). This is consistent with recent findings that ethanol can directly stimulate neurons in the VTA (Appel et al. 2003). Ethanol self-administration elevates extracellular dopamine in the NAc (Weiss et al. 1993, 1996; Weiss and Porrino 2002; Doyon et al. 2003), and ethanol self-administration can be increased or decreased by the microinjection of D2 agonists or antagonists into the NAc (Hodge et al. 1997; Cohen et al. 1998; Czachowski et al. 2001).

Short-term Activation of D2 Receptors Mimics the Effect of Ethanol by Paradoxically Increasing cAMP/PKA Signaling

The NAc/striatum contains the highest levels of A2a in the CNS (Jarvis et al. 1989; Parkinson and Fredholm 1990; Wan et al. 1990; Martinez-Mir et al. 1991; Nonaka et al. 1994; Svenningsson et al. 1997; Fredholm et al. 1998; Rosin et al. 1998; El Yacoubi et al. 2001). An intriguing observation is that medium spiny neurons in the NAc/striatum are characterized by the postsynaptic coexpression of A2a with D2 receptors on the same cells (Fink et al. 1992). The relationship between the A2 and D2 signaling systems is ordinarily antagonistic. Activated A2a receptors promote the release of $G\alpha_s$, which stimulates AC activity. In contrast, activation of D2 receptors releases $G\alpha_i$, which inhibits AC and decreases intracellular cAMP levels (Obadiah et al. 1999). The importance of A2a/D2 antagonism has been related to the close physical association and interaction of these receptors with one another (Schiffmann et al. 1991; Ferre et al. 1993, 1994; Franco et al. 2000). We investigated the relationship of D2 to ethanol/A2 activation to cAMP/PKA signaling in NG108-15/D2 cells, a stable cell line we isolated expressing functional D2 at physiological concentrations (Asai et al. 1998; Gordon et al. 2001). When studied in detail as a function of time, we found an unexpected biphasic response to D2 receptor activation in the absence of ethanol. Thus, NPA, a D2 agonist, increased cAMP production and induced PKA $C\alpha$ translocation at 10 minutes (Fig. 2) even though cAMP levels were reduced at 30 minutes (Yao et al. 2002).

A 10-minute exposure to NPA also increased cAMP-dependent gene expression 5 hours later, mimicking the response of these cells to ethanol (Yao et al. 2002). The D2 antagonist spiperone inhibited NPA-induced activation of PKA and cAMP-dependent gene expression (Yao et al. 2002). This indicates that NPA activation of cAMP/PKA signaling was specific for D2, even though D2 is a $G_{i/o}$-coupled receptor. These results were not found with the M4 muscarinic cholinergic (M4) or α2b adrenergic receptors also coupled to $G_{i/o}$ in the same cells. Activation of M4 and α2b only inhibited cAMP production (Yao et al. 2003). We do not understand why D2 receptors activate cAMP/PKA signaling and M4 and α2b receptors do not.

$G_{i/o}$-coupled D2 Receptors Increase cAMP/PKA Signaling Via $\beta\gamma$ Subunits

$G_{i/o}$ is a trimeric protein consisting of α and $\beta\gamma$ subunits. $G\alpha_{i/o}$ inhibits AC activity. However, $\beta\gamma$ dimers released from $G_{i/o}$ can potentiate cAMP production by stimulating AC isoforms II and IV (Tang and Gilman 1991; Federman et al. 1992; Inglese et al. 1994; Baker et al. 1999). ACII is widely expressed in brain, including the striatum/NAc (Mons et al. 1995; Matsuoka et al. 1997). We determined that NG108-15/D2 cells possess AC II and IV and are therefore capable of responding to $\beta\gamma$ dimers (Yao et al. 2002). Indeed, incubation with the $G_{i/o}$ inhibitor pertussis toxin (PTX) prevents D2 activation of PKA-mediated signaling (Yao et al. 2002), suggesting that D2-induced release of $\beta\gamma$ from $G\alpha_{i/o}$ probably stimulated AC (Watts and Neve 1996, 1997). This was confirmed by using $\beta\gamma$

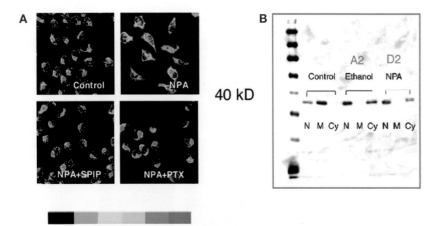

FIGURE 2. Activation of D2 promotes PKA translocation in NG108-15/D2 cells. (*A*) PKA Cα translocation detected by immunostaining. NG108-15/D2 cells were incubated with or without 50 nM of the D2-specific agonist *R*(–)-2,10,11-trihydroxy-N-propylnorapomorphine hydrobromide (NPA) for 10 min. Where indicated, cells were preincubated with the D2 antagonist spiperone (SPIP) (10 μM) for 30 min or PTX (50 ng/ml) overnight. Data represent at least three experiments. Magnification, 400x. Staining intensity is indicated by the color bar; orange indicates areas of the most intense staining. (*B*) PKA Cα translocation detected by Western blots of nuclear (N), membrane (M), and cytosolic (Cy) fractions from treated cells. (Reprinted, with permission, from Yao et al. 2002.)

dominant negative inhibitor peptides, which act as scavengers to bind free βγ dimers released from Gα$_{i/o}$ (Koch et al. 1994). We found that expression of these βγ inhibitors prevented D2 activation of PKA Cα translocation and CRE-mediated cAMP-dependent gene expression (Fig. 3) (Yao et al. 2002, 2003).

Thus, free βγ dimers released from G$_{i/o}$ appear to mediate D2-induced activation of PKA signaling. Since all G$_{i/o}$ receptors would be expected to release βγ dimers on activation, the molecular mechanisms responsible for differences between D2-mediated increases in cAMP/PKA signaling compared with M4- and α2b-mediated decreases in cAMP production remain to be determined.

Synergy between Ethanol/A2 and NPA/D2 Receptors for cAMP/PKA Signaling

The observation that ethanol or NPA activates PKA signaling through Gα$_s$ or G$_i$ βγ, respectively, suggested that ethanol/A2 activation might synergize with NPA/D2 activation. To search for evidence of synergy, we selected low concentrations of NPA (0.5 nM) or ethanol (25 mM). When added together, these subthreshold doses of ethanol and NPA, which alone would not activate PKA signaling, produced a robust synergistic activation of PKA Cα translocation and cAMP-dependent gene expression (Fig. 4) (Yao et al. 2002). Importantly, the level of activation was similar to that seen with forskolin, a potent activator of AC.

These findings have been confirmed recently in PC-12 cells where coactivation of A2 and D2 receptors synergistically increases cAMP levels (Kudlacek et al. 2003). We have found that synergy between ethanol and NPA requires adenosine and A2 receptors and

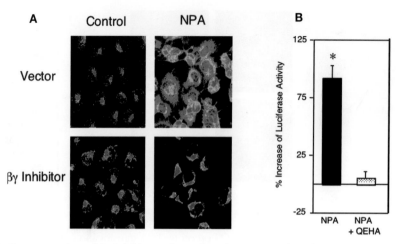

FIGURE 3. Overexpression of βγ inhibitors, the carboxyl terminus of β-adrenergic receptor kinase 1 (βARK1) and AC II QEHA inhibitor peptide, blocks PKA translocation and CRE-gene expression. (*A*) NG108-15/D2 cells were incubated in the absence or presence of NPA (50 nM) after transfection with Ad5βARK1 (βγ inhibitor) or Ad5 vector control and probed with PKA Cα antibody. Transfection efficiency was greater than 95% when visualizing cells transfected with Ad5GFP. Results are representative of six independent experiments. Magnification, 400x. (*B*) NG108-15/D2 cells were transfected with HSVCRE-luc at multiplicity of infection (MOI) 1, and with or without Ad5QEHA overnight. Cells were washed and incubated with 50 nM NPA for 10 min. Luciferase was assayed 5 hr after drug treatment and normalized for transfection efficiency as determined by β-galactosidase activity. Data are the mean ± s.e. of three experiments. (*) ($p < 0.01$) Compared with control (one-way analysis of variance and Dunnett's test). (Modified, with permission, from Yao et al. 2002.)

FIGURE 4. Subthreshold concentrations of NPA and ethanol (EtOH) act synergistically to induce PKA Cα translocation and CRE-mediated gene expression. (*A*) Synergy of PKA Cα translocation. NG108-15/D2 cells were incubated for 10 min with 0.5 nM NPA, 25 mM EtOH, or 0.5 nM NPA plus 25 mM EtOH, and localization of PKA Cα was determined. (*B*) NG108-15/D2 cells were transfected with HSVCRE-luc at MOI 1 and preincubated either with the D2 antagonist spiperone (10 μM) or A2 antagonist BW A1434U (10 μM) for 30 min, 50 ng/ml of PTX or Ad5QEHA overnight, followed by a 10-min exposure to 0.5 nM NPA or 25 mM EtOH alone or in combination, or 10 μM forskolin. Cells were then washed and cultured for up to 5 hr. Luciferase was measured as in Fig. 3B. Data are the mean ± s.e. of three experiments. (*) ($p < 0.01$) Compared with control (one-way analysis of variance and Dunnett's test). (*C*) Synergy of PKA signaling. Schematic representation for synergy between D2 and EtOH/A2 for PKA translocation and CRE-mediated gene expression. A central role for βγ subunits released from $G_{i/o}$ is proposed for D2. This diagram indicates synergy between subthreshold levels of ethanol/A2a and D2. Synergy for PKA translocation and CRE-mediated gene expression is mediated by βγ dimers from $G_{i/o}$. Adenosine A2 activation via $Gα_s$ is required for synergy. We propose that colocalization of A2 with D2 on the same neurons, e.g., in NAc, confers hypersensitivity to exogenous ethanol because of synergy. In addition, synergy promotes simultaneous hypersensitivity of postsynaptic D2 signaling, a characteristic response to ethanol. (Modified, with permission, from Yao et al. 2002.)

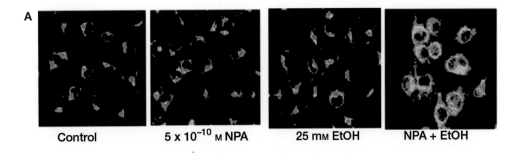

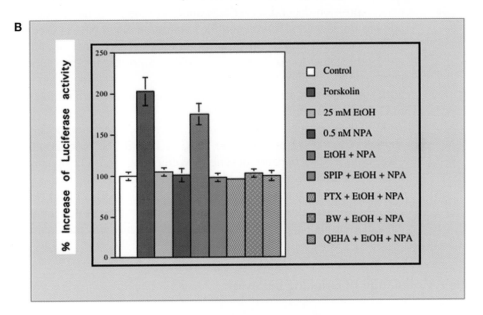

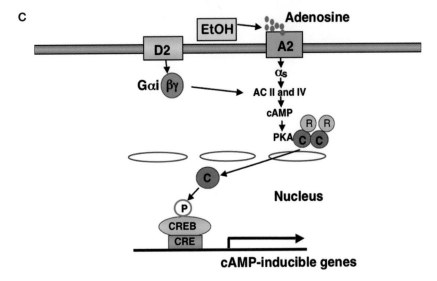

FIGURE 4. (*See facing page for legend.*)

is mediated by βγ dimers released from $G_{i/o}$ on D2 activation (Yao et al. 2002). Consistent with this observation, we also find in NAc neurons that βγ dimers mediate the increases in spike firing produced by the cooperative activation of dopamine (D1 and D2) receptors (Hopf et al. 2003), another pair of G_s- and G_i-coupled receptors implicated in responses to ethanol. This may help to explain the paradoxical observation that rats will not self-administer D1 or D2 agonists into the NAc when given separately, but will self-administer only when both agonists are administered together (Ikemoto et al. 1997). It seems likely that βγ dimers mediate this cooperative response.

Our work demonstrated that ethanol-induced PKA activation and cAMP-dependent gene expression depend on A2 activation. On the other hand, adenosine receptor blockade has no effect on D2-induced PKA activation when NPA is used at saturating concentrations (Yao et al. 2002). However, when subthreshold concentrations of ethanol and NPA are used together to produce synergy, either A2 or D2 receptor blockade prevents synergistic activation of cAMP/PKA signaling and cAMP-dependent gene expression (Yao et al. 2002). As expected, PTX and βγ scavenging peptides, which inhibit the release and action of G_i βγ dimers, also prevent synergy between ethanol/A2 and NPA/D2 (Fig. 4) (Yao et al. 2002). Taken together, these observations raise the interesting possibility that neurons that express A2a and D2 on the same cells, such as in the NAc, could be hypersensitive to exogenous ethanol because of synergy with endogenous D2 signaling. Moreover, synergy in such neurons would also be expected to confer simultaneous hypersensitivity to endogenous D2 signaling in the presence of ethanol. This might help to explain, in part, the selective activation of the NAc and the importance of dopaminergic signaling during ethanol consumption. A schematic overview of synergy is presented in Figure 4.

βγ Dimer Modulation of Drinking Behavior

Because ethanol activation of the NAc is associated with the release of dopamine in this brain region, we investigated the possibility that synergy mediated by βγ dimers between ethanol/A2 and dopamine/D2 in cell culture might have pathophysiologic significance in vivo (Yao et al. 2002). An adenoviral construct expressing a βγ inhibitor peptide was injected directly into the NAc of rats trained to consume alcohol voluntarily. Maximal expression of the βγ inhibitor in the NAc was associated with a significant reduction in voluntary ethanol consumption (Fig. 5A) compared with controls (Fig. 5B). Indeed, the time course for reduced drinking corresponded directly with the time course of adenoviral construct-induced gene expression, recovering only when gene expression had dissipated (Yao et al. 2002).

Increased action of βγ dimers during alcohol consumption could be due to upstream activation of A2a and D2 as described in cell culture systems. Indeed, it is already well established that D2 antagonists reduce alcohol consumption (Hodge et al. 1997; Cohen et al. 1998; Czachowski et al. 2001). We have tested the importance of D2 and A2a in our models of alcohol consumption and confirmed that D2 antagonists attenuate drinking behavior (Fig. 5C). Most importantly, we find that an adenosine A2a antagonist also reduces operant alcohol self-administration (Fig. 5D) (Arolfo et al. 2004). Therefore, our studies in cell culture appear to have predicted the importance of A2a signaling in ethanol self-administration.

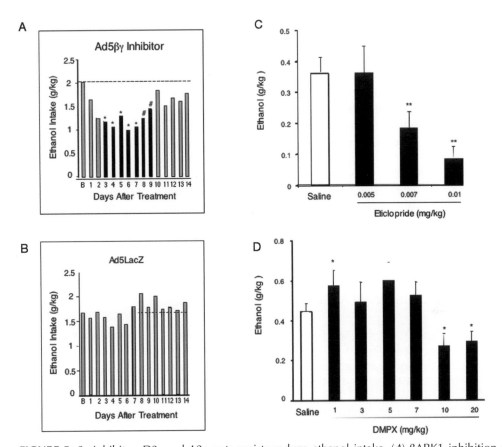

FIGURE 5. βγ inhibitor, D2, and A2a antagonists reduce ethanol intake. (*A*) βARK1 inhibition of alcohol consumption. Voluntary fluid intake and weight were measured daily in singly housed 250–300-g Long Evans male rats. Following a 4-day exposure to 10% ethanol as the only liquid source, rats were allowed to choose 10% ethanol in tap water or tap water alone. Forty-five days later, rats received permanent bilateral cannulae in the NAc. After 12–15 days to reestablish presurgery baseline drinking, rats received microinjections of Ad5βARK1 (*n* = 9), and mean daily consumption of 10% ethanol and water was measured for 14 days. Black bars represent the period of maximal gene expression (days 3–9). Dashed line indicates baseline. Mean (± s.e.m.) ethanol intake (g/kg) is presented for the baseline period (average of 7 days before vector injection) and 7 and 14 days after vector injection. (*) Significantly different from Ad5LacZ control and Ad5βARK1 day 14, *p* < 0.011; (#) significantly different from Ad5βARK1, baseline, *p* < 0.001. (*B*) Ad5LacZ (β-galactosidase) viral vector control. Mean (± s.e.m.) ethanol intake (*n* = 9). (Reprinted, with permission, from Yao et al. 2002.) (*C*) D2 antagonist inhibition of ethanol self-administration and the amount of ethanol consumed during a 30-min FR3 session of operant responding under the different eticlopride doses tested. The D2 antagonist eticlopride was administered subcutaneously 25 min before the session. Results represent means ± s.e.m. g/kg consumption. (*D*) A2 antagonist inhibition of ethanol self-administration and the amount of ethanol consumed during a 30-min FR3 session of operant responding under the different DPMX (3,7-dimethyl-1-propargylxanthine) doses tested. The A2 antagonist DPMX was administered intraperitoneally 20 min before the session. (*) Significantly different as compared with saline treatment (ANOVA with LSD post-hoc comparisons, *p* < 0.05, *n* = 8–10/group; (**) significantly different compared to saline, *p* < 0.01, *n* = 5/group. (Reprinted, with permission, from Arolfo et al. 2004.)

Studies on the role of adenosine in mediating acute ethanol-induced changes in CNS function have, in general, implicated A1 receptors (Dar 2001). In contrast, our work suggests that A2a receptors in the NAc may play a special role in reward and reinforcement of voluntary alcohol consumption. Moreover, because of the central role of the NAc in addiction and the characteristic expression of A2a and D2 receptors on medium spiny neurons in the NAc, it is possible that A2a/D2 synergy might be related to other drug-seeking behaviors in addition to alcohol. This is supported by our recent studies, discussed below, that show the presence of βγ-mediated synergy for cAMP/PKA signaling and gene expression between opioids or cannabinoids with ethanol (Yao et al. 2003). Indeed, phosphorylation of Gβ is augmented by morphine and enhances βγ stimulation of AC (Chakrabarti and Gintzler 2003). Because synergy involves A2a and D2, it is possible that new therapeutic agents directed against A2a or A2/D2 synergy may be useful in treating or attenuating alcoholism, and perhaps, other drug-seeking behaviors.

OPIATES AND CANNABINOIDS: METHODOLOGICAL CONSIDERATIONS

Opiate and Cannabinoid Activation of cAMP/PKA Signaling

We have reported in NG108-15/D2 cells and primary neurons that a 10-minute exposure to DOR and CB1 agonists increases cAMP levels, induces PKA Cα translocation (Fig. 6A,B), and activates CRE-mediated gene expression (Fig. 6C); M4 and α2b agonists do not (Yao et al. 2003). Recently, we obtained similar results after MOR activation in primary striatal neurons (Yao et al. 2005).

Nevertheless, all of these receptors inhibit cAMP production when assayed after 30 minutes of exposure, confirming their functional coupling to $G\alpha_{i/o}$. Further, we have found that overexpression of βγ scavenger peptides, PTX and the PKA inhibitor Rp-cAMPS, blocks MOR, DOR, and CB1-induced PKA Cα translocation and gene expression (Fig. 6C). These data suggest that DOR and CB1 receptors are coupled to $G_{i/o}$, releasing α_i and βγ subunits; βγ activates AC and induces PKA activation (Olianas and Onali 1999; Yao et al. 2002). Again, we have found similar results after MOR activation (Yao et al. 2005). We have begun to investigate the molecular mechanisms that facilitate opiate receptor-induced stimulation of cAMP/PKA signaling and its possible role in addictive behavior. We focused on the MOR because of its role in heroin addiction.

Recent evidence suggests that an activator of G-protein signaling 3 (AGS3) can regulate G-protein signaling. On activation, $G\alpha_i$ βγ dissociates into free $G\alpha_i$ and βγ subunits. By competing with βγ for binding to $G\alpha_{i3}$, AGS3 appears to stabilize a specific $G\alpha_{i3}$-GDP complex formed before the reassociation of free $G\alpha_{i3}$ and βγ subunits. The bound form of $G\alpha_{i3}$ does not inhibit AC, whereas free βγ subunits continue to stimulate cAMP production. This suggested that AGS3 could have a role in MOR-induced G-protein signaling and heroin addiction. We have recently identified $G\alpha_{i3}$ and AGS3 in primary striatal/NAc cultures (Yao et al. 2005). These molecular components appear to be preferentially associated with the MOR. Indeed, we find that $G\alpha_{i3}$ and AGS3 are required for MOR stimulation of PKA signaling via βγ dimers. Thus, when AGS3 levels are knocked down, MOR activation no longer stimulates CRE-mediated gene expression (Fig. 7A), apparently because AGS3 does not prevent free βγ subunits from reassociating with $G\alpha_{i3}$. In contrast, AGS3 knockdown has no effect on G_s-coupled PGE1 stimulated CRE-gene expression (Fig. 7A), as expected. On the other hand, when $G\alpha_{i3}$ is knocked down, MOR stimulation of cAMP/PKA signaling is lost,

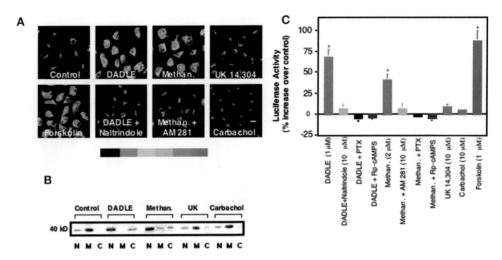

FIGURE 6. Activation of DOR and CB1 promotes PKA activation in NG108-15/D2 cells. (*A*) PKA Cα translocation detected by immunostaining. Cells were incubated with or without 1 μM DADLE, 2 μM methanandamide (Methan.), 10 μM UK 14,304 (UK), 10 μM carbachol, or 1 μM forskolin for 10 min. Where indicated, cells were preincubated with or without the DOR antagonist naltrindole (10 μM), or CB1 antagonist AM 281 (10 μM) for 30 min. Data represent at least three experiments. Scale bar, 10 μm; magnification, 400x. Staining intensity is indicated by the color bar; orange indicates areas of the most intense staining. (*B*) PKA Cα translocation detected by Western blots of nuclear (N), and membrane, including Golgi (M) and cytosolic (C) fractions from treated cells. (*C*) Gene activation measured by CRE-luciferase activity. Cells were transfected with HSVCRE-luc at MOI 1 overnight, preincubated with or without naltrindole, AM 281 for 30 min or 20 μM Rp-cAMPS (Rp) for 1 hr; or 50 ng/ml of PTX overnight and then incubated with DADLE or methanandamide for 10 min. Luciferase was assayed 5 hr after drug treatment. Data are the mean ± s.e.m. of three experiments. (*) (*p* < 0.01) Compared with control (one-way analysis of variance and Dunnett's test). (Reprinted, with permission, from Yao et al. 2003.)

apparently because a Gα$_{i3}$ βγ complex is not available to dissociate on MOR activation so that free βγ is not formed and cannot stimulate cAMP production (Yao et al. 2005).

AGS3 Antisense in the Core of the Nucleus Accumbens Abolishes Reinstatement of Heroin-seeking Behavior

Relapse, or reinstatement of heroin addiction, is a major limitation for effective treatment of human heroin addiction. In rats trained to self-administer heroin, we find that AGS3 antisense expressed in the NAc core abolishes reinstatement of heroin-seeking behavior, a valid model of human heroin craving and relapse (Fig. 7B). In contrast, knockdown of AGS3 in the NAc shell is without effect. This suggests that AGS3 may regulate craving and relapse in heroin addiction. This appears to be determined first by the predominant association of the MOR with Gα$_{i3}$ and subsequently, after Gα$_{i3}$ and βγ dimers are dissociated on MOR activation, by the selective binding interaction of Gα$_{i3}$ with AGS3 (Yao et al. 2005). Thus, Gα$_{i3}$ and AGS3 may have critical roles in the development or maintenance of heroin addiction. It remains to be determined whether this is related to βγ stimulation of PKA signaling or another effect of these G$_i$ subunits.

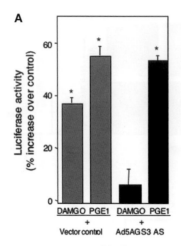
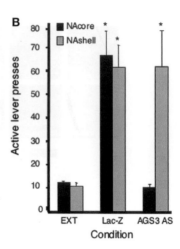

FIGURE 7. AGS3 antisense blocks MOR-induced PKA signaling and eliminates heroin-seeking behavior. (A) AGS3 antisense inhibition of CRE-luciferase expression in NAc/striatal neurons. Cells were transfected with Ad5AGS3 AS or vector control and HSVCRE-luc overnight. Cells were then treated with or without 100 nM DAMGO or 10 μM PGE1 for 10 min. Luciferase assay was done as described in Fig. 3B. Data are the mean ± S.E.M. of at least three experiments. (*) $p < 0.01$ Compared with control. (B) Ad5AGS3 AS inhibition of heroin-seeking behavior in rats. Sprague-Dawley male rats were implanted with jugular catheters and bilateral intracranial guide cannulas, and trained as previously described by McFarland et al. (2003). After animals met maintenance criteria, extinction and reinstatement were conducted. Ad5AGS3 AS or Ad5LacZ (2 μl/side, 10^{12} pfu/ml) was infused into the core or the shell of the NAc. Reinstatement testing was conducted on day 7 postinfusion. Each rat received a priming injection of heroin (0.25 mg/kg sc) before placement in the self-administration chamber for a 3-hr extinction session. Data represent mean active lever presses ± S.E.M., n = 8 in all groups. (*) ($p < 0.01$) Compared with extinction responding (EXT) (two-way ANOVA, Tukey's posttest). (Reprinted, with permission, from Yao et al. 2005.)

Synergy between Adenosine A2a and Opiate and Cannabinoid Receptors for cAMP/PKA Signaling

We have also investigated the possibility that synergy with A2a receptors also facilitates responses to opiate and cannabinoid receptors. Importantly, subthreshold concentrations of DOR or CB1 agonists, which are without effect when added separately, together with subthreshold concentrations of the D2 agonist, NPA, activate cAMP/PKA signaling synergistically (Yao et al. 2003). There is also remarkable synergy between DOR or CB1 with ethanol, another addicting agent. Synergy requires adenosine and A2a receptors, the same as synergy between ethanol and D2 (Fig. 8). Thus, our observations suggest that opiate, cannabinoid, dopamine, and ethanol share a common molecular mechanism that mediates increases in cAMP, PKA Cα translocation, and activation of gene expression. Synergy by this molecular mechanism appears to confer simultaneous hypersensitivity to opiates, cannabinoids, and D2 receptors on the same cells. This mechanism may account, in part, for drug-induced activation of medium spiny neurons in the NAc. Moreover, these results suggest that adenosinergic tone activating A2a receptors may be critical for the development of addictive behaviors.

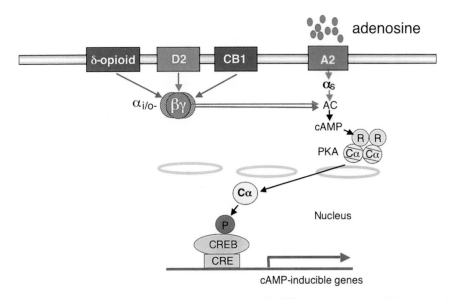

FIGURE 8. Schematic representation of postsynaptic DOR, CB1, and D2 activation-induced PKA Cα translocation and CRE-mediated gene expression via βγ dimers. A central role for βγ subunits released from $G_{i/o}$ is proposed for DOR, CB1, and D2. This diagram indicates synergy between subthreshold levels of DOR or CB1 (*blue arrows*) with D2 (*red arrows*). Synergy for PKA translocation and CRE-mediated gene expression is mediated by βγ dimers from $G_{i/o}$. Adenosine A2 activation via $Gα_s$ is required for synergy. We propose that colocalization of A2, D2, DOR, and CB1 on the same neurons, such as in NAc, confers hypersensitivity to exogenous opioids, cannabinoids, and ethanol because of synergy. In addition, synergy promotes simultaneous hypersensitivity of postsynaptic D2 signaling, characteristic of addicting drugs. (Reprinted, with permission, from Yao et al. 2003.)

IMPLICATIONS FOR ADDICTION

All addicting substances share the common property of activating the NAc and promoting dopamine release in this brain region. The NAc has the highest concentration of A2a receptors in the brain. Moreover, the NAc is characterized in part by the simultaneous expression of $G_{s/olf}$-coupled A2a with $G_{i/o}$-coupled D2 receptors on the same medium spiny neurons. Our ethanol studies suggest that a unique confluence of ENT1 and A2a and D2 receptors in these GABAergic neurons appears to modulate the regulation of operant ethanol self-administration. This may involve synergy between subthreshold concentrations of ethanol and D2 that together stimulate cAMP/PKA signaling. In the NAc, synergy appears to require A2a receptors and is mediated by unbound βγ dimers released from $G_{i/o}$ on D2 activation. Thus, expression of a dominant βγ inhibitor in the nucleus accumbens reduces voluntary alcohol consumption.

The same GABAergic medium spiny neurons in the NAc also express opiate and cannabinoid receptors on the same cells. We also find that activation of these $G_{i/o}$-coupled opiate and cannabinoid receptors, just like D2, also leads to the release of G_i βγ dimers, which stimulate cAMP/PKA signaling. Release of G_i βγ dimers appears to be determined by the preferential association of $Gα_{i3}$ with these receptors and regulation by AGS3. Thus, knockdown of $Gα_{i3}$ and AGS3 prevents MOR stimulation of cAMP/PKA signaling. In addition, expression of AGS3 antisense in the core of the nucleus accumbens, but not the shell, abolishes reinstatement of heroin-seeking behavior. This is consistent with recent evidence implicating the NAc core in the pathogenesis of addictive behavior (McFarland et al. 2003).

There is also synergy among ethanol, D2, and opiate and cannabinoid receptors for activation of cAMP/PKA signaling. As with ethanol, synergy requires adenosine A2a receptors and is mediated by $\beta\gamma$ dimers (Fig. 8). Thus, paradoxical activation of cAMP/PKA pathways could be a unique characteristic of those $G_{i/o}$-coupled receptors in NAc involved in addictive behaviors. Our hypothesis is that the cAMP/PKA pathway is activated via $\beta\gamma$ dimers in the NAc by all substances of abuse, that activation requires AGS3, and that synergy between adenosine A2a and selected $G\alpha_{i3}$-coupled receptors underlies the development and/or maintenance of craving for addicting agents.

SUMMARY AND CONCLUSIONS

The NAc is implicated in craving, reward, and reinforcement of alcohol drinking. The NAc/striatum expresses the highest levels of adenosine A2a receptors in the brain. In addition, the NAc is characterized by coexpression of A2a and dopamine D2 receptors on the same GABAergic medium spiny neurons. We have reported that ethanol promotes A2-mediated stimulation of cAMP/PKA/CRE-gene expression by inhibiting adenosine uptake, thereby increasing extracellular adenosine. Moreover, an A2a antagonist given systemically reduces operant ethanol self-administration in rats, suggesting the pathophysiologic significance of this pathway.

We also find that subthreshold concentrations of ethanol and an A2a agonist, that have no effect separately, act synergistically when added together to stimulate cAMP/PKA signaling and CRE-mediated gene expression. Synergy between A2 and D2 requires adenosine and A2a receptors, and is mediated by $\beta\gamma$ dimers released from $G_{i/o}$ by activation of D2. Thus, GABAergic medium spiny neurons in the NAc that coexpress A2a and D2 on the same cells are probably hypersensitive to ethanol in the presence of dopaminergic tone. Simultaneously, these same NAc neurons appear to exhibit an increase of D2 signaling in the presence of exogenous ethanol, a characteristic feature of ethanol consumption. Indeed, we find that synergy in the NAc appears to have a role in ethanol consumption. Expression of a $\beta\gamma$ inhibitor in the NAc reduces voluntary drinking.

The same medium spiny neurons in the NAc also coexpress opiate and cannabinoid receptors. We find that these $G_{i/o}$-coupled receptors, which are known to inhibit AC, inhibit cAMP production when assayed at 30 minutes. However, when exposed to agonists for only 10 minutes, opiate and cannabinoid receptors paradoxically increase cAMP production. This is followed by stimulation of cAMP-dependent CRE-mediated gene expression 5 hours later, long after the increased cAMP has dissipated. Indeed, our in vitro studies in cell lines and primary neuronal cultures show that stimulation of cAMP/PKA signaling occurs paradoxically only with D2, opiate, and cannabinoid receptors; other $G_{i/o}$-coupled receptors in the same cells, such as the M4 and α2b receptors, do not. Interestingly, synergy between subthreshold concentrations of receptor agonists for cAMP/PKA signaling is demonstrable between ethanol, D2, and opiate or cannabinoid receptors. As with ethanol, synergy always requires adenosine, A2a receptors, $G\alpha_{i3}$, and $\beta\gamma$ dimers, and is regulated by AGS3. These results suggest that an adenosine requirement for βγ activation of cAMP/PKA signaling may be a feature common to all addicting agents. The NAc appears to be uniquely poised to respond to adenosine and addictive agents by the coexpression of A2a with those $G_{i/o}$-coupled receptors involved in addictive behaviors. In the presence of adenosinergic tone, the presence of A2a may facilitate selective activation of the NAc during exposure to addicting drugs, as well as hypersensitvity of dopaminergic D2 signaling within this brain region. These findings may make it possible to design novel therapeutic agents to target these signaling pathways in the NAc to treat addiction.

FUTURE DIRECTIONS

Adenosine and A2a appear to be involved in all instances of synergy involving addicting agents. This suggests the feasibility of developing rationally designed A2a antagonists for alcoholism and other addictive disorders. In addition, we also need to identify the molecular mechanisms that account for selective activation of those βγ dimers responsible for paradoxical D2, opiate, and cannabinoid receptor stimulation of cAMP production or other signaling events. In addition, we need to identify the specific βγ dimers required for addictive behaviors. Then it will be possible to develop and test potential therapeutic agents designed to inhibit this pathway, from A2a antagonists to drugs interfering with βγ-mediated synergistic activation of intracellular signaling.

ACKNOWLEDGMENTS

We thank Anjlee Mahajan for her expert manuscript preparation. This research was supported by National Institutes of Health Grant No. AA-010030 (I.D.); funds provided by the State of California for medical research on ethanol and substance abuse through the University of California, San Francisco; and the Department of the Army DAMD Grant No. 17-01-1-0803 (I.D., L.Y.). The United States Army Medical Research Acquisition Activity, 820 Chandler Street, Fort Detrick, Maryland 21702-5014 is the awarding and administering acquisition office. The content of the information represented does not necessarily reflect the position or the policy of the United States Government, and no official endorsement should be inferred.

REFERENCES

Abel T., Nguyen P.V., Barad M., Deuel T.A.S., and Kandel E.R. 1997. Genetic demonstration of a role for PKA in the late phase of LTP and in hippocampus-based long-term memory. *Cell* **88:** 615–626.

Ameri A. 1999. The effects of cannabinoids on the brain. *Prog. Neurobiol.* **58:** 315–348.

Appel S.B., Liu Z., McElvain M.A., and Brodie M.S. 2003. Ethanol excitation of dopaminergic ventral tegmental area neurons is blocked by quinidine. *J. Pharmacol. Exp. Ther.* **306:** 437–446.

Arolfo M.P., Yao L., Gordon A.S., Diamond I., and Janak P.H. 2004. Ethanol operant self-administration in rats is regulated by adenosine A2 Receptors. *Alcohol. Clin. Exp. Res.* **28:** 1308–1316.

Asai K., Ishii A., Yao L., Diamond I., and Gordon A.S. 1998. Varying effects of ethanol on transfected cell lines. *Alcohol. Clin. Exp. Res.* **22:** 163–166.

Asher O., Cunningham T.D., Yao L., Gordon A.S., and Diamond I. 2002. Ethanol stimulates cAMP-responsive element (CRE)-mediated transcription via CRE-binding protein and cAMP-dependent protein kinase. *J. Pharmacol. Exp. Ther.* **301:** 66–70.

Baker L.P., Nielsen M.D., Impey S., Hacker B.M., Poser S.W.,

Chan M.Y.M., and Storm D.R. 1999. Regulation and immunohistochemical localization of βγ-stimulated adenylyl cyclases in mouse hippocampus. *J. Neurosci.* **19:** 180–192.

Barwick V.S. and Dar M.S. 1998. Adenosinergic modulation of ethanol-induced motor incoordination in the rat motor cortex. *Prog. Neuro-Psychopharmacol. Biol. Psychiatry* **22:** 587–607.

Berke J.D. and Hyman S.E. 2000. Addiction, dopamine, and the molecular mechanisms of memory. *Neuron* **25:** 515–532.

Berl S. and Frigyesi T.L. 1969. The turnover of glutamate, glutamine, aspartate and GABA labeled with [1-14C]acetate in caudate nucleus, thalamus and motor cortex (CAT). *Brain Res.* **12:** 444–455.

Berrendero F., Castane A., Ledent C., Parmentier M., Maldonado R., and Valverde O. 2003. Increase of morphine withdrawal in mice lacking A2a receptors and no changes in CB1/A2a double knockout mice. *Eur. J. Neurosci.* **17:** 315–324.

Bilecki W., Hollt V., and Przewlocki R. 2000. Acute delta-opioid receptor activation induces CREB phosphorylation in NG108-15 cells. *Eur. J. Pharmacol.* **390:** 1–6.

Buzas B., Rosenberger J., and Cox B.M. 1997. Regulation of δ opioid receptor mRNA levels by receptor-mediated and direct activation of the adenylyl cyclase-protein kinase A pathway. *J. Neurochem.* **62:** 610–615.

Campisi P., Carmichael F.J.L., Crawford M., Orrego H., and Khanna J.M. 1997. Role of adenosine in the ethanol-induced potentiation of the effects of general anesthetics in rats. *Eur. J. Pharmacol.* **325:** 165–172.

Carmichael F.J., Saldivia V., Varghese G.A., Israel Y., and Orrego H. 1988. Ethanol-induced increase in portal blood flow: Role of acetate and A_1- and A_2-adenosine receptors. *Am. J. Physiol.* **255:** G417–G423.

Carmichael F.J., Israel Y., Saldivia V., Giles H.G., Meggiorini S., and Orrego H. 1987. Blood acetaldehyde and the ethanol-induced increase in splanchnic circulation. *Biochem. Pharmacol.* **36:** 2673–2678.

Carmichael F.J., Israel Y., Crawford M., Minhas K., Saldivia V., Sandrin S., Campisi P., and Orrego H. 1991. Central nervous system effects of acetate: Contribution to the central effects of ethanol. *J. Pharm. Exp. Ther.* **259:** 403–408.

Cass C.E., Young J.D., and Baldwin S.A. 1998. Recent advances in the molecular biology of nucleoside transporters of mammalian cells. *Biochem. Cell. Biol.* **76:** 761–770.

Chakrabarti S. and Gintzler A.R. 2003. Phosphorylation of Gbeta is augmented by chronic morphine and enhances Gbetagamma stimulation of adenylyl cyclase activity. *Brain Res. Mol. Brain Res.* **119:** 144–151.

Chan J.S., Chiu T.T., and Wong Y.H. 1995. Activation of type II adenylyl cyclase by the cloned mu-opioid receptor: Coupling to multiple G proteins. *J. Neurochem.* **65:** 2682–2689.

Childers S.R. 1991. Opioid receptor-coupled second messenger systems. *Life Sci.* **48:** 1991–2003.

Choi D.S., Handa M., Young H., Gordon A.S., Diamond I., and Messing R.O. 2000. Genomic organization and expression of the mouse equilibrative, nitrobenzylthioinosine-sensitive nucleoside transporter 1 (ENT1) gene. *Biochem. Biophys. Res. Commun.* **277:** 200–208.

Choi D.S., Cascini M.G., Mailliard W., Young H., Paredes P., McMahon T., Diamond I., Bonci A., and Messing R.O. 2004. The type 1 equilibrative nucleoside transporter regulates ethanol intoxication and preference. *Nat. Neurosci.* **7:** 855–861.

Coe I.R., Yao L.N., Diamond I., and Gordon A.S. 1996a. The role of protein kinase C in cellular tolerance to ethanol. *J. Biol. Chem.* **271:** 29468–29472.

Coe I.R., Dohrman D.P., Constantinescu A., Diamond I., and Gordon A.S. 1996b. Activation of cyclic AMP-dependent protein kinase reverses tolerance of a nucleoside transporter to ethanol. *J. Pharmacol. Exp. Ther.* **276:** 365–369.

Cohen C., Perrault G., and Sanger D.J. 1998. Preferential involvement of D3 versus D2 dopamine receptors in the effects of dopamine receptor ligands on oral ethanol self-administration in rats. *Psychopharmacology* **140:** 478–485.

Constantinescu A., Diamond I., and Gordon A.S. 1999. Ethanol-induced translocation of cAMP-dependent protein kinase to the nucleus. Mechanism and functional consequences. *J. Biol. Chem.* **274:** 26985–26991.

Constantinescu A., Gordon A.S., and Diamond I. 2002. cAMP-dependent protein kinase types I and II differentially regulate cAMP response element-mediated gene expression: Implications for neuronal responses to ethanol. *J. Biol. Chem.* **277:** 18810–18816.

Constantinescu A., Wu M., Asher O., and Diamond I. 2004. cAMP-dependent protein kinase type I regulates ethanol-induced cAMP response element-mediated gene expression via activation of CREB-binding protein and inhibition of MAPK. *J. Biol. Chem.* **279:** 43321–43329.

Cornford E.M. and Oldendorf W.H. 1975. Independent blood-brain barrier transport systems for nucleic acid precursors. *Biochim. Biophys. Acta* **394:** 211–219.

Cunningham C.L., Howard M.A., Gill S.J., Rubinstein M., Low M.J., and Grandy D.K. 2000. Ethanol-conditioned place preference is reduced in dopamine D2 receptor-deficient mice. *Pharmacol. Biochem. Behav.* **67:** 693–699.

Czachowski C.L., Chappell A.M., and Samson H.H. 2001. Effects of raclopride in the nucleus accumbens on ethanol seeking and consumption. *Alcohol. Clin. Exp. Res.* **25:** 1431–1440.

Dar M.S. 2001. Modulation of ethanol-induced motor incoordination by mouse striatal A(1) adenosinergic receptor. *Brain Res. Bull.* **55:** 513–520.

Dar M.S. and Clark M. 1992. Tolerance to adenosine's accentuation of ethanol-induced motor incoordination in ethanol-tolerant mice. *Alcohol. Clin. Exp. Res.* **16:** 1138–1146.

Dassesse D., Massie A., Ferrari R., Ledent C., Parmentier M., Arckens L., Zoli M., and Schiffmann S.N. 2001. Functional striatal hypodopaminergic activity in mice lacking adenosine A(2A) receptors. *J. Neurochem.* **78:** 183–198.

Diamond I. and Gordon A.S. 1997. Cellular and molecular neuroscience of alcoholism. *Physiol. Rev.* **77:** 1–20.

Dohrman D.P., Diamond I., and Gordon A.S. 1996. Ethanol causes translocation of cAMP-dependent protein kinase catalytic subunit to the nucleus. *Proc. Natl. Acad. Sci.* **93:** 10217–10221.

Dohrman D.P., Chen H.M., Gordon A.S., and Diamond I. 2002. Ethanol-induced translocation of protein kinase A occurs in two phases: Control by different molecular mechanisms. *Alcohol. Clin. Exp. Res.* **26:** 407–415.

Doyon W.M., York J.L., Diaz L.M., Samson H.H., Czachowski C.L., and Gonzales R.A. 2003. Dopamine activity in the nucleus accumbens during consummatory phases of oral ethanol self-administration. *Alcohol. Clin. Exp. Res.* **27:** 1573–1582.

Dunwiddie T.V. and Masino S.A. 2001. The role and regulation of adenosine in the central nervous system. *Annu. Rev. Neurosci.* **24:** 31–55.

El Yacoubi M., Ledent C., Parmentier M., Ongini E., Costentin J., and Vaugeois J.M. 2001. In vivo labelling of the adenosine A2A receptor in mouse brain using the selective antagonist [3H]SCH 58261. *Eur. J. Neurosci.* **14:** 1567–1570.

Fantozzi D.A., Harootunian A.T., Wen W., Taylor S.S., Feramisco J.R., Tsien R.Y., and Meinkoth J.L. 1994. Thermostable inhibitor of cAMP-dependent protein kinase enhances the rate of export of the kinase catalytic subunit from the nucleus. *J. Biol. Chem.* **269:** 2676–2686.

Federman A.D., Conklin B.R., Schrader K.A., Reed R.R., and Bourne H.R. 1992. Hormonal stimulation of adenylyl cyclase through Gi-protein beta gamma subunits. *Nature* **356:** 159–161.

Ferre S., O'Connor W.T., Fuxe K., and Ungerstedt U. 1993. The striopallidal neuron: A main locus for adenosine-dopamine interactions in the brain. *J. Neurosci.* **13:** 5402–5406.

Ferre S., O'Connor W.T., Snaprud P., Ungerstedt U., and Fuxe K. 1994. Antagonistic interaction between adenosine A2A receptors and dopamine D2 receptors in the ventral striopallidal system. Implications for the treatment of schizophrenia. *Neuroscience* **63:** 765–773.

Fink J.S., Weaver D.R., Rivkees S.A., Peterfreund R.A., Pollack A.E., Adler E.M., and Reppert S.M. 1992. Molecular cloning of the rat A$_2$ adenosine receptor: Selective co-expression with D$_2$ dopamine receptors in rat striatum. *Mol. Brain Res.* **14:** 186–195.

Franco R., Ferre S., Agnati L., Torvinen M., Gines S., Hillion J., Casado V., Lledo P., Zoli M., Lluis C., and Fuxe K. 2000. Evidence for adenosine/dopamine receptor interactions: Indications for heteromerization. *Neuropsychopharmacology* (suppl.) **23:** S50–S59.

Fredholm B.B., Lindstrom K., Dionisotti S., and Ongini E. 1998. [3H]SCH 58261, a selective adenosine A2A receptor antagonist, is a useful ligand in autoradiographic studies. *J. Neurochem.* **70:** 1210–1216.

Funada M., Schutz C.G., and Shippenberg T.S. 1996. Role of delta-opioid receptors in mediating the aversive stimulus effects of morphine withdrawal in the rat. *Eur. J. Pharmacol.* **300:** 17–24.

Gatch M.B., Wallis C.J., and Lal H. 1999. The effects of adenosine ligands R-PIA and CPT on ethanol withdrawal. *Alcohol* **19:** 9–14.

Gatto G.J., McBride W.J., Murphy J.M., Lumeng L., and Li T.-K. 1994. Ethanol self-infusion into the ventral tegmental area by alcohol-preferring rats. *Alcohol* **11:** 557–564.

Ghozland S., Matthes H.W., Simonin F., Filliol D., Kieffer B.L., and Maldonado R. 2002. Motivational effects of cannabinoids are mediated by mu-opioid and kappa-opioid receptors. *J. Neurosci.* **22:** 1146–1154.

Glass M. and Felder C.C. 1997. Concurrent stimulation of cannabinoid CB1 and dopamine D2 receptors augments cAMP accumulation in striatal neurons: Evidence for a Gs linkage to the CB1 receptor. *J. Neurosci.* **17:** 5327–5333.

Gordon A.S., Collier K., and Diamond I. 1986. Ethanol regulation of adenosine receptor-stimulated cAMP levels in a clonal neural cell line: An in vitro model of cellular tolerance to ethanol. *Proc. Natl. Acad. Sci.* **83:** 2105–2108.

Gordon A.S., Yao L., Jiang Z., Fishburn C.S., Fuchs S., and Diamond I. 2001. Ethanol acts synergistically with a D2 dopamine agonist to cause translocation of protein kinase C. *Mol. Pharmacol.* **59:** 153–160.

Handa M., Choi D.S., Caldeiro R.M., Messing R.O., Gordon A.S., and Diamond I. 2001. Cloning of a novel isoform of the mouse NBMPR-sensitive equilibrative nucleoside transporter (ENT1) lacking a putative phosphorylation site. *Gene* **262:** 301–307.

Hassan S., Duong B., Kim K.S., and Miles M.F. 2003. Pharmacogenomic analysis of mechanisms mediating ethanol regulation of dopamine beta-hydroxylase. *J. Biol. Chem.* **278:** 38860–38869.

Hodge C.W., Samson H.H., and Chappelle A.M. 1997. Alcohol self-administration: Further examination of the role of dopamine receptors in the nucleus accumbens. *Alcohol. Clin. Exp. Res.* **21:** 1083–1091.

Hopf F.W., Cascini M.G., Gordon A.S., Diamond I., and Bonci A. 2003. Cooperative activation of dopamine D1 and D2 receptors increases spike firing of nucleus accumbens neurons via G-protein betagamma subunits. *J. Neurosci.* **23:** 5079–5087.

Howlett A.C. 1995. Pharmacology of cannabinoid receptors. *Annu. Rev. Pharmacol. Toxicol.* **35:** 607–634.

Hyman S.E. and Malenka R.C. 2001. Addiction and the brain: The neurobiology of compulsion and its persistence. *Nat. Rev. Neurosci.* **2:** 695–703.

Ikemoto S., Glazier B.S., Murphy J.M., and McBride W.J. 1997. Role of dopamine D1 and D2 receptors in the nucleus accumbens in mediating reward. *J. Neurosci.* **17:** 8580–8587.

Imperato A. and DiChiara G. 1986. Preferential stimulation of dopamine release in the nucleus accumbens of freely moving rats by ethanol. *J. Pharmacol. Exp. Ther.* **238:** 219–228.

Inglese J., Luttrell L.M., Iniguez-Lluhi J.A., Touhara K., Koch W.J., and Lefkowitz R.J. 1994. Functionally active targeting domain of the beta-adrenergic receptor kinase: An inhibitor of G beta gamma-mediated stimulation of type II adenylyl cyclase. *Proc. Natl. Acad. Sci.* **91:** 3637–3641.

Israel Y., Orrego H., and Carmichael F.J. 1994. Acetate-mediated effects of ethanol. *Alcohol. Clin. Exp. Res.* **18:** 144–148.

Jarvis M.F. and Becker H.C. 1998. Single and repeated episodes of ethanol withdrawal increase adenosine A1, but not A2A, receptor density in mouse brain. *Brain Res.* **786:** 80–88.

Jarvis M.F., Jackson R.H., and Williams M. 1989. Autoradiographic characterization of high-affinity adenosine A$_2$ receptors in the rat brain. *Brain Res.* **484:** 111–118.

Kaplan G.B., Bharmal N.H., Leite-Morris K.A., and Adams W.R. 1999. Role of adenosine A1 and A2A receptors in the alcohol withdrawal syndrome. *Alcohol* **19:** 157–162.

Khasar S.G., Wang J.F., Taiwo Y.O., Heller P.H., Green P.G., and Levine J.D. 1995. Mu-opioid agonist enhancement of prostaglandin-induced hyperalgesia in the rat: A G-protein beta gamma subunit-mediated effect? *Neuroscience* **67:** 189–195.

Koch W.J., Hawes B.E., Inglese J., Luttrell L.M., and Lefkowitz R.J. 1994. Cellular expression of the carboxyl terminus of a G protein-coupled receptor kinase attenuates G beta gamma-mediated signaling. *J. Biol. Chem.* **269:** 6193–6197.

Koob G.F. and Nestler E.J. 1997. The neurobiology of drug addiction. *J. Neuropsychiatry Clin. Neurosci.* **9:** 482–497.

Krauss S.W., Ghirnikar R.B., Diamond I., and Gordon A.S. 1993. Inhibition of adenosine uptake by ethanol is specific for one class of nucleoside transporters. *Mol. Pharmacol.* **44:** 1021–1026.

Kudlacek O., Just H., Korkhov V.M., Vartian N., Klinger M., Pankevych H., Yang Q., Nanoff C., Freissmuth M., and Boehm S. 2003. The human D(2) dopamine receptor synergizes with the A(2A) adenosine receptor to stimulate adenylyl cyclase in PC12 cells. *Neuropsychopharmacology* **28:** 1317–1327.

Kull B., Svenningsson P., and Fredholm B.B. 2000. Adenosine A$_{2A}$ receptors are colocalized with and activate G$_{olf}$ in rat striatum. *Mol. Pharmacol.* **58:** 771–777.

Latini S. and Pedata F. 2001. Adenosine in the central nervous system: Release mechanisms and extracellular concentrations. *J. Neurochem.* **79:** 463–484.

Ledent C., Vaugeois J.-M., Schiffmann S.N., Pedrazzin T., El Yacoub M., Vanderhaegen J.-J., Costentin J., Heath J.K., Vassart G., and Parmentier M. 1997. Aggressiveness, hypoalgesia and high blood pressure in mice lacking the adenosine A2a receptor. *Nature* **388:** 674–678.

Ledent C., Valverde O., Cossu G., Petitet F., Aubert J.F., Beslot F., Bohme G.A., Imperato A., Pedrazzini T., Roques B.P., Vassart G., Fratta W., and Parmentier M. 1999. Unresponsiveness to cannabinoids and reduced addictive effects of opiates in CB1 receptor knockout mice. *Science* **283:** 401–404.

Lepore M., Vorel S.R., Lowinson J., and Gardner E.L. 1995. Conditioned place preference induced by delta 9-tetrahydrocannabinol: Comparison with cocaine, morphine, and food reward. *Life Sci.* **56:** 2073–2080.

Lester L.B. and Scott J.D. 1997. Anchoring and scaffold proteins for kinases and phosphatases. *Recent Prog. Horm. Res.* **52:** 409–429; discussion 429–430.

Li J., Li Y.H., and Yuan X.R. 2003a. Changes of phosphorylation of cAMP response element binding protein in rat nucleus accumbens after chronic ethanol intake: Naloxone reversal. *Acta Pharmacol. Sin.* **24:** 930–936.

Li J., Li Y.H., Zhang X.H., Zhu X.J., Ge Y.B., and Yuan X.R. 2003b. Changes in the phosphorylation of cAMP response element binding protein in the rat nucleus accumbens after acute and chronic ethanol administration. *Shengli Xuebao* **55:** 147–152.

Mailliard W.S. and Diamond I. 2004. Recent advances in the neurobiology of alcoholism: The role of adenosine. *Pharmacol. Therapeut.* **101:** 39–46.

Malec T.S., Malec E.A., and Dongier M. 1996. Efficacy of buspirone in alcohol dependence: A review. *Alcohol. Clin. Exp. Res.* **20:** 853–858.

Martin M., Ledent C., Parmentier M., Maldonado R., and Valverde O. 2000. Cocaine, but not morphine, induces conditioned place preference and sensitization to locomotor responses in CB1 knockout mice. *Eur. J. Neurosci.* **12:** 4038–4046.

Martinez-Mir M.I., Probst A., and Palacios J.M. 1991. Adenosine A2 receptors: Selective localization in the human basal ganglia and alterations with disease. *Neuroscience* **42:** 697–706.

Matsuoka I., Suzuki Y., Defer N., Nakanishi H., and Hanoune J. 1997. Differential expression of type I, II, and V adenylyl cyclase gene in the postnatal developing rat brain. *J. Neurochem.* **68:** 498–506.

Matthes H.W.D., Maldonado R., Simonin F., Valverde O., Slowe S., Kitchen I., Befort K., Dierich A., Le Meur M., Dolle P., Tzavara E., Hanoune J., Roques B.P., and Kieffer B.L. 1996. Loss of morphine-induced analgesia, reward effect and withdrawal symptoms in mice lacking the μ-opioid-receptor gene. *Nature* **383:** 819–823.

McFarland K., Lapish C.C., and Kalivas P.W. 2003. Prefrontal glutamate release into the core of the nucleus accumbens mediates cocaine-induced reinstatement of drug-seeking behavior. *J. Neurosci.* **23:** 3531–3537.

McIntire S. and Diamond I. 2002. Alcohol neurotoxicity. In *Diseases of the nervous system: Clinical neuroscience and therapeutic principles* (ed. A.K. Asbury et al.), vol. 2, pp. 1814–1826. Saunders, Philadelphia.

Meinkoth J.L., Alberts A.S., Went W., Fantozzi D., Taylor S.S., Hagiwara M., Montminy M., and Feramisco J.R. 1993. Signal transduction through the cAMP-dependent protein kinase. *Mol. Cell. Biochem.* **127/128:** 179–186.

Melis M., Camarini R., Ungless M.A., and Bonci A. 2002. Long-lasting potentiation of GABA-ergic synapses in dopamine neurons after a single *in vivo* ethanol exposure. *J. Neurosci.* **22:** 2074–2082.

Mons N., Harry A., Dubourg P., Premont R.T., Iyengar R., and Cooper D.M.F. 1995. Immunohistochemical localization of adenylyl cyclase in rat brain indicates a highly selective concentration at synapses. *Proc. Natl. Acad. Sci.* **92:** 8473–8477.

Moore M.S., DeZazzo J., Luk A.Y., Tully T., Singh C.M., and

Heberlein U. 1998. Ethanol intoxication in *Drosophila*: Genetic and pharmacological evidence for regulation by the cAMP signaling pathway. *Cell* **93**: 997–1007.

Moriguchi M., Sakai K., Miyamoto Y., and Wakayama M. 1993. Production, purification, and characterization of D-aminoacylase from Alcaligenes xylosoxydans subsp. xylosoxydans A-6. *Biosci. Biotechnol. Biochem.* **57**: 1149–1152.

Naassila M., Ledent C., and Daoust M. 2002. Low ethanol sensitivity and increased ethanol consumption in mice lacking adenosine A2A receptors. *J. Neurosci.* **22**: 10487–10493.

Nagy L.E. 1994. Role of adenosine A1 receptors in inhibition of receptor-stimulated cyclic AMP production by ethanol in hepatocytes. *Biochem. Pharmacol.* **48**: 2091–2096.

Nagy L.E., Diamond I., and Gordon A.S. 1991. cAMP-dependent protein kinase regulates inhibition of adenosine transport by ethanol. *Mol. Pharmacol.* **40**: 812–817.

Nagy L.E., Diamond I., Casso D.J., Franklin C., and Gordon A.S. 1990. Ethanol increases extracellular adenosine by inhibiting adenosine uptake via the nucleoside transporter. *J. Biol. Chem.* **265**: 1946–1951.

Nagy L.E., Diamond I., Collier K., Lopez L., Ullman B., and Gordon A.S. 1989. Adenosine is required for ethanol-induced heterologous desensitization. *Mol. Pharmacol.* **36**: 744–748.

Negus S.S., Henriksen S.J., Mattox A., Pasternak G.W., Portoghese P.S., Takemori A.E., Weinger M.B., and Koob G.F. 1993. Effect of antagonists selective for mu, delta and kappa opioid receptors on the reinforcing effects of heroin in rats. *J. Pharmacol. Exp. Ther.* **265**: 1245–1252.

Nestler E.J. 2001. Molecular basis of long-term plasticity underlying addiction. *Nat. Rev. Neurosci.* **2**: 119–128.

Nestler E.J. and Aghajanian G.K. 1997. Molecular and cellular basis of addiction. *Science* **278**: 58–63.

Nigg E.A., Hilz H., Eppenberger H.M., and Dutly F. 1985. Rapid and reversible translocation of the catalytic subunit of cAMP-dependent protein kinase type II from the Golgi complex to the nucleus. *EMBO J.* **4**: 2801–2806.

Nonaka H., Mori A., Ichimura M., Shindou T., Yanagawa K., Shimada J., and Kase H. 1994. Binding of [3H]KF17837S, a selective adenosine A2 receptor antagonist, to rat brain membranes. *Mol. Pharmacol.* **46**: 817–822.

Obadiah J., Avidor-Reiss T., Fishburn C.S., Carmon S., Bayewitch M., Vogel Z., Fuchs S., and Levavi-Sivan B. 1999. Adenylyl cyclase interaction with the D2 dopamine receptor family; differential coupling to Gi, Gz, and Gs. *Cell. Mol. Neurobiol.* **19**: 653–664.

Olianas M.C. and Onali P. 1999. Mediation by G protein betagamma subunits of the opioid stimulation of adenylyl cyclase activity in rat olfactory bulb. *Biochem. Pharmacol.* **57**: 649–652.

Orrego H., Carmichael F.J., and Israel Y. 1988a. New insights on the mechanism of the alcohol-induced increase in portal blood flow. *Can. J. Pharmacol.* **66**: 1–9.

Orrego H., Carmichael F.J., Saldivia V., Giles H.G., Sandrin S., and Israel Y. 1988b. Ethanol-induced increase in portal blood flow: Role of adenosine. *Am. Physiol. Soc.* **254**: G495–G501.

Pandey S.C., Roy A., and Mittal N. 2001a. Effects of chronic ethanol intake and its withdrawal on the expression and phosphorylation of the CREB gene transcription factor in rat cortex. *J. Pharmacol. Exp. Ther.* **296**: 857–868.

Pandey S.C., Roy A., and Zhang H. 2003. The decreased phosphorylation of cyclic adenosine monophosphate (cAMP) response element binding (CREB) protein in the central amygdala acts as a molecular substrate for anxiety related to ethanol withdrawal in rats. *Alcohol. Clin. Exp. Res.* **27**: 396–409.

Pandey S.C., Saito T., Yoshimura M., Sohma H., and Gotz M.E. 2001b. cAMP signaling cascade: A promising role in ethanol tolerance and dependence. *Alcohol. Clin. Exp. Res.* **25**: 465–588.

Park S.K., Sedore S.A., Cronmiller C., and Hirsh J. 2000. Type II cAMP-dependent protein kinase-deficient *Drosophila* are viable but show developmental, circadian, and drug response phenotypes. *J. Biol. Chem.* **275**: 20588–20596.

Parkinson F.E. and Fredholm B.B. 1990. Autoradiographic evidence for G-protein coupled A₂-receptors in rat neostriatum using [³H]-CGS 21680 as a ligand. *Arch. Pharmacol.* **342**: 85–89.

Pertwee R.G. 1997. Pharmacology of cannabinoid CB1 and CB2 receptors. *Pharmacol. Ther.* **74**: 129–180.

Phillips T.J., Brown K.J., Burkhart-Kasch S., Wenger C.D., Kelly M.A., Rubinstein M., Grandy D.K., and Low M.J. 1998. Alcohol preference and sensitivity are markedly reduced in mice lacking dopamine D2 receptors. *Nat. Neurosci.* **1**: 610–615.

Ribeiro J.A., Sebastiao A.M., and de Mendonca A. 2003. Adenosine receptors in the nervous system: Pathophysiological implications. *Prog. Neurobiol.* **68**: 377–392.

Risinger F.O., Freeman P.A., Rubinstein M., Low M.J., and Grandy D.K. 2000. Lack of operant ethanol self-administration in dopamine D2 receptor knockout mice. *Psychopharmacology* **152**: 343–350.

Robbins T.W. and Everitt B.J. 1999. Drug addiction: Bad habits add up [news]. *Nature* **398**: 567–570.

Rodan A.R., Kiger J.A., Jr., and Heberlein U. 2002. Functional dissection of neuroanatomical loci regulating ethanol sensitivity in *Drosophila*. *J. Neurosci.* **22**: 9490–9501.

Rodriguez J.J., Mackie K., and Pickel V.M. 2001. Ultrastructural localization of the CB1 cannabinoid receptor in mu-opioid receptor patches of the rat caudate putamen nucleus. *J. Neurosci.* **21**: 823–833.

Rosin D.L., Robeva A., Woodard R.L., Guyenet P.G., and Linden J. 1998. Immunohistochemical localization of adenosine A2A receptors in the rat central nervous system. *J. Comp. Neurol.* **401**: 163–186.

Rubovitch V., Gafni M., and Sarne Y. 2003. The mu opioid agonist DAMGO stimulates cAMP production in SK-N-SH cells through a PLC-PKC-Ca⁺⁺ pathway. *Brain Res. Mol. Brain Res.* **110**: 261–266.

Sapru M.K., Diamond I., and Gordon A.S. 1994. Adenosine receptors mediate cellular adaptation to ethanol in NG108-15 cells. *J. Pharmacol. Exp. Ther.* **271**: 542–548.

Schiffmann S.N., Jacobs O., and Vanderhaeghen J.-J. 1991. Striatal restricted adenosine A2 receptor (RDC8) is expressed by enkephalin but not by substance P neurons: An *in situ* hybridization histochemistry study. *J. Neurochem.* **57:** 1062–1067.

Schwartz J.H. 2001. The many dimensions of cAMP signaling. *Proc. Natl. Acad. Sci.* **98:** 13482–13484.

Self D.W. 1998. Neural substrates of drug craving and relapse in drug addiction. *Ann. Med.* **30:** 379–389.

Snell B.J., Short J.L., Drago J., Ledent C., and Lawrence A.J. 2000a. Characterisation of central adenosine A(1) receptors and adenosine transporters in mice lacking the adenosine A(2a) receptor. *Brain Res.* **877:** 160–169.

———. 2000b. Visualisation of AMPA binding sites in the brain of mice lacking the adenosine A(2a) receptor. *Neurosci. Lett.* **291:** 97–100.

Svenningsson P., Hall H., Sedvall G., and Fredholm B.B. 1997. Distribution of adenosine receptors in the postmortem human brain: An extended autoradiographic study. *Synapse* **27:** 322–335.

Svingos A.L., Moriwaki A., Wang J.B., Uhl G.R., and Pickel V.M. 1996. Ultrastructural immunocytochemical localization of mu-opioid receptors in rat nucleus accumbens: Extrasynaptic plasmalemmal distribution and association with Leu5-enkephalin. *J. Neurosci.* **16:** 4162–4173.

Tang W.J. and Gilman A.G. 1991. Type-specific regulation of adenylyl cyclase by G protein beta gamma subunits. *Science* **254:** 1500–1503.

Thibault C., Lai C., Wilke N., Duong B., Olive M.F., Rahman S., Dong H., Hodge C.W., Lockhart D.J., and Miles M.F. 2000. Expression profiling of neural cells reveals specific patterns of ethanol-responsive gene expression. *Mol. Pharmacol.* **52:** 1593–1600.

Thiele T.E., Willis B., Stadler J., Reynolds J.G., Bernstein I.L., and McKnight G.S. 2000. High ethanol consumption and low sensitivity to ethanol-induced sedation in protein kinase A-mutant mice. *J. Neurosci.* **20:** RC75–RC80.

Thorn J.A. and Jarvis S.M. 1996. Adenosine transporters. *Gen. Pharmacol.* **27:** 613–620.

Wan W., Sutherland G.R., and Geiger J.D. 1990. Binding of the adenosine A_2 receptor ligand [^{3}H]CGS 21680 to human and rat brain: Evidence for multiple affinity sites. *J. Neurochem.* **55:** 1763–1771.

Wand G., Levine M., Zweifel L., Schwindinger W., and Abel T. 2001. The cAMP-protein kinase A signal transduction pathway modulates ethanol consumption and sedative effects of ethanol. *J. Neurosci.* **21:** 5297–5303.

Wang H. and Pickel V.M. 2001. Preferential cytoplasmic localization of delta-opioid receptors in rat striatal patches: Comparison with plasmalemmal mu-opioid receptors. *J. Neurosci.* **21:** 3242–3250.

Wang J.H., Short J., Ledent C., Lawrence A.J., and Buuse M. 2003. Reduced startle habituation and prepulse inhibition in mice lacking the adenosine A2A receptor. *Behav. Brain Res.* **143:** 201–207.

Watts V.J. and Neve K.A. 1996. Sensitization of endogenous and recombinant adenylate cyclase by activation of D2 dopamine receptors. *Mol. Pharmacol.* **50:** 966–976.

———. 1997. Activation of type II adenylate cyclase by D2 and D4 but not D3 dopamine receptors. *Mol. Pharmacol.* **52:** 181–186.

Weiss F. and Porrino L.J. 2002. Behavioral neurobiology of alcohol addiction: Recent advances and challenges. *J. Neurosci.* **22:** 3332–3337.

Weiss F., Lorang M.T., Bloom F.E., and Koob G.F. 1993. Oral alcohol self-administration stimulates dopamine release in the rat nucleus accumbens: Genetic and motivational determinants. *J. Pharmacol. Exp. Ther.* **267:** 250–258.

Weiss F., Parsons L.H., Schulteis G., Hyyti P., Lorang M.T., Bloom F.E., and Koob G.F. 1996. Ethanol self-administration restores withdrawal-associated deficiencies in accumbal dopamine and 5-hydroxytryptamine release in dependent rats. *J. Neurosci.* **16:** 3474–3485.

Wise R.A. 1998. Drug-activation of brain reward pathways. *Drug Alcohol Depend.* **51:** 13–22.

Yang X., Diehl A.M., and Wand G.S. 1996. Ethanol exposure alters the phosphorylation of cyclic AMP responsive element binding protein and cyclic AMP responsive element binding activity in rat cerebellum. *J. Pharmacol. Exp. Ther.* **278:** 338–346.

Yang X., Horn K., and Wand G.S. 1998. Chronic ethanol exposure impairs phosphorylation of CREB and CRE-binding activity in rat striatum. *Alcohol. Clin. Exp. Res.* **22:** 382–390.

Yang X., Oswald L., and Wand G. 2003. The cyclic AMP/protein kinase A signal transduction pathway modulates tolerance to sedative and hypothermic effects of ethanol. *Alcohol. Clin. Exp. Res.* **27:** 1220–1225.

Yao L., Fan P., Jiang Z., Mailliard W.S., Gordon A.S., and Diamond I. 2003. Addicting drugs utilize a synergistic molecular mechanism in common requiring adenosine and Gi-βγ dimers. *Proc. Natl. Acad. Sci.* **100:** 14379–14384.

Yao L., McFarland K., Fan P., Inoue Y., Jiang Z., and Diamond I. 2005. Activator of G-protein signaling 3 regulates opiate activation of protein kinase A signaling and relapse of heroin-seeking behavior. *Proc. Natl. Acad. Sci.* **102:** 8746–8751.

Yao L., Arolfo M.P., Dohrman D.P., Jiang Z., Fan P., Fuchs S., Janak P.H., Gordon A.S., and Diamond I. 2002. betagamma dimers mediate synergy of dopamine D2 and adenosine A2 receptor-stimulated PKA signaling and regulate ethanol consumption. *Cell* **109:** 733–743.

Yoshimura M. and Tabakoff B. 1999. Ethanol's actions on cAMP-mediated signaling in cells transfected with type VII adenylyl cyclase. *Alcohol. Clin. Exp. Res.* **23:** 1457–1461.

17 Neurotransmitter Transporters: Mechanisms and Function

Robert Edwards

Department of Neurology/Physiology, University of California at San Francisco School of Medicine, San Francisco, California 94143

ABSTRACT

Many drugs of abuse act directly on neurotransmitter transporters. The basic mechanisms involved in neurotransmitter transport across biological membranes and the role of these mechanisms in signaling between neurons will thus help us to explain how drugs of abuse influence behavior. The effect of drugs also suggests that physiological regulation of the transporters may have an important role in normal behavior and perhaps neuropsychiatric illness. In addition, recent research has begun to suggest that adaptation to drugs of abuse may involve neurotransmitter transporters.

INTRODUCTION

There are two broad categories of transporters. The ATP-binding cassette (ABC) transporters and P-type ATPases maintain ionic gradients and protect cells by extruding metabolites, toxins, or drugs. They directly use the energy of ATP hydrolysis to perform these functions. The second category, the solute carrier superfamily (SLC), provides nutrients (sugars, amino acids, fatty acids) to cells, acts as conduits (receptors) for viral entry, controls intracellular pH, and regulates neurotransmission by promoting storage and release of neurotransmitters. For active transport, these proteins rely on the ionic gradients produced by the first group.

Transporters are different from channels. Transporters move substrates and ions across membranes. Although both transporters and channels move ions from one side of the membrane to the other, transporters move their substrates much more slowly than ions, which permeate rapidly through channels. In particular, transporters generally translocate substrates on the order of 1–10^3/sec, whereas channels allow ions to move through the membrane at rates of 10^7–10^8/sec. In addition, transporters are like enzymes: They saturate with increasing concentrations of substrate and show a distinctive K_m and V_{max}. In contrast, channels are saturated only at extremely high concentrations of ions. Furthermore, transporters can generate concentration gradients, whereas channels can only let ions run down along their electrochemical gradient. However, many transporters catalyze only facilitated diffusion, and thus cannot generate a concentration gradient.

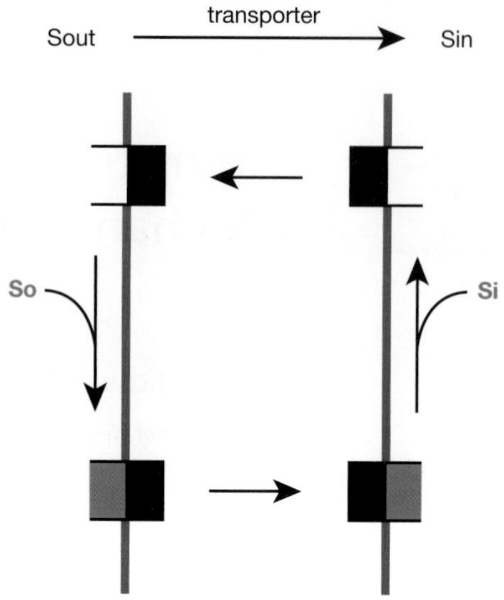

FIGURE 1. Transporters catalyze the movement of substrate (S) from one side of the membrane to the other. In contrast to channels, they involve alternating access of a substrate recognition site to the two sides of the membrane. Net flux also involves both the movement of the carrier loaded with substrate (*bottom*) and the unloaded carrier (*top*).

Therefore, the basic difference between a transporter and a channel is that a channel is simultaneously open from one side of the membrane to the other, whereas transporters show alternating access (Fig. 1). This is in fact the sine qua non of transporters: They are not open all the way through the membrane at any given time, but rather contain gates that open alternately to the two sides of the membrane.

Why do we need transporters? Transporters allow hydrophilic substrates to move across the very hydrophobic lipid bilayer, an energetically unfavorable process. By recycling and thereby conserving neurotransmitters, transporters can be considered contributors to energy conservation in the brain. Transporters thus act like enzymes to reduce the energy required for transition from one side of the membrane to the other. Channels also differ from transporters in other ways. In particular, the gating of ion channels may require substantial conformational changes, but the actual permeation of ions does not itself involve large-scale movements of the protein. Therefore, permeation of ions through channels is not very dependent on temperature. On the other hand, the alternating access mechanism requires that transporters undergo major conformational changes, making them much more temperature- and energy-dependent.

Like enzymes, transporters can mediate flux in either direction. In addition, it is important to note that the *net* movement of substrate requires the movement of *un*loaded as well as loaded transporter. Otherwise, the transporter would catalyze only the exchange of substrate on one side of the membrane for substrate on the other. Indeed, several families of transporters are obligate exchangers that do not mediate net flux in either direction. Exchange is also something channels cannot do.

In the case of a transporter that moves glucose across the plasma membrane, cells preloaded with radioactive glucose were placed in increasing concentrations of external, nonradioactive glucose. Figure 2 shows the rate of radioactive efflux as a function of external glucose concentration. Remarkably, the rate of radiolabeled glucose efflux is proportional to the concentration of extracellular glucose. Thus, efflux is stimulated by substrate on the other side of the membrane! This paradoxical result occurs because there is actually no net

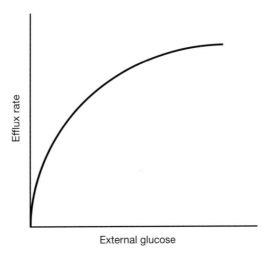

FIGURE 2. Membranes containing a glucose transporter and preloaded with radioactive glucose are incubated with increasing concentrations of external glucose, and the rates of radioactive glucose efflux are measured. Increasing nonradioactive external glucose increases the rates of efflux through the *trans*-stimulation typical of many transporters. A channel could not mediate this type of exchange reaction.

flux of substrate. The more nonradioactive substrate present on the outside, the faster the radioactive glucose moves out because the external substrate increases the amount of inwardly oriented substrate-recognition site available to carry molecules from the inside out. All transporters are thought to work through this type of mechanism, and the apparent stimulation of flux by substrate on the other side of the membrane (known as *trans*-stimulation) is one of the principal ways to distinguish a transporter from a channel.

GOALS OF THE CHAPTER

This chapter should confer the reader with an understanding of the basic principles of transport across membranes and the way in which this contributes to signaling in the nervous system.

MECHANISMS

How Do Transporters Concentrate Substrates?

Neurotransmitter transporters are secondary active transporters, i.e., they use ionic gradients made by pumps such as the plasma membrane Na^+-K^+-ATPase or, in the case of intracellular membranes such as lysosomes, endosomes, and synaptic vesicles, the vacuolar H^+-ATPase. Although many carriers mediate facilitated diffusion and primary pumps such as the Na^+ pump, vacuolar H^+-ATPase and ATPase-binding cassette (ABC) proteins have important functions, with the secondary active transporters comprising a large proportion of all transporters in the genome of many species.

Plasma membrane transporters such as the dopamine transporter couple the movement of sodium ions down their electrochemical gradient to the transport of substrate against its concentration gradient. These proteins thus function as sodium cotransporters (Fig. 3). Substrate is loaded onto the transporter and moves with the sodium ion. Ion and transmitter are both released into the cytoplasm. It is important to note that the sodium

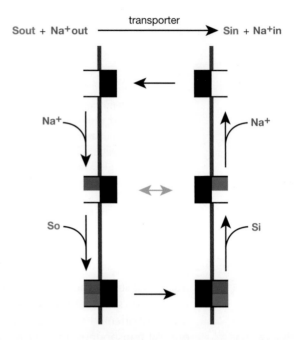

FIGURE 3. Na⁺-dependent cotransport is one form of secondary active transport. In this case, Na⁺ and substrate (S) must both occupy the transporter for translocation to the cytoplasmic face of the membrane. Binding can be ordered, and either Na⁺ or substrate alone is insufficient (*lavender double arrow* means not permitted).

cannot enter unless the substrate is also loaded. This required coupling constitutes a "rule" of transport that cannot be violated—if it were, the ionic gradient driving transport would dissipate, and the protein could not concentrate the substrate because it would leak back across the membrane. In addition, the binding and release of ion and substrate are sometimes highly ordered. After delivering substrate to the cytoplasm, the unloaded carrier reorients to the outside, a step required for net flux as noted above.

Ionic Coupling

The concentration of Na⁺ is approximately 12-fold higher on the outside of the cell than on the inside. Through the coupling mechanism, this is converted into an approximately 12-fold greater concentration of substrate on the inside of the cell. This is because at equilibrium, which may not be reached in the cell, the forward reaction must equal the reverse reaction:

$$[Na^+]_0 \times [S]_0 = [Na^+]_i \times [S]_i, \text{ or } [Na^+]_0 / [Na^+]_i = [S]_i / [S]_0$$

Some Na⁺ cotransporters move two sodium ions for each molecule of substrate. This stoichiometry confers an approximately 144-fold difference in the concentration of substrate between the inside and the outside because both forward and reverse reactions now depend on the square of the Na⁺ concentration:

$$[Na_0^+]^2[S_0] = [Na_i^+]^2[S_i]$$

The concentration of substrate achieved thus reflects the concentration gradient of ions and the stoichiometry of coupling, which is a feature of the transport protein. That is why the stoichiometry of transport reactions is very important—it makes strong predictions about the properties of transport that have direct physiological implications. Please keep in mind that this occurs at equilibrium. In terms of kinetics, the K_m of the entire reaction may differ significantly on the two sides of the membrane, consistent with binding of the substrate on one side of the membrane and release from the other. However, K_m is not the same as substrate-binding affinity. In particular, binding affinity may not differ on the two sides of the membrane, so we refer to the K_m as the "apparent" affinity.

Stoichiometry of coupling is especially important for the function of neurotransmitter transporters. Several neurotransmitter transporters have stoichiometries of two or three sodium ions for every molecule of substrate, generating steep concentration gradients of transmitter. In addition, the existence of transporter isoforms with different coupling stoichiometries enables the release of transmitter from one cell with a shallow gradient and uptake by another with a steep gradient. Another reason not to make uniformly steep gradients of all transmitters is simply that, as the stoichiometric ratio between ions and substrate increases, the rate of the reaction slows down because it takes more time to bind the additional ion. Because the ions also need to be replaced, such as by the Na^+ pump, it becomes energetically costly to the cell.

However, to our current knowlege, the cell cannot modify the stoichiometry of the transporter. The stoichiometry is an intrinsic property of protein that predicts the concentration gradients of substrate that can be reached at thermodynamic equilibrium. The cell can increase the number of transporters or change the ionic gradients that drive transport (which happens rarely), but there are few if any examples of changes in coupling. Recent work on the dopamine transporter has, however, suggested that the kinetic rates of uptake and efflux can be independently regulated by changes in phosphorylation, and this may contribute to the action of amphetamines (Gnegy et al. 2004; Khoshbouei et al. 2004).

Electrogenic Transport

So far, we have ignored the role of membrane potential in regulating the flux of substrate through transporters. According to the Nernst equation, however, for every 60-mV change in membrane potential, there will be a tenfold change in concentration of substrate if the entire reaction cycle involves the movement of one net charge. Because the cell is –60 mV or more negative than the outside, a plasma membrane neurotransmitter transporter will generate a cytoplasmic concentration 120 times greater on the inside than on the outside if the Na^+ stoichiometry is 1 and the cycle involves the inward movement of one net positive charge. Thus, the transporter depends not only on the ionic concentration gradient and coupling stoichiometry, but also on membrane potential and the amount of charge that the transporter moves. In the case where sodium and chloride anions are transported together with a neutral substrate, there would be no net movement of charge across the membrane, resulting in only a 12-fold difference in the concentration of substrate between the inside and outside of the membrane. This would be an electroneutral transporter.

In contrast, an electrogenic transporter moves net charge during the transport cycle and can therefore sense membrane potential. When these transporters are expressed in *Xenopus* oocytes or other cells, currents can be measured electrophysiologically. This is a powerful method to analyze transporter function. Figure 4 shows a plot of current versus voltage for the γ-aminobutyric acid (GABA) transporter. A current is induced by the addi-

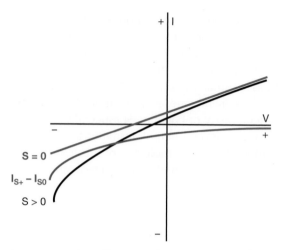

FIGURE 4. Electrogenic transport. In cells expressing a transporter for substrate S, the currents detected in the presence of S minus those in the absence of S are the currents associated with transport. However, a number of additional tests are required to show that this charge movement is stoichiometrically coupled to transport.

tion of substrate to the outside of the oocyte. The plot shows the currents in the presence and absence of GABA and the subtraction of the two plots. The addition of GABA results in an inward current. As the oocytes are held at more and more positive potentials using a voltage clamp, there is increasingly less of a GABA-induced current. This is a highly rectifying current. It can never reverse because the addition of GABA is always accompanied by a charge going inward; GABA added to the outside can never induce an outward current in the current mode.

Plasma Membrane Neurotransmitter Transporters

Na$^+$- and Cl$^-$-dependent Transporters

There are two families of plasma membrane neurotransmitter transporters. First, a relatively large family of sodium- and chloride-dependent transporters includes the glycine, GABA, norepinephrine, dopamine, and serotonin transporters. The other family of plasma membrane neurotransmitter transporters includes primarily the glutamate transporters. Although both families are sodium-dependent, the two do not share any sequence homology. In terms of function, the primary distinction is that the first depends on chloride, whereas the second does not. Let us first consider a prominent member of the Na$^+$- and Cl$^-$-dependent family of neurotransmitter transporters—the norepinephrine transporter (NET). The NET cotransports one sodium ion, one chloride anion, and one protonated norepinephrine molecule. Indeed, all of the monoamine transporters appear to recognize their substrate in the protonated state. As a result of this stoichiometry (with chloride contributing as well as sodium to the driving force) and the movement of one net charge, NET is predicted to concentrate norepinephrine approximately 1200-fold. In contrast, the dopamine transporter (DAT) moves two Na$^+$ ions, one Cl$^-$ anion, one protonated dopamine molecule, and hence two net charges, resulting in a predicted concentration gradient of approximately 1.44×10^5. However, the serotonin transporter (SERT) differs from both NET and DAT in one major way. In addition to cotransporting serotonin, Na^+, and Cl^-, SERT translocates one potassium out of the cell. As a result, SERT is electrically neutral because it does not move net charge, and so generates a more modest concentration gradient of approximately 1.2×10^3, like NET. Importantly, the serotonin transporter is indifferent to membrane potential and will not generate currents stoichiometrically coupled to transport. When the membrane is depolarized, DAT and NET (but not SERT) will also take up less neurotransmitter.

FUNCTIONAL SIGNIFICANCE OF TRANSPORTERS

What Is the Physiological Role of Plasma Membrane Neurotransmitter Transporters?

These transporters function to terminate the extracellular signal. In contrast to the enzymatic degradation of acetylcholine by acetylcholinesterase, transporters terminate signaling by most classical transmitters. Indicating the importance of reuptake for normal signaling and behavior, a number of psychoactive drugs interfere directly with monoamine reuptake. In addition to the efflux promoted by amphetamine, cocaine inhibits several of the monoamine transporters (not just DAT), and a number of antidepressants also target the monoamine transporters. Illustrating the properties of this mechanism, Marc Caron's group created DAT-null mutant mice (DAT "knockout [KO] mice") and, as expected, based on the response to cocaine, extracellular dopamine was elevated about fivefold, and the animals were hyperactive under certain conditions (Giros et al. 1996). Using electrochemistry to measure the dopamine release in striatal slices, the group showed that in wild-type animals, there is rapid removal of release dopamine. In mice heterozygous for the DAT mutation, the peak of dopamine released is reduced by half and clearance from the extracellular space is prolonged; in the homozygous KOs, decay of the dopamine signal is greatly prolonged. Indeed, clearance in this case is mediated primarily if not entirely by diffusion. However, brain homogenates from the KO animals contain dopamine levels that are 5% of wild type! In addition, dopamine continues to be released from striatal slices when evoked by electrical stimulation in slices from wild-type mice but is exhausted after two stimuli in slices from DAT KO mice. These observations indicate that the other major function of DAT, and by inference other plasma membrane neurotransmitter transporters, involves recycling the released transmitter to preserve vesicular pools. They predict that at least 95% of the dopamine released in the striatum is recycled rather than newly synthesized.

How Do Amphetamines Work?

Amphetamines act on monoamine transporters. Both dopamine and methamphetamine are substrates for the dopamine transporter. Amphetamine is thought to enter dopamine neurons through the dopamine transporter. Dopamine then binds to the cytoplasmically oriented substrate-recognition site on the transporter, driving the transporter back to the outside. This exchange reaction results in the uptake of amphetamine and the nonvesicular release of dopamine. However, amphetamines are very lipophilic and diffuse back through the membrane out of the cell. A single molecule of amphetamine thus becomes available for repeated rounds of dopamine release. This so-called exchange-diffusion mechanism is apparently responsible for the massive release of monoamines by amphetamines (Fisher and Cho 1979). However, in most monoamine neurons, the cytoplasm contains only small amounts of transmitter. Amphetamines are therefore thought to promote the release of dopamine from neurosecretory vesicles, where they are highly concentrated, as well as across the plasma membrane (Sulzer et al. 1993; Fon et al. 1997). Thus, amphetamine acts on both plasma membrane and vesicular transporters to promote the efflux of monoamine. The specificity of amphetamines for different monoamines—methamphetamine for dopamine and MDMA (ecstasy) for serotonin—indeed derives from relatively selective actions on the plasma membrane DAT and SERT, respectively.

Excitatory Amino-acid Transporters

The glutamate transporters catalyze the entry of three Na^+ ions, one H^+, as well as one glutamate molecule in exchange for the efflux of one K^+ ion (Zerangue and Kavanaugh 1995). Glutamate uptake thus involves the net movement of two positive charges and is predicted to generate gradients of at least 10^6. Importantly, the excitatory amino-acid transporters (EAATs) exist primarily on astrocytes, and when in neurons, they tend to be postsynaptic. Thus, they are not in the right place to recycle neurotransmitter. Rather, glutamine synthetase converts glutamate into glutamine after uptake into astrocytes, and the glutamine is transferred back to neurons and converted back to glutamate by the so-called phosphate-activated glutaminase before packaging into synaptic vesicles (Fig. 5).

Although recent data have suggested some reuptake of glutamate by excitatory neurons (Chen et al. 2002, 2004), glutamate is thus taken up primarily by glia. Interestingly, recent work has also shown the exocytotic release of glutamate by glia (Araque et al. 2000; Bezzi et al. 2004; Zhang et al. 2004), indicating competition between conversion to glutamine and vesicular packaging.

Because of coupling stoichiometry, EAATs are also electrogenic. However, in contrast to the highly inwardly rectifying currents usually attributable to coupled charge movement (Fig. 4), EAAT1 and EAAT3 show outward currents in response to added glutamate at depolarizing potentials (Wadiche et al. 1995). How can the addition of glutamate at positive potentials produce an outward current if glutamate uptake is stoichiometrically coupled to 2+ charge movement? In this case (and many others), the transporter shows

FIGURE 5. Glutamine-glutamate cycle. After exocytotic release from synaptic vesicles, glutamate is taken up by astrocytes, converted to glutamine by glutamine synthetase (GS), and transferred back to neurons through the action of glial system N and neuronal system A transporters (not shown) before it is converted to glutamate through the action of phosphate-activated glutaminase (PAG) and transported into vesicles.

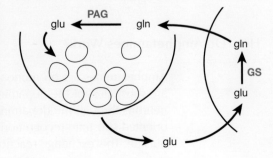

channel-like properties. However, it is not a channel for glutamate but rather for another ion, the chloride anion, that is not actually required for transport! Thus, EAAT1 and EAAT3 behave like glutamate-gated ion channels as well as glutamate transporters. The physiological significance of this dual function is not well understood. However, glutamate-gated chloride channel activity may mediate neurotransmission at a synapse in the retina (Eliasof and Werblin 1993; Arriza et al. 1997). In addition, glutamate-evoked currents mediated by EAATs in astrocytes have been used to measure the amount of glutamate released by neurons. After the induction of long-term potentiation (LTP), there was no increase in glutamate-evoked currents in astrocytes, arguing that the increased strength of the synapse after LTP must result from a postsynaptic mechanism and not from increased glutamate release (Luscher et al. 1998). Several of the monoamine transporters also show currents stoichiometrically uncoupled to flux. In particular, DAT shows small but physiologically significant uncoupled currents, and SERT expression confers a number of uncoupled currents that are essentially eliminated by its interaction with the t-SNARE protein syntaxin (Ingram et al. 2002; Quick 2003).

Vesicular Neurotransmitter Transport

In many cases, plasma membrane transport may not produce the gradients of neurotransmitter predicted by their ionic coupling. In chromaffin cells, electrochemistry indicates a cytoplasmic dopamine concentration as low as approximately 20 μM (Mosharov et al. 2003), suggesting that the plasma membrane transporter does not need to create a steep concentration gradient. The transport of monoamines into secretory vesicles that is required for their release by exocytosis presumably accounts for the relatively low cytoplasmic concentration. Plasma membrane monoamine transport may thus provide a kinetic advantage, but make a smaller than predicted contribution to net accumulation by a neuron at equilibrium. We know less about the cytoplasmic concentration of glutamate and other transmitters within neurons, although they may be considerably higher (up to 10 mM) in the case of glutamate (Shupliakov et al. 1995). In these cases, the plasma membrane transporters presumably have a larger role in equilibrium accumulation of transmitter by the cell.

Ionic Coupling and the Synaptic Vesicle Cycle

In contrast to the plasma membrane transporters, vesicular neurotransmitter transporters depend on an H^+-electrochemical gradient created by the vacuolar H^+-ATPase. Similar in structure to the mitochondrial F0-F1 ATPase that uses an H^+-electrochemical gradient to produce ATP, the vacuolar H^+ pump functions essentially in reverse, using the energy provided by ATP hydrolysis to create an H^+-electrochemical gradient (Forgac 2000). In addition to a pH gradient of 1.5–2 units, the pump creates a membrane potential, with the inside of the vesicles positive relative to the cytoplasm. The transporters then exchange lumenal H^+ for cytosolic transmitter. Protons running down their electrochemical gradient thus drive the movement of neurotransmitter into the vesicle, against its concentration gradient. In the case of vesicular monoamine (and acetylcholine) transport, which apparently couples to 2 H^+ and hence 1+ charge, the gradients approach 10^5 (Liu and Edwards 1997a).

In contrast to plasma membrane gradients of Na^+, K^+, or Cl^-, which are relatively stable, the H^+-electrochemical gradient across the membrane of a synaptic vesicle dissipates and forms again with each round of exocytosis and endocytosis, respectively. During exocytosis, the H^+-electrochemical gradient largely dissipates on fusion with the plasma membrane, which shows no substantial pH gradient. However, synaptic vesicle membrane can become available for a second round of release as quickly as 20 seconds after exocytosis (Ryan et al. 1993). Within 5 seconds after endocytosis, a pH gradient reemerges, allowing the vesicle to fill (Sankaranarayanan and Ryan 2000). Remarkably, up to 5–10,000 molecules of transmitter are predicted to enter the vesicle within the next approximately 15 seconds. If each vesicle contains a single vesicular transporter, the maximum turnover rates measured in vitro would not suffice to complete filling, at least in the case of vesicular monoamine transport, where the turnover number has actually been measured (~5–20/sec at 29°C) (Peter et al. 1994). After endocytosis, large Na^+ and Cl^- gradients present at the plasma membrane are presumably dissipated, but how this occurs remains unclear. In addition, the small size of the vesicle makes it likely that any pH gradient is dissipated by transmitter entry almost as soon as it forms, resulting in a very dynamic process that eventually fills the vesicle because of coordination of H^+ entry and exit with neurotransmitter packaging. There is yet another problem related to charge imbalance in the case of vesicular monoamine transport: two protons but only one charge exit because of the positive charge on the substrate. The vesicle should then accumulate positive charge at the expense of the pH gradient, and presumably this charge must be dissipated for filling with transmitter. One possibility is that the entry of an anion dissipates this positive charge. Although it remains unclear how ATP enters neurosecretory vesicles, ATP is stored at high concentrations and released at many synapses to activate postsynaptic purinergic receptors. A third problem concerns the osmolarity of the vesicle. In chromaffin cells, the neurotransmitter norepinephrine is out of solution in dense core vesicles. If it were in solution, the concentrations would be molar and the vesicle would explode because of osmotic forces. By removing the neurotransmitter from solution, the vesicle is thus able to store extremely high concentrations of transmitter. Interestingly, these osmotic forces may also drive exocytosis. A fusion pore is generated when the vesicle undergoes exocytosis. Because the chromaffin granule is hypertonic, water will rush in to dilate the fusion pore and produce full fusion of the vesicle (Troyer and Wightman 2002).

Can Quantal Size Be Altered Presynaptically?

Although considered invariant, quantal size can change under a number of conditions. Physiologists generally assume that changes in quantal size result from changes in the number or activity of postsynaptic receptors, and a change in AMPA (α-amino-3-hydroxy-5-methyl-4-isoazole) receptors does appear to underlie LTP in stratum radiatum of CA1 hippocampus (Malinow and Malenka 2002; Bredt and Nicoll 2003). However, it is clear that altering the amount of neurotransmitter produced or packaged can also change quantal size. Some of the first evidence for this came from the study of amphetamines that reduce quantal size by depleting vesicular monoamine stores (Sulzer et al. 1995). In particular, amphetamines act as weak bases to dissipate the H^+-electrochemical gradient across the vesicle membrane (Sulzer and Rayport 1990). This eliminates the driving force for vesicular monoamine transport and may even promote flux reversal. In addition, administration of L-Dopa increases and the VMAT (vesicle-membrane-associated transporter) inhibitor reserpine reduces quantal size measured amperometrically (Pothos et al. 1996, 1998). Surprisingly, the concentration of neurotransmitter within the vesicle does not appear to change. Rather, chromaffin granules appear to enlarge in response to L-Dopa and shrink in the presence of reserpine (Colliver et al. 2000). Genetic manipulation has also indicated that increases as well as decreases in the vesicular transporter can regulate quantal size (Fon et al. 1997; Pothos et al. 2000). Additional work has indicated the potential for presynaptic changes in quantal size at the neuromuscular junction (Van der Kloot 1990; Song et al. 1997). Thus, the amount of transporter present under physiological conditions appears to limit vesicle filling. There may also be a sensor for filling vesicles. One potential sensor is osmotic force, which will constrain the amount of transmitter stored if it remains soluble in the lumen of the vesicle. Reinhard Jahn's group has observed that the distribution of vesicle sizes visualized by electron microscopy is highly correlated with the quantal size distribution measured by amperometry (Bruns et al. 2000). The only force that could make the vesicles so similar in concentration is osmotic. Nonetheless, other investigators have proposed a receptor on the surface of the vesicles that senses the concentration of neurotransmitter in the vesicle (Ahnert-Hilger et al. 1998; Höltje et al. 2000; Pahner et al. 2003). Although there is evidence for G proteins on neurosecretory vesicles, and they may regulate transport activity, the receptors to which they couple would have to sense molar but not millimolar concentrations of transmitter, something they would not normally do, but might do in the acidic environment of the vesicle. In addition, recent work has suggested postfusional control of quantal size, by regulation of the fusion pore (Staal et al. 2004). Although not strictly related to vesicle filling, the large quanta observed under these conditions may reflect continued action of the vesicle transporter even after exocytosis. Transmitter release by kiss-and-run, which would presumably involve postfusional control of exocytosis, would also require high rates of filling because the vesicle becomes available for a second round of fusion much more quickly after the first than in the case of clathrin-dependent endocytosis.

Vesicular Amino-acid Transport

In the case of GABA and glycine, there is no net charge on the substrate—they are both zwitterions. Although we do not know the precise stoichiometry of proton exchange mediated by their uptake into synaptic vesicles, the entry of GABA in exchange for protons results in the efflux of protons equal in number to the efflux of charge. In this case, the transport activity depends more equally than vesicular monoamine transport on the pH gradient and membrane potential. Naturally, transport activity will reflect the conditions under which it is measured—the presence of only ΔpH (no $\Delta\Psi$) will make transport depend on only ΔpH, for example. In the case of glutamate, which is anionic, the efflux of one proton (we do not know the actual stoichiometry) would result in the efflux of two charges. The glutamate transporter thus depends more on $\Delta\Psi$ than on ΔpH. Ideally, the vesicular monoamine and acetylcholine transporters that depend principally on ΔpH should therefore reside in vesicles with a large pH gradient. Similarly, those dependent primarily on charge should reside on vesicles with a large $\Delta\Psi$. A key question thus is: How do secretory vesicles regulate ΔpH and $\Delta\Psi$? This is accomplished primarily by chloride channels (Fig. 6). Chloride channels enable chloride entry into vesicles where they

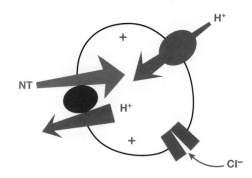

FIGURE 6. Transport of all classical neurotransmitters into synaptic vesicles depends on a proton electrochemical gradient, but different transmitters vary in the extent of dependence on the chemical (ΔpH) and electrical ($\Delta\Psi$) components. Chloride influx can influence the relative proportion of these two components: Chloride will neutralize the positive charge inside the vesicle, dissipate $\Delta\Psi$, and so increase the activity of the H$^+$ pump, increasing ΔpH.

can neutralize the charge on protons. This dissipates $\Delta\Psi$ and permits the proton pump to increase the number of protons pumped into the vesicle, making secretory vesicles, endosomes, and lysosomes more acidic. Chloride channels thus enhance the filling of vesicles that depend on pH gradients. However, chloride channels would dissipate the membrane potential required for transport of glutamate into synaptic vesicles.

The Vesicular Monoamine Transporter, Parkinson's Disease, and Dopamine Metabolism

Molecular cloning has now shown that the three classes of vesicular transport activities (dependent primarily on pH gradient, membrane potential, or both) reflect three distinct gene families with no obvious sequence similarity. The first of these families was identified by selection in a neurotoxin.

MPTP (1-methyl-4-phenyl-1,2,3,6-tetrahydropyridine) produces a model of Parkinson's disease (PD). MPTP causes selective degeneration of dopamine neurons in the substantia nigra and to a lesser extent the ventral tegmental area, very similar to idiopathic PD. MPTP has thus provided a way to study the pathogenesis of PD and so develop treatment aimed at disease progression as well as symptoms. MPTP crosses the blood–brain barrier readily because it is lipophilic. Outside neurons, monoamine oxidase (MAO) B in glia converts MPTP to MPP$^+$ (1-methyl-4-phenylpyridium), the active metabolite. MPP$^+$ is then taken up by plasma membrane monoamine transporters and enters mitochondria where it probably both inhibits respiration and activates the apoptotic pathway of cell death. The selectivity for dopamine neurons derives from the cell-specific expression of plasma membrane monoamine transporters. However, a number of monoamine neurons and adrenal chromaffin cells take up large amounts of MPP$^+$ without showing much toxicity. Taking advantage of these MPP$^+$-resistant cells, we identified a sequence that would protect against MPP$^+$ after transfer into sensitive cells (Liu et al. 1992a). This sequence protects cells by sequestering MPP$^+$ inside secretory vesicles, lowering its concentration in the cytoplasm, and so preventing its entry into mitochondria. We then showed that this protein transports monoamines into vesicles with all of the properties previously described for vesicular monoamine transport (Liu et al. 1992b). However, it is important to point out that serotonergic, noradrenergic neurons, and chromaffin cells all take up large amounts of MPP$^+$ and express the neuronal isoform vesicular monoamine transporter 2 (VMAT2). (VMAT1 is expressed by non-neuronal cells outside the central nervous system, such as adrenal chromaffin cells.) It is thus likely that there are other mechanisms to prevent or reduce the toxicity of MPP⁺. Although inhibiting VMAT pharmacologically or genetically greatly potentiates MPTP toxicity, we still do not fully understand the basis for dopamine cell selectivity.

What is the relevance of vesicular monoamine transport for PD when MPP⁺ has been implicated in only a few cases? Many investigators have speculated that normal neurotransmitter dopamine acts as an endogenous toxin. Dopamine oxidizes very rapidly to produce free radicals and damages many cultured cells more potently than MPP⁺. We therefore proposed that VMAT functions to lower the concentration of dopamine in the cytoplasm, where it may oxidize and damage cell constituents. To test the hypothesis that VMAT protects against toxicity of endogenous dopamine, we and other investigators knocked out the neuronal isoform VMAT2 in mice (Fon et al. 1997; Takahashi et al. 1997; Wang et al. 1997). However, mice lacking VMAT2 do not live very long because they are unable to release dopamine, serotonin, and norepinephrine. The pups do not suckle and hence die within days after birth, so it is not possible to assess neural degeneration.

On the other hand, the brains of newborn VMAT2 KOs contain essentially no dopamine, norepinephrine, or serotonin. This means that 97–99% of monoamines are loaded into vesicles, and if not packaged, they do not accumulate elsewhere. This makes sense if dopamine is toxic, because cytoplasmic accumulation would be harmful. But what accounts for the loss of monoamines in the KO animals? One possibility is that monoamines not loaded into vesicles are degraded by MAO. To test this hypothesis, Ted Fon administered MAO inhibitor to the VMAT2 KO mice. He observed that levels of serotonin in MAO inhibitor-treated VMAT2 KOs exceeded the levels of untreated wild-type mice (Fon et al. 1997). In contrast, dopamine levels did not increase at all in the KO animals treated with MAO inhibitor. This means that serotonin levels are regulated by the VMAT and MAO, but additional mechanisms must exist to limit cytoplasmic dopamine accumulation. Importantly, the decreased levels of dopamine do not result from a decreased number of neurons synthesizing the neurotransmitter. The same number of tyrosine-hydroxylase-positive cells is seen in VMAT2 KO and wild-type mice. Dopaminergic projections in the mice also remain intact. We presume that the other mechanism leading to a decrease in intracellular levels of dopamine is feedback inhibition of tyrosine hydroxylase, by either L-Dopa or dopamine. Thus, midbrain neurons stop making dopamine if it starts to accumulate in the cytoplasm. Tryptophan hydroxylase, on the other hand, is not subject to feedback inhibition. This presumably explains why serotonin but not dopamine accumulates in the presence of MAO inhibitors in VMAT2 KO animals. Because cytoplasmic dopamine levels are tightly regulated, it is difficult to manipulate them—inhibiting both VMAT2 and MAO appears to have relatively little effect. Dopamine levels are only elevated if VMAT and MAO inhibition are combined with exogenous L-Dopa, bypassing feedback inhibition of tyrosine hydroxylase.

Somatodendritic Monoamine Release

Although midbrain dopamine neurons are considered to release dopamine primarily from axon terminals in the striatum, they also release transmitter at their cell bodies and dendrites in the midbrain. Importantly, somatodendritic dopamine release has been implicated in the development of behavioral sensitization, which is generally considered a model for drug craving. Behavioral sensitization involves the augmented response to a single dose of psychostimulant that results from previous drug treatment and can persist for long periods after drug administration (Robinson and Berridge 2000; Vezina 2004). Originally described in terms of the increased locomotor activity that can be elicited for up to 1 year after drug exposure (Kalivas and Stewart 1991), sensitization is accompanied by an elevation of extracellular dopamine in the striatum (Robinson and Becker 1982; Kolta et al. 1989; Paulson and Robinson 1995). Although amphetamine injected directly into the midbrain does not affect acute locomotor activity, behavioral sensitization can only be elicited by systemic administration or direct midbrain administration. Indeed, injection into terminal fields such as the nucleus accumbens does not produce sensitization (Dougherty and Ellinwood 1981; Kalivas and Weber 1988). In addition, sensitization to amphetamine generally requires the activation of D1 but not D2 dopamine receptors (Stewart and Vezina 1989; Bjijou et al. 1996). In contrast, D2 antagonists generally have no effect. Because dopamine neurons do not express D1 receptors, dopamine released in the midbrain must activate receptors on afferent terminals.

Although amphetamine releases dopamine by promoting flux reversal, considerable evidence also supports the somatodendritic release of dopamine, norepinephrine, and serotonin in response to physiological activity (Geffen et al. 1976; Robertson et al. 1991; Heeringa and Abercrombie 1995; Jaffe et al. 1998). However, the mechanism by which physiological activity releases dopamine in the midbrain remains unclear. Sensitivity to reserpine and dependence on Ca^{2+} suggest an exocytotic mechanism (Björklund and Lindvall 1975; Cheramy et al. 1981; Rice et al. 1997; Beckstead et al. 2004). On the other hand, somatodendritic release can be Ca^{2+}-independent (Chen and Rice 2001) and insensitive to the clostridial toxins that cleave SNARE proteins required for exocytosis (Bergquist et al. 2002). In addition, activation of excitatory afferents to the midbrain stimulates dopamine release that is blocked by a more specific inhibitor of the DAT, albeit over a longer timescale (Falkenburger et al. 2001), suggesting that physiological stimulation may also promote flux reversal mediated by DAT.

VMAT Trafficking

The subcellular location of VMAT2 determines which vesicles are capable of exocytotic monoamine release. Supporting an exocytotic mechanism for somatodendritic release, VMAT2 localizes prominently to the cell body and dendrites of monoamine neurons in the brain stem (Peter et al. 1995). However, the protein resides on a poorly characterized population of tubulovesicular membranes at this location (Nirenberg et al. 1996). To understand the mechanism and indeed the role of physiological dopamine release from cell bodies and dendrites, we became interested in the trafficking of VMAT2.

Rat pheochromocytoma PC12 cells contain both large, dense, core vesicles (LDCVs) and small, synaptic-like microvesicles (SLMVs). Previous work had shown that these cells store monoamines in LDCVs and acetylcholine in SLMVs (Bauerfeind et al. 1993), suggesting that the VMATs would localize to LDCVs, whereas the closely related vesicular acetylcholine transporter (VAChT) (Alfonso et al. 1993; Roghani et al. 1994; Varoqui and Erickson 1996) would localize to SLMVs. Subsequent work confirmed the differential localization of these two transport proteins in PC12 cells (Liu et al. 1994; Weihe et al.

1996; Liu and Edwards 1997b). We then used heterologous expression to identify the sequences required for these differences in trafficking.

Dileucine Motif

In non-neural as well as neural cells, VMAT2 and VAChT internalize on reaching the cell surface, and a carboxy-terminal dileucine-like motif is responsible for the endocytosis of both proteins (Tan et al. 1998). The differential sorting in PC12 cells must therefore require more specific sorting determinants. Interestingly, the VMATs and VAChT diverge in sequence just upstream of the dileucine-like motif, and these residues are highly conserved across species within the two subfamilies:

VMAT2	K *E E* K M A **I L**
VAChT	R *S E* R D V **L L**

Indeed, we have found that glutamates four and five residues upstream of the dileucine (italicized) are required for the sorting of VMAT2 to dense core vesicles (Krantz et al. 2000), and the serine five residues upstream in VAChT are required for sorting to SLMVs. Furthermore, phosphorylation of this serine (underlined) confers the negative charge required to direct this protein to dense core vesicles (Krantz et al. 2000). In the absence of phosphorylation, VAChT targets to synaptic vesicles. The sorting decision appears to be made in the *trans*-Golgi network (TGN) (Fig. 7).

These results raise the possibility that the same sequence directs VMAT2 to somatodendritic vesicles in neurons. Although it has not been tested directly, the results also predict that when phosphorylated, VAChT targets somatodendritic vesicles. Interestingly, the cell bodies and dendrites of cholinergic neurons show strong labeling for VAChT (Gilmor et al. 1996), raising the possibility of somatodendritic release of ACh as well as monoamines.

Acidic Cluster

Many proteins enter dense core vesicles as they bud from the TGN but are removed in the course of a maturation process that follows budding. Among the best studied, furin is a protease that normally processes peptides in the TGN but still enters dense core vesi-

FIGURE 7. A dileucine-like signal directs VMAT2 (*red*) to LDCVs at the level of the TGN. VAChT targets to constitutive secretory vesicles (CSVs), but when phosphorylated on a serine upstream of its dileucine motif, VAChT also targets to LDCVs. After constitutive delivery to the plasma membrane, VAChT enters synaptic vesicles by endocytosis.

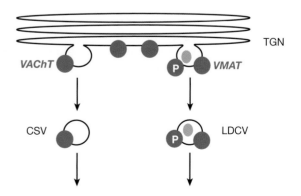

FIGURE 8. VMAT2 remains on maturing LDCVs, whereas furin undergoes removal. Immature LDCVs (iLDCVs) contain both proteins, whereas mature LDCVs (mLDCVs) contain only VMAT2. Furin is removed in a budding process that requires the adaptor proteins AP-1 and PACS as well as clathrin.

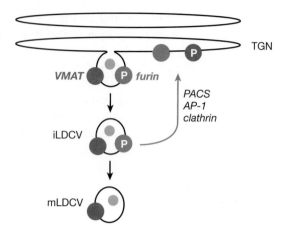

cles at substantial levels. Furin is removed as dense core vesicles mature and phosphorylation by casein kinase of an acidic patch at the carboxyl terminus promotes the retrieval of furin (Molloy et al. 1999) (Fig. 8). Two serines in an acidic patch at the carboxyl terminus of VMAT2 also undergo phosphorylation by casein kinase 2 (Krantz et al. 1997). However, VMAT2 remains on dense core vesicles as they mature, whereas furin is removed. How does the same sequence result in retrieval of furin but retention of VMAT2 on maturing LDCVs? Presumably, VMAT2 is dephosphorylated by a phosphatase, whereas furin is not. Surprisingly, deletion of the acidic cluster at the carboxyl terminus of VMAT2 actually promotes removal during LDCV maturation, further suggesting that phosphorylation inactivates a signal for retention (Waites et al. 2001). And what is the relationship between the dileucine-based motif and this acidic cluster? Recent work has begun to suggest that the extended dileucine motif promotes sorting to LDCVs at the level of the TGN, whereas the acidic cluster operates afterward, on maturation of the vesicle. Because sorting to LDCVs in the TGN has been proposed to rely on lumenal rather than cytoplasmic sequences (Arvan and Castle 1998), these results provide some of the first evidence for specific cytosolic machinery in the biogenesis of LDCVs. We are now in the process of determining whether the mechanism of trafficking for VMAT and VAChT that we have observed in PC12 cells is similar in neurons, and in particular, whether they contribute to somatodendritic localization and transmitter release. We would also be very interested in determining whether the trafficking of the transporters changes as a function of drug exposure; recent work has shown changes in the distribution of VMAT2 in vivo in response to cocaine and amphetamines (Brown et al. 2001; Riddle et al. 2002; Sandoval et al. 2003; Truong et al. 2004), changes that may influence toxicity as well as the acute behavioral response. The results in PC12 cells may begin to suggest a mechanism for these observations.

Vesicular GABA Transport, Systems N and A

Corresponding to the different dependencies on pH and membrane potential, three distinct families of proteins transport neurotransmitters into synaptic vesicles. VMAT and

VAChT, primarily dependent on ΔpH, belong to the first family identified. The second family, dependent more equally on ΔpH and ΔΨ, includes the vesicular GABA transporter (VGAT). This protein family also includes the System N and A transporters implicated in the glutamine-glutamate cycle as well as many other metabolic pathways (Chaudhry et al. 2002). The third family includes the vesicular glutamate transporters, which are required for excitatory transmission. Indeed, every cell has high concentrations of glutamate, but only cells that release it as an extracellular signal require vesicular glutamate transport. Identification of the transport protein would thus provide an unequivocal marker for glutamatergic neurons and help to resolve a number of controversies.

Vesicular Glutamate Transport

Grown as single cells on glial islands in culture, serotonin and dopamine neurons form excitatory glutamatergic synapses onto themselves (Johnson 1994; Sulzer et al. 1998). This suggested that monoamine neurons release glutamate as well as monoamine, but many considered this a tissue culture artifact of dedifferentiation in vitro and argued that the release of glutamate had little physiological significance in vivo. Similarly, astrocytes have been reported to release glutamate in culture (Araque et al. 2000; Bezzi et al. 2004; Montana et al. 2004; Zhang et al. 2004). Identification of the vesicular glutamate transporters would help to establish the physiological significance of these observations.

Several years ago, genetic studies in *Caenorhabditis elegans* identified the *eat-4* mutation as important for eating, foraging, and chemotaxis (Avery 1993; Lee et al. 1999). These behaviors all involve glutamate, and the EAT-4 protein was found to be expressed only in glutamate neurons, suggesting a presynaptic role in glutamate release. The sequence of the polytopic membrane protein showed similarity to a large number of inorganic phosphate transporters in mammals. In fact, one member of this family, the brain-specific Na^+-dependent phosphate inorganic transporter (BNPI) was closely related in sequence to *eat-4* (Ni et al. 1994). BNPI was considered to be important in ATP synthesis and energy metabolism at the nerve terminal, but its expression in a subset of glutamate neurons suggested a more specific role in glutamate release. Remarkably, the last step of the glutamine-glutamate cycle involves the conversion of glutamine to glutamate by glutaminase, also known as phosphate-activated glutaminase (PAG) because it is activated by inorganic phosphate. We therefore considered that this transporter might contribute to glutamate release by pumping inorganic phosphate into the nerve terminal where it could locally activate PAG to provide a source of glutamate for synaptic vesicles.

To test this hypothesis, we raised antibodies to BNPI and found that the protein indeed localizes to nerve terminals at asymmetric (excitatory) synapses by immunoelectron microscopy (Bellocchio et al. 1998). However, the labeling was not found at the plasma membrane but rather over synaptic vesicles in the cytoplasm. We then considered that perhaps exocytosis delivers BNPI to the plasma membrane where it can take up phosphate and hence activate PAG to replace the glutamate just released. This regulatory mechanism would couple glutamate production to glutamate release. On the other hand, the location on vesicles and several other observations raised the alternative possibility that BNPI acted as a vesicular glutamate transporter. Supporting this possibility, a number of other so-called type-I phosphate transporters were found to transport organic anions with higher affinity than phosphate (Busch et al. 1996; Verheijen et al. 1999). Indeed, we found that BNPI expression conferred vesicular glutamate transport with all of the properties previously observed in native synaptic vesicles from brain, including specificity for glutamate (it does not recognize aspartate), an unusual biphasic dependence on chloride concentration, and particular reliance on membrane potential rather than pH gradient (Bellocchio et al. 2000). At around the same time, Reinhard Jahn's groups showed that when expressed in inhibitory neurons, BNPI enables them to release glutamate (Takamori et al. 2000). Thus, the protein is sufficient to make neurons excitatory.

The vesicular glutamate transporter has been difficult to study. No specific drugs activate or inhibit this transporter, and its basic properties remain poorly understood. In addition, BNPI (renamed VGLUT1) is expressed by only a subset of known glutamate neurons. More recently, we and other investigators have identified a second isoform, also originally identified as a phosphate transporter (Fremeau et al. 2001; Herzog et al. 2001; Takamori et al. 2001). These two transporters have a complementary distribution in brain. VGLUT1 is expressed in the cortex, hippocampus, and cerebellar cortex. The deep cerebellar and brain stem nuclei express VGLUT2 but not VGLUT1. Although many presynaptic proteins involved in transmitter release exist in multiple isoforms, their expression is highly overlapping, and this complementary pattern is unique for the VGLUTs. Together, the distribution of these transporters accounts for glutamate release by essentially all established excitatory neurons. However, they are not expressed by dopamine or serotonin neurons, suggesting that these cells either do not release glutamate in vivo or do not use VGLUT1 or VGLUT2 to store glutamate.

In the hippocampus, VGLUT1[+] puncta localize to dendritic fields, whereas VGLUT2[+] boutons terminate on cell bodies in the pyramidal layer. In the molecular layer of the cerebellum, parallel fibers express VGLUT1, whereas climbing fibers express VGLUT2. What properties of these synapses correlate with expression of the two isoforms? In general, VGLUT1 is expressed at synapses where there is a low probability of transmitter release. Vesicle release at these synapses occurs only rarely or after multiple stimulations, whereas VGLUT2[+] climbing fiber synapses show a high probability of release. In addition, parallel fiber synapses show more plasticity than climbing fibers. The expression of VGLUT1 and VGLUT2 may therefore correlate with the potential for plasticity as well as the probability of release. But how do the two isoforms differ? VGLUT1 and VGLUT2 transport glutamate with similar properties (they have the same intrinsic transport activity). Rather, they appear to differ in trafficking, at least in PC12 cells (Fremeau et al. 2001). More recent work suggests that they also show major differences in trafficking in neurons.

We and other investigators have also identified an interesting third isoform, VGLUT3. The mRNA for this isoform is expressed at much lower levels in brain than VGLUT1 and VGLUT2 (Fremeau et al. 2002; Gras et al. 2002; Takamori et al. 2002). In addition, it is expressed in liver and kidney as well as brain, whereas VGLUT1 and VGLUT2 are essentially brain-specific. Within the brain, VGLUT3 mRNA is expressed in the striatum, cortex, hippocampus, ventral tegmental area, substantia nigra, raphe, and cerebellum. This suggests that VGLUT3 is the isoform enabling glutamate release by monoamine neurons. Indeed, VGLUT3 does not appear to be expressed by neurons generally accepted to release glutamate. In the cortex and hippocampus, VGLUT3 is associated with GABAergic interneurons. In the striatum, cholinergic interneurons express VGLUT3, and VGLUT3 is also expressed by astrocytes. By immunoelectron microscopy, VGLUT3 in the hippocampus is found to be expressed at symmetrical inhibitory synapses (Fremeau et al. 2002). However, the protein is also detectable at dendrites, where glutamate presumably acts as a retrograde signal released by the postsynaptic cell. There is evidence in the hippocampus and cortex that glutamate can indeed be used as a retrograde signal (Zilberter 2000; Harkany et al. 2004). In the retina, antibodies to VGLUT3 label amacrine cells that are not stained for glutamic acid decarboxylase (GAD), suggesting that VGLUT3 marks a subset of amacrine cells that is excitatory.

Finally, VGLUT3 partially colocalizes with VMAT2 but not TH, consistent with expression by serotonergic neurons. It remains unclear whether dopamine neurons express VGLUT3 or another isoform, possibly VGLUT2 (Dal Bo et al. 2004). However, it is possible that within certain monoamine populations, the monoamine and glutamate release sites may be distinct, accounting for the lack of VGLUT3 colocalization with many monoamine markers (Sulzer et al. 1998). Distinct synaptic release sites could indeed contribute to the different roles of glutamate and monoamines in signaling.

CONCLUSIONS

Neurotransmitter transporters have fundamental roles in both normal signaling and in drug abuse. In normal signaling, they recycle released transmitter so that the nerve terminal can keep up with high rates of release. At the same time, they control extracellular concentrations of transmitter and may contribute to synaptic plasticity by influencing quantal size. With regard to drug abuse, many psychostimulants act directly on monoamine transporters. However, recent work has also begun to indicate important influences on drug abuse beyond simple acute inhibition. In the case of the dopamine transporter, phosphorylation appears to control the direction of flux and may thus contribute to long-term adaptation. Somatodendritic dopamine release, through either a vesicular or nonvesicular mechanism, is required to induce behavioral sensitization to amphetamines. The corelease of glutamate by monoamine neurons suggests further mechanisms for plasticity. Implicated in drug abuse through the direct action of psychostimulants, neurotransmitter transporters thus also appear to have a role in the long-term changes central to the addicted state.

THE FUTURE

The high incidence of substance abuse and escalating need for improved medications and strategies to treat drug addiction provide a compelling case to clarify physiological and pathophysiological mechanisms governing neurotransmitter transport. In contrast to a role confined simply to drug action, neurotransmitter transporters now appear to serve as a major contributor to adaptive processes that underlie addiction. Transporter function can be modulated by physiological changes in response to drugs of abuse or by pathological processes engendered by toxic effects of drugs (e.g., amphetamines). The regulatory mechanisms remain poorly understood.

REFERENCES

Ahnert-Hilger G., Nurnberg B., Exner T., Schafer T., and Jahn R. 1998. The heterotrimeric G protein GO2 regulates catecholamine uptake by secretory vesicles. *EMBO J.* **17:** 406–413.

Alfonso A., Grundahl K., Duerr J.S., Han H.-P., and Rand J.B. 1993. The *Caenorhabditis elegans unc-17* gene: A putative vesicular acetylcholine transporter. *Science* **261:** 617–619.

Araque A., Li N., Doyle R.T., and Haydon P.G. 2000. SNARE protein-dependent glutamate release from astrocytes. *J. Neurosci.* **20:** 666–673.

Arriza J.L., Eliasof S., Kavanaugh M.P., and Amara S.G. 1997. Excitatory amino acid transporter 5, a retinal glutamate transporter coupled to a chloride conductance. *Proc. Natl. Acad. Sci.* **94:** 4155–4160.

Arvan P. and Castle D. 1998. Sorting and storage during secretory granule biogenesis: Looking backward and looking forward. *Biochem. J.* **332:** 593–610.

Avery L. 1993. The genetics of feeding in *Caenorhabditis elegans*. *Genetics* **133:** 897–917.

Bauerfeind R., Regnier-Vigouroux A., Flatmark T., and Huttner W.B. 1993. Selective storage of acetylcholine, but not catecholamines, in neuroendocrine synaptic-like microvesicles of early endosomal origin. *Neuron* **11:** 105–121.

Beckstead M.J., Grandy D.K., Wickman K., and Williams J.T. 2004. Vesicular dopamine release elicits an inhibitory postsynaptic current in midbrain dopamine neurons. *Neuron* **42:** 939–946.

Bellocchio E.E., Reimer R.J., Fremeau R.T.J., and Edwards R.H. 2000. Uptake of glutamate into synaptic vesicles by an inorganic phosphate transporter. *Science* **289:** 957–960.

Bellocchio E.E., Hu H., Pohorille A., Chan J., Pickel V.M., and Edwards R.H. 1998. The localization of the brain-specific inorganic phosphate transporter suggests a specific presynaptic role in glutamatergic transmission. *J. Neurosci.* **18:** 8648–8659.

Bergquist F., Niazi H.S., and Nissbrandt H. 2002. Evidence for different exocytosis pathways in dendritic and terminal dopamine release in vivo. *Brain Res.* **950:** 245–253.

Bezzi P., Gundersen V., Galbete J.L., Seifert G., Steinhauser C., Pilati E., and Volterra A. 2004. Astrocytes contain a vesicular compartment that is competent for regulated exocytosis of glutamate. *Nat. Neurosci.* **7:** 613–620.

Bjijou Y., Stinus L., Le Moal M., and Cador M. 1996. Evidence for selective involvement of dopamine D1 receptors of the ventral tegmental area in the behavioral sensitization induced by intra-ventral tegmental area injections of D-amphetamine. *J. Pharmacol. Exp. Ther.* **277:** 1177–1187.

Björklund A. and Lindvall O. 1975. Dopamine in dendrites of substantia nigra neurons: Suggestions for a role in dendritic terminals. *Brain Res.* **83:** 531–537.

Bredt D.S. and Nicoll R.A. 2003. AMPA receptor trafficking at excitatory synapses. *Neuron* **40:** 361–379.

Brown J.M., Hanson G.R., and Fleckenstein A.E. 2001. Regulation of the vesicular monoamine transporter-2: A novel mechanism for cocaine and other psychostimulants. *J. Pharmacol. Exp. Ther.* **296:** 762–767.

Bruns D., Riedel D., Klingauf J., and Jahn R. 2000. Quantal release of serotonin. *Neuron* **28:** 205–220.

Busch A.E., Schuster A., Waldegger S., Wagner C.A., Zempel G., Broer S., Biber J., Murer H., and Lang F. 1996. Expression of a renal type I sodium/phosphate transporter (NaPi-1) induces a conductance in *Xenopus* oocytes permeable for organic and inorganic anions. *Proc. Natl. Acad. Sci.* **93:** 5347–5351.

Chaudhry F.A., Reimer R.J., and Edwards R.H. 2002. The glutamine commute: Take the N line and transfer to the A. *J. Cell Biol.* **157:** 349–355.

Chen B.T. and Rice M.E. 2001. Novel Ca^{++} dependence and time course of somatodendritic dopamine release: Substantia nigra versus striatum. *J. Neurosci.* **21:** 7841–7847.

Chen W., Aoki C., Mahadomrongkul V., Gruber C.E., Wang G.J., Blitzblau R., Irwin N., and Rosenberg P.A. 2002. Expression of a variant form of the glutamate transporter GLT1 in neuronal cultures and in neurons and astrocytes in the rat brain. *J. Neurosci.* **22:** 2142–2152.

Chen W., Mahadomrongkul V., Berger U.V., Bassan M., DeSilva T., Tanaka K., Irwin N., Aoki C., and Rosenberg P.A. 2004. The glutamate transporter GLT1a is expressed in excitatory axon terminals of mature hippocampal neurons. *J. Neurosci.* **24:** 1136–1148.

Cheramy A., Leviel V., and Glowinski J. 1981. Dendritic release of dopamine in the substantia nigra. *Nature* **289:** 537–542.

Colliver T.L., Pyott S.J., Achalabun M., and Ewing A.G. 2000. VMAT-mediated changes in quantal size and vesicular volume. *J. Neurosci.* **20:** 5276–5282.

Dal Bo G., St-Gelais F., Danik M., Williams S., Cotton M., and Trudeau L.E. 2004. Dopamine neurons in culture express VGLUT2 explaining their capacity to release glutamate at synapses in addition to dopamine. *J. Neurochem.* **88:** 1398–1405.

Dougherty G.G., Jr. and Ellinwood E.H., Jr. 1981. Chronic D-amphetamine in nucleus accumbens: Lack of tolerance or reverse tolerance of locomotor activity. *Life Sci.* **28:** 2295–2298.

Eliasof S. and Werblin F. 1993. Characterization of the glutamate transporter in retinal cones of the tiger salamander. *J. Neurosci.* **13:** 402–411.

Falkenburger B.H., Barstow K.L., and Mintz I.M. 2001. Dendrodendritic inhibition through reversal of dopamine transport. *Science* **293:** 2465–2470.

Fisher J.F. and Cho A.K. 1979. Chemical release of DA from striatal homogenates: Evidence for an exchange diffusion model. *J. Pharm. Exp. Ther.* **208:** 203–209.

Fon E.A., Pothos E.N., Sun B.-C., Killeen N., Sulzer D., and Edwards R.H. 1997. Vesicular transport regulates monoamine storage and release but is not essential for amphetamine action. *Neuron* **19:** 1271–1283.

Forgac M. 2000. Structure, mechanism and regulation of the clathrin-coated vesicle and yeast vacuolar H(+)-ATPases. *J. Exp. Biol.* **203:** (Pt 1) 71–80.

Fremeau R.T., Jr., Troyer M.D., Pahner I., Nygaard G.O., Tran C.H., Reimer R.J., Bellocchio E.E., Fortin D., Storm-Mathisen J., and Edwards R.H. 2001. The expression of vesicular glutamate transporters defines two classes of excitatory synapse. *Neuron* **31:** 247–260.

Fremeau R.T., Jr., Burman J., Qureshi T., Tran C.H., Proctor J., Johnson J., Zhang H., Sulzer D., Copenhagen D.R., Storm-Mathisen J., Reimer R.J., Chaudhry F.A., and Edwards R.H. 2002. The identification of vesicular glutamate transporter 3 suggests novel modes of signaling by glutamate. *Proc. Natl. Acad. Sci.* **99:** 14488–14493.

Geffen L.B., Jessell T.M., Cuello A.C., and Iversen L.L. 1976. Release of dopamine from dendrites in rat substantia nigra. *Nature* **260:** 258–260.

Gilmor M.L., Nash N.R., Roghani A., Edwards R.H., Yi H., Hersch S.M., and Levey A.I. 1996. Expression of the putative vesicular acetylcholine transporter in rat brain and localization in cholinergic synaptic vesicles. *J. Neurosci.* **16:** 2179–2190.

Giros B., Jaber M., Jones S.R., Wightman R.M., and Caron M.G. 1996. Hyperlocomotion and indifference to cocaine and amphetamine in mice lacking the dopamine transporter. *Nature* **379:** 606–612.

Gnegy M.E., Khoshbouei H., Berg K.A., Javitch J.A., Clarke W.P., Zhang M., and Galli A. 2004. Intracellular Ca^{2+} regulates amphetamine-induced dopamine efflux and currents mediated by the human dopamine transporter. *Mol. Pharmacol.* **66:** 137–143.

Gras C., Herzog E., Bellenchi G.C., Bernard V., Ravassard P., Pohl M., Gasnier B., Giros B., and El Mestikawy S. 2002. A third vesicular glutamate transporter expressed by cholinergic and serotoninergic neurons. *J. Neurosci.* **22:** 5442–5451.

Harkany T., Holmgren C., Hartig W., Qureshi T., Chaudhry F.A., Storm-Mathisen J., Dobszay M.B., Berghuis P., Schulte G., Sousa K.M., et al. 2004. Endocannabinoid-independent retrograde signaling at inhibitory synapses in layer 2/3 of neocortex: Involvement of vesicular glutamate transporter 3. *J. Neurosci.* **24:** 4978–4988.

Heeringa M.J. and Abercrombie E.D. 1995. Biochemistry of somatodendritic dopamine release in substantia nigra: An in vivo comparison with striatal dopamine release. *J. Neurochem.* **65:** 192–200.

Herzog E., Bellenchi G.C., Gras C., Bernard V., Ravassard P., Bedet C., Gasnier B., Giros B., and El Mestikaway S. 2001. The existence of a second vesicular glutamate transporter specifies subpopulations of glutamatergic neurons. *J. Neurosci.* **21:** RC181.

Höltje M., von Jagow B., Pahner I., Lautenschlager M., Hörtnagl H., Nürnberg B., Jahn R., and Ahnert-Hilger G. 2000. The neuronal monoamine transporter VMAT2 is regulated by the trimeric GTPase Go(2). *J. Neurosci.* **20:** 2131–2141.

Ingram S.L., Prasad B.M., and Amara S.G. 2002. Dopamine transporter-mediated conductances increase excitability of midbrain dopamine neurons. *Nat. Neurosci.* **5:** 971–978.

Jaffe E.H., Marty A., Schulte A., and Chow R.H. 1998. Extrasynaptic vesicular transmitter release from the somata of substantia nigra neurons in rat midbrain slices. *J. Neurosci.* **18:** 3548–3553.

Johnson M.D. 1994. Synaptic glutamate release by postnatal rat serotonergic neurons in microculture. *Neuron* **12:** 443–442.

Kalivas P.W. and Stewart J. 1991. Dopamine transmission in the initiation and expression of drug- and stress-induced sensitization of motor activity. *Brain Res. Brain Res. Rev.* **16:** 223–244.

Kalivas P.W. and Weber B. 1988. Amphetamine injection into the ventral mesencephalon sensitizes rats to peripheral amphetamine and cocaine. *J. Pharmacol. Exp. Ther.* **245:** 1095–1102.

Khoshbouei H., Sen N., Guptaroy B., Johnson L., Lund D., Gnegy M.E., Galli A., and Javitch J.A. 2004. N-terminal phosphorylation of the dopamine transporter is required for amphetamine-induced efflux. *PLoS Biol.* **2:** E78.

Kolta M.G., Shreve P., and Uretsky N.J. 1989. Effect of pretreatment with amphetamine on the interaction between amphetamine and dopamine neurons in the nucleus accumbens. *Neuropharmacology* **28:** 9–14.

Krantz D.E., Peter D., Liu Y., and Edwards R.H. 1997. Phosphorylation of a vesicular monoamine transporter by casein kinase II. *J. Biol. Chem.* **272:** 6752–6759.

Krantz D.E., Waites C., Oorschot V., Liu Y., Wilson R.I., Tan P.K., Klumperman J., and Edwards R.H. 2000. A phosphorylation site in the vesicular acetylcholine transporter regulates sorting to secretory vesicles. *J. Cell Biol.* **149:** 379–395.

Lee R.Y., Sawin E.R., Chalfie M., Horvitz H.R., and Avery L. 1999. EAT-4, a homolog of a mammalian sodium-dependent inorganic phosphate cotransporter, is necessary for glutamatergic neurotransmission in *Caenorhabditis elegans*. *J. Neurosci.* **19:** 159–167.

Liu Y. and Edwards R.H. 1997a. The role of vesicular transport proteins in synaptic transmission and neural degeneration. *Annu. Rev. Neurosci.* **20:** 125–156.

———. 1997b. Differential localization of vesicular acetylcholine and monoamine transporters in PC12 cells but not CHO cells. *J. Cell Biol.* **139:** 907–916.

Liu Y., Roghani A., and Edwards R.H. 1992a. Gene transfer of a reserpine-sensitive mechanism of resistance to MPP+. *Proc. Natl. Acad. Sci.* **89:** 9074–9078.

Liu Y., Schweitzer E.S., Nirenberg M.J., Pickel V.M., Evans C.J., and Edwards R.H. 1994. Preferential localization of a vesicular monoamine transporter to dense core vesicles in PC12 cells. *J. Cell Biol.* **127:** 1419–1433.

Liu Y., Peter D., Roghani A., Schuldiner S., Prive G.G., Eisenberg D., Brecha N., and Edwards R.H. 1992b. A cDNA that suppresses MPP+ toxicity encodes a vesicular amine transporter. *Cell* **70:** 539–551.

Luscher C., Malenka R.C., and Nicoll R.A. 1998. Monitoring

glutamate release during LTP with glial transporter currents. *Neuron* **21:** 335–441.

Malinow R. and Malenka R.C. 2002. AMPA receptor trafficking and synaptic plasticity. *Annu. Rev. Neurosci.* **25:** 103–126.

Molloy S.S., Anderson E.D., Jean F., and Thomas G. 1999. Bicycling the furin pathway: From TGN localization to pathogen activation and embryogenesis. *Trends Cell Biol.* **9:** 28–35.

Montana V., Ni Y., Sunjara V., Hua X., and Parpura V. 2004. Vesicular glutamate transporter-dependent glutamate release from astrocytes. *J. Neurosci.* **24:** 2633–2642.

Mosharov E.V., Gong L.W., Khanna B., Sulzer D., and Lindau M. 2003. Intracellular patch electrochemistry: Regulation of cytosolic catecholamines in chromaffin cells. *J. Neurosci.* **23:** 5835–5845.

Ni B., Rosteck P.R., Nadi N.S., and Paul S.M. 1994. Cloning and expression of a cDNA encoding a brain-specific Na$^+$-dependent inorganic phosphate cotransporter. *Proc. Natl. Acad. Sci.* **91:** 5607–5611.

Nirenberg M.J., Chan J., Liu Y., Edwards R.H., and Pickel V.M. 1996. Ultrastructural localization of the vesicular monoamine transporter-2 in midbrain dopaminergic neurons: Potential sites for somatodendritic storage and release of dopamine. *J. Neurosci.* **16:** 4135–4145.

Pahner I., Holtje M., Winter S., Takamori S., Bellocchio E.E., Spicher K., Laake P., Numberg B., Ottersen O.P., and Ahnert-Hilger G. 2003. Functional G-protein heterotrimers are associated with vesicles of putative glutamatergic terminals: Implications for regulation of transmitter uptake. *Mol. Cell. Neurosci.* **23:** 398–413.

Paulson P.E. and Robinson T.E. 1995. Amphetamine-induced time-dependent sensitization of dopamine neurotransmission in the dorsal and ventral striatum: A microdialysis study in behaving rats. *Synapse* **19:** 56–65.

Peter D., Jimenez J., Liu Y., Kim J., and Edwards R.H. 1994. The chromaffin granule and synaptic vesicle amine transporters differ in substrate recognition and sensitivity to inhibitors. *J. Biol. Chem.* **269:** 7231–7237.

Peter D., Liu Y., Sternini C., de Giorgio R., Brecha N., and Edwards R.H. 1995. Differential expression of two vesicular monoamine transporters. *J. Neurosci.* **15:** 6179–6188.

Pothos E.N., Davila V., and Sulzer D. 1998. Presynaptic recording of quanta from midbrain dopamine neurons and modulation of the quantal size. *J. Neurosci.* **18:** 4106–4118.

Pothos E.N., Desmond M., and Sulzer D. 1996. L-3,4-dihydroxyphenylalanine increases the quantal size of exocytotic dopamine release in vitro. *J. Neurochem.* **66:** 629–636.

Pothos E.N., Larsen K.E., Krantz D.E., Liu Y.-J., Haycock J.W., Setlik W., Gershon M.E., Edwards R.H., and Sulzer D. 2000. Synaptic vesicle transporter expression regulates vesicle phenotype and quantal size. *J. Neurosci.* **20:** 7297–7306.

Quick M.W. 2003. Regulating the conducting states of a mammalian serotonin transporter. *Neuron* **40:** 537–549.

Rice M.E., Cragg S.J., and Greenfield S.A. 1997. Characteristics of electrically evoked somatodendritic dopamine release in

substantia nigra and ventral tegmental area in vitro. *J. Neurophysiol.* **77:** 853–862.

Riddle E.L., Topham M.K., Haycock J.W., Hanson G.R., and Fleckenstein A.E. 2002. Differential trafficking of the vesicular monoamine transporter-2 by methamphetamine and cocaine. *Eur. J. Pharmacol.* **449:** 71–74.

Robertson G.S., Damsma G., and Fibiger H.C. 1991. Characterization of dopamine release in the substantia nigra by in vivo microdialysis in freely moving rats. *J. Neurosci.* **11:** 2209–2216.

Robinson T.E. and Becker J.B. 1982. Behavioral sensitization is accompanied by an enhancement in amphetamine-stimulated dopamine release from striatal tissue in vitro. *Eur. J. Pharmacol.* **85:** 253–254.

Robinson T.E. and Berridge K.C. 2000. The psychology and neurobiology of addiction: An incentive-sensitization view. *Addiction* (suppl. 2) **95:** S91–S117.

Roghani A., Feldman J., Kohan S.A., Shirzadi A., Gundersen C.B., Brecha N., and Edwards R.H. 1994. Molecular cloning of a putative vesicular transporter for acetylcholine. *Proc. Natl. Acad. Sci.* **91:** 10620–10624.

Ryan T.A., Reuter H., Wendland B., Schweizer F.E., Tsien R.W., and Smith S.J. 1993. The kinetics of synaptic vesicle recycling measured at single presynaptic boutons. *Neuron* **11:** 713–724.

Sandoval V., Riddle E.L., Hanson G.R., and Fleckenstein A.E. 2003. Methylphenidate alters vesicular monoamine transport and prevents methamphetamine-induced dopaminergic deficits. *J. Pharmacol. Exp. Ther.* **304:** 1181–1187.

Sankaranarayanan S. and Ryan T.A. 2000. Real-time measurements of vesicle-SNARE recycling in synapses of the central nervous system. *Nat. Cell Biol.* **2:** 197–204.

Shupliakov O., Atwood H.L., Ottersen O.P., Storm-Mathisen J., and Brodin L. 1995. Presynaptic glutamate levels in tonic and phasic motor axons correlate with properties of synaptic release. *J. Neurosci.* **15:** 7168–7180.

Song H.-J., Ming G.-l., Fon E., Bellocchio E., Edwards R.H., and Poo M.-M. 1997. Expression of a putative vesicular acetylcholine transporter facilitates quantal transmitter packaging. *Neuron* **18:** 815–826.

Staal R.G., Mosharov E.V., and Sulzer D. 2004. Dopamine neurons release transmitter via a flickering fusion pore. *Nat. Neurosci.* **7:** 341–346.

Stewart J. and Vezina P. 1989. Microinjections of Sch-23390 into the ventral tegmental area and substantia nigra pars reticulata attenuate the development of sensitization to the locomotor activating effects of systemic amphetamine. *Brain Res.* **495:** 401–406.

Sulzer D., and Rayport S. 1990. Amphetamine and other psychostimulants reduce pH gradients in midbrain dopaminergic neurons and chromaffin granules: A mechanism of action. *Neuron* **5:** 797–808.

Sulzer D., Maidment N.T., and Rayport S. 1993. Amphetamine and other weak bases act to promote reverse transport of dopamine in ventral midbrain neurons. *J. Neurochem.* **60:** 527–535.

Sulzer D., Chen T.-K., Lau Y.Y., Kristensen H., Rayport S., and Ewing A. 1995. Amphetamine redistributes dopamine from synaptic vesicles to the cytosol and promotes reverse transport. *J. Neurosci.* **15:** 4102–4108.

Sulzer D., Joyce M.P., Lin L., Geldwert D., Haber S.N., Hattori T., and Rayport S. 1998. Dopamine neurons make glutamatergic synapses in vitro. *J. Neurosci.* **18:** 4588–4602.

Takahashi N., Miner L.L., Sora I., Ujike H., Revay R.S., Kostic V., Jackson-Lewis V., Przedborski S., and Uhl G.R. 1997. VMAT2 knockout mice: Heterozygous display reduced amphetamine-conditioned reward, enhanced amphetamine locomotion and enhanced MPTP toxicity. *Proc. Natl. Acad. Sci.* **94:** 9938–9943.

Takamori S., Malherbe P., Broger C., and Jahn R. 2002. Molecular cloning and functional characterization of human vesicular glutamate transporter 3. *EMBO Rep.* **3:** 798–803.

Takamori S., Rhee J.S., Rosenmund C., and Jahn R. 2000. Identification of a vesicular glutamate transporter that defines a glutamatergic phenotype in neurons. *Nature* **407:** 189–194.

———. 2001. Identification of differentiation-associated brain-specific phosphate transporter as a second vesicular glutamate transporter. *J. Neurosci.* **21:** RC182.

Tan P.K., Waites C., Liu Y., Krantz D.E., and Edwards R.H. 1998. A leucine-based motif mediates the endocytosis of vesicular monoamine and acetylcholine transporters. *J. Biol. Chem.* **273:** 17351–17360.

Troyer K.P. and Wightman R.M. 2002. Temporal separation of vesicle release from vesicle fusion during exocytosis. *J. Biol. Chem.* **277:** 29101–29107.

Truong J.G., Newman A.H., Hanson G.R., and Fleckenstein A.E. 2004. Dopamine D2 receptor activation increases vesicular dopamine uptake and redistributes vesicular monoamine transporter-2 protein. *Eur. J. Pharmacol.* **504:** 27–32.

Van der Kloot W. 1990. The regulation of quantal size. *Prog. Neurobiol.* **36:** 93–130.

Varoqui H., and Erickson J.D. 1996. Active transport of acetyl-choline by the human vesicular acetylcholine transporter. *J. Biol. Chem.* **271:** 27229–27232.

Verheijen F.W., Verbeek E., Aula N., Beerens C.E., Havelaar A.C., Joosse M., Peltonen L., Aula P., Galjaard H., van der Spek P.J., and Mancini G.M. 1999. A new gene, encoding an anion transporter, is mutated in sialic acid storage diseases. *Nat. Genet.* **23:** 462–465.

Vezina P. 2004. Sensitization of midbrain dopamine neuron reactivity and the self-administration of psychomotor stimulant drugs. *Neurosci. Biobehav. Rev.* **27:** 827–839.

Wadiche J.I., Amara S.G., and Kavanaugh M.P. 1995. Ion fluxes associated with excitatory amino acid transport. *Neuron* **15:** 721–728.

Waites C.L., Mehta A., Tan P.K., Thomas G., Edwards R.H., and Krantz D.E. 2001. An acidic motif retains vesicular monoamine transporter 2 on large dense core vesicles. *J. Cell Biol.* **152:** 1159–1168.

Wang Y.-M., Gainetdinov R.R., Fumagalli F., Xu F., Jones S.R., Bock C.B., Miller G.W., Wightman R.M., and Caron M.G. 1997. Knockout of the vesicular monoamine transporter 2 gene results in neonatal death and supersensitivity to cocaine and amphetamine. *Neuron* **19:** 1285–1296.

Weihe E., Tao-Cheng J.-H., Schafer M.K.-H., Erickson J.D., and Eiden L.E. 1996. Visualization of the vesicular acetylcholine transporter in cholinergic nerve terminals and its targeting to a specific population of small synaptic vesicles. *Proc. Natl. Acad. Sci.* **93:** 3547–3552.

Zerangue N. and Kavanaugh M.P. 1995. Flux coupling in a neuronal glutamate transporter. *Nature* **383:** 634–637.

Zhang Q., Pangrsic T., Kreft M., Krzan M., Li N., Sul J.Y., Halassa M., Van Bockstaele E., Zorec R., and Haydon P.G. 2004. Fusion-related release of glutamate from astrocytes. *J. Biol. Chem.* **279:** 12724–12733.

Zilberter Y. 2000. Dendritic release of glutamate suppresses synaptic inhibition of pyramidal neurons in rat neocortex. *J. Physiol.* **528:** 489–496.

PART 4
Synaptic Plasticity and Addiction

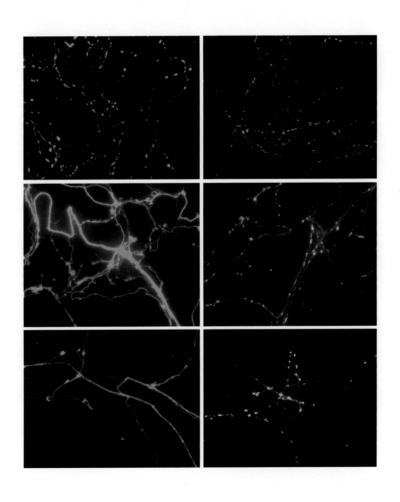

18

Synaptic Vesicle Trafficking and Drug Addiction in Synapsin Triple Knockout Mice

Daniel Gitler,[1] Jian Feng,[2,3] Yoshiko Takagishi,[4] Vladimir M. Pogorelov,[5] Ramona M. Rodriguiz,[5] B. Jill Venton,[6] Paul E.M. Phillips,[6,7] Yong Ren,[3] Hung-Teh Kao,[2,8] Mark Wightman,[6] Paul Greengard,[2] William C. Wetsel,[5] and George J. Augustine[1]

[1]Department of Neurobiology, Duke University Medical Center, Durham, North Carolina 27710; [2]Laboratory of Molecular and Cellular Neuroscience, The Rockefeller University, New York, New York 10021; [3]Department of Physiology and Biophysics, State University of New York, Buffalo, New York 14214; [4]Research Institute of Environmental Medicine, Nagoya University, Nagoya 464-8601, Japan; [5]Departments of Psychiatry and Behavioral Sciences, and Cell Biology, Mouse Behavioral and Neuroendocrine Analysis Core Facility, Duke University Medical Center, Durham, North Carolina 27710; [6]Department of Chemistry and Neuroscience Center, University of North Carolina at Chapel Hill, Chapel Hill, North Carolina 27599; [7]Department of Psychology, University of North Carolina at Chapel Hill, Chapel Hill, North Carolina 27599; [8]Center for Dementia Research, Nathan Kline Institute and Department of Psychiatry, New York University School of Medicine, Orangeburg, New York 10962

ABSTRACT

Synapsins, a family of synaptic vesicle proteins, are the most abundant phosphoproteins in the brain. We have studied the functions of these proteins by deleting all three known synapsin genes in mice. Synapsin triple knockout (TKO) mice are viable but have a number of behavioral defects, most notably increased sensitivity to psychostimulants such as cocaine and amphetamine. Studies of synaptic transmission in neurons from synapsin TKO mice reveal that synapsins have different roles at different types of synapses. At glutamatergic synapses, synapsins are not required for basal transmission, but are required for maintaining a reserve pool of glutamatergic vesicles that is mobilized during high rates of synaptic activity. At GABAergic synapses, synapsins both regulate the amount of basal GABA release and influence the number of vesicles in the reserve pool. At dopaminergic synapses, synapsins maintain vesicles in a reserve pool that is mobilized during cocaine treatment; knockout of synapsins perturbs this relationship so that cocaine releases less dopamine from TKO mice. Hypersensitization of these mice to cocaine is not caused by this impairment in vesicle trafficking, but appears attributable to a compensatory increase in the responsiveness of postsynaptic neurons to dopamine.

INTRODUCTION

It is clear that synaptic transmission is the main target of drugs of abuse (Nestler and Aghajanian 1997; Koob et al. 1998). To date, most studies of the mechanism of action of these drugs have focused on postsynaptic targets. For example, it is well established that opioid compounds work on specific postsynaptic opioid receptors and that alcohol works, at least in part, via actions on postsynaptic GABA receptors (Nestler 2002). However, it is likely that the presynaptic mechanisms involved in release of synaptic neurotransmitters are also important targets of drugs of abuse (Roberto et al. 2003). For example, the actions of cocaine appear to arise largely from blockade of dopamine transporters on presynaptic terminals (Ritz et al. 1987; Laakso et al. 2002).

Presynaptic terminals rely on a complex cycle of membrane trafficking to release neurotransmitters. After synaptic vesicles are originally synthesized, they undergo repeated cycles of exocytosis and endocytosis within the presynaptic terminal (Augustine et al. 1999; Südhof 2004). During this cycling, at least two distinguishable pools of vesicles can be discerned: (1) a readily releasable pool (RRP) of vesicles that actively participates in exocytosis and endocytosis and (2) a reserve pool (RP) of vesicles that can be mobilized during periods of high synaptic activity, when demand for transmitter-filled vesicles exceeds the supply available in the RRP (Rizzoli and Betz 2005). This trafficking of synaptic vesicle membrane requires the actions of dozens of proteins that are found on the vesicles, in the presynaptic cytoplasm, and in the plasma membrane of the nerve terminal (Augustine et al. 1999). These proteins, as well as the components of the calcium signaling cascade that triggers exocytotic discharge of transmitters from the synaptic vesicle, provide a wealth of possible molecular targets for drugs of abuse.

GOALS OF THE CHAPTER

This chapter focuses on the synapsins, the first synaptic vesicle proteins discovered (Greengard et al. 1993). The goals are to familiarize readers with synapsins and to explore the synaptic functions of these proteins. Furthermore, we consider the roles of synapsins at the behavioral level, with special emphasis on their involvement in behavioral responses to psychostimulant drugs.

Synapsins Are a Family of Phosphoproteins Found on Synaptic Vesicles

Three mammalian synapsin genes have been identified (Südhof et al. 1989; Kao et al. 1999). Each is subject to alternative splicing, producing at least ten isoforms: synapsins Ia, Ib, IIa, IIb, IIIa, IIIb, IIIc, IIId, and the atypical IIIe and IIIf (Porton et al. 1999). These isoforms arise from the combinatorial arrangement of numerous domains (Fig. 1; Südhof et al. 1989; Kao et al. 1999). The amino-terminal portions of synapsins are highly conserved and consist of domains A–C, whereas the carboxy-terminal portions are variable owing to heterogeneous combinations of domains D–J.

Synapsins are expressed throughout the central and peripheral nervous systems, although their expression varies across synapse types and brain regions (Fried et al. 1982;

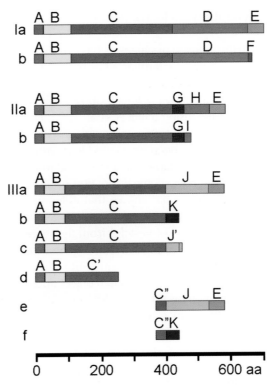

FIGURE 1. Domain structure of the three synapsin genes. Domains are indicated by different colors.

Südhof et al. 1989; Mandell et al. 1992; Matus-Leibovitch et al. 1997). In contrast to synapsins I and II, which are expressed in adult presynaptic terminals, synapsin III expression is developmentally controlled and not confined to synaptic terminals (Ferreira et al. 2000; Pieribone et al. 2002).

Synapsins exhibit many intriguing biochemical properties. They are peripherally associated with synaptic vesicles (SVs), attributable to binding to both phospholipid and protein constituents of the SV (Greengard et al. 1993; Hilfiker et al. 1999). Synapsins also bind to cytoskeletal elements such as actin, tubulin, and spectrin, and it is thought that these connections between synaptic vesicles and the cytoskeleton may be important for synapsin function (Hilfiker et al. 1999). Synapsins are the most abundant phosphoproteins in the brain and serve as substrates for several protein kinases, including protein kinase A (PKA), calcium-calmodulin kinase I (CaMKI), CaMKII, and mitogen-activated protein (MAP) kinase (Matsubara et al. 1996; Hilfiker et al. 1999). Phosphorylation by CaMKII causes synapsins to dissociate from synaptic vesicles (Benfenati et al. 1992). Synapsins also interact with other elements of cellular signaling pathways, such as phosphatidylinositol 3-kinase and regulators of G-protein signaling (RGS proteins; Cousin et al. 2003; Tu et al. 2003). Synapsins bind to ATP, and structural similarities to ATP-utilizing enzymes suggest that synapsins may possess an as-yet-unidentified enzymatic function (Esser et al. 1998; Hosaka and Südhof 1998a,b).

Synapsins Regulate Synaptic Vesicle Trafficking

Although the function of synapsins has been studied for more than 20 years, the role of these proteins in presynaptic terminals has been somewhat elusive. The best-established function of synapsins is to maintain the reserve pool of synaptic vesicles (Li et al. 1995; Hilfiker et al. 1999; Gitler et al. 2004a). Other proposed roles for synapsins include regulation of the kinetics of membrane fusion (Hilfiker et al. 1998; Humeau et al. 2001), stabilization of synaptic vesicles (Rosahl et al. 1995), and regulation of late steps of endocytosis (Bloom et al. 2003). Synapsins also have been implicated in neuronal development, synaptogenesis, and maintenance of mature synapses (Ferreira and Rapoport 2002; Kao et al. 2002).

To define the physiological function of these proteins, we turned to genetic studies in mice. Previous work has shown that deleting individual synapsin genes in mice causes varied effects on synaptic transmission. Synaptic depression, a form of plasticity that requires mobilization of vesicles from the reserve pool (Zucker and Regehr 2002), is not affected by deletion of the synapsin I gene (Rosahl et al. 1993, 1995; Li et al. 1995), but is accelerated by loss of synapsin II (Rosahl et al. 1995; Samigullin et al. 2004) and is slowed by deletion of synapsin III (Feng et al. 2002). Although these studies begin to identify different functions for the three synapsin genes, they do not define physiological roles because synapsins may have redundant functions, so that remaining synapsins compensate for the loss of a given synapsin gene. In addition, deleting individual genes does not define the importance of the various isoforms produced from each gene by alternative splicing. Thus, deleting all three synapsin genes is a necessary starting point for defining the roles that synapsins have in synaptic function.

Synapsin-deficient Mice Are Viable

Synapsin TKO mice were generated by crossing preexisting lines of mice harboring null mutations in single synapsin genes (Chin et al. 1995; Ferreira et al. 1998; Feng et al. 2002). To serve as a genetically similar control, a line of matching wild-type mice, termed synapsin triple wild-type (TWT) mice, was derived in parallel. The synapsin TKO mice have a normal life span and are fertile, although to a lower degree than the TWT mice. The TKO mice do not express obvious signs of morbidity, except for transient, mild seizures. Gross brain morphology is not noticeably affected by deleting the synapsins, indicating that these proteins do not have a central role in the development of the nervous system. However, the TKO mice have various deficiencies in basic neurophysiological performance such as muscle strength, posture, and reflexes. In addition, the mice also show impairments on learning and memory tests, as exemplified by defects in performance in the radial arm maze (Gitler et al. 2004a) and the Morris water maze (R.M. Rodriguiz et al., unpubl.).

Synapsins Regulate the Dynamics of Glutamatergic Synaptic Transmission

To determine the role of synapsins in synaptic function, we compared synaptic transmission in TKO and TWT neurons. Synaptic function was studied in single hippocampal neurons grown on microislands (Segal and Furshpan 1990; Bekkers and Stevens 1991). Neurons cultured in this manner form synapses on themselves and these "autapses" offer several advantages for quantifying neurotransmitter release. Most importantly, all autaptic responses originate from the same cell, so that the sources of incoming synaptic input are homogeneous and defined. Thus, both spontaneous synaptic currents and currents evoked by presynaptic action potentials arise from the same population of synapses, allowing characterization of the quantal properties of synaptic function (del Castillo and Katz 1954). In addition, because neurotransmitter release occurs in response to stimulation of a single neuron, synaptic transmission is unitary and resulting synaptic currents are largely independent of the stimulus properties. The ability of synapsin TKO neurons to form synapses in culture did not differ from that of TWT neurons (Gitler et al. 2004a), which allows us to attribute any differences in synaptic transmission to changes in the properties of release from synapses, rather than changes in the number of synapses.

We first considered the role of synapsins in release of glutamate from pyramidal neurons, the main type of neuron in the hippocampus. We used stimulus paradigms that report on the properties of the RRP and RP of synaptic vesicles. Spontaneous transmitter release and release evoked by single stimuli draw vesicles from the RRP. In contrast, trains of stimuli elicit synaptic depression, which results from the depletion of synaptic vesicles from the RRP, resulting in the mobilization of vesicles from the RP (Zucker and Regehr 2002; Rizzoli and Betz 2005).

Properties of synaptic transmission that are related to the size of the RRP such as the amplitude of evoked excitatory postsynaptic currents (EPSCs) (Fig. 2A,B), the frequency of spontaneous events (Fig. 2C), and the quantal content of evoked EPSCs (Fig. 2D) did not differ in the TKO and TWT neurons. This indicates that loss of synapsins does not affect the RRP of glutamatergic neurons. In addition, stimulation with pairs of stimuli elicited synaptic facilitation that was unaltered between TKO and TWO neurons, suggesting that synapsins do not affect the probability of release from the RRP (Dobrunz and Stevens 1997).

On the other hand, the rate of synaptic depression was accelerated by the deletion of synapsins (Fig. 2E–G). Importantly, the difference between TWT and TKO neurons was evident only after the neurons had been stimulated 30 times (Fig. 2F), which would be expected to release most of the RRP (Murthy and Stevens 1999; Wesseling and Lo 2002). Thus, it appears that the RP of glutamatergic vesicles is reduced in the TKO neurons. To further evaluate this effect, the total amount of synaptic charge produced during the stimulus train, calculated by integrating each synaptic current, was determined to quantify the number of release events that were evoked (Fig. 2H). TKO neurons released much less synaptic charge than TWT neurons, indicating that loss of synapsins caused release of approximately half the normal number of vesicles from the reserve pool (Fig. 2I).

Deleting Synapsins Reduces the Number of Synaptic Vesicles

To follow up on these electrophysiological indications of defects in the RP of TKO neurons, we used several approaches to examine the number and distribution of synaptic vesicles within presynaptic terminals. We first quantified the levels of synaptic proteins in the brains of TKO and TWT mice. The relative quantities of synaptic vesicle proteins (synaptobrevin 2, synaptotagmin I, and synaptophysin I) were reduced in homogenates of brains from TKO mice. In contrast, levels of representative postsynaptic proteins (PSD95 and NMDA receptor subunits NR1 and NR2B), structural proteins (α-tubulin), or non-vesicular presynaptic proteins (syntaxin I) were not affected by deletion of the synapsins

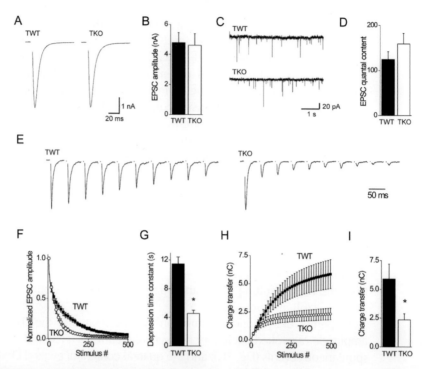

FIGURE 2. Characteristics of glutamatergic neurotransmission in cultured TWT and TKO neurons. (*A*) Averaged excitatory postsynaptic currents (EPSCs) recorded from cultured glutamatergic autaptic neurons (*n* = 29 and 35 neurons, respectively). (*B*) Mean amplitude of evoked EPSCs of TWT and TKO neurons. (*C*) Representative traces of spontaneous synaptic activity in TWT and TKO neurons. (*D*) Mean quantal content of evoked EPSCs of TWT and TKO neurons. (*E*) Representative traces showing every 15th event in a train of EPSCs evoked at a frequency of 10 Hz in TWT and TKO neurons. (*F*) Mean EPSC amplitudes during the 500-event trains. Symbols represent EPSC amplitudes normalized to the first response in the train and every 15th response is plotted (*n* = 20 TWT neurons and *n* = 22 TKO neurons). (*G*) Mean time constants for synaptic depression in TWT and TKO neurons, determined after the 30th stimulus. (*H*) Cumulative EPSC charge transfer in TWT and TKO neurons. (*I*) Mean values for total EPSC charge transfer after 500 stimuli were applied to TWT and TKO neurons. (*G,I*) Asterisks denote a statistically significant difference at a 5% confidence level (Student's *t*-test). All error bars represent ±s.e.m. (Modified, with permission, from Gitler et al. 2004a [©Society for Neuroscience].)

(Gitler et al. 2004a). These results indicate specific reductions in the amount of synaptic vesicle proteins in the terminals of TKO mice.

We next assessed the density of synaptic vesicles by examining the localization of a synaptic vesicle marker in live neurons. This was done by measuring the synaptic targeting of synaptobrevin 2, an integral protein of the synaptic vesicle membrane (Baumert et al. 1989). This protein was tagged with green fluorescent protein (GFP) to permit visualization of synaptic vesicle clusters (Gitler et al. 2004a). We observed that synaptic targeting of GFP-synaptobrevin 2 was significantly reduced in TKO neurons, again suggesting a reduction in the number of synaptic vesicles.

Finally, we used high-resolution electron microscopy to determine the absolute number and distribution of synaptic vesicles (Fig. 3A). Ultrastructural quantification revealed that the total number of vesicles in TKO terminals was reduced to approximately half that

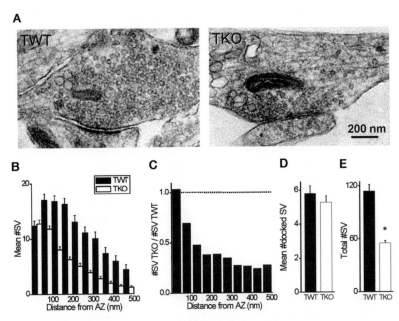

FIGURE 3. Spatial distribution of synaptic vesicles in glutamatergic presynaptic terminals at the ultrastructural level. (*A*) Representative electron micrographs of excitatory terminals of TWT and TKO neurons. (*B*) Mean number of synaptic vesicles located within 50-nm-wide concentric compartments centered at the active zone of TWT and TKO neurons ($n = 47$ and 66 terminals, respectively). (*C*) Ratio of the number of vesicles in the various compartments, calculated from the data shown in *B* by dividing TKO values by TWT values. Dotted line marks equality between TWT and TKO data. (*D*) Mean number of morphologically docked vesicles in TWT and TKO terminals. (*E*) Mean total number of vesicles up to 500 nm from the active zone in TWT and TKO terminals. (Reprinted, with permission, from Gitler et al. 2004a [©Society for Neuroscience].)

in TWT neurons (Fig. 3E). Analysis of the spatial distribution of vesicles showed that although the number of vesicles docked at the presynaptic active zone was unchanged (Fig. 3D), a specific reduction in the number of the vesicles located far from the active zone occurred in TKO neurons (Fig. 3B,C). Thus, we conclude that synapsins determine the ability of the neuron to preserve the vesicles that belong to its functional RP. In the absence of synapsins, the size of this pool is diminished. This result can explain the inability of TKO neurons to maintain glutamatergic transmission during periods of intense synaptic activity.

Synapsin Isoforms Have Different Functions

Although our observations regarding the synapsin TKO mice indicate that synapsins determine the dynamics of glutamatergic synaptic transmission, it is not clear why so many different synapsin genes and protein isoforms exist. Previous experimental approaches using mice in which single synapsin genes are deleted (Rosahl et al. 1993, 1995; Li et al. 1995; Feng et al. 2002; Samigullin et al. 2004) suggest that the synapsin genes are not equivalent in their ability to support the RP of glutamatergic synaptic vesicles. However, knocking out entire genes cannot address the possible variations among the various isoforms formed by each gene. Furthermore, such studies cannot take into account the compensatory influences of remaining synapsin gene products.

We took advantage of the TKO mice to determine the functions of individual synapsin isoforms. On this null background, individual synapsin isoforms were transfected into cultured TKO neurons and electrophysiological methods were used to examine the effect of the introduced synapsins on synaptic depression. Surprisingly, we found that the products of different synapsin genes were not equivalent in function; indeed, even isoforms derived from the same gene were different from one another (D. Gitler et al., in prep.).

Synapsin IIa slowed synaptic depression when expressed in TKO neurons, illustrating that synapsin IIa functions to support the reserve pool of glutamatergic synaptic vesicles (Fig. 4A). In contrast, synapsin IIb had no effect on synaptic depression, indicating that isoforms from the same gene are not equivalent in function even though these differ by only their carboxy-terminal tails. In fact, synapsin IIa was the only synapsin among the several that we examined (synapsins Ia, Ib, IIa, IIb, and IIIa) that was able to slow synaptic depression when expressed in TKO neurons.

Synapsin isoforms are capable of forming heterodimers (Hosaka and Südhof 1999). To examine whether such interactions have a role in synapsin function, we also expressed several synapsin isoforms in TWT neurons. As in the case of TKO neurons, we found that synapsin IIa further slowed synaptic depression beyond the already-slow rate normally observed in TWT neurons. This result indicates that synapsin IIa can increase the size of the RP. In contrast, synapsin Ia had no effect, comparable to its lack of effect in TKO neurons. Remarkably, synapsin Ib, the other isoform of synapsin I, reduced the reserve pool, again illustrating differences among close isoforms of the same gene. When overexpressed in TWT neurons, synapsin Ib significantly enhanced the rate of synaptic depression, so that depression kinetics in synapsin-Ib-transfected TWT neurons was indistinguishable from TKO neurons (Fig. 4B). Because synapsin Ib had no effect when expressed in TKO neurons, we conclude that an interaction between synapsin Ib and endogenous synapsins was responsible for the inhibitory effects of synapsin Ib.

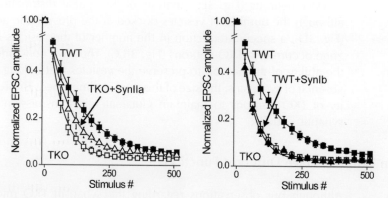

FIGURE 4. Effect of introduction of individual synapsin isoforms on synaptic depression. (A) Normalized EPSC amplitudes during trains of 500 stimuli delivered at 10 Hz. EPSCs are normalized by the amplitude of the first EPSC in the train, and every 30th EPSC is shown. The solid line represents a fitted exponential decay curve, starting from the 30th EPSC in the train. Shown are mean amplitudes for untransfected TWT and TKO neurons, and for TKO neurons transfected with GFP-synapsin IIa (n = 7 transfected neurons). (B) As in A, except data are from untransfected TWT and TKO neurons, and from TWT neurons transfected with GFP-synapsin Ib (n = 6 transfected neurons).

Synapsins Differentially Target to Presynaptic Terminals

Hints about the molecular mechanisms responsible for the ability of synapsin Ib to regulate synaptic depression came from studies of the targeting of synapsins to presynaptic terminals (Gitler et al. 2004b). GFP-tagged versions of any synapsin isoform, including synapsin Ib, properly target to presynaptic terminals in TWT neurons. However, synapsin Ib, unlike the other synapsins, was not able to target in TKO neurons (Fig. 5A–C). This suggests that targeting of synapsin Ib requires the presence of other synapsin isoforms. Previous in vitro work shows that synapsin I is capable of dimerizing, either with itself or with synapsin II (Hosaka and Südhof 1999). The fact that cotransfection of synapsin Ia or IIa rescued the targeting of synapsin Ib (Fig. 5D) indicates that targeting of this isoform is mediated by heterodimerization with other synapsin isoforms.

Determination of the structural requirements for targeting of synapsins elucidated the basis for the lack of targeting of synapsin Ib. By transfecting synapsin I fragments into TKO

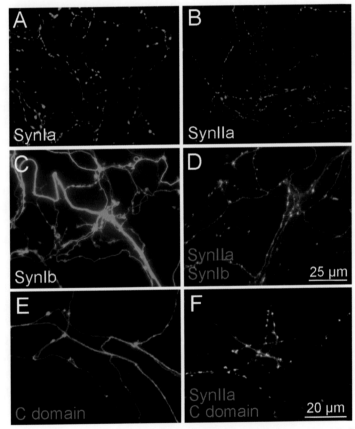

FIGURE 5. Synaptic targeting capabilities of synapsins and synapsin fragments. (A) GFP-synapsin Ia and (B) GFP-synapsin IIa localize to presynaptic terminals in cultured TKO neurons. (C) GFP-synapsin Ib does not localize to the presynaptic terminals, but rather concentrates in the axons of transfected neurons. (D) Upon cotransfection with YFP-synapsin IIa (*red*), CFP-synapsin Ib (*green*) colocalizes (*yellow*) to presynaptic terminals. (E) A CFP-tagged C domain fragment of synapsin I does not localize to synapses on its own. (F) The same fragment does localize to synapses when cotransfected with GFP-synapsin IIa.

and TWT neurons, the central C domain was found to be the site of the interaction underlying targeting (Fig. 5F; Gitler et al. 2004b). This domain is also sufficient for dimerization in vitro (Esser et al. 1998). However, additional domains are involved in the targeting of synapsin Ia because the C domain alone could not target to synaptic terminals of TKO neurons (Fig. 5E). Combinations of either domains B and C or domains C and E were minimally sufficient to produce targeting. In addition, domain D contains a segment that inhibits targeting, with this inhibition overcome by the presence of the E domain (Gitler et al. 2004b). The difference in the targeting of synapsins Ia and Ib in TKO neurons, therefore, arises because synapsin Ia, but not synapsin Ib, possesses this regulatory E domain.

Dimerization Is Important for Synapsin Functions

The results presented above indicate that dimerization with other synapsins is involved in both the acceleration of synaptic depression by synapsin Ib and the targeting of this isoform. Given that synapsin Ib alone cannot target to presynaptic terminals, these results suggest that this isoform could form heterodimers that interfere with the targeting and subsequent presynaptic function of the other synapsin isoforms that dimerize with synapsin Ib.

We first considered this hypothesis by examining the influence of synapsin Ib on the synaptic targeting of cotransfected synapsins in TKO neurons. Synapsin Ib reduced the targeting of synapsins, particularly that of synapsin IIa, the isoform most important for determining the kinetics of synaptic depression (Fig. 6A). Likewise, synapsin Ib reduced the presynaptic targeting of endogenous synapsins. This was established by immunolabeling cultured TWT neurons with an antibody that recognizes synapsins Ia, IIa, and IIIa, yet does not recognize synapsin Ib. Overexpressing synapsin Ib caused a reduction in the amount of these endogenous synapsins in the synaptic terminals and an uncharacteristic increase in their abundance in axons (D. Gitler et al., in prep.). Thus, synapsin Ib does interfere with targeting of other synapsin isoforms.

Given that synapsin Ib interferes with the targeting of other synapsins and that reducing the amount of synapsins reduces the size of the RP (in the synapsin TKO neurons), it is possible that synapsin Ib reduces the RP. To examine this possibility, we evaluated the effect of synapsin Ib on the RP, as assayed by targeting GFP-synaptobrevin 2. Synapsin Ib caused a substantial reduction in the targeting of synaptobrevin 2 (Fig. 6B). This reduction in the RP seems to be responsible for the acceleration of depression by synapsin Ib, because there was a very good correlation between RP size and the rate of depression in neurons expressing synapsin Ib (Fig. 6C). In summary, synapsin Ib appears to serve as a dominant negative isoform by heterodimerizing with other synapsin isoforms and consequently reducing the targeting of these isoforms, thereby reducing the size of the RP of synaptic vesicles (SVs).

Synapsin Dimerization May Control Synaptic Dynamics

On the basis of our observations, we propose that dimerization of synapsins controls the cross-linking RP by SVs in the presynaptic terminal (Fig. 7). This model is based on proposed differences in the ability of different synapsin isoforms to cross-link SVs (Fig. 7A).

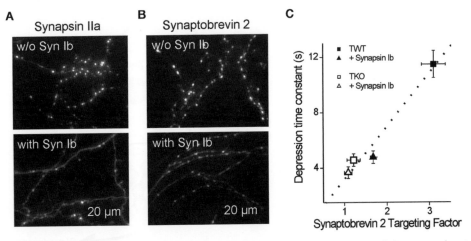

FIGURE 6. Synapsin Ib interferes with synaptic targeting of other synapsins and decreases the size of the reserve pool. (*A*) Synaptic localization of GFP-synapsin IIa when expressed in TKO neurons, in the presence (*bottom*) or absence (*top*) of coexpressed CFP-synapsin Ib (*not seen in these images*). (*B*) Synaptic localization of GFP-synaptobrevin 2 when expressed in TWT neurons, in the presence (*bottom*) or absence (*top*) of coexpressed GFP-synapsin Ib (*not seen in images*). (*C*) Correlation of time constant of synaptic depression with synaptic targeting of GFP-synaptobrevin 2 in TKO and TWT neurons. Slow synaptic depression is correlated with a larger vesicle pool in TWT neurons. Expression of synapsin Ib accelerates synaptic depression and reduces the size of the vesicle pool in TWT neurons to the level of the TKO neurons.

Synapsin IIa acts as the prototypical synapsin (Greengard et al. 1993; Hilfiker et al. 1999) by clustering SVs in the terminals through formation of divalent dimers that can interact with more than one vesicle, either directly with the vesicle or through interactions with cytoskeletal elements (Hirokawa et al. 1989; Thiel et al. 1990; Hilfiker et al. 1999). However, synapsin Ib negates this effect, by forming heterodimers with other synapsins that are monovalent for interaction with vesicles and, thus, do not allow cross-linking and vesicle clustering (Gitler et al. 2004b).

This model predicts that the relative abundance of the synapsin isoforms determines the net behavior of the RP. An increase in the abundance of synapsin IIa will shift the balance toward divalent dimers, thus enlarging the reserve pool and slowing synaptic depression (Fig. 7B). Conversely, a relative increase in synapsin Ib will raise the proportion of ineffective monovalent dimers, thus dispersing the reserve pool and accelerating synaptic depression (Fig. 7C). Hence, synapsins can bidirectionally fine-tune the characteristics of synaptic depression. Future work will be directed toward testing the predictions of this model.

Synapsins Have Different Functions at GABAergic Synapses

The data summarized indicate that synapsins regulate the RP of glutamatergic synaptic vesicles. However, synapsins have other functions at other synapses. At GABAergic synapses, loss of synapsins also reduces the number of synaptic vesicles, but this effect is

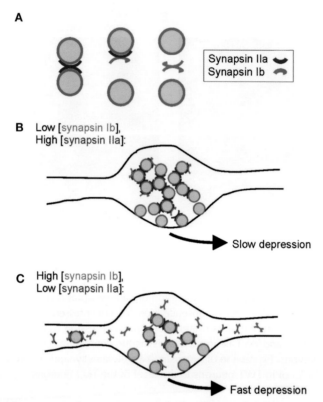

FIGURE 7. Model for synapsin function in control of vesicle pool distribution in presynaptic terminals. (*A*) Synapsin IIa forms a divalent dimer that can bind and cross-link vesicles, whereas heterodimerization of synapsin IIa with synapsin Ib disrupts this clustering ability. (*B*) A predominance of synapsin IIa in the presynaptic terminal favors an increase in the size of the reserve pool, whereas a predominance of synapsin Ib favors a dispersal of this pool (*C*).

not restricted to the RP (Gitler et al. 2004a). This homogenous decrease in the number of GABAergic vesicles throughout the terminal also reduces the number of docked vesicles. As a result, basal-evoked release of GABA is diminished, whereas the rate of synaptic depression is unaffected. Thus, loss of synapsins has different structural and physiological effects at glutamatergic and GABAergic neurons. It will be interesting to determine the molecular mechanisms underlying these differences, as well as to examine the functions of synapsins at synapses that use other neurotransmitters.

Synapsins Regulate Behavioral Responses to Psychostimulants

While examining the behavior of synapsin TKO mice, we observed that these mice were more active than TWT mice. To quantify this difference, animals were placed into an open field and spontaneous locomotor activity was monitored for 40 minutes. When initially placed into the open field, TKO mice traveled greater horizontal distances and more frequently reared, as indicated by vertical beam-breaks. The levels of stereotypic behavior, as revealed by repeated beam-breaks occurring less than 1 second apart, was also

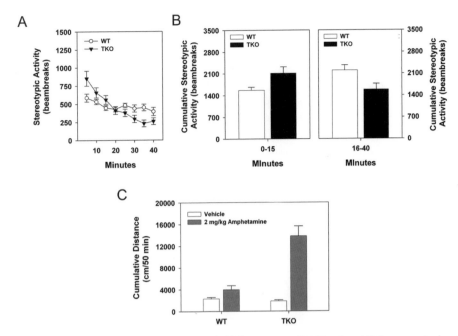

FIGURE 8. Spontaneous activity in the open field for TWT and TKO mice. (*A*) Stereotypical activity (repetitive beam breaks <1 sec apart) of the mice over 40 min of testing. (*B*) Stereotypical activity of TWT and TKO mice during the first 15 min in the open field (*left*) and from 16–40 min (*right*). (*C*) Effects of psychostimulants on locomotor activity of TWT and TKO mice. Animals were given vehicle (saline) or 2 mg/kg amphetamine (intraperitoneally) and immediately placed into the open field for 1 hr. Cumulative locomotor activities under vehicle and psychostimulant treatments are depicted for both genotypes.

enhanced in TKO mice (Fig. 8A,B). However, within 20 minutes, the activity of TKO mice was reduced relative to that of the TWT controls (Fig. 8A,B). These findings indicate that TKO mice are initially more responsive to the locomotor-stimulating effects of a novel environment and also habituate more rapidly to their surroundings.

Increased activity in the open field is typically associated with increased dopaminergic tone (Pijnenburg et al. 1976; Gainetdinov et al. 1999; Carlsson et al. 2001). To examine this possibility, we treated TKO and TWT mice with psychostimulants to enhance dopaminergic transmission (Robinson and Berridge 1993; Nestler and Aghajanian 1997; Koob et al. 1998; Kopinksy and Hyman 2002). Animals were given amphetamine or cocaine, or vehicle as a control, and immediately placed into the open field. The two genotypes of mice exhibited remarkably different sensitivities to the psychostimulants: For example, locomotor activity was stimulated by amphetamine (2 mg/kg, intraperitoneally) to a much greater extent in TKO mice than in TWT controls (Fig. 8C). TKO mice were similarly hypersensitive to cocaine. Collectively, these results indicate that the locomotor-stimulating effects of psychostimulants are enhanced in TKO mice.

Because the TKO mice were more responsive than TWT mice to acute administration of psychostimulants, we decided to determine whether they could become sensitized more readily to these drugs. We did this by injecting TKO and TWT mice over 5 consecutive days with saline or with cocaine (20 mg/kg, intraperitoneally). Following repeated administration of psychostimulants, mice were withdrawn from these drugs for 5 days and subsequently tested for their response to cocaine. The TWT mice displayed sensitization to the drug because locomotor responses to cocaine were much higher at the time of challenge than on the first day of exposure. In contrast, repeated administration of cocaine to TKO animals provided no augmentation of locomotion. Hence, although TWT mice exhibit behavioral sensitization to the psychostimulant, TKO animals appear to be fully sensitized to the drug even though they have not been exposed to it previously.

Psychostimulants, such as cocaine and amphetamine, block plasma membrane transporters that remove catecholamines and serotonin from the extracellular space (Woolverton and Johnson 1992; Seiden et al. 1993; Kuczenski et al. 1995; Paulson and Robinson 1995; Heidbreder et al. 1996). As a result, psychostimulants induce hyperlocomotion by increasing extracellular dopamine (DA) in the ventral striatum (Paulson and Robinson 1995; Heidbreder et al. 1996). For this reason, microdialysis was used to measure striatal DA levels in TWT and TKO mice. Following collection of baseline samples for 1 hour, animals were injected with cocaine (20 mg/kg, intraperitoneally) and samples were collected over another 2 hours. Cocaine potentiated extracellular DA levels in the striata of both TWT and TKO animals. However, DA concentrations were augmented by cocaine to a greater extent in TWT than in TKO mice. Because extracellular DA levels were less responsive to cocaine in TKO than in TWT mice, whereas locomotor activity in TKO animals showed the opposite response, these data suggest that the enhanced locomotor activity must result from additional mechanisms.

It is well known that changes in postsynaptic receptor sensitivity can mediate sensitization to cocaine and other psychostimulants that act as indirect agonists at DA receptors (M. Xu et al. 1994, 1997; Rubinstein et al. 1997; F. Xu et al. 2000; Chausmer et al. 2002; Elliot et al. 2003). Five DA receptors have been cloned and they are assigned to two receptor classes termed the D1- and D2-like receptors (Civelli et al. 1993). D1 receptors comprise the D1 and D5 subtypes, whereas the D2 receptor class is composed of D2, D3, and D4 receptors. To analyze DA receptor responses, TWT and TKO animals were administered apomorphine, an agonist at both D1- and D2-like receptors, and changes in behavior were assessed. For this purpose, locomotor activity was measured in an open field (Bailey and Jackson 1978) and climbing behavior was assessed in a cage (Wilcox et al. 1980). In both behavioral tests, TKO mice were more responsive to a given dose of apomorphine than were the TWT controls. These findings suggest an increased sensitivity of D1- and D2-like DA receptors in TKO animals.

To investigate whether the enhanced apomorphine response was the result of changes in D1- and/or D2-like receptors, mice were treated with the selective D1 receptor agonist, SKF 81297, or the selective D2 receptor agonist, quinpirole. In the open field, TKO animals were more responsive to lower doses of either drug than were the TWT controls. Hence, these findings suggest that the enhanced behavioral responses of TKO mice to psychostimulants and apomorphine may be attributable to enhanced responsiveness of both D1- and D2-like receptors.

We also have examined presynaptic mechanisms underlying the reduced amount of cocaine-induced DA release in TKO mice. Levels of DA were measured by dissecting the striata from TWT and TKO mice. Monoamines were extracted from these tissue samples, separated by high-pressure liquid chromatography, and then measured by electrochemical detection. No differences in striatal DA levels in TKO and TWT mice were found. To study whether biosynthesis of catecholamines was different, animals were treated with the drug NSD-1015 (100 mg/kg, intraperitoneally) to block conversion of L-dihydroxy-phenylalanine (L-Dopa) to DA. Accumulation of L-Dopa was enhanced by approximately 50% in TKO mice compared to that in TWT animals. Together, these findings show that tissue levels of DA are not increased in striata of TKO mice, whereas synthesis rates are augmented; neither of these observations can account for the reduction in cocaine-induced DA release in TKO mice.

As is the case for the glutamatergic and GABAergic neurons described above, separate pools of DA have been described in dopaminergic neurons (Javoy and Glowinski 1971; McMillen et al. 1980; Ewing et al. 1983; Justice et al. 1988). Although many such pools have been postulated, at the minimum these can be grouped into a storage (reserve) pool and a newly synthesized pool that is rapidly released in response to stimulation. Given that loss of synapsins impairs the RP in glutamatergic and GABAergic synapses, it is possible that the reduction in cocaine-induced DA release in TKO mice arises from a defect in the DA storage pool. To test this hypothesis, we took advantage of the fact that newly synthesized DA is preferentially released on stimulation (Groppetti et al. 1977). Thus, we could eliminate this component by blocking synthesis of DA by treating mice with α-methyl-p-tyrosine (AMPT; 200 mg/kg, intraperitoneally) and then examine the effects of cocaine on the residual release of DA from the storage pool. TWT and TKO mice were implanted with a stimulating electrode in the medial forebrain bundle and a cyclic voltametry electrode was positioned in the caudate-putamen to measure DA levels with subsecond time resolution (B.J. Venton et al., in prep.).

Electrical stimulation of the medial forebrain bundle (0.5-sec trains at 10–50 Hz) increased extracellular levels of DA in the caudate-putamen of both TWT and TKO mice. After treatment with AMPT, trains of 600 electrical stimuli (60 Hz) were given every 20 minutes to eliminate the rapidly releasable pool of DA. This procedure reduced releasable DA by approximately 75% in both TWT and TKO striata. Subsequent administration of cocaine was sufficient to restore extracellular concentrations of DA to pre-AMPT levels in TWT animals. However, in TKO mice, cocaine only increased DA levels to 40% of their original values. These results indicate that cocaine acts, at least in part, by mobilizing DA from the reserve pool in TWT mice. Furthermore, these results indicate that the reduced ability of cocaine to increase extracellular DA in TKO mice arises because of the loss of a reserve pool of DA vesicles that requires synapsins.

In summary, synapsin TKO mice are highly responsive to psychostimulants. In fact, these animals seem to be fully sensitized on their first exposure to these drugs. Disruption of all three synapsin genes in these mice appears to produce both presynaptic and postsynaptic alterations in dopaminergic neurotransmission. Although tissue levels of catecholamines in TKO mice are similar to those of the TWT controls, mobilization of stored DA from the reserve pool to the readily releasable pool is diminished in TKO mice. As stimulation preferentially mobilizes DA in the readily releasable pool (Javoy and Glowinski 1971; McMillen et al. 1980; Ewing et al. 1983; Justice et al. 1988), DA neurons appear to adapt to these circumstances by increasing the rate of catecholamine biosynthesis, thereby providing additional transmitter for stimulation conditions. In this way, more transmitter may be available for rapid release from DA terminals in TKO mice when they are initially exposed to a novel environment (e.g., open field) and/or psychostimulants. In addition to these presynaptic changes, TKO mice also have increased responsiveness of both D1- and D2-like DA receptors. These changes in postsynaptic DA receptors may serve to compensate for the reduced DA release resulting from loss of the synapsin-dependent reserve pool. The enhanced sensitivity of DA receptors in TKO mice, coupled with the presynaptic changes in DA homeostasis, apparently is responsible for the high sensitivity of these animals to psychostimulant drugs (Clark and White 1987; Rubinstein et al. 1988; Waddington and Daly 1993).

SUMMARY AND CONCLUSIONS

Our studies with TKO mice devoid of all three synapsin genes reveal that these proteins have a variety of presynaptic functions that depend on the type of neurotransmitter used. At hippocampal glutamatergic synapses, synapsins are responsible for maintaining a reserve pool of synaptic vesicles that are mobilized during synaptic depression, enhancing the duration and frequency range of glutamatergic synaptic transmission. Synapsin IIa seems to be mainly responsible for this action, with synapsin Ib serving as a dominant negative isoform that reduces the reserve pool by heterodimerizing with synapsin IIa and reducing vesicle cross-linking. At hippocampal GABAergic synapses, synapsins influence both the readily releasable and reserve pools of synaptic vesicles. The synapsin isoforms involved in these functions are not yet clear. At striatal dopaminergic synapses, synapsins also appear to be responsible for maintaining a reserve vesicle pool that is mobilized by cocaine and other psychostimulant drugs.

These presynaptic roles for synapsins can also partially explain the behavioral phenotypes of the synapsin TKO mice. The deficits in learning and memory presumably arise from the accelerated synaptic depression at hippocampal excitatory synapses: The more rapid loss of synaptic excitation during repetitive activity should make it more difficult to elicit long-term potentiation, long-term depression, and other long-lasting forms of synaptic plasticity that are thought to be responsible for learning and memory. The reduction in basal transmission at inhibitory synapses may account for the propensity of the synapsin TKO mice to undergo seizures, given that seizures may result from an imbalance between synaptic excitation and inhibition (Dichter and Ayala 1987). Although apparent defects in the release of dopamine exist, these presynaptic defects cannot explain the observed enhancements in locomotor activity and hypersensitization to psychostimulants observed in the synapsin TKO mice. Instead, these changes apparently result from enhanced postsynaptic sensitivity to dopamine. These postsynaptic changes may arise to compensate for the impaired release of dopamine at these synapses.

GAPS AND FUTURE DIRECTIONS

Although the work summarized above provides a new and coherent picture of the functions of the synapsin protein family, much remains to be understood. First, our work has examined transmission at only a very small number of synapses and it will be important to determine whether the conclusions derived from these initial studies can be generalized to the thousands of other types of synapses found within the brain. Second, our studies indicate that synapsins have transmitter-specific functions and it will be important to determine whether these proteins serve additional functions at synapses using transmitters other than glutamate, GABA, and dopamine. Third, among the ten known synapsin isoforms, our studies have identified roles for only two, namely, synapsins Ib and IIa, which serve opposing functions in regulating the reserve pool of glutamatergic synaptic vesicles. Future studies certainly will be needed to identify the function of the eight other isoforms, as well as to consider whether synapsins Ib and IIa have additional roles. Finally, although the synaptic defects in the synapsin TKO mice provide plausible explanations for some of the behavioral phenotypes of these mice, future work is required to actually establish cause/effect relationships in these cases.

In conclusion, the development of mice defective in all known forms of synapsins has provided an important new resource for identifying the functions of synapsins at both the synaptic and behavioral levels. Future work should be able to continue exploiting this resource to learn more about the diverse functions of the intriguing, but elusive, synapsin family.

ACKNOWLEDGMENTS

This work is supported by a European Molecular Biology Organization postdoctoral fellowship and a Pfizer fellowship of the Life Science Research Foundation (D.G.), a Theodore and Vada Stanley Foundation Research Award (P.G. and J.F.), the Nagoya University Foundation (Y.T.), and National Institutes of Health (Grant Nos. MH-39327 and DA-10044 (P.G.), MH-67044 (G.J.A.), and NS-047209 (H.-T.K.).

REFERENCES

Augustine G.J., Burns M.E., DeBello W.M., Hilfiker S., Morgan J.R., Schweizer F.E., Tokumaru H., and Umayahara K. 1999. Proteins involved in synaptic vesicle trafficking. *J. Physiol.* **520:** 33–41.

Bailey R.C. and Jackson D.M. 1978. A pharmacological study of changes in central nervous system receptor responsiveness after long-term dexamphetamine and apomorphine administration. *Psychopharmacology* **56:** 317–326.

Baumert M., Maycox P.R., Navone F., De Camilli P., and Jahn R. 1989. Synaptobrevin: An integral membrane protein of 18,000 daltons present in small synaptic vesicles of rat brain. *EMBO J.* **8:** 379–384.

Bekkers J. and Stevens C. 1991. Excitatory and inhibitory autaptic currents in isolated hippocampal neurons maintained in cell culture. *Proc. Natl. Acad. Sci.* **88:** 7834–7838.

Benfenati F., Valtorta F., Rubenstein J.L., Gorelick F.S., Greengard P., and Czernik A.J. 1992. Synaptic vesicle-associated Ca^{2+}/calmodulin-dependent protein kinase II is a binding protein for synapsin I. *Nature* **359:** 417–420.

Bloom O., Evergren E., Tomilin N., Kjaerulff O., Low P., Brodin L., Pieribone V.A., Greengard P., and Shupliakov O. 2003. Colocalization of synapsin and actin during synaptic vesicle recycling. *J. Cell Biol.* **161:** 737–747.

Carlsson A., Waters N., Holm-Waters S., Tedroff J., Nilsson M., and Carlsson M.L. 2001. Interactions between monoamines, glutamate, and GABA in schizophrenia: New evidence. *Annu. Rev. Pharmacol. Toxicol.* **41:** 237–260.

Chausmer A.L., Elmer G.I., Rubinstein M., Low M.J., Grandy D.K., and Katz J.L. 2002. Cocaine-induced locomotor activity and cocaine discrimination in dopamine D2 receptor mutant mice. *Psychopharmacology* **163:** 54–61.

Chin L.S., Li L., Ferreira A., Kosik K.S., and Greengard P. 1995. Impairment of axonal development and of synaptogenesis in hippocampal neurons of synapsin I-deficient mice. *Proc. Natl. Acad. Sci.* **92:** 9230–9234.

Civelli O., Bunzow J.R., and Grandy D.K. 1993. Molecular diversity of the dopamine receptors. *Annu. Rev. Pharmacol. Toxicol.* **33:** 281–307.

Clark D. and White F.J. 1987. D1 dopamine receptor—The search for a function: A critical evaluation of the D1/D2 dopamine receptor classification and its functional implications. *Synapse* **1:** 347–388.

Cousin M.A., Malladi C.S., Tan T.C., Raymond C.R., Smillie K.J., and Robinson P.J. 2003. Synapsin I-associated phosphatidylinositol 3-kinase mediates synaptic vesicle delivery to the readily releasable pool. *J. Biol. Chem.* **278:** 29065–29071.

del Castillo J. and Katz B. 1954. Quantal components of the end plate potential. *J. Physiol.* **124:** 560–573.

Dichter M.A. and Ayala G.F. 1987. Cellular mechanisms of epilepsy: A status report. *Science* **237:** 157–164.

Dobrunz L.E. and Stevens C.F. 1997. Heterogeneity of release probability, facilitation, and depletion at central synapses. *Neuron* **18:** 995–1008.

Elliot E.E., Sibley D.R., and Katz J.L. 2003. Locomotor and dis-criminative-stimulus effects of cocaine in dopamine D5 receptor knockout mice. *Psychopharmacology* **169:** 161–168.

Esser L., Wang C.R., Hosaka M., Smagula C.S., Südhof T.C., and Deisenhofer J. 1998. Synapsin I is structurally similar to atp-utilizing enzymes. *EMBO J.* **17:** 977–984.

Ewing A.G., Bigelow J.C., and Wightman R.M. 1983. Direct in vivo monitoring of dopamine released from two striatal com-partments in the rat. *Science* **221:** 169–171.

Feng J., Chi P., Blanpied T.A., Xu Y., Magarinos A.M., Ferreira A., Takahashi R.H., Kao H.T., McEwen B.S., Ryan T.A., Augustine G.J., and Greengard P. 2002. Regulation of neurotransmitter release by synapsin III. *J. Neurosci.* **22:** 4372–4380.

Ferreira A. and Rapoport M. 2002. The synapsins: Beyond the reg-ulation of neurotransmitter release. *Cell. Mol. Life Sci.* **59:** 589–595.

Ferreira A., Kao H.T., Feng J., Rapoport M., and Greengard P. 2000. Synapsin III: Developmental expression, subcellular localization, and role in axon formation. *J. Neurosci.* **20:** 3736–3744.

Ferreira A., Chin L.S., Li L., Lanier L.M., Kosik K.S., and Greengard P. 1998. Distinct roles of synapsin I and synapsin II during neuronal development. *Mol. Med.* **4:** 22–28.

Fried G., Nestler E.J., De Camilli P., Stjarne L., Olson L., Lundberg J.M., Hokfelt T., Ouimet C.C., and Greengard P. 1982. Cellular and subcellular localization of protein I in the peripheral nerv-ous system. *Proc. Natl. Acad. Sci.* **79:** 2717–2721.

Gainetdinov R.R., Wetsel W.C., Jones S.R., Levin E.D., Jaber M., and Caron M.G. 1999. Role of serotonin in the paradoxical calming effect of psychostimulants on hyperactivity. *Science* **283:** 397–401.

Gitler D., Takagishi Y., Feng J., Ren Y., Rodriguiz R.M., Wetsel W.C., Greengard P., and Augustine G.J. 2004a. Different presynaptic roles of synapsins at excitatory and inhibitory synapses. *J. Neurosci.* **24:** 11368–11380.

Gitler D., Xu Y., Kao H.-T., Lin D., Lim S., Feng J., Greengard P., and Augustine G.J. 2004b. Molecular determinants of synapsin targeting to presynaptic terminals. *J. Neurosci.* **24:** 3711–3720.

Greengard P., Valtorta F., Czernik A.J., and Benfenati F. 1993. Synaptic vesicle phosphoproteins and regulation of synaptic function. *Science* **259:** 780–785.

Groppetti A., Algeri S., Cattabeni F., Di Giulio A.M., Galli C.L., Ponzio F., and Spano P.F. 1977. Changes in specific activity of dopamine metabolites as evidence of a multiple compartmen-tation of dopamine in striatal neurons. *J. Neurochem.* **28:** 193–197.

Heidbreder C.A., Thompson A.C., and Shippenberg T.S. 1996. Role of extracellular dopamine in the initiation and long-term expression of behavioral sensitization to cocaine. *J. Pharmacol. Exp. Ther.* **278:** 490–502.

Hilfiker S., Pieribone V.A., Czernik A.J., Kao H.T., Augustine G.J., and Greengard P. 1999. Synapsins as regulators of neurotrans-mitter release. *Philos. Trans. R. Soc. Lond. B Biol. Sci.* **354:** 269–279.

Hilfiker S., Schweizer F.E., Kao H.T., Czernik A.J., Greengard P.,

and Augustine G.J. 1998. Two sites of action for synapsin domain E in regulating neurotransmitter release. *Nat. Neurosci.* **1:** 29–35.

Hirokawa N., Sobue K., Kanda K., Harada A., and Yorifuji H. 1989. The cytoskeletal architecture of the presynaptic terminal and molecular structure of synapsin 1. *J. Cell Biol.* **108:** 111–126.

Hosaka M. and Südhof T.C. 1998a. Synapsin III, a novel synapsin with an unusual regulation by Ca^{2+}. *J. Biol. Chem.* **273:** 13371–13374.

——. 1998b. Synapsins I and II are ATP-binding proteins with dif-ferential Ca^{2+} regulation. *J. Biol. Chem.* **273:** 1425–1429.

——. 1999. Homo- and heterodimerization of synapsins. *J. Biol. Chem.* **274:** 16747–16753.

Humeau Y., Doussau F., Vitiello F., Greengard P., Benfenati F., and Poulain B. 2001. Synapsin controls both reserve and releasable synaptic vesicle pools during neuronal activity and short-term plasticity in Aplysia. *J. Neurosci.* **21:** 4195–4206.

Javoy F. and Glowinski J. 1971. Dynamic characteristic of the 'functional compartment' of dopamine in dopaminergic ter-minals of the rat striatum. *J. Neurochem.* **18:** 1305–1311.

Justice J.B., Jr., Nicolaysen L.C., and Michael A.C. 1988. Modeling the dopaminergic nerve terminal. *J. Neurosci. Methods* **22:** 239–252.

Kao H.T., Porton B., Hilfiker S., Stefani G., Pieribone V.A., DeSalle R., and Greengard P. 1999. Molecular evolution of the synapsin gene family. *J. Exp. Zool.* **285:** 360–377.

Kao H.T., Song H.J., Porton B., Ming G.L., Hoh J., Abraham M., Czernik A.J., Pieribone V.A., Poo M.M., and Greengard P. 2002. A protein kinase A-dependent molecular switch in synapsins regulates neurite outgrowth. *Nat. Neurosci.* **5:** 431—437.

Koob G.F., Sanna P.P., and Bloom F.E. 1998. Neuroscience of addiction. *Neuron* **21:** 467–476.

Kopinsky K.L. and Hyman S.E. 2002. Molecular and cellular biology of addiction. In *Neuropharmacology: The fifth gener-ation of progress* (ed. K.L. Davis et al.), pp. 1367–1380. Lippincott Williams & Wilkins, Philadelphia.

Kuczenski R., Segal D.S., Cho A.K., and Melega W. 1995. Hippocampus norepinephrine, caudate dopamine and sero-tonin, and behavioral responses to the stereoisomers of amphetamine and methamphetamine. *J. Neurosci.* **15:** 1308–1317.

Laakso A., Mohn A.R., Gainetdinov R.R., and Caron M.G. 2002. Experimental genetic approaches to addiction. *Neuron* **36:** 213–228.

Li L., Chin L.S., Shupliakov O., Brodin L., Sihra T.S., Hvalby O., Jensen V., Zheng D., McNamara J.O., Greengard P., et al. 1995. Impairment of synaptic vesicle clustering and of synap-tic transmission, and increased seizure propensity, in synapsin I-deficient mice. *Proc. Natl. Acad. Sci.* **92:** 9235–9239.

Mandell J.W., Czernik A.J., De Camilli P., Greengard P., and Townes-Anderson E. 1992. Differential expression of synapsins I and II among rat retinal synapses. *J. Neurosci.* **12:** 1736–1749.

Matsubara M., Kusubata M., Ishiguro K., Uchida T., Titani K., and Taniguchi H. 1996. Site-specific phosphorylation of synapsin I by mitogen-activated protein kinase and cdk5 and its effects on physiological functions. *J. Biol. Chem.* **271:** 21108–21113.

Matus-Leibovitch N., Nevo I., and Vogel Z. 1997. Differential distribution of synapsin IIa and IIb mRNAs in various brain structures and the effect of chronic morphine administration on the regional expression of these isoforms. *Brain Res. Mol. Brain Res.* **45:** 301–316.

McMillen B.A., German D.C., and Shore P.A. 1980. Functional and pharmacological significance of brain dopamine and norepinephrine storage pools. *Biochem. Pharmacol.* **29:** 3045–3050.

Murthy V.N. and Stevens C.F. 1999. Reversal of synaptic vesicle docking at central synapses. *Nat. Neurosci.* **2:** 503–507.

Nestler E.J. 2002. From neurobiology to treatment: Progress against addiction. *Nat. Neurosci.* **5:** 1076–1079.

Nestler E.J. and Aghajanian G.K. 1997. Molecular and cellular basis of addiction. *Science* **278:** 58–63.

Paulson P.E. and Robinson T.E. 1995. Amphetamine-induced time-dependent sensitization of dopamine neurotransmission in the dorsal and ventral striatum: A microdialysis study in behaving rats. *Synapse* **19:** 56–65.

Pieribone V.A., Porton B., Rendon B., Feng J., Greengard P., and Kao H.T. 2002. Expression of synapsin III in nerve terminals and neurogenic regions of the adult brain. *J. Comp. Neurol.* **454:** 105-114.

Pijnenburg A.J., Honig W.M., Van der Heyden J.A., and Van Rossum J.M. 1976. Effects of chemical stimulation of the mesolimbic dopamine system upon locomotor activity. *Eur. J. Pharmacol.* **35:** 45–58

Porton B., Kao H.T., and Greengard P. 1999. Characterization of transcripts from the synapsin III gene locus. *J. Neurochem.* **73:** 2266–2271.

Ritz M.C., Lamb R.J., Goldberg S.R., and Kuhar M.J. 1987. Cocaine receptors on dopamine transporters are related to self-administration of cocaine. *Science* **237:** 1219–1223.

Rizzoli S.O. and Betz W.J. 2005. Synaptic vesicle pools. *Nat. Rev. Neurosci.* **6:** 57–69.

Roberto M., Madamba S.G., Moore S.D., Tallent M.K., and Siggins G.R. 2003. Ethanol increases GABAergic transmission at both pre- and postsynaptic sites in rat central amygdala neurons. *Proc. Natl. Acad. Sci.* **100:** 2053–2058.

Robinson T.E. and Berridge K.C. 1993. The neural basis of drug craving: An incentive-sensitization theory of addiction. *Brain Res. Brain Res. Rev.* **18:** 247–291.

Rosahl T.W., Geppert M., Spillane D., Herz J., Hammer R.E., Malenka R.C., and Südhof T.C. 1993. Short-term synaptic plasticity is altered in mice lacking synapsin I. *Cell* **75:** 661–670.

Rosahl T.W., Spillane D., Missler M., Herz J., Selig D.K., Wolff J.R., Hammer R.E., Malenka R.C., and Südhof T.C. 1995. Essential functions of synapsins I and II in synaptic vesicle regulation. *Nature* **375:** 488–493.

Rubinstein M., Gershanik O., and Stefano F.J. 1988. Postsynaptic bimodal effect of sulpiride on locomotor activity induced by pergolide in catecholamine-depleted mice. *Naunyn Schmiedebergs Arch. Pharmacol.* **337:** 115–117.

Rubinstein M., Phillips T.J., Bunzow J.R., Falzone T.L., Dziewczapolski G., Zhang G., Fang Y., Larson J.L., McDougall J.A., Chester J.A., Saez C., Pugsley T.A., Gershanik O., Low M.J., and Grandy D.K. 1997. Mice lacking dopamine D4 receptors are supersensitive to ethanol, cocaine, and methamphetamine. *Cell* **90:** 991–1001.

Samigullin D., Bill C.A., Coleman W.L., and Bykhovskaia M. 2004. Regulation of transmitter release by synapsin II in the mouse motor terminal. *J. Physiol.* **23:** 23.

Segal M.M. and Furshpan E.J. 1990. Epileptiform activity in microcultures containing small numbers of hippocampal neurons. *J. Neurophysiol.* **64:** 1390–1399.

Seiden L.S., Sabol K.E., and Ricaurte G.A. 1993. Amphetamine: Effects on catecholamine systems and behavior. *Annu. Rev. Pharmacol. Toxicol.* **33:** 639–677.

Südhof T.C. 2004. The synaptic vesicle cycle. *Annu. Rev. Neurosci.* **27:** 509–547.

Südhof T.C., Czernik A.J., Kao H.T., Takei K., Johnston P.A., Horiuchi A., Kanazir S.D., Wagner M.A., Perin M.S., De Camilli P., et al. 1989. Synapsins: Mosaics of shared and individual domains in a family of synaptic vesicle phosphoproteins. *Science* **245:** 1474–1480.

Thiel G., Südhof T.C., and Greengard P. 1990. Synapsin II. Mapping of a domain in the NH2-terminal region which binds to small synaptic vesicles. *J. Biol. Chem.* **265:** 16527–16533.

Tu Y., Nayak S.K., Woodson J., and Ross E.M. 2003. Phosphorylation-regulated inhibition of the Gz GAP activity of RGS proteins by synapsin I. *J. Biol. Chem.* **278:** 52273–52281.

Waddington J.L. and Daly S.A. 1993. Regulation of unconditioned motor behaviour by D1:D2 interactions. In *D1:D2 dopamine receptor interactions* (ed. J.L. Waddington), pp. 51–78. Academic Press, San Diego.

Wesseling J.F. and Lo D.C. 2002. Limit on the role of activity in controlling the release-ready supply of synaptic vesicles. *J. Neurosci.* **22:** 9708–9720.

Wilcox R.E., Smith R.V., Anderson J.A., and Riffee W.H. 1980. Apomorphine-induced stereotypic cage climbing in mice as a model for studying changes in dopamine receptor sensitivity. *Pharmacol. Biochem. Behav.* **12:** 29–33.

Woolverton W.L. and Johnson K.M. 1992. Neurobiology of cocaine abuse. *Trends Pharmacol. Sci.* **13:** 193–200.

Xu F., Gainetdinov R.R., Wetsel W.C., Jones S.R., Bohn L.M., Miller G.W., Wang Y.M., and Caron M.G. 2000. Mice lacking the norepinephrine transporter are supersensitive to psychostimulants. *Nat. Neurosci.* **3:** 465–471.

Xu M., Moratalla R., Gold L.H., Hiroi N., Koob G.F., Graybiel A.M., and Tonegawa S. 1994. Dopamine D1 receptor mutant mice are deficient in striatal expression of dynorphin and in dopamine-mediated behavioral responses. *Cell* **79:** 729–742.

Xu M., Koeltzow T.E., Santiago G.T., Moratalla R., Cooper D.C., Hu X.T., White N.M., Graybiel A.M., White F.J., and Tonegawa S. 1997. Dopamine D3 receptor mutant mice exhibit increased behavioral sensitivity to concurrent stimulation of D1 and D2 receptors. *Neuron* **19:** 837–848.

Zucker R.S. and Regehr W.G. 2002. Short-term synaptic plasticity. *Annu. Rev. Physiol.* **64:** 355–405.

19

Synaptic Plasticity in the Mesolimbic Dopamine System and Addiction

Daniel Saal[1] and Robert C. Malenka[2]

[1]*Department of Psychiatry and Behavioral Science, Emory University, Atlanta, Georgia;*
[2]*Nancy Pritzker Laboratory, Department of Psychiatry and Behavioral Sciences,*
Stanford University School of Medicine, Palo Alto, California

ABSTRACT

Long-lasting changes in synaptic strength are thought to contribute to the modifications in neural circuitry that mediate all forms of experience-dependent plasticity, including learning and memory. The most well-established models for such experience-dependent synaptic plasticity are long-term potentiation (LTP) and long-term depression (LTD) of synaptic transmission, which can be elicited at most excitatory synapses in the mammalian brain. Here we review the mechanisms and role of plasticity at excitatory synapses in the mesolimbic dopamine (DA) system. Specifically, we examine LTP and LTD processes in the ventral tegmental area and nucleus accumbens, and the role that they may have in the induction and maintenance of addiction and related behaviors.

INTRODUCTION

A major challenge for neuroscience is to elucidate how in vivo experiences modify brain function and, thereby, behavior. A leading hypothesis is that the neural circuit changes underlying learning and memory involve strengthening or weakening the synaptic connections within the relevant network of neurons. These synaptic changes, like memories, are long lasting and subject to further modification. Such long-lasting increases in strength at excitatory synapses are termed LTP, whereas decreases in strength are classified as LTD (Malenka and Nicoll 1999; Malenka and Bear 2004). Given the great complexity of neural circuits in the mammalian brain and the problems inherent in determining which specific synapses are affected by an experience, we have attempted to use the relatively well-defined circuits involved in mediating addiction to understand the role of synaptic plasticity in experience-dependent modifications of behavior. Although historically, the dominant focus of the addiction field has not been on excitatory synapses, a large and growing body of evidence shows that modulation of excitatory neural circuitry contributes to many of the neural adaptations that underlie addiction (Kim et al. 1996; Tzschentke 1998; Wolf 1998; Sanchez et al. 2003; Suto et al. 2003; Kauer 2004).

Addiction Is a Maladaptive Form of Learning and Memory

Given that cues associated with past drug use often initiate drug craving and are associated with relapse (O'Brien et al. 1998), it was natural to predict that mechanisms that underlie adaptive forms of learning, such as LTP and LTD, may also contribute importantly to addiction (Hyman and Malenka 2001; Everitt and Wolf 2002; Kauer 2004). Historically, LTP in particular was thought to be the mechanism of so-called Hebbian learning, in which two coincident events are required for synaptic strengthening. Thus, the strong link between drug-associated cues and the drug experience that is critical for many addiction-related behaviors suggests that this kind of synaptic process may be important. In fact, a number of studies have shown an important role for modulation of excitatory synapses of the mesolimbic DA system in initiation and maintenance of addiction behaviors in animals (Wolf 1998; Hyman and Malenka 2001; Carlezon and Nestler 2002; Wolf et al. 2004). These sorts of findings have led to the conceptualization of addiction as a maladaptive form of learning and memory (Kelley 2004). In other words, the addictiveness of a substance may stem from its ability to co-opt natural learning and memory processes and make the brain greatly overvalue the drug experience.

GOALS OF THE CHAPTER

In this chapter, we examine some of the mechanisms and potential behavioral roles of synaptic plasticity at excitatory synapses in specific brain areas and circuits that have a role in addiction. The basic mechanisms underlying the major forms of LTP and LTD described here provide the requisite background for the subsequent discussion of the relevance of these synaptic phenomena to addiction. We specifically focus on the similarities and differences between the mechanisms of LTP and LTD in the ventral tegmental area (VTA) and nucleus accumbens (NAc), the two main components of the mesolimbic DA system. Finally, we show that in vivo exposure to cocaine and other addictive substances do, in fact, modify excitatory synapses in the VTA and NAc in ways that mimic LTP and LTD and that at least some of these changes appear to have an important role in mediating the behavioral effects of drugs of abuse.

MECHANISMS OF LTP AND LTD

To provide a framework in which to consider the role in addiction of drug-induced plasticity at excitatory synapses in the mesolimbic DA system, we first briefly describe our current understanding of the basic mechanisms underlying LTP and LTD. It is now clear that there are multiple forms of both LTP and LTD, which have been described in many different regions of the mammalian brain (Malenka and Bear 2004). The most extensively studied and well-characterized forms of LTP and LTD are found at excitatory synapses in the CA1 region of the hippocampus and are triggered by activation of the *N*-methyl-D-aspartate (NMDA) subtype of glutamate receptor. It is these forms of LTP and LTD that we review because it appears that many, although certainly not all, of the principles learned from the study of the hippocampus are generalizable to circuits throughout the brain.

AMPA and NMDA Glutamate Receptors' Role in LTP and LTD

The neurotransmitter used at excitatory synapses is glutamate, which activates two different subtypes of ionotropic receptors that flux positively charged ions. These two glutamate receptor subtypes are termed AMPA (alpha-amino-3-hydroxy-5-methyl-4-isoxazole propionate) and NMDA receptors and are colocalized in the postsynaptic membrane of most excitatory synapses. AMPA receptors are considered the workhorses of excitatory synapses and carry most of the synaptic current during the routine functioning of excitatory neural circuits. Changes in synaptic strength during LTP and LTD are largely thought to be caused by alteration in the number of AMPA receptors at individual synapses. AMPA receptors are composed of individual subunits termed GluR1, GluR2, GluR3, and GluR4, which are likely combined as tetramers (Contractor and Heinemann 2002). The subunit composition of AMPA receptors influences their biophysical properties and perhaps their trafficking to and away from synapses during LTP and LTD (Contractor and Heinemann 2002; Malinow and Malenka 2002).

NMDA receptors have two very important properties that distinguish them from AMPA receptors. They exhibit a strong voltage dependence such that even when they bind glutamate, they do not pass much current when the cell's membrane potential is hyperpolarized (i.e., the cell is not very active). This is because the NMDA receptor channel pore contains a binding site for magnesium ions that blocks the flux of other ions through the channel pore. The magnesium ions are held in place by the relatively negative voltage across the plasma membrane when the cell is not very active. When the postsynaptic membrane is depolarized, however, the magnesium ions are expelled out of the channel pore, allowing current to flow. Thus, NMDA receptors only pass current when two conditions are met: They are activated by synaptically released glutamate and the cell is relatively depolarized by, for example, strong afferent activity. The second important property of NMDA receptors is that unlike AMPA receptors, they are highly permeable to calcium, which can activate a host of intracellular signaling molecules. It is this NMDA receptor-mediated increase in postsynaptic calcium concentration that is the critical trigger for LTP and LTD (Malenka and Nicoll 1999; Malenka and Bear 2004).

How Can an Increase in Postsynaptic Calcium Trigger Both LTP and LTD?

Current evidence suggests that the different patterns of afferent activity that elicit LTP or LTD result in differences in the level and perhaps the time course of the postsynaptic calcium signal. At low levels of calcium, high-affinity proteins are bound and activated, resulting in LTD. These are thought to include protein phosphatase 2B (also known as calcineurin), which in turn indirectly activates another protein phosphatase termed protein phosphatase 1 (Lisman 1989; Morishita et al. 2001; Mulkey et al. 1993). These enzymes, along with others, ultimately promote the endocytosis (i.e., removal) of synaptic AMPA receptors, resulting in a decrease in synaptic strength (Fig. 1) (Carroll et al. 2001). On the other hand, higher levels of calcium entry through NMDA receptors activate calcium-/calmodulin-dependent protein kinase II (CaMKII), which is clearly critical for triggering LTP, as well as a number of other signaling enzymes that may also contribute to LTP (Malenka and Nicoll 1999; Lisman 2001; Malenka and Bear 2004). Activation of CaMKII ultimately promotes the insertion of GluR1-containing AMPA receptors into synapses,

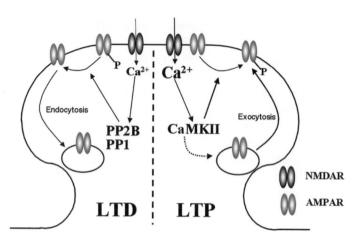

FIGURE 1. The mechanisms of hippocampal LTP and LTD: The figure is a depiction of a postsynaptic dendritic spine. The left side illustrates the mechanisms underlying LTD. Calcium ions enter the cell via NMDA receptors and activate protein phosphotase 1 and 2B (PP1 and PP2B). These enzymes remove phosphate groups on AMPA receptors (as well as other proteins that are not shown) and this leads to the removal (endocytosis) of these receptors from the plasma membrane. The right side of the figure illustrates the processes that underlie LTP. A relatively larger number of calcium ions enter the cell through NMDA receptors. The resulting elevated calcium concentration activates calcium-/calmodulin-dependent protein kinase II (CaMKII). This enzyme adds phosphate groups to AMPA receptors (and other proteins that are not shown), leading to the insertion of AMPA receptors into the synaptic plasma membrane.

resulting in LTP (Fig. 1) (Malinow and Malenka 2002; Malenka and Bear 2004). This requirement for GluR1 has been demonstrated in a number of ways (Malinow and Malenka 2002), including the lack of LTP in GluR1 knockout mice (Zamanillo et al. 1999). As we will see later, important aspects of this model for LTP are that, although required for activation of the process, presynaptic processes and NMDA receptors do not change in a sustained way during LTP.

THE MESOLIMBIC DA SYSTEM IS INVOLVED IN ADDICTION

A large body of evidence implicates the so-called mesolimbic DA system in the induction and maintenance of addiction. These anatomical substrates were defined using structural mapping techniques (Overton and Clark 1997; Carr et al. 1999; Sesack et al. 2003) and functional studies in which the impact of stimulating or lesioning regions affected many different addiction-related behaviors (Koob et al. 1998; Hyman and Malenka 2001; Nestler 2001). The two major components of the mesolimbic DA system are the VTA and the NAc (Fig. 2). The VTA contains GABAergic neurons and DA-producing cells that provide important inputs to the prefrontal cortex and NAc (Oades and Halliday 1987; Carr and Sesack 2000; Sesack et al. 2003). Indeed, the increase in DA release from VTA projections to the NAc is believed to be a key common feature of all drugs of abuse (Wise

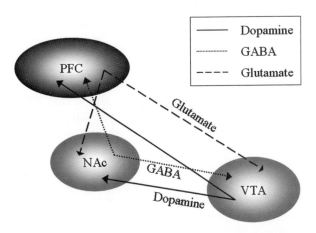

FIGURE 2. A highly simplified circuit diagram of some of the key components of the mesolimbic DA system. The prefrontal cortex (PFC) projects excitatory glutamatergic afferents to the nucleus accumbens (NAc) and ventral tegmental area (VTA). The NAc projects inhibitory GABAergic afferents to the PFC and VTA. The VTA sends dopamine-releasing afferents to many targets including the PFC and NAc.

1996; Koob et al. 1998; Nestler 2001). The major excitatory afferent inputs to the VTA that control DA cell firing come from the prefrontal cortex, the pedunculopontine nucleus, and the bed nucleus of the stria terminalis (Overton and Clark 1997). In the NAc, the major cell subtype is the medium spiny neuron, which receives important excitatory projections from the prefrontal cortex, hippocampus, and amygdala. These GABAergic cells in turn project to the globus pallidus and back to the VTA.

Drug-induced Synaptic Changes in the Mesolimbic DA System Lead to Behavioral Sensitization

In terms of examining the role of synaptic plasticity in the mesolimbic DA system in addiction, a particularly important form of experience-dependent plasticity has been the phenomenon of behavioral sensitization (Robinson and Becker 1986; Vanderschuren and Kalivas 2000a; Robinson and Berridge 2001). In the routine version of this protocol, rodents are given repeated injections of psychostimulant drugs and their motor activity is determined. Typically, the degree of motor activity increases with each injection. After a withdrawal period, lasting anywhere from days to months, animals are given a test dose of the psychostimulant. Animals that were previously exposed to psychostimulants still show an enhanced (i.e., sensitized) motor response. This increased locomotor response correlates with enhanced responsiveness to psychostimulants in several other addiction-related behavioral assays (Robinson and Becker 1986; Tzschentke 1998; Robinson and Berridge 2003; Vezina 2004) and is believed to model aspects of human addiction including drug craving (Wise 1988; Robinson and Berridge 1993, 2003).

Drug-induced Adaptations in the VTA and NAc Are Critical for Behavioral Sensitization

Much evidence suggests that drug-induced adaptations in the VTA are critical for the induction or triggering of behavioral sensitization, whereas adaptations in the NAc are critical for its expression (Robinson and Becker 1986; Kalivas and Weber 1988; Perugini and Vezina 1994; Wolf 1998; Vanderschuren and Kalivas 2000a). Of particular relevance was the finding that injection of NMDA receptor antagonists into the VTA blocked the development of sensitization, but not its expression (Wolf 1998; Vanderschuren and Kalivas 2000a). Furthermore, infusion of cocaine or amphetamine into the VTA was sufficient to induce sensitization to a subsequent systemic injection of cocaine (Kalivas and Weber 1988; Wolf 1998). On the other hand, injection of cocaine into the NAc was sufficient to evoke expression of sensitization, whereas injection of glutamate receptor antagonists into this structure could prevent its expression in response to systemically applied drugs (Bell and Kalivas 1996; Wolf 1998; Vanderschuren and Kalivas 2000a).

LTP AND LTD IN THE MESOLIMBIC DA SYSTEM

The results from experiments studying behavioral sensitization as well as other addiction-related behaviors suggested that synaptic plasticity mechanisms may be important (Wolf 1998; Overton et al. 1999; Carlezon and Nestler 2002; Wolf et al. 2004). However, before addressing the question of whether drugs of abuse actually elicit LTP or LTD in the VTA and NAc, it was important to determine whether synaptic plasticity occurred in these structures and if so, whether the properties of synaptic plasticity are the same in these structures as in the hippocampus. Experiments using acute brain slices did in fact demonstrate that excitatory synapses on VTA DA neurons can undergo both LTP (Bonci and Malenka 1999) and LTD (Jones et al. 2000; Thomas et al. 2000).

Comparing LTP and LTD in VTA with Hippocampus

LTP in these cells, like in the hippocampus, requires activation of NMDA receptors, whereas LTD does not but instead appears to be caused by activation of voltage-dependent calcium channels (Jones et al. 2000; Thomas et al. 2000). Also, unlike LTD in the hippocampus, which requires protein phosphatase activity (Malenka and Bear 2004), cAMP-dependent protein kinase activation is required for LTD in the VTA (Gutlerner et al. 2002). However, both hippocampal LTD and VTA LTD involve a decrease in the number of synaptic AMPA receptors (Carroll et al. 1999, 2001; Gutlerner et al. 2002). These results suggest that in different cell types (CA1 pyramidal cells and VTA DA cells) different signaling cascades can result in the same net synaptic effect. A further interesting difference between hippocampal and VTA LTD is the finding that bath application of d-amphetamine blocks LTD only in the VTA (Jones et al. 2000). This, as we will see later, may have a role in the drug-induced synaptic adaptations that contribute to the addictive process. The key implication of these results is that some of the mechanisms of both LTP and LTD may be quite different in the VTA compared with the rest of the brain. Thus, if the assumption that these forms of synaptic plasticity are important for addiction is correct, these differences may present important targets for intervention. On the other hand, those aspects of the mechanisms of plasticity that are conserved across different brain regions would likely not be practical therapeutic targets.

LTP and LTD in the NAc

Excitatory synapses on NAc medium spiny neurons can also express both LTP and LTD (Kombian and Malenka 1994; Thomas et al. 2000; Baldi et al. 2004) that are triggered by NMDA receptors, but little else is known about their underlying mechanisms. Recently, a distinct and potentially important form of LTD was observed in the NAc as well as the dorsal striatum (Gerdeman et al. 2002; Robbe et al. 2002). This form of LTD is triggered by strong postsynaptic activation of metabotropic glutamate receptors, another subtype of glutamate receptor that is directly coupled to second messenger systems via G proteins. This in turn causes the release of endocannabinoids that feed back on presynaptic terminals to cause a long-lasting depression of glutamate release. Interestingly, this form of LTD in the NAc was blocked by administration of a single dose of THC (Mato et al. 2004) or cocaine (Fourgeaud et al. 2004), suggesting that its modulation may be important for mediating some of the effects of these drugs of abuse.

DRUGS OF ABUSE TRIGGER SYNAPTIC PLASTICITY IN THE MESOLIMBIC DA SYSTEM

Ventral Tegmental Area

We have presented evidence that excitatory synapses in the VTA and NAc are important for addiction-related behaviors and can undergo LTP and LTD. However, two critical questions remain: Do drugs of abuse administered in vivo cause analogous synaptic modifications? Are these changes important for the development and maintenance of addiction? As mentioned above, injection of NMDA receptor antagonists into the VTA blocks the induction of behavioral sensitization (Wolf 1998) as well as conditioned place preference (CPP) in response to cocaine (Kim et al. 1996; Harris and Aston-Jones 2003). Furthermore, repeated cocaine or amphetamine exposures caused VTA neurons to be more responsive to iontophoresed AMPA (White et al. 1995). These data can be interpreted as evidence for an LTP-like event in the VTA. Molecular studies provide additional correlative evidence that synaptic plasticity in the VTA contributes to the development of addiction-related behaviors. Because LTP, at least in the hippocampus, involves insertion of GluR1-containing AMPA receptors into synapses, a long-lasting increase in AMPA receptor expression might imply that an LTP-like process has occurred. Indeed, repeated exposure to cocaine, morphine (Fitzgerald et al. 1996), and ethanol (Ortiz et al. 1995) results in an increase of GluR1 protein in the VTA. Of note, however, is that there is controversy in the literature over this point. Other groups found that there was no change in GluR1 levels following cocaine and amphetamine administration (Lu et al. 2002). These differences may be because of differences in drug administration protocol. To determine if this correlative, drug-induced change in a key synaptic protein has functional consequences, in a set of technically demanding experiments GluR1 was directly overexpressed in the VTA using viral-mediated gene transfer. This caused a significant enhancement in morphine-induced CPP, whereas overexpression of GluR2, an AMPA receptor subunit not required for LTP, did not (Carlezon et al. 1997; Carlezon and Nestler 2002).

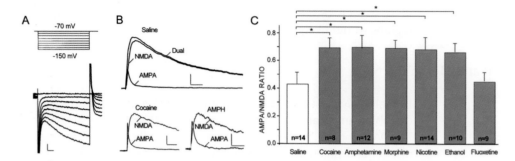

FIGURE 3. (*A*) Family of I_h currents: Dopamine cells in the midbrain are identified by the presence of an inward current evoked by hyperpolarizing voltage steps cells. (*B*) Representative synaptic currents mediated by AMPA and NMDA receptors from dopamine neurons in saline-, cocaine-, and amphetamine-treated animals. (*C*) The ratio of peak AMPA-receptor-mediated to peak NMDA-receptor-mediated current was used as an assay of synaptic strength. The bar graph shows the AMPA to NMDA ratio in dopamine cells from saline-, cocaine-, amphetamine-, morphine-, nicotine-, ethanol-, and fluoxetine-treated animals.

Measuring LTP and LTD in the VTA Following In Vivo Administration of Drugs of Abuse

Taken together, these results provided motivation to determine directly whether an LTP-like process occurred in the VTA in response to in vivo administration of drugs of abuse. This was initially accomplished by administering animals single doses of cocaine and then recording synaptic responses (excitatory postsynaptic currents, EPSCs) 24 hours later from individual DA cells in acute slices of the VTA and substantia nigra that were prepared from these animals (Ungless et al. 2001). As in previous work (Bonci and Malenka 1999), DA cells were identified by their morphological appearance and a characteristic hyperpolarization-activated current, termed the I_h current (see Fig. 3A). Differences in synaptic strength between cocaine- and saline-treated animals were assayed by taking a ratio of the relative amount of synaptic currents mediated by AMPA receptors versus NMDA receptors, a procedure that offers many technical advantages (Ungless et al. 2001). This AMPA/NMDA ratio was significantly increased in the cocaine-treated animals, a finding consistent with the dominant model for hippocampal LTP in which the magnitude of AMPA-receptor-mediated, but not NMDA-receptor-mediated, currents are increased. In addition, the ability to further enhance AMPA-receptor-mediated EPSCs by induction of LTP was diminished in DA cells from animals that had been treated with cocaine, whereas the generation of LTD was enhanced. These results strongly imply that the synapses that had been previously exposed to cocaine in vivo had already been subject to an LTP-like process. Additional electrophysiological assays indicated that the cocaine-induced "LTP" at excitatory synapses in the VTA was caused by an enhancement of AMPA receptor function and/or the number at individual synapses.

Further Characterization of Drug-induced LTP in the VTA

The demonstration that in vivo cocaine administration induces a form of LTP at excitatory synapses on midbrain DA neurons raised a number of important questions. Is this synaptic modification specific to cocaine or is it a component of the neural adaptations elicited by all addictive substances? How long does this change last? Is this LTP important for the behavioral changes elicited by cocaine or other drugs of abuse? What might the natural function of this synaptic plasticity be? Subsequent work has in fact demonstrated that multiple different classes of drugs of abuse including amphetamine, morphine, nicotine, and ethanol (Fig. 3) caused the same change in the AMPA/NMDA ratio following in vivo administration (Saal et al. 2003). Importantly, nonabused psychoactive substances, specifically carbamazepine, which like cocaine blocks sodium channels, and fluoxetine, which also shares properties with cocaine in that it is a monoamine reuptake inhibitor, both failed to cause an increase in the AMPA/NMDA ratio. The drug-induced LTP can be detected in as little as 2 hours after in vivo drug administration, at least for amphetamine (Faleiro et al. 2004). Furthermore, it lasts for at least 5 but less than 10 days following a single injection of cocaine (Ungless et al. 2001). Surprisingly, however, 7 days of repeated cocaine injections did not prolong the duration of the drug-induced synaptic enhancement; it still lasted at least 5 days but less than 10 days after the last dose (Borgland et al. 2004). This resetting to baseline is consistent with the notion that neural adaptations in the VTA function as a "gate" for the initiation of addictive processes but are not involved in the long-term maintenance of the addicted state.

What Might Be One of the Adaptive Functions of LTP on Midbrain DA Cells?

A possible clue comes from the effects that stress can have on addiction-related behaviors (Marinelli and Piazza 2002; Shaham et al. 2002). Stress can potentiate acquisition of self-administration (Kabbaj et al. 2001), trigger reinstatement (Piazza and Le Moal 1998), potentiate CPP, and induce reinstatement of CPP after it has extinguished (Will et al. 1998; Sanchez and Sorg 2001; Sanchez et al. 2003). Furthermore, like drugs of abuse, various stressors result in increased release of DA in the NAc (Rouge-Pont et al. 1998). To test whether acute stress can elicit LTP on midbrain DA cells, like drugs of abuse, animals were subjected to the Porsolt forced swim test (Porsolt et al. 1977). This did indeed cause a large increase in the AMPA/NMDA ratio in midbrain DA neurons measured 24 hours later (Saal et al. 2003). Although this result showed that a more "natural" event could cause the same synaptic adaptation as drugs of abuse, it also raised the possibility that this effect was caused by some acute stress that the drugs had generated. To examine this possibility, the glucocorticoid receptor antagonist mifepristone (RU-486) was administered. This prevented the LTP in response to the forced swim test but had no effect on the cocaine-induced LTP (Saal et al. 2003). These results indicate that the cocaine effect was not attributable to any unexpected distress that it caused the animals.

The differential effect of mifepristone on the cocaine- and stress-induced LTP in DA cells also suggested that the initial triggering mechanisms for these synaptic modifications differed. Consistent with this conclusion, in vivo administration of the D1 DA receptor antagonist SCH-23390 blocked the cocaine-induced LTP but not that elicited by the acute stress (Dong et al. 2004). On the other hand, the two processes do have some common mechanistic features. Animals exposed to stress and then given cocaine do not have higher AMPA/NMDA ratios than animals that were stressed and then given a saline injection (Dong et al. 2004). This so-called occlusion between the stress- and cocaine-induced synaptic enhancement demonstrates that the two processes do share at least one common element. Consistent with this conclusion, like the cocaine effect, the stress-induced LTP involves an up-regulation of AMPA receptors (Dong et al. 2004).

Investigating the Role of GLUR1 in LTP Using a Mouse Knockout

Because LTP in the hippocampus, which involves insertion of AMPA receptors into synapses (Malinow and Malenka 2002), is absent in adult mutant mice in which the GluR1 gene has been deleted (Zamanillo et al. 1999), an interesting question is whether the stress- and cocaine-induced LTP in midbrain DA neurons is similarly affected. Indeed, GluR1 knockout (KO) mice exposed to either stress or cocaine did not express the synaptic enhancements that their wild-type littermates showed (Dong et al. 2004). These findings provided the motivation to examine several cocaine-induced behaviors in these mice to see if they were still present and normal. Surprisingly, despite the evidence suggesting an important role for LTP in the VTA in behavioral sensitization (Overton et al. 1999; Vanderschuren and Kalivas 2000b; Wolf et al. 2004), cocaine-induced sensitization still occurred in the GluR1 KO mice (Dong et al. 2004), a result consistent with that from a previous study on morphine sensitization (Vekovischeva et al. 2001). Although this result clearly demonstrates that cocaine-induced LTP is not required for sensitization in the GluR1 KO mouse, because of the significant possibility of developmental compensations, such a negative result does not rule out that LTP in the VTA normally has an important role in this phenomenon.

Indeed, two other cocaine-induced behaviors, that assay the ability to associate and remember the context in which the drug experience occurred, were abnormal in the GluR1 KO mice (Dong et al. 2004). First, standard CPP in response to cocaine was absent in these mice. Second, wild-type mice exhibit an enhanced locomotor response when placed in an open field apparatus in which they had experienced cocaine 24 hours previously. The GluR1 KO mice did not exhibit such a context-dependent enhancement of activity. Additional correlative evidence supporting an important functional role for the cocaine-induced LTP in DA cells is provided by the findings that a D1 receptor antagonist, which blocked LTP, also prevented CPP (Nazarian et al. 2004) as well as the enhanced motor activity normally exhibited by animals placed in a context in which they had previously experienced cocaine (Dong et al. 2004). Furthermore, the enhancement of morphine CPP by administration of a priming dose of cocaine has several properties that are very similar to those of the cocaine-induced LTP in DA cells (Kim et al. 2004). Both are blocked by an NMDA receptor antagonist and both last 5, but not 10, days. On the basis of this combination of correlative behavioral findings, we suggest that the cocaine-induced enhancement of excitatory synapses on DA neurons may contribute either to learning the association between context and drug experience or to attributing motivational significance to the experience. More generally, because increasing excitatory synaptic strength on midbrain DA neurons will result in increased DA cell firing in response to afferent input (Overton and Clark 1997; Grillner and Mercuri 2002) and a consequent increase in DA release in target regions such as the NAc, experiences (such as an acute stress) that elicit LTP in the VTA may prepare or prime the animal such that subsequent experiences have greater incentive or motivational value.

Nucleus Accumbens

As in the VTA, modulation of excitatory synaptic transmission in the NAc appears to have an important role in addiction-related behaviors. For example, injecting glutamate receptor antagonists into the NAc disrupts several drug-evoked behaviors, including behavioral sensitization (Vanderschuren and Kalivas 2000a; Everitt and Wolf 2002). Furthermore, overexpressing GluR1 in the NAc facilitates extinction of cocaine-seeking responses (Sutton et al. 2003) and also makes cocaine aversive, rather than rewarding, in a CPP assay (Kelz et al. 1999). Relatively less work, however, has been performed on the synaptic modifications elicited by drugs of abuse in the NAc.

Drug-induced Synaptic Modification in the NAc

In contrast to the VTA, single in vivo doses of cocaine had no detectable effect on excitatory synaptic strength in the NAc (Thomas et al. 2001). However, repetitive cocaine administration in a manner that elicited robust behavioral sensitization caused a decrease in the AMPA/NMDA ratio when measured 10–14 days after the end of the chronic drug administration protocol (Thomas et al. 2001). Interestingly, this decrease in synaptic strength was observed in the NAc shell but not in the NAc core. The generation of LTD was greatly reduced in the cocaine-treated animals, suggesting that their NAc shell synapses were already depressed. Additional electrophysiological assays suggested that the cocaine-induced LTD in the NAc likely involved down-regulation of AMPA receptors. Consistent with this model, in vivo recording in animals previously treated with repeated doses of cocaine or amphetamine demonstrated that NAc neurons are less responsive to iontophoresed AMPA (White et al. 1995).

Currently, little data is available to suggest a functional role for this cocaine-induced synaptic plasticity in the NAc or whether other drugs of abuse will cause the same change. The depression of synaptic strength in the NAc elicited by cocaine is the exact opposite of what would be expected to occur in animals in which GluR1 was overexpressed in the NAc, a manipulation that made cocaine aversive (Kelz et al. 1999). This could be taken to suggest that LTD in the NAc contributes to the enhancement of the incentive/motivational value of cocaine following repeated exposure. If this NAc LTD contributes to behavioral sensitization, drugs that block LTD should also prevent sensitization. As reviewed earlier, the protein phosphatase calcineurin is important for LTD in the hippocampus. If LTD in the NAc also requires calcineurin, it may be of significant interest that rats administered the calcineurin antagonist FK506 fail to sensitize (Tsukamoto et al. 2001). Clearly, these points are highly speculative and indicate the need for additional work on the synaptic modifications in the NAc elicited by drugs of abuse and their functional consequences.

EFFECTS OF DRUGS OF ABUSE ON SYNAPTIC PLASTICITY

The data reviewed thus far suggest that drugs of abuse can elicit synaptic modifications at excitatory synapses in the mesolimbic DA system and that these adaptations may have functional consequences. Because various forms of synaptic plasticity (i.e., LTP and LTD) likely have very important functional roles in adaptive behaviors, of equal importance for the development of addiction may be the ability of drugs of abuse to modulate or interfere with "normal" LTP and LTD. For example, applying amphetamine to slices of the VTA blocks the generation of LTD in this structure (Jones et al. 2000). This may importantly contribute to the generation of the LTP in the VTA, which is elicited by in vivo administration of psychostimulants. Nicotine also facilitates the generation of LTP at excitatory synapses on DA cells but via a very different mechanism—by enhancing the presynaptic release of glutamate (Mansvelder and McGehee 2000; Mansvelder et al. 2002). This enhanced glutamate release might shift the probability of induction of LTP and importantly contribute to the generation of LTP observed in the VTA following in vivo exposure to nicotine (Saal et al. 2003). Although the effects on VTA synapses of in vivo THC administration have not been tested, in vivo THC or cocaine blocks that form of LTD in the NAc that requires postsynaptic release of endocannabinoids (Hoffman et al. 2003; Fourgeaud et al. 2004; Mato et al. 2004). These findings cannot yet be tied to any behavioral effects of the drugs. However, given that LTP and LTD are widely believed to be important for adaptive forms of experience-dependent plasticity, including learning and memory, it seems reasonable to suggest that the ability of drugs of abuse to interfere with synaptic plasticity may contribute to some of their detrimental cognitive effects.

SUMMARY

A growing body of literature shows that the mechanisms of synaptic plasticity, which have been defined in other regions of brain, are relevant to the neural circuit adaptations that mediate addiction. Here we have focused on the mechanisms of LTP and LTD in the VTA and NAc, the key components of the mesolimbic DA system, and have presented evidence that drugs of abuse may usurp these mechanisms. Although some of the underlying mechanisms of LTP and LTD in these structures appear to be homologous to those underlying prototypic LTP and LTD in the hippocampus, there is also evidence that excitatory synapses in the VTA and possibly the NAc have unique properties.

GAPS AND OPPORTUNITIES

We have reviewed literature that shows that excitatory synapses in the VTA and NAc are subject to modulation following exposure to cocaine as well as to other drugs of abuse. It is our belief that further elucidation of the molecular differences between the mechanisms of synaptic plasticity in mesolimbic DA system structures and other brain regions should not only further our understanding of the pathophysiology of addiction but also yield molecules that may provide targets for the development of novel and more efficacious pharmacotherapies for the treatment of addiction. For example, little is known about the detailed mechanisms by which the drug-induced synaptic modifications in the VTA and NAc occur and any number of different models can be envisioned. These need to be tested because understanding such detailed molecular mechanisms may point toward novel molecular targets for the development of drugs that will disrupt the addictive process while sparing normal, adaptive cognitive processes. Another interesting avenue for investigation will be to determine why the drug-induced synaptic potentiation in the VTA appears to reverse after approximately 5 days. This may provide clues as to how to achieve such a reversal with therapeutic interventions. Clearly, much is still to be learned about the synaptic and neural circuit modifications that lead to addiction. However, progress is being made and we are optimistic that this will indeed lead to a more sophisticated understanding of the neurobiology of addiction and eventually to better treatments.

REFERENCES

Baldi E., Lorenzini C.A., and Bucherelli C. 2004. Footshock intensity and generalization in contextual and auditory-cued fear conditioning in the rat. *Neurobiol. Learn. Mem.* **81:** 162–166.

Bell K. and Kalivas P.W. 1996. Context-specific cross-sensitization between systemic cocaine and intra-accumbens AMPA infusion in the rat. *Psychopharmacology* **127:** 377–383.

Bonci A. and Malenka R.C. 1999. Properties and plasticity of excitatory synapses on dopaminergic and GABAergic cells in the ventral tegmental area. *J. Neurosci.* **19:** 3723–3730.

Borgland S.L., Malenka R.C., and Bonci A., 2004. Acute and chronic cocaine-induced potentiation of synaptic strength in the ventral tegmental area: Electrophysiological and behavioral correlates in individual rats. *J. Neurosci.* **24:** 7482–7490.

Carlezon W.A., Jr., and Nestler E.J. 2002. Elevated levels of GluR1 in the midbrain: A trigger for sensitization to drugs of abuse? *Trends Neurosci.* **25:** 610–615.

Carlezon W.A., Jr., Boundy V.A., Haile C.N., Lane S.B., Kalb R.G., Neve R.L., and Nestler E.J. 1997. Sensitization to morphine induced by viral-mediated gene transfer. *Science* **277:** 812–814.

Carr D.B. and Sesack S.R. 2000. GABA-containing neurons in the rat ventral tegmental area project to the prefrontal cortex. *Synapse* **38:** 114–123.

Carr D.B., O'Donnell P., Card J.P., and Sesack S.R. 1999. Dopamine terminals in the rat prefrontal cortex synapse on pyramidal cells that project to the nucleus accumbens. *J. Neurosci.* **19:** 11049–11060.

Carroll R.C., Beattie E.C., von Zastrow M., and Malenka R.C. 2001. Role of AMPA receptor endocytosis in synaptic plasticity. *Nat. Rev. Neurosci.* **2:** 315–324.

Carroll R.C., Lissin D.V., von Zastrow M., Nicoll R.A., and Malenka R.C. 1999. Rapid redistribution of glutamate receptors contributes to long-term depression in hippocampal cultures. *Nat. Neurosci.* **2:** 454–460.

Contractor A. and Heinemann S.F. 2002. Glutamate receptor trafficking in synaptic plasticity. *Sci. STKE* **2002:** RE14.

Dong Y., Saal D., Thomas M., Faust R., Bonci A., Robinson T., and Malenka R.C. 2004. Cocaine-induced potentiation of synaptic strength in dopamine neurons: Behavioral correlates in GluRA(–/–) mice. *Proc. Natl. Acad. Sci.* **101:** 14282–14287.

Everitt B.J. and Wolf M.E. 2002. Psychomotor stimulant addiction: A neural systems perspective. *J. Neurosci.* **22:** 3312–3320.

Faleiro L.J., Jones S., and Kauer J.A. 2004. Rapid synaptic plasticity of glutamatergic synapses on dopamine neurons in the ventral tegmental area in response to acute amphetamine injection. *Neuropsychopharmacology* **29:** 2115–2125.

Fitzgerald L.W., Ortiz J., Hamedani A.G., and Nestler E.J. 1996. Drugs of abuse and stress increase the expression of GluR1 and NMDAR1 glutamate receptor subunits in the rat ventral tegmental area: Common adaptations among cross-sensitizing agents. *J. Neurosci.* **16:** 274–282.

Fourgeaud L., Mato S., Bouchet D., Hemar A., Worley P.F., and Manzoni O.J. 2004. A single in vivo exposure to cocaine abolishes endocannabinoid-mediated long-term depression in the nucleus accumbens. *J. Neurosci.* **24:** 6939–6945.

Gerdeman G.L., Ronesi J., and Lovinger D.M. 2002. Postsynaptic endocannabinoid release is critical to long-term depression in the striatum. *Nat. Neurosci.* **5:** 446–451.

Grillner P. and Mercuri N.B. 2002. Intrinsic membrane properties and synaptic inputs regulating the firing activity of the dopamine neurons. *Behav. Brain Res.* **130:** 149–169.

Gutlerner J.L., Penick E.C., Snyder E.M., and Kauer J.A. 2002. Novel protein kinase A-dependent long-term depression of excitatory synapses. *Neuron* **36:** 921–931.

Harris G.C. and Aston-Jones G. 2003. Critical role for ventral tegmental glutamate in preference for a cocaine-conditioned environment. *Neuropsychopharmacology* **28:** 73–76.

Hoffman A.F., Oz M., Caulder T., and Lupica C.R. 2003. Functional tolerance and blockade of long-term depression at synapses in the nucleus accumbens after chronic cannabinoid exposure. *J. Neurosci.* **23:** 4815–4820.

Hyman S.E. and Malenka R.C. 2001. Addiction and the brain: The neurobiology of compulsion and its persistence. *Nat. Rev. Neurosci.* **2:** 695–703.

Jones S., Kornblum J.L., and Kauer J.A. 2000. Amphetamine blocks long-term synaptic depression in the ventral tegmental area. *J. Neurosci.* **20:** 5575–5580.

Kabbaj M., Norton C.S., Kollack-Walker S., Watson S.J., Robinson T.E., and Akil H. 2001. Social defeat alters the acquisition of cocaine self-administration in rats: Role of individual differences in cocaine-taking behavior. *Psychopharmacology* **158:** 382–387.

Kalivas P.W. and Weber B. 1988. Amphetamine injection into the ventral mesencephalon sensitizes rats to peripheral amphetamine and cocaine. *J. Pharmacol. Exp. Ther.* **245:** 1095–1102.

Kauer J.A. 2004. Learning mechanisms in addiction: Synaptic plasticity in the ventral tegmental area as a result of exposure to drugs of abuse. *Annu. Rev. Physiol.* **66:** 447–475.

Kelley A.E. 2004. Memory and addiction: Shared neural circuitry and molecular mechanisms. *Neuron* **44:** 161–179.

Kelz M.B., Chen J., Carlezon W.A., Jr., Whisler K., Gilden L., Beckmann A.M., Steffen C., Zhang Y.J., Marotti L., Self D.W., Tkatch T., Baranauskas G., Surmeier D.J., Neve R.L., Duman R.S., Picciotto M.R., and Nestler E.J. 1999. Expression of the transcription factor deltaFosB in the brain controls sensitivity to cocaine. *Nature* **401:** 272–276.

Kim H.S., Park W.K., Jang C.G., and Oh S. 1996. Inhibition by MK-801 of cocaine-induced sensitization, conditioned place preference, and dopamine-receptor supersensitivity in mice. *Brain Res. Bull.* **40:** 201–207.

Kim J.A., Pollak K.A., Hjelmstad G.O., and Fields H.L. 2004. A single cocaine exposure enhances both opioid reward and aversion through a ventral tegmental area-dependent mechanism. *Proc. Natl. Acad. Sci.* **101:** 5664–5669.

Kombian S.B. and Malenka R.C. 1994. Simultaneous LTP of non-NMDA- and LTD of NMDA-receptor-mediated responses in the nucleus accumbens. *Nature* **368:** 242–246.

Koob G.F., Sanna P.P., and Bloom F.E. 1998. Neuroscience of addiction. *Neuron* **21:** 467–476.

Lisman J. 1989. A mechanism for the Hebb and the anti-Hebb processes underlying learning and memory. *Proc. Natl. Acad. Sci.* **86:** 9574–9578.

——. 2001. Three Ca^{2+} levels affect plasticity differently: The LTP zone, the LTD zone and no man's land. *J. Physiol.* **532:** 285.

Lu W., Monteggia L.M., and Wolf M.E. 2002. Repeated administration of amphetamine or cocaine does not alter AMPA receptor subunit expression in the rat midbrain. *Neuropsychopharmacology* **26:** 1–13.

Malenka R.C. and Bear M.F. 2004. LTP and LTD: An embarrassment of riches. *Neuron* **44:** 5–21.

Malenka R.C. and Nicoll R.A. 1999. Long-term potentiation—A decade of progress? *Science* **285:** 1870–1874.

Malinow R. and Malenka R.C. 2002. AMPA receptor trafficking and synaptic plasticity. *Annu. Rev. Neurosci.* **25:** 103–126.

Mansvelder H.D. and McGehee D.S. 2000. Long-term potentiation of excitatory inputs to brain reward areas by nicotine. *Neuron* **27:** 349–357.

Mansvelder H.D., Keath J.R., and McGehee D.S. 2002. Synaptic mechanisms underlie nicotine-induced excitability of brain reward areas. *Neuron* **33:** 905–919.

Marinelli M. and Piazza P.V. 2002. Interaction between glucocorticoid hormones, stress and psychostimulant drugs. *Eur. J. Neurosci.* **16:** 387–394.

Mato S., Chevaleyre V., Robbe D., Pazos A., Castillo P.E., and Manzoni O.J. 2004. A single in-vivo exposure to delta 9THC blocks endocannabinoid-mediated synaptic plasticity. *Nat. Neurosci.* **7:** 585–586.

Morishita W., Connor J.H., Xia H., Quinlan E.M., Shenolikar S., and Malenka R.C. 2001. Regulation of synaptic strength by protein phosphatase 1. *Neuron* **32:** 1133–1148.

Mulkey R.M., Herron C.E., and Malenka R.C. 1993. An essential role for protein phosphatases in hippocampal long-term depression. *Science* **261:** 1051–1055.

Nazarian A., Russo S.J., Festa E.D., Kraish M., and Quinones-Jenab V. 2004. The role of D1 and D2 receptors in the cocaine conditioned place preference of male and female rats. *Brain Res. Bull.* **63:** 295–299.

Nestler E.J. 2001. Molecular basis of long-term plasticity underlying addiction. *Nat. Rev. Neurosci.* **2:** 119–128.

Oades R.D. and Halliday G.M. 1987. Ventral tegmental (A10) system: Neurobiology. 1. Anatomy and connectivity. *Brain Res.* **434:** 117–165.

O'Brien C.P., Childress A.R., Ehrman R., and Robbins S.J. 1998. Conditioning factors in drug abuse: Can they explain compulsion? *J. Psychopharmacol.* **12:** 15–22.

Ortiz J., Fitzgerald L.W., Charlton M., Lane S., Trevisan L., Guitart X., Shoemaker W., Duman R.S., and Nestler E.J., 1995. Biochemical actions of chronic ethanol exposure in the mesolimbic dopamine system. *Synapse* **21:** 289–298.

Overton P.G. and Clark D. 1997. Burst firing in midbrain dopaminergic neurons. *Brain Res. Brain Res. Rev.* **25:** 312–334.

Overton P.G., Richards C.D., Berry M.S., and Clark D. 1999. Long-term potentiation at excitatory amino acid synapses on midbrain dopamine neurons. *Neuroreport* **10:** 221–226.

Perugini M. and Vezina P. 1994. Amphetamine administered to the ventral tegmental area sensitizes rats to the locomotor effects of nucleus accumbens amphetamine. *J. Pharmacol. Exp. Ther.* **270:** 690–696.

Piazza P.V. and Le Moal M. 1998. The role of stress in drug self-administration. *Trends Pharmacol. Sci.* **19:** 67–74.

Porsolt R.D., Bertin A., and Jalfre M. 1977. Behavioral despair in mice: A primary screening test for antidepressants. *Arch. Int. Pharmacodyn. Ther.* **229:** 327–336.

Robbe D., Bockaert J., and Manzoni O.J. 2002. Metabotropic glutamate receptor 2/3-dependent long-term depression in the nucleus accumbens is blocked in morphine withdrawn mice. *Eur. J. Neurosci.* **16:** 2231–2235.

Robinson T.E. and Becker J.B. 1986. Enduring changes in brain and behavior produced by chronic amphetamine administration: A review and evaluation of animal models of amphetamine psychosis. *Brain Res.* **396:** 157–198.

Robinson T.E. and Berridge K.C. 1993. The neural basis of drug craving: An incentive-sensitization theory of addiction. *Brain Res. Brain Res. Rev.* **18:** 247–291.

——. 2001. Incentive-sensitization and addiction. *Addiction* **96:** 103–114.

——. 2003. Addiction. *Annu. Rev. Psychol.* **54:** 25–53.

Rouge-Pont F., Deroche V., Le Moal M., and Piazza P.V. 1998. Individual differences in stress-induced dopamine release in the nucleus accumbens are influenced by corticosterone. *Eur. J. Neurosci.* **10:** 3903–3907.

Saal D., Dong Y., Bonci A., and Malenka R.C. 2003. Drugs of abuse and stress trigger a common synaptic adaptation in dopamine neurons. *Neuron* **37:** 577–582.

Sanchez C.J. and Sorg B.A. 2001. Conditioned fear stimuli reinstate cocaine-induced conditioned place preference. *Brain Res.* **908:** 86–92.

Sanchez C.J., Bailie T.M., Wu W.R., Li N., and Sorg B.A. 2003. Manipulation of dopamine d1-like receptor activation in the rat medial prefrontal cortex alters stress- and cocaine-induced reinstatement of conditioned place preference behavior. *Neuroscience* **119:** 497–505.

Sesack S.R., Carr D.B., Omelchenko N., and Pinto A. 2003. Anatomical substrates for glutamate-dopamine interactions: Evidence for specificity of connections and extrasynaptic actions. *Ann. N.Y. Acad. Sci.* **1003:** 36–52.

Shaham Y., Shalev U., Lu L., De Wit H., and Stewart J. 2002. The reinstatement model of drug relapse: History, methodology and major findings. *Psychopharmacology* **26:** 26.

Suto N., Tanabe L.M., Austin J.D., Creekmore E., and Vezina P. 2003. Previous exposure to VTA amphetamine enhances cocaine self-administration under a progressive ratio schedule in an NMDA, AMPA/kainate, and metabotropic glutamate receptor-dependent manner. *Neuropsychopharmacology* **28:** 629–639.

Sutton M.A., Schmidt E.F., Choi K.H., Schad C.A., Whisler K., Simmons D., Karanian D.A., Monteggia L.M., Neve R.L., and Self D.W. 2003. Extinction-induced upregulation in AMPA receptors reduces cocaine-seeking behaviour. *Nature* **421:** 70–75.

Thomas M.J., Malenka R.C., and Bonci A. 2000. Modulation of long-term depression by dopamine in the mesolimbic system. *J. Neurosci.* **20:** 5581–5586.

Thomas M.J., Beurrier C., Bonci A., and Malenka R.C. 2001. Long-term depression in the nucleus accumbens: A neural correlate of behavioral sensitization to cocaine. *Nat. Neurosci.* **4:** 1217–1223.

Tsukamoto T., Iyo M., Tani K., Sekine Y., Hashimoto K., Ohashi Y., Suzuki K., Iwata Y., and Mori N. 2001. The effects of FK506, a specific calcineurin inhibitor, on methamphetamine-induced behavioral change and its sensitization in rats. *Psychopharmacology* **158:** 107–113.

Tzschentke T.M. 1998. Measuring reward with the conditioned place preference paradigm: A comprehensive review of drug effects, recent progress and new issues. *Prog. Neurobiol.* **56:** 613–672.

Ungless M.A., Whistler J.L., Malenka R.C., and Bonci A. 2001. Single cocaine exposure in vivo induces long-term potentiation in dopamine neurons. *Nature* **411:** 583–587.

Vanderschuren L.J. and Kalivas P.W. 2000a. Alterations in dopaminergic and glutamatergic transmission in the induction and expression of behavioral sensitization: A critical review of preclinical studies. *Psychopharmacology* **151:** 99–120.

——. 2000b. Alterations in dopaminergic and glutamatergic transmission in the induction and expression of behavioral sensitization: A critical review of preclinical studies. *Psycho-pharmacology* **151:** 99–120.

Vekovischeva O.Y., Zamanillo D., Echenko O., Seppala T., Uusi-Oukari M., Honkanen A., Seeburg P.H., Sprengel R., and Korpi E.R. 2001. Morphine-induced dependence and sensitization are altered in mice deficient in AMPA-type glutamate receptor-A subunits. *J. Neurosci.* **21:** 4451–4459.

Vezina P. 2004. Sensitization of midbrain dopamine neuron reactivity and the self-administration of psychomotor stimulant drugs. *Neurosci. Biobehav. Rev.* **27:** 827–839.

White F.J., Hu X.T., Zhang X.F., and Wolf M.E. 1995. Repeated administration of cocaine or amphetamine alters neuronal responses to glutamate in the mesoaccumbens dopamine system. *J. Pharmacol. Exp. Ther.* **273:** 445–454.

Will M.J., Watkins L.R., and Maier S.F. 1998. Uncontrollable stress potentiates morphine's rewarding properties. *Pharmacol. Biochem. Behav.* **60:** 655–664.

Wise R.A. 1988. The neurobiology of craving: Implications for the understanding and treatment of addiction. *J. Abnorm. Psychol.* **97:** 118–132.

——. 1996. Neurobiology of addiction. *Curr. Opin. Neurobiol.* **6:** 243–251.

Wolf M.E. 1998. The role of excitatory amino acids in behavioral sensitization to psychomotor stimulants. *Prog. Neurobiol.* **54:** 679–720.

Wolf M.E., Sun X., Mangiavacchi S., and Chao S.Z. 2004. Psychomotor stimulants and neuronal plasticity. *Neuropharmacology* (suppl. 1) **47:** 61–79.

Zamanillo D., Sprengel R., Hvalby O., Jensen V., Burnashev N., Rozov A., Kaiser K.M., Koster H.J., Borchardt T., Worley P., Lubke J., Frotscher M., Kelly P.H., Sommer B., Andersen P., Seeburg P.H., and Sakmann B. 1999. Importance of AMPA receptors for hippocampal synaptic plasticity but not for spatial learning [see comments]. *Science* **284:** 1805–1811.

20

Long-term Memory Storage in *Aplysia*

James H. Schwartz

Center for Neurobiology and Behavior, Columbia University, New York, New York 10032

ABSTRACT

Every instance of learning—memory, drug addiction, art appreciation—is likely to proceed through changes in gene expression mediated by second messengers or electrical depolarization. Gene expression involves the formation of a proper promoter complex and the decondensation of DNA, processes now beginning to be characterized in *Aplysia*. In addition to early gene expression, the functions of downstream late proteins need to be explained. These mechanisms are best examined in *Aplysia*. Because most of these processes have been studied at the level of the sensory-to-motor synapse isolated in cell culture, the mechanisms are not complicated by the difficulties of network interactions and other problems that occur with higher animals.

INTRODUCTION

Aplysia as a Model System

Aplysia has been used as an experimental animal in the neurosciences since the 1930s, first in France for the chemistry of acetylcholine as a neurotransmitter (Fig. 1). The animal has a simple nervous system (Eales 1921); its neurons can be seen with the naked eye, and single cell bodies can be dissected out with a forceps (Schwartz and Swanson 1987). The name *Aplysia* comes from the Greek, meaning "it won't wash" (Pliny; Thompson 1947) When frightened, the animal emits purple ink and a sticky white substance. Species of *Aplysia* have a worldwide distribution in semitropical waters: the Mediterranean, Indian Ocean, Sea of Japan, and the waters from Venice to Baja, California. Early in the 20th century, an important strategy in biological research was to identify experimentally convenient systems for research, for example, the squid giant axon to study the electrical properties of neural membranes, *Drosophila* for genetics, and bacteriophage for studying viruses. Like several other invertebrates, *Aplysia* is ideal for studying the cellular basis of behavior. The *Aplysia* nervous system contains about 20,000 neurons, many large and identifiable. During the 1960s, discrete behaviors were found to be controlled by the actions of a small number of specific identified neurons. As a result, modern molecular-biological techniques

FIGURE 1. *Aplysia*. Woodcut from S. Rang's *Aplysia, Histoire Naturelle des Aplysiens*, published in 1828. The animal is seen advancing to the right. The two oral tentacles followed by the two rhineophores suggest a resemblance to the rabbit. Hence, the common name, sea hare (*lepus marinus*), for *Aplysia*. (Reprinted, with permission, from Kandel 1979 [©W.H. Freeman].)

can be applied directly to single neurons to determine whether a molecular mechanism actually operates to produce the observed changes in synaptic action (Kandel 2001).

When we first began to study *Aplysia* neurons, we often had to rationalize the relevance of our findings to vertebrates. It is astonishing, however, that over the years molecular-biological techniques have revealed how similar the molecular components that regulate synaptic plasticity in *Aplysia* and other invertebrates are when compared with those of higher animals. Perhaps this should have been expected, because second messenger pathways, receptor mechanisms, and regulation of gene expression are all fundamental and universal processes likely to be conserved throughout phylogeny.

GOALS OF THE CHAPTER

In this chapter, the behavior that is described is called sensitization, a form of memory in which a stimulus enables a neuron to give an enhanced response for 15–20 minutes (short term) or for days to weeks (long term). The duration of the change in synaptic plasticity depends on the intensity and extent of the stimulus. Underlying this behavior is an electrophysiological process called facilitation. Both forms of facilitation are the result of a change in synaptic strength: The synaptic connection between an identified sensory neuron and a motor neuron becomes approximately twice as strong because twice as much neurotransmitter has been released. The molecular mechanisms underlying long-term sensitization in *Aplysia* are likely to be similar to the processes that underlie other forms of memory storage, in particular, drug addiction (see Wolf and Heberlein 2003). Both sensitization and addiction are thought to involve changes in synaptic activity produced by new gene expression. In both sensitization and drug addiction, the induction of new proteins is regulated by the cyclic AMP (cAMP)-dependent protein phosphorylation of the cAMP–response element binding protein (CREB) (Frank and Greenberg 1994; Belvin et al. 1999; Park et al. 2000; Hou et al. 2004; Nestler 2004). Therefore, an analysis of the pathway that produces long-term sensitization in *Aplysia* should be instructive for understanding the molecular processes underlying drug addiction.

SHORT-TERM SENSITIZATION IS PRODUCED BY cAMP

Behavioral Sensitization Is Mediated by Serotonin, Increasing Transmitter Release from the Presynaptic Neuron

By 1982, Eric Kandel and I had worked out the mechanism for the increased release of transmitter that results in short-term sensitization (Kandel and Schwartz 1982). This understanding emerged from examining how synaptic action can be regulated by modulatory neurotransmitters. Thus, it became clear that the strength of a given synapse could be modified and that a mediating synapse could be modulated by certain neurotransmitters. Most important for facilitation in *Aplysia* was the discovery that the synthesis of cAMP is stimulated in brain slices by the application of serotonin (5-HT) (Mansour et al. 1960; Rall 1985). We were then able to show that 5-HT, a neurotransmitter plentiful in *Aplysia* ganglia, could enhance transmitter release.

5-HT Activates Adenylyl Cyclase, Resulting in Transient Covalent Modifications of Proteins

Brief noxious inputs activate serotonergic interneurons that in turn activate adenylyl cyclase in mechanosensory neurons through a G protein to produce the second messenger cAMP. In the sensory neuron, cAMP binds to regulatory subunits of cAMP-dependent protein kinase (PKA), releasing catalytic subunits to phosphorylate proteins involved in neurotransmitter vesicle mobilization and release (Fig. 2). Because of the action of phosphodiesterases, protein phosphorylation is transient, lasting only as long as the concentration of cAMP remains elevated. Because the inputs are brief, the activation of the cyclase is soon finished, and no additional cAMP is produced. Thus, the enhanced ability to release neurotransmitter is transient and reversible, ending within 15–20 minutes.

Long-term Memory Storage Requires New Protein Synthesis But Is Not Specific to Memory Type

The chief difference between short- and long-term synaptic plasticity is that long-term memory storage requires new protein synthesis. This dependence was shown pharmacologically in the 1960s in experiments with drugs that block protein synthesis, for example, anisomycin and puromycin (Davis and Squire 1984). What sort of proteins need to be made for memory storage? Is the protein an "engram," an informational molecule that encodes a particular memory and that differs for different memories, or do the proteins have similar functions in processing different memories? As we shall see, the change from short- to long-term memory has been shown to be a molecular cascade involving PKA phosphorylation and not a specific informational molecular substance. Ultimately, the required protein molecules are cellular components needed for the formation of new synapses (Fig. 3). Thus, the proteins induced are not specific to memory type: The cAMP-CREB pathway operates in *Aplysia* sensitization, drug addiction, and several other forms of memory storage.

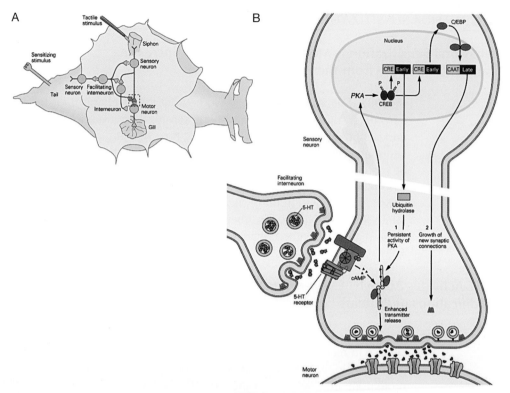

FIGURE 2. (*A*) Sensitization of the gill withdrawal reflex is produced by a prolonged noxious stimulus to the tail. This stimulus activates sensory neurons that excite facilitatory interneurons. These interneurons, which innervate the siphon, form axo-axonic synapses on terminals of mechanosensory neurons. Release of 5-HT from the interneurons enhances transmitter release from the terminals of the mechanosensory neurons onto motor neurons. This process is called presynaptic facilitation. (*B*) 5-HT released from the facilitatory interneuron activates a 5-HT receptor, producing cAMP, which in turn activates PKA. Proteins involved in the release of synaptic vesicles are phosphorylated, enhancing the release of transmitter from mechanosensory neurons onto the motor neurons. PKA is also imported into the neuron's nucleus where it phosphorylates CREB. Phosphorylated CREB, bound to the CRE site, causes the induction of two immediate early genes, ubiquitin carboxy-terminal hydrolase and C/EBP. The ubiquitin-mediated degradation of R subunits of PKA results in a persistently active protein kinase (*1*). C/EBP, a transcription activator for late genes, induces proteins needed for growth of new synaptic connections (*2*). It should be noted that in *Aplysia* Bartch et al. (1995) called the constitutively active CREB, CREB1, and an inhibitory factor, CREB2. These two factors are not related biochemically. (Reprinted, with permission of the McGraw-Hill Companies, from Kandel et al. 2000.)

cAMP-dependent Phosphorylation Induces the Formation of Long-term Facilitation

New proteins are made during the formation of long-term facilitation when the stimulus is prolonged (2 hr) and the cAMP-dependent protein kinase is activated for some time by the persistent synthesis of cAMP. Gene expression begins when catalytic subunits, dissociated from regulatory subunits, are imported into the sensory neuron's nucleus to initiate transcription with the constitutive transcription activator, CREB (Montminy 1997). CREB is bound to promoter regions capable of being activated by PKA. Activation occurs

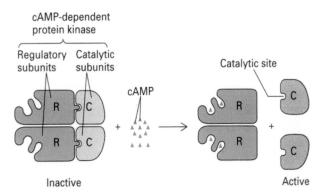

FIGURE 3. The cAMP-dependant protein kinase is a tetramer consisting of two regulatory (R) subunits and two catalytic (C) subunits. R subunits bind to the C subunits at pseudosubstrate regions. The kinase becomes active with the production of the second messenger, cAMP, that binds at two points in each R subunit, releasing active C subunits. The R subunits are bound to one another at their amino-terminal ends as a dimer that reassociates with the two C subunits when the concentration of cAMP decreases. The two types of PKA are type I and II; type is determined by the R subunits. Because R subunits exist only as dimers that therefore always reassociate with C subunits together, the tetramer is always either type I or II; no mixing occurs. On the other hand, a variety of C subunits exist that are used by both types. (Reprinted, with permission, from Lodish et al. 1995 [©W.H. Freeman].)

when CREB is phosphorylated and the CREB binding protein (CBP) is recruited to the promoter to produce active promoter complexes (Guan et al. 2002). CBP is a histone acetylase, and acetylation, described below, promotes gene expression.

MANY LATE PROTEINS SYNTHESIZED ARE BUILDING BLOCKS OF NEW SYNAPSES

Regulatory and Structural Genes Are Induced during Formation of Long-term Facilitation

What genes are induced during the formation of long-term facilitation? Three types of late proteins are made: regulatory, translational, and structural.

To date, we have identified two immediate early genes (early response genes), *Aplysia* CCAAT/enhancer binding protein (Ap-C/EBP) (Alberini et al. 1994) and *Aplysia* ubiquitin carboxy-terminal hydrolase (Ap-uch) (Hedge et al. 1997). C/EBP is a transcription factor whose synthesis is activated by the phosphorylation of CREB. Ap-C/EBP, in turn, activates downstream induction of late genes, for example, the translation factor eEFIA. The other immediate early gene, Ap-uch, is a hydrolase that binds to proteasomes to enhance the degradation of ubiquitinated proteins, including regulatory subunits of PKA. As described below, ubiquitin-proteasome-mediated degradation of R subunits results in the production of an autonomously active PKA (Greenberg et al. 1987; Müller and Carew 1998; Chain et al. 1999). 5-HT–mediated synthesis of cAMP is complete a few minutes after the last stimulation (2 hr). Nevertheless, in the sensory neuron, PKA continues to be active for at least 24 hours, thus continuing to have a role in the consolidation of memory long after the initial burst of cAMP is over.

How do the induced proteins operate to produce long-term facilitation? It is likely that sensorin and Tolloid (BMP-1) act as growth factors for synaptogenesis. The translational proteins facilitate new protein synthesis, probably mostly locally at nerve endings to which mRNAs have been transported from the cell body (Steward and Schuman 2001; et Giustetto al. 2003; Moccia et al. 2003). Finally, actin, intermediary filament (IF), and tubulin are obvious building blocks for new synapses. The others are needed for vesicle metabolism (calreticulin, calmodulin, clathrin, and syntaxin) (e.g., see Liu et al. 1997; Zwartjes et al. 1998; Giustetto et al. 2003). The induction of the RII regulatory subunit as a late protein is of special interest and is discussed below.

LONG-TERM MEMORY STORAGE REQUIRES CHROMATIN REMODELING AT THE C/EBP PROMOTER

Chromatin Remodeling and Decondensation of Double-stranded DNA Is Required for CREB to Initiate Transcription

Synthesis of new proteins requires chromatin remodeling, specifically to decondense the double-stranded DNA around proper promoter regions. Direct evidence for remodeling during long-term facilitation in *Aplysia* was obtained with chromatin immunoprecipitation (Chip) assays using antibodies against C/EPB, CREB-CBP, and the TATA box–binding protein (TBP)—all components of the C/EBP promoter complex (Fig. 4). We found that exposure to 5-HT, which induces C/EBP, recruits CBP to the C/EBP promoter to form the CREB-CBP complex necessary for gene espression. Correlated with the induction of C/EBP, the formation of the promoter complex was complete when TBP was also recruited to the promoter after treatment with 5-HT (Guan et al. 2003).

Chromatin Remodeling Is Mediated by Acetylation and Deacetylation of Histones

Chromatin remodeling involves the acetylation and deacetylation of histones. Acetylation opens up the DNA for transcription. We therefore tested whether the induction of C/EBP is regulated by histone acetylation. Treatment with 5-HT increases acetylation of both histone H3 and H4 at the C/EBP promoter. The histone acetylase of the CBP is important in

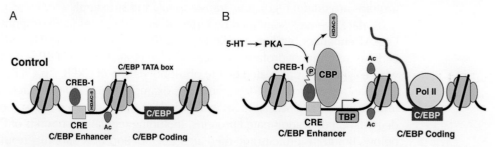

FIGURE 4. Histone acetylation and chromatin remodeling that results when CREB is phosphorylated. (*A*) In the basal state (uninduced), CREB resides at the CRE site in the C/EBP promoter. Some lysine residues of histones are acetylated, but the DNA is not opened up for transcription. (*B*) PKA, activated through 5-HT, phosphorylates the CREB bound to the C/EBP promoter. CBP and the TATA box–binding protein (TBP), both components of the promoter essential for induction, are then recruited. Because CBP is a histone acetylase, more lysine residues are acetylated, and the DNA becomes decondensed to permit transcription by RNA polymerase pol II. (Reprinted, with permission, from Guan et al. 2002 [©Elsevier].)

chromatin remodeling. Chip assays revealed that the recruitment of CBP to the C/EBP promoter is correlated with the acetylation of histones and the induction of C/EBP. This indicates that histone acetylation is a key process in induction.

In the basal state, CREB binds to the promoter of C/EBP, the immediate response gene critical for switching from short- to long-term facilitation. In this state, the DNA is condensed and the configuration of chromatin does not favor gene expression. Changes in chromatin structure are therefore required for transcription. This is achieved by the release of 5-HT during learning, which activates CREB through PKA phosphorylation. Phosphorylated CREB recruits CBP, which in turn leads to the acetylation of specific histone residues at the C/EBP promoter, allowing the recruitment of TBN and other regulatory proteins needed for transcribing C/EBP. Experiments using Chip assays also explain the block with the inhibitory neuropeptide FMRFamide that produces long-term synaptic depression. FRMFamide activates an inhibitory transcription factor (CREB2) through p38 MAP kinase. CREB2 displaces CREB from the C/EBP promoter and recruits HDACS, a histone deacetylase. This represses C/EBP, presumably by packaging the promoter region more densely with chromatin (Guan et al. 2002).

PolyADP RIBOSYLATION: ANOTHER METHOD FOR DECONDENSATION

We have seen that the development of long-term memory storage requires specific chromatin remodeling and the decondensation of DNA in preparation for a cascade of transcription and synthesis of new proteins. Another indication that chromatin remodeling is required for long-term facilitation is the observation that treatments that produce long-term sensitization activate the enzyme polyADP-ribosylase-1 (PARP-1; Cohen-Armon et al. 2004). PARP-1 can produce decondensation of DNA by modifying many of the proteins in chromatin (D'Amours et al. 1999; Kraus and Lis 2003). PolyADP ribosylation is a transient and reversible modification of nuclear proteins by PARP-1, an abundant and conserved enzyme that can modify histones, transcription factors, RNA polymerases, and topoisomerases by polyADP ribosylation (Fig. 5). PARP-1 has been shown to be activated by stressful stimuli that damage DNA (nicks and breaks) and has therefore been considered primarily as part of the mechanism of apoptosis. It is now known that the enzyme can be activated by other, less damaging stimuli, e.g., by depolarization of rat cortical neurons in culture (Homburg et al. 2000).

PARP-1 Is also Activated during Sensitization in *Aplysia*

We tested whether exposure to 5-HT can activate PARP-1 in isolated ganglia. We found that the treatment with 5-HT that produces long-term facilitation activates PARP-1, but treatments with 5-HT that produce only short-term facilitation did not. Interestingly, PARP-1 was not activated during long-term depression, which is produced by the application of the inhibitory neuropeptide FMRFamide, even though new protein synthesis is also needed for long-term depression (Guan et al. 2002). Thus, the activation of PARP-1 appears to be selective, because its activation was not correlated with all forms of long-term synaptic plasticity.

Activation of PARP-1 specifically in ganglia-mediating memory storage can be shown in the following way. Four spaced noxious stimuli to one side of an intact *Aplysia* result in the long-term sensitization of the withdrawal reflex in response to ipsilateral stimulation with-

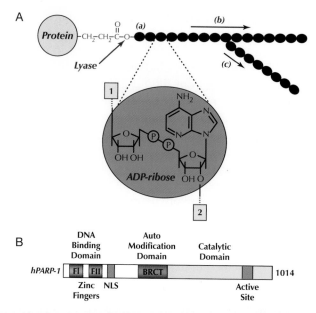

FIGURE 5. Structure of polyADP-ribosyl (PAR) polymers. (*A*) The polymers are produced by PARP-1 step-wise by (*a*) initiation, (*b*) elongation, and (*c*) branching. PAR polymers are attached to glutamate residues in target proteins at the position labeled *1* in the expanded view of the ADP ribose unit. The ADP ribose units within the polymer are linked by glycosidic ribose 1''-2' bonds at the positions labeled *1* and *2* in the expanded view. (*B*) PARP-1 is highly conserved from *Drosophila* to humans. The molecule contains an amino-terminal DNA-binding domain, an automodification domain, and a carboxy-terminal NAD+-binding catalytic domain. The DNA-binding domain has two zinc finger motifs and a nuclear localization signal (NLS). The automodification domain, which contains a BRCT motif, functions as a break on the polymerization reaction. During the reaction, the enzyme adds PAR to this domain. At some point during the reaction, the polymer on the enzyme, because of its bulk, separates the enzyme from the substrate proteins. (Reprinted, with permission, from Kraus and Lis 2003 [©Elsevier].)

out affecting the response to stimulation on the other side. Underlying this behavioral sensitization is the long-term facilitation of sensory-to-motor synapses in the ipsilateral pleural-pedal ganglia, indicating that the activation occurs only in ganglia on the stimulation side.

PARP-1 appears to operate through polyADP ribosylation of histones. For example, histone H1 was polyADP-ribosylated in pleural-pedal ganglia as a result of long-term treatment with 5-HT (but not short term). PolyADP ribosylation of H1 has been shown to cause the fast and transient relaxation of the highly condensed structure of chromatin, rendering DNA accessible to transcription (D'Amours et al. 1999). PolyADP ribosylation would provide quick access to DNA, enabling the transcription required for long-term memory. A role in learning is strengthened by Satchell et al.'s (2003) observation that polyADP ribosylation contributes to memory as measured in the Morris water maze.

UBIQUITIN CARBOXY-TERMINAL HYDROLASE PROLONGS THE ACTION OF PKA

I have described in some detail the molecular processes that occur just after the application of 5-HT or natural stimuli. We have seen how the early components of the induction process operate, i.e., opening up the DNA around the promoter of the immediate-early

Table 1. Removing Ubiquitin Chains Facilitates Degradation: Ap-uch Enhances the In Vitro Degradation of Substrate Proteins by the Proteasome

Addition	Degradation (%)			
	Proteasome	Ap-uch + proteasome	Ap-uch (S90) + proteasome	Ap-uch alone
Lysozyme	44 ± 4	89 ± 3	48 ± 3	8 ± 3
R subunit (N4)	41 ± 2	84 ± 3	43 ± 2	6 ± 2

[125]I-lysozyme-ubiquitin conjugates and 8-N_3-cAMP ^{32}P-labeled R1 subunit N4 of *Aplysia* (Hedge et al. 1993; Chain et al. 1999) were assayed for degradation. For these experiments, we titrated the amount of proteasome needed for submaximal (4–50%) degradation of substrates. Wild-type and mutant Ap-uchs were tested. After SDS-PAGE and autoradiography, the extent of degradation was determined by densitometry, and the percent of degradation relative to controls was calculated. With both lysozyme and N4, significant ($p < 0.01$, paired t-test) enhancement of degradation was seen with wild-type Ap-uch. The mutant produced by site-directed mutagenesis at the catalytic site (S90) had no effect. No degradation was seen with wild-type Ap-uch alone. Values are mean ± s.e.m., $n = 4$.

gene, C/EBP, allowing transcription, and resulting in the induction of downstream late proteins.

The other immediate-early gene encodes a ubiquitin carboxy-terminal hydrolase. The action of this second early response gene was unexpected: The hydrolase normally facilitates degradation of ubiquitin-conjugated small peptides during proteolysis through the proteasome (Table 1).

One of the important substrates for the ubiquitin pathway is a PKA regulatory subunit. Treatments that produce long-term facilitation result in the limited degradation of RI regulatory subunits (Fig. 6). During the brief period from when the hydrolase is induced, the degradation of R subunits does not go to completion, rather only about 20% is hydrolyzed. This extent of degradation results in some of the catalytic subunits being unregulated, or persistently activated, even in the absence of cAMP. This mechanism has been shown to operate in spinal dorsal horn neurons to cause neuropathic pain (Moss et al. 2002). Persistence is characteristic of PKA, but also occurs with other persistently active kinases (cognitive kinases such as PKC, Ca^{2+}-/calmodulin-dependent protein kinases that are thought to mediate memory storage in neurons; Schulman and Hyman 1999; Schwartz 2001).

It is tempting to speculate that the function of the persistent kinase in *Aplysia* extends the action of PKA past the period of 2 hours when second messenger concentrations are high. It is a fact that PKA activity in the sensory neurons persists for at least 24 hours, and it is attractive to think that the persistent PKA serves to maintain CREB in a state of persistent phosphorylation. If this is so, the phenomenon of persistent phosphorylation is a form of reiterative induction.

ENHANCED SYNAPTIC ACTION THROUGH THE INDUCTION OF TYPE II PKA

All of the mechanisms described so far can be thought of as belonging to the induction phase in the formation of long-term memory storage, presumably leading to the synthesis of late proteins. The list of downstream components that need to be made to obtain long-term memory storage can be seen in Table 2. The late genes fall into three categories: regulatory, translational, and structural. For example, the translation factor eEF1A is essential for the formation of polypeptide bonds on ribosomes, and BiP is a chaperone necessary for the correct folding of nasent proteins. On the other hand, actin and intermediate filament protein are clearly building blocks for new synapses. The messenger

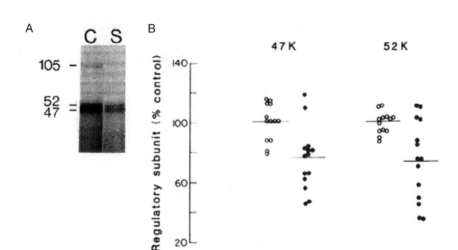

FIGURE 6. The effect of long-term sensitization on PKA R subunits. *Aplysia* were trained to sensitize the gill-siphon withdrawal reflex (Frost et al. 1985). Sensory neurons were dissected out. At 24 hr, proteins were extracted for labeling with [^{32}P]8-N$_3$-cAMP (which binds irreversibly to the cAMP sites in R subunits when exposed to ultraviolet light). (*A*) Autoradiograph of an electropherogram containing labeled samples of sensory neurons with equal amounts of proteins applied, one from control (*C*; untreated) *Aplysia*, the other from sensitized (*S*) animals. (*B*) Summary of results. Each point is a radiolabeled R subunit from a trained (*closed circle*) or naïve (*open circle*) animal. Means are marked by horizontal lines. Values for the two major separated components of R subunits from trained animals were significantly lower than those from controls. Catalytic subunits were not affected. (Reprinted, with permission, from Greenberg et al. 1987 [©Nature Publishing Group].)

RNAs (mRNAs) for many of these late genes are transported down the axon to the nerve endings presumably to be expressed at these nerve endings for construction of new synapses (Moccia et al. 2003).

Is RII a Regulatory or Structural Protein?

How does RII, which is induced as a late gene, fit into the two categories of regulatory and structural? It could be argued that RII is regulatory because the subunit regulates protein phosphorylation. On the other hand, complexes with channels and receptors in the cytoskeleton, as well as with other second messenger enzymes, might suggest that RII constitutes a substantial bulk of the protein in synapses.

Both PKAs are tetramers comprised of homodimers of two R and two C subunits. Animals have two types of the kinase, differing in their R subunits. Although the open reading frames for RI and RII are similar, the 5′-untranslated regions (UTRs) are quite different. The 5′ UTR of RI has no signal elements and is therefore likely to be noninducible. In contrast, the 5′ UTR of RII has a CRE (cAMP response element), therefore likely to be inducible by phosphorylated CREB (Fig. 7). Another difference is that type II operates together with A-kinase-anchoring proteins (AKAPs). Although encoded by different genes, both types of R are conservatively built: The carboxy-terminal two thirds of the molecule enhance two cAMP-binding sites connected to the amino terminal by a hinge region con-

TABLE 2. Proteins Induced by 5-HT in Sensory Neurons

Genes	mRNA transported
Early response genes	
Ap-C/EBP	No
Ap-uch	No
Late Genes	
Regulatory	
Sensorin	Yes
Tolloid (BMP-1)	?
RII	Yes
Translational	
Staufen	Yes
CPEB	Yes
BiP	?
eEF1A	Yes
Calreticulin	?
Ribosomal proteins	Yes
Structural	
Actin	Yes
β-thymosin	Yes
IF	?
α-tubulin	Yes
Calmodulin	?
Syntaxin	?
Synapsis	?
Clathrin	?
Fasciclin	?
Other	
Phosphoglycerate kinase	?
Glutamate transporter	?

The effector late genes so far identified can be characterized as regulatory, translational, or structural. The regulatory proteins are sensorin, a sensory neuron-specific neuropeptide akin to BDNF (Moccia et al. 2003); Tolloid (BMP-1), a growth factor (Liu et al. 1997); and the RII subunit of PKA. The proteins involved in translation are two RNA-binding proteins, Staufen and CPEB (cytoplasmic polyadenylation element binding protein); eEF1A, a translation factor; two chaperones, BiP and calreticulin; and 13 ribosomal proteins. Structural proteins are actin, β-thymosin (an actin-binding protein), IF (intermediary filament protein), α-tubulin, calmodulin (the Ca^{2+}-binding protein), syntaxin (Hu et al. 2003), synapsin (Angiers et al. 2002), and clathrin (three proteins that operate in vesicle metabolism), and fasciclin (a cell-adhesion molecule that has a role in the guidance of growth cones). Other proteins are phosphoglycerate kinase, the enzyme that catalyzes the first ATP-generating reaction in glycolysis, and a glutamate transporter, involved in the uptake and restoration of the sensory neuron's neurotransmitter (Khabour et al. 2004). Note that many of the proteins' mRNAs are transported to the nerve endings for local protein synthesis at synapses (Eberwine et al. 2001; Steward and Schuman 2001; Moccia et al. 2003).

taining the inhibiting (regulatory) binding site for carboxyl, and in RIIs, an autophosphorylation site. Of special interest, in RII, the amino-terminal 30 amino-acid residues contain a binding site for members of a functional group of binding proteins, AKAPs. Although type-I PKA is mainly cytosolic, most of type II is bound to membranes and cytoskeleton through association of RII with AKAPs (Michel and Scott 2002). Thus, a key biochemical difference between the two PKA types is AKAP binding, which targets the kinase to a variety of organelles and subcellular components. These molecular complex-

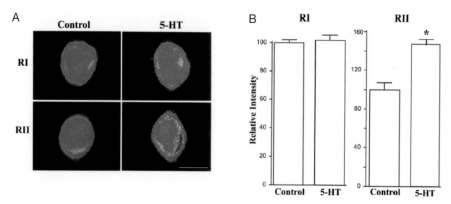

FIGURE 7. RII mRNA is induced during long-term facilitation. (*A*) In situ hybridization of *Aplysia* RI and RII in sensory neurons cocultured with the motor neuron. Cell bodies were fixed 6 hr after 5-HT treatment. Artificial colors (*light red/yellow*) indicate the induction of RII after treatment with 5-HT. (*B*) Quantitation of the changes in the in situ hybridization signals during the development of LTF. The amount of RII mRNA increased significantly; no changes occurred in RI mRNA. (Reprinted, with permission, from Liu et al. 2004 [©Society for Neuroscience].)

es can be quite large, functioning by bringing together second messenger enzymes as well as associating membranes, channels, and receptors. These complexes have been appropriately called "transducisomes."

During long-term facilitation of sensory neuron synapses, C subunits remain constant, RII is induced, and RI decreases through ubiquitin-mediated proteolysis. Proteolysis of R subunits results in some C that is unregulated. The new synthesis of RII subunits may be induced through phosphorylation of CREB by this persistently active form of the kinase. These experiments provide evidence for the novel idea that two isoforms with distinctive functions within a cell cooperate to produce a common physiological end (facilitation). What is the function of RII? As we shall see, anchoring of RII at nerve terminals is essential for neurotransmitter release. RII is a crucial synaptic component.

AKAP BINDING IS CRUCIAL TO MEMORY STORAGE

Inhibition of AKAP binding abolishes both short- and long-term facilitation (Fig. 8). The binding is blocked by Ht31, a peptide isolated from human thyroid AKAP. That both short and long term are blocked suggests that the release of neurotransmitter is affected. The site of action is therefore likely to be nerve terminals. In accord with this idea, we found that type-II PKA is enriched in neurites and synaptosomes. The amount of RII in synaptosomes is 2.5 times greater than that in the homogenate from which the synaptosomes were isolated. In the hippocampus, type-II PKA is targeted to synaptic glutamate receptors, and phosphorylation of the Glu1 receptor subunit by type-II PKA regulates internalization and turnover of AMPA receptors (Colledge et al. 2000; Esteban 2003). In *Aplysia*, 5-HT stimulates the phosphorylation of synapsin at nerve terminals, presumably to facilitate synaptic vesicle release (Angers et al. 2002). It can be concluded that RII, enriched at synapses, phosphorylates synaptic components selectively. Presumably, the increase of type-II PKA during long-term facilitation is to stock the newly formed synapses.

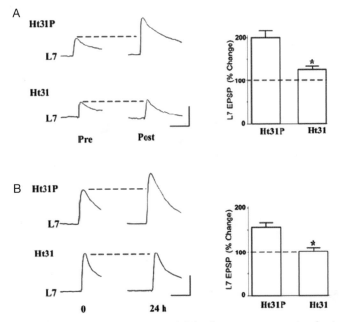

FIGURE 8. Binding of PKA II to AKAPs is essential for long-term synaptic plasticity. Intracellular injection of the inhibitory Ht31 peptide into the sensory neuron blocks the long-term facilitation produced by 5-HT. Excitatory postsynaptic potential (EPSP) traces were recorded in L7, the motor neuron, before (*A*) and 24 hr after (*B*) treatment with 5-HT. (Reprinted, with permission, from Liu et al. 2004 [©The American Society for Biochemistry & Molecular Biology].) The inhibitory peptide is based on a human thyroid AKAP and blocks the interaction between RII subunits and AKAPs (Vijayaraghavan et al. 1997).

CONCLUSIONS, GAPS, AND OPPORTUNITIES

Learning and memory appear to be an example of neuronal differentiation, changes induced by the induction of gene expression. At present, CREB-induced transcription has been studied the most; in the nervous system, it has been shown to be the core mechanism underlying both memory storage and drug addiction. Of course, other second messenger pathways are activated during the formation of synaptic plasticity, for example, protein kinase C (Sacktor and Schwartz 1990), p42–44 MAP kinase (Martin et al. 1997; Michael et al. 1998; Yamamoto et al. 1999), and p38 MAP kinase (Guan et al. 2003).

Three important opportunities exist for further study in both *Aplysia* and other invertebrates (see Wolf and Heberlein 2003). The first is to determine how other forms of long-term synaptic plasticity are induced. What are the transcription factors that govern these other inductions? Second, it is of obvious importance to identify the protein products and functions of the genes induced. Only with this information will we be able to understand the molecular mechanisms that produce memory storage. Finally, our studies of the critical proteins, although incomplete, have resulted in the discovery of pharmacological targets that block the development of long-term memory storage in *Aplysia*. Effective are inhibitors of proteasomes, inhibitors of PARP-1, and inhibitors of the interaction of PKA with AKAPs. We might expect that characterization of the molecular steps that produce drug addiction will also identify useful therapeutic targets.

ACKNOWLEDGMENTS

This work was supported by National Institutes of Health Grant No. NS-29255 and National Institute of Mental Health Grant No. MH-48850.

REFERENCES

Alberini C.M., Ghirardi M., Metz R., and Kandel E.R. 1994. C/EBP is an immediate-early gene required for the consolidation of long-term facilitation in *Aplysia. Cell* **76:** 1099–1114.

Angers A., Fioravante D., Chin J., Cleary L.J., Bean A.J., and Byrne J.H. 2002. Serotonin stimulates phosphorylation of *Aplysia* synapsin and alters its subcellular distribution in sensory neurons. *J. Neurosci.* **22:** 5412–5422.

Bartsch D., Ghirardi M., Skehel P.A., Karl K.A., Herder S.P., Chen M., Bailey C.H., and Kandel E.R. 1995. *Aplysia* CREB2 represses long-term facilitation: Relief of repression converts transient facilitation into long-term functional and structural change. *Cell* **83:** 979–992.

Chain D.G., Casadio A., Schacher S., Hegde A.N., Valbrun M., Yamamoto N., Goldberg A.L., Bartsch D., Kandel E.R., and Schwartz J.H. 1999. Mechanisms for generating the autonomous cAMP-dependent protein kinase required for long-term facilitation in *Aplysia. Neuron* **22:** 147–156.

Cohen-Armon M., Visochek L., Katzoff A., Levitan D., Susswein A.J., Klein R., Valbrun M., and Schwartz J.H. 2004. Long-term memory requires polyADP-ribosylation. *Science* **304:** 1820–1822.

Colledge M., Dean R.A., Scott G.K., Langeberg L.K., Huganir R.L., and Scott J.D. 2000. Targeting of PKA to glutamate receptors through a MAGUK-AKAP complex. *Neuron* **27:** 107–119.

Davis H.P. and Squire L.R. 1984. Protein synthesis and memory: A review. *Psychol. Bull.* **96:** 518–559.

D'Amours D., Desnoyers S., D'Silva I., and Poirier G.G. 1999. Poly(ADP-ribosyl)ation reactions in the regulation of nuclear functions. *Biochem. J.* **342:** 249–268.

Eales N.B. 1921. *Aplysia.* LMBC Memoirs XXIV, pp. 183–273. Transactions of the Liverpool Biological Society.

Eberwine J., Miyashiro K., Kacharmina J.E., and Job C. 2001. Local translation of classes of mRNAs that are targeted to neuronal dendrites. *Proc. Natl. Acad. Sci.* **98:** 7080–7085.

Esteban J.A. 2003. AMPA receptor trafficking: A road map for synaptic plasticity. *Mol. Interv.* **3:** 375–385.

Frank D.A. and Greenberg M.E. 1994. CREB: A mediator of long-term memory from mollusks to mammals. *Cell* **79:** 5–8.

Frost W.N., Castellucci V.F., Hawkins R.D., and Kandel E.R. 1985. Monosynaptic connections made by the sensory neurons of the gill- and siphon-withdrawal reflex in *Aplysia* participate in the storage of long-term memory for sensitization. *Proc. Natl. Acad. Sci.* **82:** 8266–8269.

Giustetto M., Hegde A.N., Si K., Casadio A., Inokuchi K., Pei W.,

Kandel E.R., and Schwartz J.H. 2003. Axonal transport of eukaryotic translation elongation factor 1 alpha mRNA couples transcription in the nucleus to long-term facilitation at the synapse. *Proc. Natl. Acad. Sci.* **100:** 13680–13685.

Greenberg S.M., Castellucii V.F., Bayley H., and Schwartz J.H. 1987. A molecular mechanism for long-term sensitization in *Aplysia. Nature* **329:** 62–65.

Guan Z., Kim J.H., Lomvardas S., Holick K., Xu S., Kandel E.R., and Schwartz J.H. 2003. p38 MAP kinase mediates both short-term and long-term synaptic depression in *Aplysia. J. Neurosci.* **23:** 7317–7325.

Guan Z., Giustetto M., Lomvardas S., Kim J.H., Miniaci M.C., Schwartz J.H., Thanos D., and Kandel E.R. 2002. Integration of long-term-memory-related synaptic plasticity involves bidirectional regulation of gene expression and chromatin structure. *Cell* **111:** 483–493.

Hegde A.N, Inokuchi K., Pei W., Casadio A., Ghirardi M., Chain D.G., Martin K.C., Kandel E.R., and Schwartz J.H. 1997. Ubiquitin C-terminal hydrolase is an immediate-early gene essential for long-term facilitation in *Aplysia. Cell* **89:** 115–126.

Homburg S., Visochek L., Moran N., Dantzer F., Priel E., Asculai E., Schwartz D., Rotter V., Dekel N., and Cohen-Armon M. 2000. A fast signal-induced activation of Poly(ADP-ribose) polymerase: A novel downstream target of phospholipase C. *J. Cell Biol.* **150:** 293–307.

Hu J.Y., Meng X., and Schacher S. 2003. Redistribution of syntaxin mRNA in neuronal cell bodies regulates protein expression and transport during synapse formation and long-term synaptic plasticity. *J. Neurosci.* **23:** 1804–1815.

Kandel E.R. 1979. *Behavioral biology of* Aplysia, p. 28. W.H. Freeman, San Francisco.

———. 2001. The molecular biology of memory storage: A dialog between genes and synapses. *Biosci. Rep.* **5:** 565–611.

Kandel E.R. and Schwartz J.H. 1982. Molecular biology of learning: Modulation of transmitter release. *Science* **218:** 433–443.

Kandel E.R., Schwartz J.H., and Jessel T. 2000. *Principles of neural science,* 4th edition, pp. 1251, 1254. McGraw-Hill, New York.

Khabour O., Levenson J., Lyons L.C., Kategaya L.S., Chin J., Byrne J.H., and Eskin A. 2004. Coregulation of glutamate uptake and long-term sensitization in *Aplysia. J. Neurosci.* **24:** 8829–8837.

Kraus W.L. and Lis J.T. 2003. PARP goes transcription. *Cell* **113:** 677–683.

Lee J.A., Kim H., Lee Y.S., and Kaang B.K. 2003. Overexpression and RNA interference of Ap-cyclic AMP-response element binding protein-2, a repressor of long-term facilitation, in *Aplysia kurodai* sensory-to-motor synapses. *Neurosci. Lett.* **337:** 9–12.

Liu J., Hu J.Y., Schacher S., and Schwartz J.H. 2004. The two regulatory subunits of *Aplysia* cAMP-dependent protein kinase mediate distinct functions in producing synaptic plasticity. *J. Neurosci.* **24:** 2465–2474.

Liu Q.R., Hattar S., Endo S., MacPhee K., Zhang H., Cleary L.J., Byrne J.H., and Eskin A. 1997. A developmental gene (Tolloid/BMP-1) is regulated in *Aplysia* neurons by treatments that induce long-term sensitization. *J. Neurosci.* **17:** 755–764.

Lodish H., Baltimore D., Berk A., Zipursky S.L., Matsudaira P., and Darnell J. 1995. *Molecular cell biology*. W.H. Freeman, New York.

Mansour T.E., Sutherland E.W., Rall T.W., and Bueding E. 1960. The effect of serotonin (5-hydroxytryptamine) on the formation of adenosine 3′,5′-phosphate by tissue particles from the liver fluke, *Fasciola hepatica*. *J. Biol. Chem.* **235:** 466–470.

Martin K.C., Michael D., Rose J.C., Barad M., Casadio A., Zhu H., and Kandel E.R. 1997. MAP kinase translocates into the nucleus of the presynaptic cell and is required for long-term facilitation in *Aplysia*. *Neuron* **18:** 899–912.

Michael D., Martin K.C., Seger R., Ning M.M., Baston R., and Kandel E.R. 1998. Repeated pulses of serotonin required for long-term faciliation activate mitogen-activated protein kinase in sensory neurons of *Aplysia*. *Proc. Natl. Acad. Sci.* **95:** 1864–1869.

Michel J.J. and Scott J.D. 2002. AKAP mediated signal transduction. *Annu. Rev. Pharmacol. Toxicol.* **42:** 235–257.

Moccia R., Chen D., Lyles V., Kapuya E., Kalachikov S., Spahn C.M., Frank J., Kandel E.R., Barad M., and Martin K.C. 2003. An unbiased cDNA library prepared from isolated *Aplysia* sensory neuron processes is enriched for cytoskeletal and translational mRNAs. *J. Neurosci.* **23:** 9409–9417.

Montminy M. 1997. Transcriptional regulation by cyclic AMP. *Annu. Rev. Biochem.* **66:** 807–822.

Moss A., Blackburn-Munro G., Garry E.M., Blakemore J.A., Dickinson T., Rosie R., Mitchell R,. and Fleetwood-Walker S.M. 2002. A role of the ubiquitin-proteasome system in neuropathic pain. *J. Neurosci.* **22:** 1363–1372.

Nestler E.J. 2004. Historical review: Molecular and cellular mechanisms of opiate and cocaine addiction. *Trends Pharmacol. Sci.* **25:** 210–218.

Pliny. *Natural History*, **IX:** 150–155.

Rall T.W. 1975. On the importance of cyclic AMP in neurobiology: An essay. *Metabolism* **24:** 241–247.

Sacktor T.C. and Schwartz J.H. 1990. Sensitizing stimuli cause translocation of protein kinase C. *Proc. Natl. Acad. Sci.* **87:** 2036–2039.

Satchell M.A., Zhang X., Kochanek P.M., Dixon C.E., Jenkins L.W., Melick J., Szabo C., and Clark R.S. 2003. A dual role for poly-ADP-ribosylation in spatial memory acquisition after traumatic brain injury in mice involving NAD$^+$ depletion and ribosylation of 14-3-3 gamma. *J. Neurochem.* **85:** 697–708.

Schulman H. and Hyman S.E. 1999. Intracellular signaling. In *Fundamental neuroscience* (ed. M.J. Zigmond et al.), pp. 269–316. Academic Press, New York.

Schwartz J.H. 2001. The many dimensions of cAMP signaling. *Proc. Natl. Acad. Sci.* **98:** 13482–13484 .

Schwartz J.H. and Swanson M.E. 1987. Dissection of tissues for characterizing nucleic acids from *Aplysia:* Isolation of the structural gene encoding calmodulin. *Methods Enzymol.* **139:** 277–290.

Steward O. and Schuman E.M. 2001. Protein synthesis at synaptic sites on dendrites. *Annu. Rev. Neurosci.* **24:** 299–325.

Thompson D'Arcy W. 1947. *A glossary of green fishes*, pp. 142–144. Oxford University Press, United Kingdom.

Vijayaraghavan S., Goueli S.A., Davey M.P., and Carr D.W. 1997. Protein kinase A anchoring inhibitor peptides arrest mammalian sperm motility. *J. Biol. Chem.* **272:** 4747–4752.

Yamamoto N., Hedge A.N., Chain D.G., and Schwartz J.H. 1999. Activation and degradation of the transcription factor C/EBP during long-term facilitation in *Aplysia*. *J. Neurochem.* **73:** 2415–2423.

Zwartjes R.E., West H., Hattar S., Ren X., Noel F., Nunez-Regueiro M., MacPhee K., Homayouni R., Crow M.T., Byrne J.H., and Eskin A. 1998. Identification of specific mRNAs affected by treatments producing long-term facilitation in *Aplysia*. *Learn. Mem.* **4:** 478–495.

21

Molecular Mechanisms of Drug Addiction

Eric J. Nestler

Department of Psychiatry and Center for Basic Neuroscience, The University of Texas Southwestern Medical Center, Dallas, Texas 75390-9070

ABSTRACT

Addiction can be viewed as a form of drug-induced neural plasticity. One of the best-established molecular mechanisms of addiction is up-regulation of the cAMP second messenger pathway, which occurs in many neuronal cell types in response to chronic administration of opiates or other drugs of abuse. This up-regulation, and the resulting activation of the transcription factor CREB (cAMP response element binding protein), appear to mediate aspects of tolerance and dependence. In contrast, induction of another transcription factor, termed ΔFosB (a Fos family protein), exerts the opposite effect and may contribute to sensitized responses to drug exposure. Knowledge of these and many other molecular mechanisms of addiction could one day lead to more effective treatments for addictive disorders.

INTRODUCTION

Drug addiction is a complex phenomenon, involving many psychological and social factors; it is also, at its core, a biological process caused by the long-lasting effects of a physical substance (a drug of abuse) on a physical substrate (a vulnerable brain).

Acute and Chronic Drug Exposure Modifies Synaptic Function

When an individual is exposed to a drug, the drug enters the bloodstream and then the brain. Once in the brain, it binds to an initial protein target, usually located at the synapse. Such perturbation of transmission at certain synapses in the brain mediates the acute effects of drug action—for example, a high, sedation, or activation.

These acute drug actions, however, are not sufficient to explain addiction, which requires repeated drug exposure. For that reason, addiction can be viewed as an example of drug-induced neural plasticity (Nestler et al. 1993). That is, repeated drug exposure, through repeated interactions with its acute target and repeated perturbation of synaptic transmission, causes changes in nerve cells, at the molecular and cellular level, that are ultimately responsible for the behavioral abnormalities that define addiction.

Biological Markers Are Needed for Mental Disorders such as Addiction

As an aside, one of the awkward facts about addiction, and psychiatry in general, is that all mental disorders are defined solely on the basis of behavioral abnormalities. This is in striking contrast to other fields of medicine, where behavioral symptoms, such as a cough or shortness of breath (these are behaviors, too), are correlated very closely with underlying biological processes that can be readily measured in patients. We do not have that luxury yet in psychiatry. When a patient presents with a mental symptom, we do not have objective laboratory tests—brain scans, genetic tests, and the like—to anchor behavioral abnormalities to their underlying biology. The development of such tests are a major goal for the field, but until they are available, behavioral abnormalities are all we have to define the syndrome of addiction.

CNS Targets for Drugs of Abuse

During the past 20 years, addiction research has succeeded in defining the acute targets of most drugs of abuse including neurotransmitter receptor agonists and antagonists, inhibitors of monoamine transporters, and activators and inhibitors of ligand-gated ion channels (Table 1). Increasingly over the past decade, the field has also provided a cataloguing of large numbers of molecular and cellular changes that occur in particular parts of the brain after chronic drug exposure. As will be seen below, it has even been possible to relate certain molecular and cellular adaptations to the behavioral abnormalities that characterize addiction. The major gap in the field is the lack of understanding of how changes at the level of individual nerve cells actually mediate these complex behaviors. This is a challenge throughout the neurosciences and requires highly sophisticated knowledge of the neural circuits involved, something that is not yet available.

All known acute targets of drugs of abuse are located at the extracellular aspect of the synapse. But to understand how drugs cause addiction after repeated exposure, it is necessary to consider the complex web of intracellular signals that regulate synaptic transmission. According to this view, repeated exposure to a drug, by repeatedly perturbing synaptic transmission, would cause robust perturbation of intracellular signaling cascades. Because many of the behavioral abnormalities that characterize addiction are very stable—for example, people can remain at increased risk for relapse despite a lifetime of abstinence—there must be very stable changes that drugs cause in the brain. For that reason, the field has considered a role for gene expression in the genesis of addiction. Accordingly, a drug, acting extracellularly, would repeatedly perturb the cell's intracellular cascades, which would ultimately cause changes in the cell nucleus at the level of transcription factors and other nuclear proteins that regulate the expression of individual genes. It is hypothesized that the altered expression of particular genes underlies some of the stable changes in neural function and behavior that represent addiction (Hyman and Malenka 2001; Nestler 2001).

Neural Substrates for Addiction

To investigate changes in gene expression that contribute to addiction, it is critical to first know where to look for these changes in the brain. Several decades of research, per-

TABLE 1. Initial Targets of Drugs of Abuse

Drug	Target
Opiates	Agonist at μ-, δ-, and κ-opioid receptors[a]
Cocaine	Indirect agonist at dopamine receptors by inhibiting dopamine transporters[b]
Amphetamine and related stimulants[c]	Indirect agonist at dopamine receptors by stimulating dopamine release[b]
Ethanol	Facilitates $GABA_A$ and inhibits NMDA glutamate receptor function[d]
Nicotine	Agonist at nicotinic acetylcholine receptors
Cannabinoids	Agonist at CB_1 and CB_2 cannabinoid receptors[e]
Phencyclidine (PCP)	Antagonist at NMDA glutamate receptors
Hallucinogens	Partial agonist at $5\text{-}HT_{2A}$ serotonin receptors
Inhalants	Unknown

Adapted, with permission, from Nestler 2001, 2004 [©Elsevier].

[a]Activity at μ (and possibly δ) receptors mediates the addicting actions of opiates; κ receptors mediate aversive actions.

[b]Cocaine and amphetamine exert analogous actions on serotonergic and noradrenergic systems, which may also contribute to the addicting effects of these drugs.

[c]For example, methamphetamine and methylphenidate.

[d]Ethanol affects several other ligand-gated channels, and at higher concentrations, voltage-gated channels as well. In addition, ethanol is reported to influence many other neurotransmitter systems.

[e]Activity at CB_1 receptors mediates the addicting actions of cannabinoids; CB_2 receptors are expressed in the periphery only. Proposed endogenous ligands for the CB_1 receptor include the arachidonic-acid metabolites, anandamide, and 2-arachidonylglycerol.

formed in hundreds of laboratories, has very convincingly established a role for so-called brain reward regions in mediating various aspects of the addiction process.

Mesolimbic Dopamine System

The circuit that has received the most attention is called the mesolimbic dopamine system (Koob et al. 1998; Wise 1998; Everitt and Wolf 2002; Chao and Nestler 2004). This circuit arises in dopamine-containing nerve cells in the ventral tegmental area (VTA) and includes the many regions of the limbic forebrain to which these dopamine cells project, in particular, the nucleus accumbens (NAc). Compelling evidence shows that this mesolimbic dopamine system is an integral mediator of the ability of all drugs of abuse to elicit an acute rewarding response. Likewise, it is widely believed that all drugs of abuse change this reward circuit to cause certain aspects of addiction.

Drugs of Abuse "Hijack" the Reward System

Although the field has focused a large majority of its effort on this one circuit, this likely overstates the role of the mesolimbic dopamine system in the addiction process. Thus, many other brain areas are also believed to contribute to critical features of addiction, however, the precise roles of each region are not yet as well established as that for the

VTA-NAc circuit. Such areas include several regions of the frontal cerebral cortex, amygdala, hippocampus, and hypothalamus, to name just a few (Koob et al. 1998; Everitt et al. 2001; Georgescu et al. 2003; Chao and Nestler 2004; Kalivas 2004).

Under normal conditions, brain reward regions are thought to regulate an individual's responses to natural rewards such as food, sex, and social interaction (Kelley and Berridge 2002). Drugs of abuse perturb these reward circuits with a power and persistence not seen with natural rewards. Such powerful effects likely explain the longer-term changes in these circuits that result from addiction. The idea is that drugs commandeer these reward circuits; they corrupt them in a way that directs the person's entire being (in the extreme case of addiction) to the drug. A striking feature of these brain reward regions is their age, from an evolutionary point of view. Dopamine neurons serve as an interface between a rewarding stimulus and motor function to consume that reward that shows substantial conservation down to organisms such as worms and flies, suggesting that these systems are about 2 billion years old (McClung and Hirsch 1998; Sawin et al. 2000). These considerations highlight the extremely powerful nature of the biological forces that drive an addiction.

The above discussion assumes that all drugs of abuse, despite their very different acute targets, converge at the level of the mesolimbic dopamine system and related reward circuits in brain to cause the similar changes that underlie addiction. Increasing evidence now shows that drugs of abuse do indeed share many common mechanisms of addiction. On the other hand, it is also likely that each drug induces its own unique changes that give rise to drug-specific features of an addiction.

GOALS OF THE CHAPTER

The above discussion provides an overview of how drugs of abuse affect the mesolimbic dopamine circuit and other brain reward circuits initially, but it does not answer the question of how they cause addiction. Recall that addiction requires repeated drug exposure, which somehow changes these neurons to alter an individual's reward mechanisms and related memories, and such changes somehow lead to addiction. Major clues to these molecular and cellular processes have come from the study of the locus coeruleus, a brain region not critically involved in drug reward per se, but which nonetheless has sparked several advances in the field. The remainder of this chapter provides a detailed description of some of the molecular mechanisms believed to have a major role in drug addiction. The methodologies developed and used to uncover and study these molecular targets are described. As will be seen, drug regulation of the cAMP pathway and the transcription factor CREB appears to be a mechanism of tolerance and dependence, whereas drug regulation of the transcription factor ΔFosB appears to be a mechanism for sensitization. Both types of adaptations ultimately summate to cause addiction.

MOLECULAR MECHANISMS OF ADDICTION

Up-regulation of the cAMP Pathway by Opiates

One of the best-established molecular mechanisms of addiction is up-regulation of the cAMP second messenger and protein phosphorylation pathway. This work arose from studies first performed by Sharma, Klee, and Nirenberg (1975) (Fig. 1). The investigators

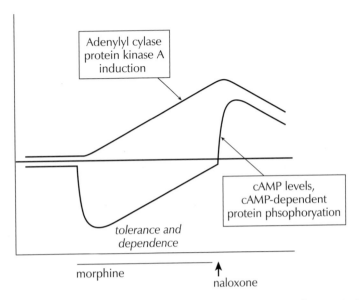

FIGURE 1. Up-regulation of the cAMP pathway as a mechanism of opiate tolerance and dependence. Opiates acutely inhibit the functional activity of the cAMP pathway (e.g., as indicated by cellular levels of cAMP or cAMP-dependent protein phosphorylation). With continued opiate exposure, functional activity of the cAMP pathway gradually recovers and increases far above control levels on removal of the opiate (e.g., by administration of the opioid receptor antagonist naloxone). These changes in the functional state of the cAMP pathway are mediated via the induction of adenylyl cyclase and protein kinase A (PKA) in response to chronic opiate administration. Induction of these enzymes accounts for the gradual recovery in the functional activity of the cAMP pathway seen during chronic opiate exposure (tolerance and dependence) and for the activation of the cAMP pathway seen on removal of the opiate (withdrawal). (Reprinted, with permission, from Nestler 2004 [©Elsevier].)

added morphine to cultured neuroblastoma x glia cells and found that morphine acutely decreased cAMP levels in the cells. Continued morphine exposure was accompanied by a gradual recovery of cAMP levels and, after a few days, cAMP levels recovered to normal. The investigators further showed that, on addition of the opioid receptor antagonist naloxone, cAMP levels rose dramatically above control. This was proposed as a cellular model of opiate tolerance and dependence. It depicts tolerance because cAMP levels gradually recover through the course of continued opiate exposure, and it depicts dependence because cAMP levels increase to above normal during "withdrawal." Collier and Francis (1975) independently provided similar lines of evidence for involvement of the cAMP pathway in opiate dependence.

Mechanism of Tolerance and Dependence in the Locus Coeruleus

On the basis of these findings in cultured cells, my laboratory determined whether similar phenomena may occur within the brain in vivo. For these initial studies, we focused on a small brain region called the locus coeruleus, the largest noradrenergic region in brain. We found the same basic phenomena as found earlier for the cultured cells (Nestler and Aghajanian 1997; Nestler 2004). But we went one step further by providing the mechanisms underlying at least part of these responses by understanding some of the molecular changes that opiates induce in locus coeruleus neurons (Fig. 1). We found that, during a course of chronic morphine exposure, locus coeruleus neurons express higher levels of adenylyl cyclase (the enzyme that synthesizes cAMP) and protein kinase A (PKA, the enzyme that mediates most of the functional effects of cAMP). Consequently, the combination of the up-regulated cAMP pathway and the continued presence of morphine results in functional tolerance. When the morphine is removed, the up-regulated cAMP pathway is unopposed, and this reveals the full functional effect of that up-regulated pathway. Considerable evidence now supports this hypothesis, that the up-regulated cAMP pathway in the locus coeruleus is one mechanism underlying the cellular signs of tolerance and dependence showed electrophysiologically by these nerve cells, as well as signs of physical opiate withdrawal that are mediated by locus coeruleus neuronal activation (Nestler and Aghajanian 1997; Punch et al. 1997; Ivanov and Aston-Jones 2001).

We now know a great deal more about the molecular details of these morphine responses in the locus coeruleus (Fig. 2). We and others have identified subtypes of adenylyl cyclase, ACVIII and ACI, and subunits of PKA that are increased in the locus coeruleus by chronic morphine. We also have implicated the transcription factor CREB in mediating some of these adaptations to morphine (Nestler 2001; Chao and Nestler 2004). This is consistent with the observation that mice lacking the predominant forms of CREB show attenuated morphine dependence and withdrawal (Maldonado et al. 1996).

Drugs of Abuse Up-regulate the cAMP Pathway in Other Brain Regions

Since these initial discoveries, we and many other groups have shown a similar up-regulation of the cAMP pathway in many other types of brain regions and even in peripheral nerve cells (Nestler 2001, 2004; Williams et al. 2001). For example, we showed that after a course of chronic morphine exposure, the changes described in the locus coeruleus also occur in the NAc. The notion is that this up-regulated cAMP pathway alters the excitability of NAc neurons in a way that contributes to changes in reward mechanisms related to addiction (I return to that notion below). Additional examples for the involvement of an

TABLE 2. Up-regulation of the cAMP Pathway in Opiate Addiction

Site of up-regulation	Functional consequence
Locus coeruleus[a]	Physical dependence and withdrawal
Ventral tegmental area[b]	Dysphoria during early withdrawal periods
Periaqueductal gray[b]	Dysphoria during early withdrawal periods, and physical dependence and withdrawal
Nucleus accumbens	Dysphoria during early withdrawal periods
Amygdala	Conditioned aspects of addiction
Dorsal horn of spinal cord	Tolerance to opiate-induced analgesia
Myenteric plexus of gut	Tolerance to opiate-induced reductions in intestinal motility and increases in motility during withdrawal

Adapted, with permission, from Nestler 2001.

[a]The cAMP pathway is up-regulated within the principal noradrenergic neurons located in this region.

[b]Indirect evidence suggests that the cAMP pathway is up-regulated within GABAergic neurons that innervate the dopaminergic VTA and serotonergic (periaqueductal gray) neurons located in these regions. During withdrawal, the up-regulated cAMP pathway would become fully functional and could contribute to a state of dysphoria by increasing the activity of the GABAergic neurons, which would then inhibit the dopaminergic and serotonergic neurons (see Bonci and Williams 1997; Nestler 2001).

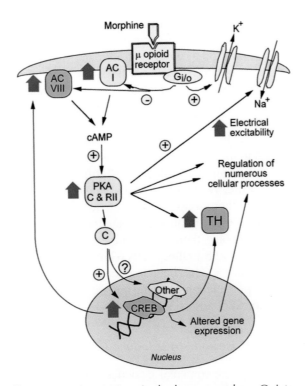

FIGURE 2. Scheme illustrating opiate actions in the locus coeruleus. Opiates acutely inhibit locus coeruleus (LC) neurons by increasing the conductance of an inwardly rectifying K+ channel via coupling with subtypes of $G_{i/o}$, and possibly by decreasing a Na+-dependent inward current via coupling with $G_{i/o}$ and the consequent inhibition of adenylyl cyclase (AC), reduced levels of PKA activity, and reduced phosphorylation of the responsible channel or pump. Inhibition of the cAMP pathway also decreases phosphorylation of numerous other proteins and thereby affects many additional processes in the neuron. For example, it reduces the phosphorylation state of CREB, which may initiate some of the longer-term changes in LC function. Upward arrows summarize effects of chronic morphine in the LC: increased levels of ACI and ACVIII, PKA catalytic (C) and regulatory type-II (RII) subunits, and several phosphoproteins including CREB and tyrosine hydroxylase (TH), the rate-limiting enzyme in norepinephrine biosynthesis. These changes contribute to the altered phenotype of the drug-addicted state. For example, the intrinsic excitability of LC neurons is increased via enhanced activity of the cAMP pathway and Na+-dependent inward current, which contributes to the tolerance, dependence, and withdrawal shown by these neurons. Up-regulation of ACVIII and TH is mediated via CREB, whereas up-regulation of ACI and the PKA subunits appears to occur via a CREB-independent mechanism not yet identified. (Adapted, with permission, from Nestler and Aghajanian 1997.)

up-regulated cAMP pathway in other brain areas are given in Table 2. Not surprisingly, the functional role played by cAMP pathway up-regulation depends on how the affected neurons are wired in the brain. The locus coeruleus is wired in such a way that when its functional activity is affected by opiates and opiate withdrawal through the cAMP pathway, physical dependence and withdrawal symptoms occur. Perturbations in the NAc affect drug reward, perturbations in the myenteric plexus of gut affect intestinal motility, etc. We now turn to a more detailed discussion of this neuroadaptation in the NAc.

Up-regulation of the cAMP Pathway and Induction of CREB by Many Drugs of Abuse

We and several other groups found many years ago that morphine was not alone in producing up-regulation of the cAMP pathway, including induction of CREB activity, in the NAc. Other drugs of abuse including cocaine, amphetamine, and alcohol exert the same effect (Nestler 2001; see below). It is proposed that the up-regulated cAMP pathway, including the induction of CREB activity, in these cells represents part of a common molecular adaptation that mediates changes in reward mechanisms related to aspects of addiction to many drugs. Several lines of evidence support the hypothesis that up-regulation of the cAMP pathway and CREB mediates a form of tolerance and dependence to drug exposure. We focus on available evidence for CREB.

We and other groups concentrated our efforts on CREB originally on the basis of the notion that if drugs of abuse activate the cAMP pathway in NAc nerve cells during a sustained time course, then it would be expected that the drugs would also lead to a sustained activation of CREB in these cells (Guitart et al. 1992). Moreover, because CREB is a transcription factor and has been implicated in a wide range of examples of neural plasticity—long-term potentiation, learning and memory, etc.—it was an interesting target to examine in the context of addiction.

Drugs of Abuse and Stress Induce CREB Activity

Two types of experimental evidence support the conclusion that drugs of abuse induce CREB activity. First, the drugs have been shown to increase levels of phospho-CREB (CREB phosphorylated at Ser133) in the NAc (Cole et al. 1995; Turgeon et al. 1997; Self et al. 1998) (phospho-CREB is the active form of the protein). Although we presume that induction of phospho-CREB is mediated by the up-regulated cAMP pathway in the NAc, it could be mediated by several other protein kinases also known to phosphorylate and activate CREB. Second, several drugs of abuse have been shown to induce CREB-mediated transcription in the NAc in CRE (cAMP response element)-LacZ transgenic mice (Asher et al. 2002; Barrot et al. 2002; Shaw-Lutchman et al. 2002, 2003). The transgene is comprised of CRE's driving expression of the LacZ reporter gene, which encodes β-galactosidase.

One of the interesting features of the mesolimbic dopamine system is that it is also highly responsive to stress. The fact that a key brain reward region also responds to stress has been somewhat of a surprise to investigators, and the functional implications remain a topic of some debate. Regardless, it was of interest to determine whether stress also activates CREB function in the NAc. As might be expected, we found that several forms of stress did indeed induce CREB activity in the NAc (Pliakas et al. 2001; Barrot et al. 2002).

Localization of CREB Regulation within the NAc

The NAc is not a homogeneous issue; it is part of the ventral striatum. The dorsal striatum is also called the caudate-putamen. Ninety to ninety-five percent of the cells in the NAc and dorsal striatum are medium spiny neurons that can be divided into two major subtypes (Gerfen 2000; Graybiel 2000). Direct projecting neurons, which express the neuropeptides dynorphin and substance P, project directly back to the midbrain. Indirect projecting neurons, which express enkephalin, project to the pallidum, which in turn projects to the midbrain and elsewhere. The NAc and dorsal striatum also contain small numbers of interneurons, including several types of GABAergic interneurons and large cholinergic interneurons. Using a variety of double-labeling techniques, we were able to show that the induction of CREB activity in the NAc seen with chronic drugs of abuse occurs in both major subtypes of medium spiny neurons, as well as sparsely in certain GABAergic interneurons (Shaw-Lutchman et al. 2002, 2003). No induction in non-neural cells was observed. Studies aimed at examining whether stress induces CREB activity in the same types of neurons are underway.

CREB Activation: Compensatory Response to Drugs and Stress, and its Role in Emotional States

As stated earlier, we have hypothesized that CREB activation by drugs of abuse or by stress represents a classic negative-feedback mechanism. In other words, CREB activation is a compensatory mechanism that decreases the animal's responsiveness to subsequent exposures to drug or stress. This hypothesis is based on a series of experiments in which we have increased or decreased CREB function selectively in the NAc by use of viral-mediated gene transfer. In these studies, we inject a herpes simplex virus (HSV) vector encoding CREB or a dominant negative mutant (mCREB) directly into the NAc. We have shown that such injections produce highly localized increases and decreases, respectively, of CREB activity within this brain region. Animals are then subjected to a wide battery of behavioral tests to assess the functional consequences of such localized changes in CREB activity. Our data to date all point to the conclusion that increased CREB function in the NAc decreases the animal's responses to cocaine and morphine, to natural rewards such as sucrose, and also to aversive stimuli such as anxiety, stress, and pain (Carlezon et al. 1998; Pliakas et al. 2001; Barrot et al. 2002). Conversely, in each case, decreased CREB function exerts the opposite effects.

Together, this work indicates that CREB activation reduces the behavioral responsiveness of an animal to both rewarding and aversive stimuli. Our hypothesis is that CREB, in the NAc, may be part of a molecular gating mechanism that weakens the sensitivity of an animal to emotional stimuli, whether that emotional stimulus is aversive or rewarding. Under normal conditions, this would appear to be a positive, coping response. With low levels of activation, there might be an advantage to be less responsive to emotional stimuli in the environment to facilitate dealing with the situation. Under pathological conditions of excessive, prolonged CREB activation, however, excessive emotional numbing could contribute to part of the addiction phenotype as well as the depression phenotype (Nestler et al. 2002). For example, in the context of chronic drug exposure or prolonged, severe stress, up-regulation of CREB could underlie part of the negative emotional state (emotional numbing, anhedonia, etc.) seen in many patients during early phases of withdrawal, or with depression or other stress-related disorders.

Concordant and Discrepant Results for CREB: Methodological Considerations

The above conclusions are based on studies involving viral-mediated gene transfer. Although this method offers a powerful means to alter levels of CREB (or some other protein of interest) within localized brain regions of adult animals, it nevertheless involves intracranial surgery and viral infection. It is, therefore, significant that we have been able to replicate aspects of the CREB and mCREB phenotype in a second experimental system: inducible transgenic mice in which CREB or mCREB is overexpressed with some selectivity in the NAc and dorsal striatum of adult animals (Newton et al. 2002; Sakai et al. 2002; McClung and Nestler 2003). Replication of this behavioral phenotype, in both rats and mice with viral-mediated gene transfer and inducible transgenesis, provides much greater support for the hypothesized role of CREB in the NAc outlined above.

Nevertheless, it is important to consider findings from a CREB knockdown mouse; data from this third experimental system agree with our overexpression data in some respects but differ in others (Walters and Blendy 2001). This CREB mouse lacks the two predominant forms of CREB expressed in brain. Consistent with our overexpression data, these mice show greater sensitivity to the behavioral effects of cocaine. However, inconsistent with our overexpression data, the mice are *less* sensitive to the behavioral effects of morphine. An important question is what accounts for this apparent discrepancy with morphine. One possibility is that the overexpression systems produce some artifactual results on the basis of supraphysiological levels of CREB or mCREB expression. A second possibility is that although mCREB would block the actions of all CREB family proteins, the CREB knockdown mice lack CREB only. A third possibility is that the overexpression systems influence CREB function within the NAc preferentially and in adult animals, whereas the CREB knockdown mice are deficient in CREB ubiquitously from early stages of development. In fact, some recent evidence shows that the reduced sensitivity to morphine seen in the CREB knockdown mice may be caused by the loss of CREB from the VTA, not the NAc (Walters et al. 2003).

The important message is that every experimental system has its advantages and disadvantages, and the best approach is to bring to bear diverse types of methodologies to understand something as complicated as the role of a single transcription factor, such as CREB, in a phenotype as complicated as addiction.

CREB Regulates Target Genes

As a transcription factor, one can presume that the behavioral effects of CREB at the level of the NAc are mediated through the regulation of specific target genes. Considerable evidence now shows that dynorphin, an opioid peptide, is one of those genes. This is depicted in Figure 3. Recall from an earlier discussion that dynorphin is expressed in one major subset of medium spiny neurons in the NAc that project directly back to VTA dopamine (DA) neurons. Dynorphin appears to mediate a feedback loop in this VTA-NAc circuit. When these medium spiny neurons fire, they release dynorphin onto VTA neuron cell bodies and their terminals in the NAc. Dynorphin activates κ-opioid receptors that inhibit the dopamine cells. This inhibition dampens the sensitivity of this reward circuit (Shippenberg and Rea 1997; Kreek 2001). The dynorphin gene is known to be activated by CREB in vitro and in the NAc in vivo (Cole et al. 1995; Carlezon et al. 1998). Hence, drug or stress activation of CREB in the NAc leads to induction of dynorphin and to a dampening of the reward circuit. Consistent with this hypothesis are the findings that blockade of κ-opioid receptors antagonizes the ability of CREB in the NAc to dampen responses to rewarding and stressful stimuli (Carlezon et al. 1998; Pliakas et al. 2001). Also, κ-opioid antagonists have been shown to induce antidepressant-like effects in animal models (Mague et al. 2003).

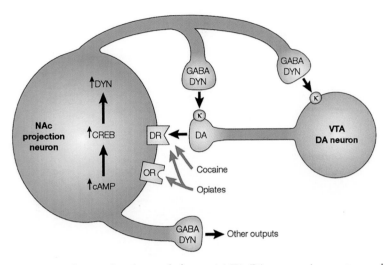

FIGURE 3. Regulation of CREB by drugs of abuse. A VTA DA neuron innervates a class of NAc GABAergic projection neuron that expresses dynorphin (DYN). Dynorphin serves a negative-feedback mechanism in this circuit: Released from terminals of the NAc neurons, dynorphin acts on κ-opioid receptors located on nerve terminals and cell bodies of the DA neurons to inhibit their functioning. Chronic exposure to cocaine or opiates, and possibly other drugs of abuse, up-regulates the activity of this negative-feedback loop via up-regulation of the cAMP pathway, activation of CREB, and induction of dynorphin.

Dynorphin is just one of many genes through which CREB exerts its complex effects in the NAc. Several other putative targets of CREB in this region have been identified and now require more detailed investigation (see McClung and Nestler 2003).

ΔFosB: A MOLECULAR SWITCH FOR ADDICTION

ΔFosB is a second transcription factor that we have investigated for its role in addiction. It provides a very interesting contrast to CREB, with respect to its temporal properties and functional role in addiction.

Our interest in ΔFosB began in the early 1990s when several groups showed that acute administration of cocaine or amphetamine triggers the very rapid induction of several Fos family proteins in the NAc and dorsal striatum (Graybiel et al. 1990; Young et al. 1991; Hope et al. 1992). Other drugs of abuse elicited a very similar phenomenon.

Drug Exposure Induces Several Fos Family Proteins including ΔFosB

This is shown in Figure 4. c-Fos is by far the most rapidly induced. Within 2 hours, it is already maximal and then back to normal, undetectable levels within 4–6 hours. Several other Fos family proteins including FosB, Fra1, and Fra2 are induced in a somewhat delayed fashion, but are also back to normal within 8–12 hours. We found a third wave of induction that was later shown to represent ΔFosB (Hope et al. 1994; Hiroi et al. 1997).

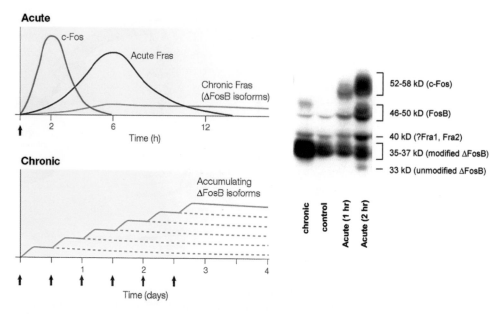

FIGURE 4. Scheme showing the gradual accumulation of ΔFosB versus the rapid and transient induction of other Fos family proteins. (*Top*) Several waves of Fos family proteins (comprised of c-Fos, FosB, ΔFosB [33-kD isoform], Fra-1, and Fra-2) are induced in a region-specific manner in brain by acute administration of any of several stimuli (see Table 1). Also induced are biochemically modified isoforms of ΔFosB (35–37 kD); they, too, are induced (although at low levels) following an acute stimulus, but persist in brain for long periods because of their stability. (*Bottom*) With repeated (e.g., twice daily) stimulation, each acute stimulus induces a low level of the stable ΔFosB isoforms. This is illustrated by the lower set of overlapping lines, which indicate ΔFosB induced by each acute stimulus. The result is a gradual increase in the total levels of ΔFosB with repeated stimuli during a course of chronic treatment, illustrated by the increasing stepped line in the graph. The gel inset (*right*) shows an example of this phenomenon: the induction of c-Fos, FosB, Fra1, Fra2, and the 33-kD isoform of ΔFosB in the nucleus accumbens after an acute cocaine exposure, as well as the switch to the predominant and selective induction of 35–37-kD isoforms of ΔFosB after chronic cocaine administration. (Adapted, with permission, from Hope et al. 1994 [©Elsevier].)

(ΔFosB is a truncated product of the FosB gene generated by alternative splicing.) ΔFosB is different from the other Fos family proteins for two reasons. First, after acute drug exposure, it is induced to low levels only. Second, ΔFosB is highly stable. Unlike all other Fos family proteins, which are induced rapidly but decay rapidly, once ΔFosB is induced, it persists in brain because it is a highly stable protein (Chen et al. 1995, 1997). Consequently, during a course of repeated drug exposure, ΔFosB gradually accumulates and eventually becomes, by far, the predominant Fos family protein in these cells. At the same time, most of the other Fos family proteins become desensitized with repeated drug exposure (Hope et al. 1992; Daunais et al. 1993; Persico et al. 1993). Thus, in a very real way, induction of ΔFosB represents a molecular switch that develops during the course of chronic drug administration and lasts for a relatively long time (weeks–months) after the last drug exposure (Nestler et al. 2001). This longevity is in striking contrast to the CREB signal, which dissipates within days of drug withdrawal.

Recent work has indicated two major mechanisms explaining ΔFosB's unique stability (see McClung et al. 2004). First, ΔFosB is phosphorylated on at least one serine residue, and this phosphorylation appears to increase the stability of the protein. Second, ΔFosB lacks the carboxy-terminal 101 amino acids present in FosB, and this carboxyl terminus contains a conserved motif present in all other Fos family proteins, which targets the proteins to rapid degradation in proteasomes. ΔFosB is unique among the Fos family in lacking this motif.

The second feature of ΔFosB that makes it an interesting substrate is the range of stimuli that induces it. ΔFosB is induced by every type of drug of abuse that has been investigated to date including cocaine and other stimulants, opiates, nicotine, alcohol, cannabinoids, and phencyclidine (see Hope et al. 1994; Moratalla et al. 1996; Pich et al. 1997; McClung et al. 2004). Moreover, the induction of ΔFosB by a drug appears to be greater in adolescent animals (Ehrlich et al. 2002).

Induction of ΔFosB Is Associated with Natural Rewards

Recall from our earlier discussion that the NAc circuit is important in regulating responses to natural rewards. Accordingly, increasing interest has generated in the field to determine whether the mechanisms underlying so-called natural addictions (e.g., pathological gambling, pathological overeating, etc.) may be similar to mechanisms underlying drug addiction. It is obviously difficult to study these phenomena in animal models, although the field has made progress. Interestingly, we have found that induction of ΔFosB is seen in two such models: compulsive running and compulsive sucrose drinking (Werme et al. 2002; McClung et al. 2004). Thus, ΔFosB is induced not only by chronic exposure to drugs of abuse, but also chronic, excessive consumption of natural rewards.

Functional Consequences of ΔFosB Induction

Drugs of abuse induce ΔFosB both in the NAc and dorsal striatum, whereas natural rewards seem to induce ΔFosB with greater selectivity in the NAc. There is also accruing evidence that the induction of ΔFosB by both drugs and natural rewards occurs predominantly in the dynorphin subset of medium spiny neurons. This has been shown for cocaine and compulsive running and is currently under investigation for other stimuli (Werme et al. 2002; McClung et al. 2004).

To study the functional consequences of ΔFosB induction in the NAc (and dorsal striatum), we made three lines of inducible transgenic mice: two that overexpress ΔFosB and one that overexpresses a mutant form of cJun (ΔcJun) that antagonizes transcription mediated by all Fos and Jun family proteins. One of the ΔFosB-expressing lines, the 11A line, expresses ΔFosB solely in dynorphin+ medium spiny neurons, whereas the other line, 11B, expresses ΔFosB preferentially in enkephalin+ medium spiny neurons. The ΔcJun line expresses the protein equally in both cell types. In all three lines, transgene expression is seen predominantly in the NAc and dorsal striatum, with much lower levels of expression seen in some of the lines in hippocampus and frontal cortex. Transgene expression in each of these lines is mediated by the neuron-specific enolase promoter; the differing cellular patterns of expression seen in the lines presumably reflect the different insertion sites of the transgenes.

TABLE 3. Behavioral Phenotype of ΔFosB in Dynorphin+ Neurons of NAc and Dorsal Striatum[a]

Cocaine	Increased locomotor responses to acute drug administration[b]
	Increased locomotor sensitization to repeated drug administration[b]
	Increased conditioned place preference at lower drug doses[c,d]
	Increased acquisition of cocaine self-administration at lower drug doses
	Increased incentive for drug in progressive ratio procedure
Morphine	Increased conditioned place preference at lower drug doses[c,d]
	Increased development of physical dependence and withdrawal[c]
	Decreased analgesic responses; enhanced tolerance[c]
Alcohol	Increased anxiolytic responses
Wheel running	Increased[c,d]
Sucrose	Increased incentive for food in progressive ratio procedure

[a]The listed phenotypes occur on inducible overexpression of ΔFosB in line A bitransgenic NSE-tTA x TetOp-ΔFosB mice. ΔFosB expression in this line predominates in NAc and dorsal striatum (where it occurs solely in dynorphin+/substance P+ neurons).

[b]A similar phenotype is observed in line B bitransgenic NSE-tTA x TetOp-ΔFosB mice who overexpress ΔFosB predominantly in enkephalin+ NAc and dorsal striatal neurons.

[c]The opposite phenotype is observed in NSE-tTA x Tetop-ΔcJun mice.

[d]The phenotype is not observed in line B bitransgenic NSE-tTA x TetOp-ΔFosB mice.

We are currently in the process of analyzing the behavioral phenotypes of these three lines of mice in animal models of addiction. Work to date, summarized in Table 3, suggests these main phenotypes: (1) ΔFosB expression in dynorphin+ cells of the NAc and dorsal striatum increases an animal's sensitivity to the locomotor and rewarding effects of cocaine; (2) such expression also appears to increase incentive drive for cocaine; (3) ΔFosB expression increases an animal's sensitivity to the rewarding effects of morphine and also makes the animal more prone to physical dependence and analgesic tolerance to morphine; (4) ΔFosB expression increases responses to natural rewards, including running and sucrose; (5) most of these effects are not seen when ΔFosB is expressed predominantly in enkephalin+ neurons and, in most cases, opposite effects are seen in mice overexpressing ΔcJun (Kelz et al. 1999; Werme et al. 2002; Colby et al. 2003; Peakman et al 2003; McClung et al. 2004; V. Zachariou et al., in prep.).

Together, these data suggest that ΔFosB induction in the NAc (and in some cases in the dorsal striatum) by drugs of abuse and natural rewards represents a mechanism for relatively prolonged sensitization. In addition, although this remains highly inferential, we hypothesize that ΔFosB induction also increases drive or incentive motivation for the drugs and natural rewards. If this is the case, it would support the notion that ΔFosB induction could be a persistent molecular switch that, once thrown, first initiates and then maintains a state of addiction for a relatively prolonged period of time (Nestler et al. 2001).

Differences between Transcription Factors ΔFosB and CREB

Comparison of ΔFosB with CREB highlights several interesting differences. The CREB signal is relatively specific to the NAc, but is expressed in several major cell types within this region. The ΔFosB signal is seen more broadly throughout the NAc and dorsal striatum, but shows greater cellular specificity to dynorphin+ cells. The CREB signal is relatively short lived, lasts a few days only, and mediates a negative emotional state. The ΔFosB signal is much more long lived, persists for up to 6–8 weeks, and seems to mediate a sensitized state. A state of drug addiction would involve both of these transcription factors or a summation of these two competing functional effects. In fact, addiction is much more complicated, because it likely involves hundreds of mechanisms in addition to CREB and ΔFosB. A major goal of future research is to understand how these numerous factors integrate and summate over time to produce the complex behaviors that characterize a state of addiction.

Gene Targets for CREB and ΔFosB

As with CREB, ΔFosB's behavioral phenotype is presumably mediated via the altered expression of specific sets of target genes in the NAc. We have used Affymetrix chips in an open-ended way to search for such genes. We used the various lines of inducible transgenic mice described above: those that overexpress CREB, mCREB, ΔFosB (the 11A line), or ΔcJun (McClung and Nestler 2003). The first conclusion from this work is that the effects of CREB are relatively simple: For the preponderance of genes examined, expression of CREB and mCREB produces opposite effects. In contrast, the effects of ΔFosB are more complex: Only a small majority of genes regulated by ΔFosB show reciprocal regulation by ΔcJun, whereas a smaller subset of genes shows similar regulation by the two proteins. These results teach us that, for the former genes, ΔFosB acts as a transcriptional activator, but for the latter genes, it acts as a transcriptional repressor. We are interested in studying what it is about those target genes that determines whether ΔFosB will act as an activator or repressor. The other interesting observation from these microarray studies is the identification of a subset of genes, which show opposite regulation by ΔFosB and CREB. This is interesting because the two proteins generally exert opposite effects on behavior, as discussed previously. Further characterization of the genes that show this pattern of regulation is now warranted.

We also compared the patterns of regulation of gene expression by ΔFosB and CREB to those seen with cocaine (Fig. 5). We gave one group of wild-type mice a brief period of cocaine exposure (5 days) and another group a longer exposure (4 weeks). Both cocaine treatments regulated a subset of genes, although the two subsets overlapped by 25% only. This illustrates that patterns of gene expression shift during a course of chronic cocaine administration. We then determined which of the genes regulated by cocaine could be attributed to CREB or ΔFosB. We found that CREB seems to be more important during early phases of cocaine use, but ΔFosB becomes more important after prolonged cocaine exposure, at which point close to 30% of all genes regulated by cocaine could be accounted for by this single transcription factor. This finding supports a dominant role for ΔFosB in mediating the genomic effects of prolonged cocaine exposure (McClung and Nestler 2003).

Characterizing each of these distinct subsets of genes is an enormous undertaking, but is expected to provide new insight into the molecular mechanisms underlying the long-term effects of cocaine on the brain. This can be illustrated by consideration of one novel substrate for ΔFosB, CDK5 (cyclin-dependent kinase-5).

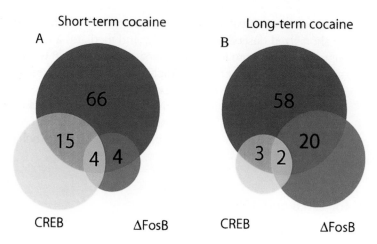

FIGURE 5. Regulation of gene expression by cocaine: role of CREB and ΔFosB. Mice were treated with cocaine for 5 days (*A*) or 4 weeks (*B*), and RNA isolated from the nucleus accumbens was subjected to DNA microarray analysis. The Venn diagrams show the number of genes whose regulation by short- or longer-term cocaine administration (*blue circles*) could be accounted for by ΔFosB (*red*) or CREB (*yellow*). The figure illustrates an important role for CREB in earlier phases of cocaine exposure and an increasingly dominant role for ΔFosB during more prolonged periods of drug administration. (Reprinted, with permission, from McClung and Nestler 2003.)

CDK5: A Target Gene for ΔFosB

CDK5 is a serine/threonine protein kinase that has been implicated in regulating the growth and survival of neural cells. We first identified CDK5 as a target for ΔFosB in microarray studies of our inducible transgenic mice (Chen et al. 2000; Bibb et al. 2001). On the basis of this finding, we confirmed that CDK5 is induced in the NAc by chronic cocaine, an effect blocked by overexpression of ΔcJun. Moreover, using the method of chromatin immunoprecipitation, we have recently been able to show that ΔFosB binds to the CDK5 gene promoter in the NAc in vivo, and that this binding is increased by chronic cocaine administration and is associated with chromatin remodeling (e.g., changes in histone acetylation) that reflects gene activation (A. Kumar et al., in prep.). In fact, an inhibitor of histone deacetylation, that would be expected to promote the action of ΔFosB, potentiates the behavioral effects of cocaine, as would be expected. These data are important because they provide the first direct evidence for the regulation of a gene in vivo by chronic cocaine administration, and they begin to elaborate the detailed molecular mechanisms involved.

The induction of a growth-regulatory protein by cocaine was interesting in light of a series of studies by Terry Robinson and colleagues (Robinson and Kolb 2004) that showed that chronic exposure to cocaine increases the dendritic arborization and the density of dendritic spines on NAc medium spiny neurons. We therefore asked whether the induction of CDK5 by ΔFosB and cocaine could represent one molecular mechanism underlying this effect. Indeed, local infusion of a CDK5 inhibitor into the NAc blocked the ability of cocaine to increase the density of dendritic spines on NAc neurons (Norrholm et al. 2003).

It is possible that the observed expansion of the dendritic arbor of NAc neurons could contribute to a state of sensitization and enhanced motivation for drug and related cues (Nestler 2001). However, it is imperative to emphasize that this hypothesis remains highly speculative. No direct evidence is available to show that more dendrites means greater responsiveness. We also have no idea what is happening to the presynaptic inputs to these cells. It will be important to learn whether drug-induced alterations exist in the glutamatergic terminals from frontal cortex, amygdala, and hippocampus that impinge on these NAc neurons. Nevertheless, these studies highlight the new lines of investigation that we can expect will result from open-ended studies of gene expression changes in the brain's reward pathways.

SUMMARY

This chapter presents an overview of the mechanisms by which repeated exposure to drugs of abuse alters the brain's reward pathways to cause a state of addiction. Evidence is presented that shows that up-regulation of the cAMP pathway and CREB by opiates and several other drugs of abuse is a general mechanism of tolerance and dependence, and contributes to different aspects of addiction in the various regions of the central and peripheral nervous system where these adaptations occur. In contrast, induction of ΔFosB in the nucleus accumbens seems to cause an opposite phenotype, namely, one of drug sensitization. Both types of adaptations, along with drug-induced changes in numerous other molecular pathways, ultimately cause the complex behavioral phenotype of addiction.

GAPS AND OPPORTUNITIES

Several major challenges remain. One need in the field is to analyze molecular adaptations to drugs of abuse not in isolation, but together, to investigate mechanisms by which hundreds of adaptations interact antagonistically or synergistically to cause a state of addiction. In a related sense, we need to characterize drug-induced adaptations throughout the brain's emotional circuitry and understand how changes in these various regions contribute to the addiction process. Currently, the tools of molecular biology and the advanced animal models of addiction have made it possible to implicate numerous proteins and biochemical pathways in the abnormal behaviors associated with addiction. However, this current information is limited in two key respects. First, we do not understand with certainty how most drug-induced adaptations alter the electrophysiological properties of the neurons involved. Second, we have a very poor understanding of how any given brain region actually leads to addiction. This requires a much more sophisticated appreciation of neural networks as mediators of complex behavior and represents perhaps the greatest remaining challenge for the neurosciences today. Finally, the ultimate challenge is to use this evolving knowledge of the molecular basis of addictive states to develop more effective treatments and preventive measures for addiction disorders.

REFERENCES

Asher O., Cunningham T.D., Tao L., Gordon A.S., and Diamond I. 2002. Ethanol stimulates cAMP-responsive element (CRE)-mediated transcription via CRE-binding protein and cAMP-dependent protein kinase. *J. Pharmacol. Exp. Ther.* **301**: 66–70.

Barrot M., Olivier J.D.A., Perrotti L.I., Impey S., Storm D.R., Neve R.L., Zachariou V., and Nestler E.J. 2002. CREB activity in the nucleus accumbens shell controls gating of behavioral responses to emotional stimuli. *Proc. Natl. Acad. Sci.* **99**: 11435–11440.

Bibb J.A., Chen J.S., Taylor J.R., Svenningsson P., Nishi A., Snyder G.L., Yan Z., Sagawa Z.K., Ouimet C.C., Nairn A.C., Nestler E.J., and Greengard P. 2001. Effects of chronic exposure to cocaine are regulated by the neuronal protein Cdk5. *Nature* **410**: 376–380.

Bonci A. and Williams J.T. 1997. Increased probability of GABA release during withdrawal from morphine. *J. Neurosci.* **17**: 796–803.

Carlezon W.A., Jr., Thome J., Olson V.G., Lane-Ladd S.B., Brodkin E.S., Hiroi N., Duman R.S., Neve R.L., and Nestler E.J. 1998. Regulation of cocaine reward by CREB. *Science* **282**: 2272–2275.

Chao J. and Nestler E.J. 2004. Molecular neurobiology of drug addiction. *Annu. Rev. Med.* **55**: 113–132.

Chen J., Kelz M.B., Hope B.T., Nakabeppu Y., and Nestler E.J. 1997. Chronic FRAs: Stable variants of ΔFosB induced in brain by chronic treatments. *J. Neurosci.* **17**: 4933–4941.

Chen J.S., Nye H.E., Kelz M.B., Hiroi N., Nakabeppu Y., Hope B.T., and Nestler E.J. 1995. Regulation of ΔFosB and FosB-like proteins by electroconvulsive seizure (ECS) and cocaine treatments. *Mol. Pharmacol.* **48**: 880–889.

Chen J.S., Zhang Y.J., Kelz M.B., Steffen C., Ang E.S., Zeng L., and Nestler E.J. 2000. Induction of cyclin-dependent kinase 5 in hippocampus by chronic electroconvulsive seizures: Role of ΔFosB. *J. Neurosci.* **20**: 8965–8971.

Colby C.R., Whisler K., Steffen C., Nestler E.J., and Self D.W. 2003. ΔFosB enhances incentive for cocaine. *J. Neurosci.* **23**: 2488–2493.

Cole R.L., Konradi C., Douglass J., and Hyman S.E. 1995. Neuronal adaptation to amphetamine and dopamine: Molecular mechanisms of prodynorphin gene regulation in rat striatum. *Neuron* **14**: 813–823.

Collier H.O. and Francis D.L. 1975. Morphine abstinence is associated with increased brain cyclic AMP. *Nature* **255**: 159–162

Daunais J.B., Roberts D.C., and McGinty J.F. 1993. Cocaine self-administration increases preprodynorphin, but not c-fos, mRNA in rat striatum. *Neuroreport* **4**: 543–546.

Ehrlich M.E., Sommer J., Canas E., and Unterwald E.M. 2002. Periadolescent mice show enhanced DeltaFosB upregulation in response to cocaine and amphetamine. *J. Neurosci.* **22**: 9155–9159.

Everitt B.J. and Wolf M.E. 2002. Psychomotor stimulant addiction: A neural systems perspective. *J. Neurosci.* **22**: 3312–3320.

Everitt B.J., Dickinson A., and Robbins T.W. 2001. The neuropsychological basis of addictive behaviour. *Brain Res. Rev.* **36**: 129–138.

Georgescu D., Zachariou V., Barrot M., Mieda M., Willie J.T., Eisch A.J., Yanagisawa M., Nestler E.J., and DiLeone R.J. 2003. Involvement of the lateral hypothalamic peptide orexin in morphine dependence and withdrawal. *J. Neurosci.* **23**: 3106–3111.

Gerfen C.R. 2000. Molecular effects of dopamine on striatal-projection pathways. *Trends Neurosci.* (suppl.) **23**: S64–S70.

Graybiel A.M. 2000. The basal ganglia. *Curr. Biol.* **10**: R509–R511.

Graybiel A.M., Moratalla R., and Robertson H.A. 1990. Amphetamine and cocaine induce drug-specific activation of the c-fos gene in striosome-matrix compartments and limbic subdivisions of the striatum. *Proc. Natl. Acad. Sci.* **87**: 6912–6916.

Guitart X., Thompson M.A., Mirante C.K., Greenberg M.E., and Nestler E.J. 1992. Regulation of CREB phosphorylation by acute and chronic morphine in the rat locus coeruleus. *J. Neurochem.* **58**: 1168–1171.

Hiroi N., Brown J., Haile C., Ye H., Greenberg M.E., and Nestler E.J. 1997. FosB mutant mice: Loss of chronic cocaine induction of Fos-related proteins and heightened sensitivity to cocaine's psychomotor and rewarding effects. *Proc. Natl. Acad. Sci.* **94**: 10397–10402.

Hope B., Kosofsky B., Hyman S.E., and Nestler E.J. 1992. Regulation of IEG expression and AP-1 binding by chronic cocaine in the rat nucleus accumbens. *Proc. Natl. Acad. Sci.* **89**: 5764–5768.

Hope B.T., Nye H.E., Kelz M.B., Self D.W., Iadarola M.J., Nakabeppu Y., Duman R.S., and Nestler E.J. 1994. Induction of a long-lasting AP-1 complex composed of altered Fos-like proteins in brain by chronic cocaine and other chronic treatments. *Neuron* **13**: 1235–1244.

Hyman S.E. and Malenka R.C. 2001. Addiction and the brain: The neurobiology of compulsion and its persistence. *Nat. Rev. Neurosci.* **2**: 695–703

Ivanov A. and Aston-Jones G. 2001. Local opiate withdrawal in locus coeruleus neurons in vitro. *J. Neurophysiol.* **85**: 2388–2397.

Kalivas P.W. 2004. Glutamate systems and cocaine addiction. *Curr. Opin. Pharmacol.* **4**: 23–29.

Kelley A.E. and Berridge K.C. 2002. The neuroscience of natural rewards: Relevance to addictive drugs. *J. Neurosci.* **22**: 3306–3311

Kelz M.B., Chen J.S., Carlezon W.A., Whisler K., Gilden L., Beckmann A.M., Steffen C., Zhang Y.-J., Marotti L., Self D.W., Tkatch R., Baranauskas G., Surmeier D.J., Neve R.L., Duman R.S., Picciotto M.R., and Nestler E.J. 1999. Expression of the transcription factor ΔFosB in the brain controls sensitivity to cocaine. *Nature* **401**: 272–276.

Koob G.F., Sanna P.P., and Bloom F.E. 1998. Neuroscience of

addiction. *Neuron* **21**: 467–476.

Kreek M.J. 2001. Drug addictions: Molecular and cellular end-points. *Ann. N.Y. Acad. Sci.* **937**: 27–49.

Mague S.D., Pliakas A.M., Todtenkopf M.S., Tomasiewica H.C., Zhang Y., Stevens W.C., Jr., Jones R.M., Portoghese P.S., and Carlezon W.A., Jr. 2003. Antidepressant-like effects of kappa-opioid receptor antagonists in the forced swim test in rats. *J. Pharmacol. Exp. Ther.* **305**: 323–330.

Maldonado R., Blendy J.A., Tzavara E., Gass P., Roques B.P., Hanoune J., and Schutz G. 1996. Reduction of morphine abstinence in mice with a mutation in the gene encoding CREB. *Science* **273**: 657–659.

McClung C. and Hirsh J. 1998. Stereotypic behavioral responses to free-base cocaine and the development of behavioral sensitization in *Drosophila*. *Curr. Biol.* **8**: 109–112.

McClung C.A. and Nestler E.J. 2003. Regulation of gene expression and cocaine reward by CREB and ΔFosB. *Nat. Neurosci.* **11**: 1208–1215.

McClung C.A., Ulery P.G., Perrotti L.I., Zachariou V., Berton O., and Nestler E.J. 2004. ΔFosB: A molecular switch for long-term adaptation. *Mol. Brain Res.* **132**: 146–154.

Moratalla R., Elibol B., Vallejo M., and Graybiel A.M. 1996. Network-level changes in expression of inducible Fos-Jun proteins in the striatum during chronic cocaine treatment and withdrawal. *Neuron* **17**: 147–156.

Nestler E.J. 2001. Molecular basis of long-term underlying addiction. *Nat. Rev. Neurosci.* **2**: 119–128.

———. 2004. Historical review: Molecular and cellular mechanisms of opiate and cocaine addiction. *Trends Pharmacol. Sci.* **25**: 210–218.

Nestler E.J. and Aghajanian G.K. 1997. Molecular and cellular basis of addiction. *Science* **278**: 58–63.

Nestler E.J., Barrot M., and Self D.W. 2001. ΔFosB: A molecular switch for addiction. *Proc. Natl. Acad. Sci.* **98**: 11042–11046.

Nestler E.J., Hope B.T., and Widnell K.L. 1993. Drug addiction: A model for the molecular basis of neural plasticity. *Neuron* **11**: 995–1006.

Nestler E.J., Barrot M., DiLeone R.J., Eisch A.J., Gold S.J., and Monteggia L.M. 2002. Neurobiology of depression. *Neuron* **34**: 13–25.

Newton S.S., Thome J., Wallace T., Shirayama Y., Dow A., Schlesinger L., Duman C.H., Sakai N., Chen J.S., Neve R., Nestler E.J., and Duman R.S. 2002. Inhibition of CREB or dynorphin in the nucleus accumbens produces an antidepressant-like effect. *J. Neurosci.* **22**: 10883–10890.

Norrholm S.D., Bibb J.A., Nestler E.J., Ouimet C.C., Taylor J.R., and Greengard P. 2003. Cocaine-induced proliferation of dendritic spines in nucleus accumbens is dependent on the activity of the neuronal kinase Cdk5. *Neuroscience* **116**: 19–22.

Peakman M.C., Colby C., Perrotti L.I., Tekumalla P., Carle T., Ulery P., Chao J., Duman C., Steffen C., Monteggia L., Allen M.R., Stock J.L., Duman R.S., McNeish J.D., Barrot M., Self D.W., Nestler E.J., and Schaeffer E. 2003. Inducible, brain region specific expression of a dominant negative mutant of c-Jun in transgenic mice decreases sensitivity to cocaine. *Brain Res.* **970**: 73–86.

Persico A.M., Schindler C.W., O'Hara B.F., Brannock M.T., and Uhl G.R. 1993. Brain transcription factor expression: Effects of acute and chronic amphetamine and injection stress. *Mol. Brain Res.* **20**: 91–100.

Pich E.M., Pagliusi S.R., Tessari M., Talabot-Ayer D., hooft van Huijsduijnen R., and Chiamulera C. 1997. Common neural substrates for the addictive properties of nicotine and cocaine. *Science* **275**: 83–86.

Pliakas A.M., Carlson R.R., Neve R.L., Konradi C., Nestler E.J., and Carlezon W.A., Jr. 2001. Altered responsiveness to cocaine and increased immobility in the forced swim test associated with elevated CREB expression in the nucleus accumbens. *J. Neurosci.* **21**: 7397–7403.

Punch L., Self D.W., Nestler E.J., and Taylor J.R. 1997. Opposite modulation of opiate withdrawal behaviors upon microinfusion of a protein kinase A inhibitor versus activator into the locus coeruleus or periaqueductal gray. *J. Neurosci.* **17**: 8520–8527.

Robinson T.E. and Kolb B. 2004. Structural plasticity associated with exposure to drugs of abuse. *Neuropharmacology* (suppl.) **47**: 33–46.

Sakai N., Thome J., Chen J.S., Kelz M.B., Steffen C., Nestler E.J., and Duman R.S. 2002. Inducible and brain region specific CREB transgenic mice. *Mol. Pharmacol.* **61**: 1453–1464.

Sawin E.R., Ranganathan R., and Horvitz H.R. 2000. *C. elegans* locomotory rate is modulated by the environment through a dopaminergic pathway and by experience through a serotonergic pathway. *Neuron* **26**: 619–631

Self D.W., Genova L.M., Hope B.T., Barnhart W.J., Spencer J.J., and Nestler E.J. 1998. Involvement of cAMP-dependent protein kinase in the nucleus accumbens in cocaine self-administration and relapse of cocaine-seeking behavior. *J. Neurosci.* **18**: 1848–1859.

Sharma S.K., Klee W.A., and Nirenberg M. 1975. Dual regulation of adenylate cyclase accounts for narcotic dependence and tolerance. *Proc. Natl. Acad. Sci.* **72**: 3092–3096.

Shaw-Lutchman T.Z., Impey S., Storm D., and Nestler E.J. 2003. Regulation of CRE-mediated transcription in mouse brain by amphetamine. *Synapse* **48**: 10–17.

Shaw-Lutchman T.Z., Barrot M., Wallace T., Gilden L., Zachariou V., Impey S., Duman R.S., Storm D., and Nestler E.J. 2002. Regional and cellular mapping of CRE-mediated transcription during naltrexone-precipitated morphine withdrawal. *J. Neurosci.* **22**: 3663–3672.

Shippenberg T.S. and Rea W. 1997. Sensitization to the behavioral effects of cocaine: Modulation by dynorphin and kappa-opioid receptor agonists. *Pharmacol. Biochem. Behav.* **57**: 449–455

Turgeon S.M., Pollack A.E., and Fink J.S. 1997. Enhanced CREB phosphorylation and changes in c-Fos and FRA expression in striatum accompany amphetamine sensitization. *Brain Res.* **749**: 120–126.

Walters C.L. and Blendy J.A. 2001. Different requirements for cAMP response element binding protein in positive and negative reinforcing properties of drugs of abuse. *J. Neurosci.* **21:** 9438–9444.

Walters C.L., Kuo Y.C., and Blendy J.A. 2003. Differential distribution of CREB in the mesolimbic dopamine reward pathway. *J. Neurochem.* **87:** 1237–1244.

Werme M., Messer C., Olson L., Gilden L., Thorén P., Nestler E.J., and Brené S. 2002. ΔFosB regulates wheel running. *J. Neurosci.* **22:** 8133–8138.

Williams J.T., Christie M.J., and Manzoni O. 2001. Cellular and synaptic adaptations mediating opioid dependence. *Physiol. Rev.* **81:** 299–343.

Wise R.A. 1998. Drug-activation of brain reward pathways. *Drug Alcohol Depend.* **51:** 13–22.

Young S.T., Porrino L.J., and Iadarola M.J. 1991. Cocaine induces striatal c-fos-immunoreactive proteins via dopaminergic D1 receptors. *Proc. Natl. Acad. Sci.* **88:** 1291–1295.

PART 5

Systems Analysis of Drug Abuse

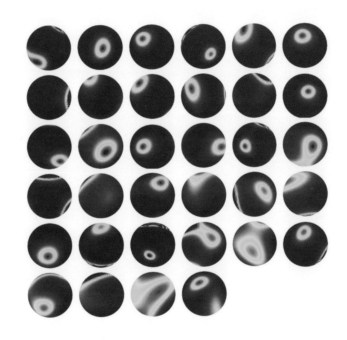

PART 5

Systems Analysis of Drug Abuse

22

Dynamic Analyses of Neural Representations Using the State-space Modeling Paradigm

Emery N. Brown and Riccardo Barbieri

Neuroscience Statistics Research Laboratory, Department of Anesthesia and Critical Care, Massachusetts General Hospital, Division of Health Sciences and Technology, Harvard Medical School/Massachusetts Institute of Technology, Boston, Massachusetts 02114-2696

ABSTRACT

Understanding how neurons represent biological signals in their ensemble spiking activity is a fundamental challenge in neuroscience. A key property of the neural representations is that they are plastic. That is, with repeated exposure to the same stimulus, a neuron changes its firing properties. Understanding these fundamental properties of neurons is essential for understanding how the nervous system responds to any stimulus including, light, sound, temperature, contact, a cognitive task, or pharmacological substances. This suggests that methods used to analyze neurophysiological data should be designed specifically to characterize ensemble spiking activity and neural plasticity. In this chapter, we describe how a broad class of neural spike train decoding algorithms for studying ensemble representations and adaptive filter algorithms for analyzing neural plasticity can be derived in a unified way using the state-space modeling paradigm. We show the decoding algorithm in the analysis of position decoding from the ensemble spiking activity of rat hippocampal neurons. In addition, we illustrate the adaptive filter algorithm by studying the temporal dynamics of the spatial receptive field of a hippocampal neuron. In the final section, we discuss directions for future research.

INTRODUCTION

Understanding how neurons represent a biological signal in their joint spiking activity is a fundamental challenge in neuroscience (Georgopoulos et al. 1986; Bialek et al. 1991; Wilson and McNaughton 1993; Stanley et al. 1999; Brown et al. 2004; Wu et al. 2004). During the past 12 years, the development of multiunit electrode arrays that allow the simultaneous recording of a large number of neurons is an important experimental advance that now makes it possible to study these ensemble representations (Wilson and McNaughton 1993). In addition, neural representations are plastic in that, with experience, neurons in many brain regions change their spiking responses to relevant stimuli or biological signals (Kaas et al. 1983; Merzenich et al. 1984; Jog et al. 1999; Gandolfo et al. 2000; Mehta et al. 2000). Deciphering how groups of neurons represent information in their ensemble spiking activity and the dynamics of neural plasticity are key elements to understanding how the nervous system responds to any stimulus—whether it be light, sound, temperature, contact, a cognitive task, or a pharmacological substance. These characterizations are crucial for defining both the normal physiological states of neural systems as well as pathological conditions that occur in diseases such as drug addiction. These observations suggest that methods used to analyze neurophysiological data should be designed specifically to characterize ensemble spiking activity and the plasticity of the individual neurons (Brown et al. 2004).

Signal Processing Algorithms for Neural Spike Train Data Analysis

Neural spike train decoding algorithms are commonly used with signal processing methods for analyzing how neural systems represent biological signals (Brown et al. 1998; Barbieri et al. 2004; Brockwell et al. 2004; Wu et al. 2004). These algorithms use a wide range of methods to estimate from neural spiking activity the most likely value of a biological signal. More recently, these algorithms have been one of several strategies used to design controls for neural prosthetic devices and other brain-machine interfaces (Chapin et al. 1999; Wessberg et al. 2000; Donoghue 2002; Serruya et al. 2002; Taylor et al. 2002; Mussallam et al. 2004). Developing optimal strategies for constructing and testing decoding algorithms is therefore an important challenge in computational neuroscience. Similarly, adaptive filtering signal processing algorithms offer an approach to analyzing the dynamics of neural receptive fields (Brown et al. 2001; Frank et al. 2002, 2004; Wirth et al. 2003; Eden et al. 2004). Both the neural spike train decoding algorithms and the adaptive filtering algorithms can be constructed using the state-space modeling paradigm. State-space modeling is an established framework used to study dynamical processes in engineering, computer science, and statistics (Kitagawa and Gersh 1996; Smith and Brown 2003).

GOALS OF THE CHAPTER

In this chapter, we derive a class of broadly applicable neural spike train decoding algorithms and adaptive filter algorithms using the state-space modeling paradigm. This approach provides a unified framework for analyzing both neural ensemble representations of biological signals and neural plasticity. First, we review the relevant neurophysiology of the rat hippocampus. Second, we review the state-space paradigm and derive the neural spike train decoding algorithms. Third, we illustrate the decoding algorithm in the analysis of position decoding from ensemble neural spiking activity of simultaneously recorded rat hippocampal neurons. Fourth, we show how the adaptive filter algorithm can be derived as a special case of the decoding algorithm, and we illustrate its use with an analysis of the temporal dynamics of hippocampal place receptive fields. In the final section, we present a summary and discuss gaps and opportunities.

DYNAMIC PROPERTIES OF RAT HIPPOCAMPAL NEURONS

The Hippocampus: A Model for Neural Information Processing

The rat hippocampus is an experimental system widely used to study how neural systems represent information. Located in the temporal lobes, the hippocampus is a bilateral region critical for the formation and storage of both short- and long-term memories (Scoville and Milner 1957; Squire 1982; Cohen and Eichenbaum 1993; Rudy and Sutherland 1995). It is now appreciated that, within the rat hippocampus, specialized pyramidal neurons known as place cells tune their receptive field properties to the animal's current environment thereby allowing the animal to develop a spatial representation (O'Keefe and Dostrovsky 1971). That is, as an animal moves through its environment, each one of a subset of the hippocampal place cells demarcates, within a few minutes, its own region in the environment by firing spikes only when the animal is within that region. The region of the environment in which the cell fires is termed its place field. Large numbers of hippocampal place cells tile each environment with overlapping place fields. The recorded shapes of these fields depend on the structure of the environment and the task the animal is required to perform. When the animal forages randomly in an open environment, these fields resemble approximately two-dimensional Gaussian surfaces (Fig. 1), whereas when the animal runs on a linear track, the fields resemble a one-dimensional Gaussian curve. The hippocampal θ rhythm with its frequency range of 6–14 Hz, the animal's running velocity, and its direction of motion are additional factors known to modulate place cell spiking activity (O'Keefe and Reece 1993; Wilson and McNaughton 1993). The spatial information carried by hippocampal place cells is believed to be an important component of the rat's spatial navigation system (McNaughton et al. 1996).

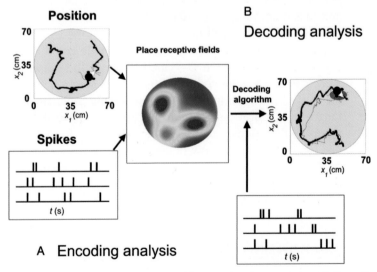

FIGURE 1. Decoding of position from rat ensemble neural spiking activity. (A) Encoding analysis in which the relationship between the biological signal (trajectory of the rat in the environment, *solid black line in the Position panel*) and spiking activity (*Spikes panel*) is estimated as place receptive fields for three neurons. (B) Decoding analysis in which the estimated place receptive fields are used in the point process filter decoding algorithm to compute the predicted position (*thin black line*) of the rat in the environment from new spiking activity of the neural ensemble recorded during the decoding stage. The predicted position is compared with the observed position (*thick black line*) during the decoding stage. The blue oval defines a 95% confidence region centered at that location. (Reprinted, with permission, from Brown et al. 2004.)

Multiunit Electrode Arrays for Recording Ensemble Neural Activity

During the past 12 years, the development of multiunit electrode arrays that allow the simultaneous recording of a large number of individual neurons represents an important experimental advance in neuroscience (Wilson and McNaughton 1993; Buzsaki 2004). Once implanted, a multiunit electrode array can be used in some cases to study population neural activity for up to 1 year. The technical capability to simultaneously record the spiking activity of large numbers of hippocampal place cells—between 20 and 100—along with the animal's position in its environment has made feasible formal quantitative study of how rats represent spatial information in short-term memory. We show below how the state-space paradigm may be used to analyze the representation of the animal's position in its environment maintained by the ensemble spiking activity of the hippocampal place cell neurons recorded from multiunit electrode arrays.

Dynamics of Hippocampal Place Receptive Fields

Once formed, the place receptive fields of hippocampal neurons are not static. That is, rat hippocampal neurons have been shown to change their spatial receptive fields as the animal executes either a novel or familiar behavioral task (Mehta et al. 2000; Frank et al. 2002, 2004). For example, when the experimental environment is a linear track, these spatial receptive fields have been shown to migrate in the direction opposite to the cell's preferred direction of firing relative to the animal's movement, and to increase in scale and maximum firing rate (Mehta et al. 2000). This behavior can be seen in the spiking activity of the place cell neuron recorded on the U-shaped track shown in Figure 2. To display all of the experimental data on a single graph in space and time, we show a linear representation of the U-shaped track. At the outset of the experiment, the spiking activity is sparse, centered near 90 cm, and ranges between 75 and 100 cm, whereas by the end of the experiment, it is appreciably more intense, centered near 70 cm, and ranges between 50 and 90 cm. Because receptive field plasticity is a characteristic of many neural systems, analysis of these dynamics from experimental measurements is crucial for understanding how different brain regions learn and adapt their representations of relevant biological information. Therefore, we show below how the state-space paradigm may be used to track the dynamics of hippocampal place receptive field properties on a millisecond timescale.

THE STATE-SPACE MODELING PARADIGM

The State Equation

We assume that neural representations can be studied with the state-space framework (Kitagawa and Gersh 1996; Smith and Brown 2003). The state-space model consists of two equations: a state equation and an observation equation. The state equation defines an unobservable process whose evolution is tracked across time. Such state models with unobservable processes are often referred to as hidden Markov or latent process models

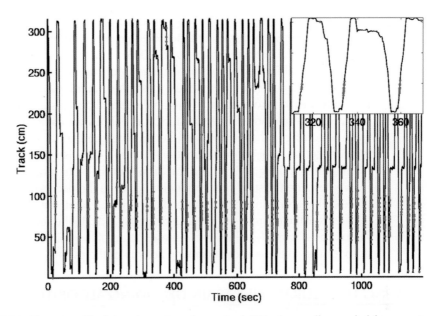

FIGURE 2. Place-specific firing dynamics of an actual CA1 place cell recorded from a rat running back and forth on a 300-cm U-shaped track for 1200 sec. The track was linearized to display the entire experiment in a single graph. The vertical lines show the animal's position and the red dots indicate the time at which a spike was recorded. The inset is an enlargement of the display from 320 to 360 sec to show the cell's unidirectional firing, i.e., spiking only when the animal runs from the bottom to the top of the track. (Reprinted, with permission, from Brown et al. 2001 [©National Academy of Sciences, U.S.A.].)

(Fahrmeir and Tutz 2001; Smith and Brown 2003). In the analyses we present, the state process represents either the biological signal or the parameters of the neural receptive field. For an ensemble of rat hippocampal neurons, the state process would be the position of the animal in its environment (Barbieri et al. 2004). In the case of a hippocampal neuron whose receptive field is changing with time, the state equation defines the dynamics of the parameters in the model describing the receptive field (Brown et al. 2001; Eden et al. 2004). In other examples, for an ensemble of cat lateral geniculate neurons, the state equation could be the natural image being projected onto the animal's visual field (Stanley et al. 1999), whereas for monkey M1 neurons, it could be the velocity profile of an intended hand movement (Wu et al. 2004).

The Observation Equation

The observation equation completes the state-space model paradigm and defines how the observed data relate to the unobservable state process. In all of the cases that we consider, the observed data from the neurophysiological experiments represent neural spike trains. For these analyses, we consider the neural spike trains as point processes, time series that take on the values of 0 or 1 in continuous time (Brown et al. 2003; Daley and

Vere-Jones 2003; Brown 2005). For a neural spike train modeled as a point process, the relationship between the spiking activity and the state process is defined by the conditional intensity function (Brown et al. 2003; Daley and Vere-Jones 2003; Brown 2005). The conditional intensity function is the rate function of each neuron. Like the rate function of an inhomogeneous Poisson process, the conditional intensity function can depend on a time-varying process that in our analyses will be the state process. For any small time interval, the conditional intensity function defines for each neuron the probability of a spike occurring in that small time interval. The conditional intensity function generalizes the concept of a Poisson rate function in that it can also be dependent on the history of the neurons' spiking activity and the history of the state process.

We describe the state-space model in terms of the decoding problem for hippocampal place cell neurons. In the first part of the analysis, termed the encoding analysis, the receptive fields of the neurons are characterized (Fig. 1A). In the second part, the decoding algorithm is used to analyze how the representation of the animal's position in its environment is performed in the ensemble spiking activity of the pyramidal neurons (Fig. 1B).

FORMULATING THE NEURAL SPIKE TRAIN DECODING ALGORITHM

We denote the decoding interval as $(0, T]$. We define the updating lattice by choosing K large and dividing $(0, T]$ into K intervals of equal width $\Delta = K^{-1}T$. In the decoding analysis, we choose $\Delta = 3.3$ msec, whereas in the neural receptive field analysis, we choose $\Delta = 1$ msec.

State and Observation Equations

To define the state-space model, we assume that the position of the rat during foraging obeys a bivariate first-order linear Gaussian autoregressive $AR(1)$ model defined as

$$x_k = \mu_x + Fx_{k-1} + \varepsilon_k \tag{1}$$

where $x_k = (x_1(k\Delta), x_2(k\Delta))$ is the animal's position at time $k\Delta$, F is a 2 x 2 matrix of system parameters, μ_x is a 2 x 1 vector of mean parameters, and ε_k is a 2 x 1 Gaussian random variable with mean zero and a 2 x 2 covariance matrix W_ε. The ε_k's are assumed to be independent.

The observation model for the ensemble neural spike train is the discretized version of the point process joint probability density defined as (Brown 2005)

$$p(n_k \mid x_k) = \exp\left(\sum_{c=1}^{C}[n_k^c \log \lambda_k^c - \lambda_k^c \Delta]\right) \tag{2}$$

where n_k^c is the number of spikes in the interval $((k-1)\Delta, k\Delta]$, from neurons $c = 1,...,C$; λ_k^c is the conditional intensity function of neuron c at time $k\Delta$; and $n_k = (n_k^1,...,n_k^c)$. For each problem, we give the explicit form of the conditional intensity function. We let $N_k = [n_1, ..., n_k]$ be the sequences of spike train observations in the interval $(0, k\Delta]$ for $k = 1,...,K$. Equation (2) assumes that the spiking activities of the individual neurons are independent given the value of the position at time $k\Delta$. In the subsequent discussions, we denote time $k\Delta$ as time k.

The Bayes' Rule and Chapman-Kolmogorov Equations

The objective of our state-space modeling analysis is to construct a recursive filter algorithm to estimate the state x_k at time k from the spiking activity N_k. The standard approach to deriving such a filter is to define recursively the probability density of the state given the observations. For the state model defined in Eq. (1) and the observation process defined in Eq. (2), the Bayes' rule formula for the posterior probability density of the state-given spike train observations is (Brown et al. 1998; Barbieri et al. 2004; Eden et al. 2004)

$$p(x_k \mid N_k) = \frac{p(x_k \mid N_{k-1})p(n_k \mid x_k)}{p(n_k \mid N_{k-1})} \tag{3}$$

and the associated one-step prediction probability density or Chapman-Kolmogorov equation is

$$p(x_k \mid N_{k-1}) = \int p(x_{k-1} \mid N_{k-1})p(x_k \mid x_{k-1})dx_{k-1} \tag{4}$$

Before presenting the explicit form of the recursive filter algorithm, we explain the logic behind Eqs. (3) and (4) (Fig. 3). The first term on the right side of Eq. (3), $p(x_k \mid N_{k-1})$, is the one-step prediction probability density from Eq. (4). Before recording the spike trains at time k, it defines the predictions of the most likely values of the state at time k given the observations up through time $k-1$. Equation (4) predicts the most likely values of the state at time k by integrating ("averaging") over the most likely values of the state at time $k-1$ given the data up through $k-1$ defined by $p(x_{k-1} \mid N_{k-1})$ and the most likely set of changes in state between times $k-1$ to k defined by $p(x_k \mid x_{k-1})$. The most likely value of the state at time $k-1$ given the data up through $k-1$, $p(x_{k-1} \mid N_{k-1})$ is the first term of the integrand in Eq. (4), and this value defines the posterior probability density at $k-1$. The most likely change in state from time $k-1$ to k, $p(x_k \mid x_{k-1})$, is defined by the autoregressive probabil-

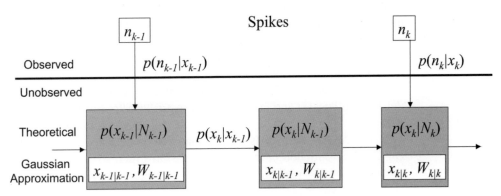

FIGURE 3. Schematic representation of the state-space paradigm showing the relationship between the observation process $p(n_k \mid x_k)$ (Eq. [2]), the state process $p(x_k \mid x_{k-1})$ defined by Eq. (1), the one-step prediction $p(x_k \mid N_{k-1})$ (Eq. [4]), its Gaussian approximation from the point process filter algorithm with mean $x_{k \mid k-1}$ (Eq. [5]) and variance $W_{k \mid k-1}$ (Eq. [6]), and the posterior probability density $p(x_k \mid N_k)$ (Eq. [3]) and its Gaussian approximation with mean $x_{k \mid k}$ (Eq. [7]) and variance $W_{k \mid k}$ (Eq. [8]) from the point process filter algorithm.

ity model for the state in Eq. (1). The second term on the right-hand side of Eq. (3), $p(n_k \mid x_k)$, is the point process probability mass function at time k given the state x_k defined in Eq. (2). The probability density $p(n_k \mid N_{k-1})$ in the denominator of Eq. (3) is the integral of the numerator and defines the normalizing constant, which ensures that the posterior probability density integrates to 1.

Together, Eqs. (3) and (4) define a recursion that can be used to compute the probability of the state given the observations (Fig. 3). The formulae in Eqs. (3) and (4) are recursive because Eq. (4) uses the posterior probability density at time $k - 1$, $p(x_{k-1} \mid N_{k-1})$, to generate the one-step prediction probability density at time k, $p(x_k \mid N_{k-1})$, which in turn, makes it possible to compute the new posterior probability density at time k, $p(x_k \mid N_k)$, given in Eq. (3). In other words, at each time k, the recursion relations fuse information from two sources (Fig. 3). The first is the prediction of the state at time k given the model, N_{k-1}, and the observations up through time $k - 1$, as given by Eq. (4), and the second is the spike train's observations recorded in the interval $((k - 1)\Delta, k\Delta]$ as given by Eq. (2).

Evaluating the Bayes' Rule and Chapman-Kolmogorov Equations

Given the state model in Eq. (1) and the observation process in Eq. (2), the challenge of estimating the unobserved states from the observed neural spike trains is simply the computational problem of evaluating Eqs. (3) and (4). For systems with low-dimensional state and observation models, Eqs. (3) and (4) can be evaluated numerically (Kitagawa and Gersh 1996). As these dimensions increase, numerical computation becomes less feasible. A standard approach, and the one we apply here, is to compute Gaussian approximations to Eqs. (3) and (4) (Brown et al. 1998; Barbieri et al. 2004; Eden et al. 2004). This approach, also termed *maximum a posteriori estimation*, amounts to finding the maximum and the second derivative of Eq. (3) as a function of the state x_k. The maximum corresponds to the mode or mean of the posterior density (Eq. [3]) and the curvature defines its covariance matrix. These two quantities are sufficient to compute the Gaussian approximation to Eq. (3) because a Gaussian probability density is defined completely by its mean and covariance matrix (Fig. 3).

Assuming that a Gaussian approximation to Eq. (3) was computed at time $k - 1$, then the integral in Eq. (4) can be computed analytically using standard formulae for computing the mean and covariance of the sum of two Gaussian random variables (Barbieri et al. 2004). Under the Gaussian approximation, the recursion defined by the probability densities in Eqs. (3) and (4) becomes a recursive filter because it simplifies to computing recursively just the means and covariance of these probability densities. In the special case that the state process is a linear Gaussian system and the observation model is a linear Gaussian function of the state process, this recursive computation of the means and covariances is the well-known Kalman filter algorithm (Kitagawa and Gersh 1996).

Neural Spike Train Decoding Algorithm

Under the assumption that the neurons are conditionally independent, the solutions to these equations using Gaussian approximations to Eqs. (3) and (4) for the state model in

Eq. (1) and the point process observation model in Eq. (2) yield the following recursive filter algorithm (Barbieri et al. 2004):

One-step prediction: $x_{k\,|\,k-1} = \mu_x + Fx_{k-1\,|\,k-1}$ (5)

One-step prediction variance: $W_{k\,|\,k-1} = FW_{k-1\,|\,k-1}F' + W_\varepsilon$ (6)

Posterior mode: $x_{k\,|\,k} = x_{k\,|\,k-1} + W_{k\,|\,k-1} \sum_{c=1}^{C} \nabla \log \lambda_k^c [n_k^c - \lambda_k^c \Delta]$ (7)

Posterior variance: $W_{k\,|\,k-1}^{-1} = \left[W_{k\,|\,k-1}^{-1} - \sum_{c=1}^{C} \left[\nabla^2 \log \lambda_k^c [n_k^c - \lambda_k^c \Delta] - \nabla \log \lambda_k^c [\nabla \lambda_k^c \Delta]' \right] \right]$ (8)

for $k = 1,\ldots,K$, where λ_k^c is the conditional intensity function for neuron c; F and W_ε are, respectively, the transition matrix and the white noise covariance matrix for the $AR(1)$ model in Eq. (1); ∇ (∇^2) denotes the first (second) derivative of the indicated function with respect to x_k; and the notation $j|k$ denotes the estimate at time j given the spiking activity in the interval $(0,k\Delta]$.

We termed the algorithm in Eqs. (5)–(8) a point process filter because it takes point process observations as inputs at each time k, which, in this case, are the neural spike train observations n_k, and it provides a recursive system of Gaussian approximations with which to compute Eqs. (3) and (4) (Fig. 3). Equations (7) and (8) give, respectively, the mean and covariance matrix for the Gaussian approximation to $p(x_k|N_k)$ (Eq. [3]), whereas Eqs. (5) and (6) give, respectively, the mean and covariance matrix for the Gaussian approximation to $p(x_k|N_{k-1})$ (Eq. [4]). As mentioned above, because we compute Gaussian approximations to $p(x_k|N_k)$ and $p(x_k|N_{k-1})$, it suffices to compute recursively only the respective means and covariance matrices (Fig. 3).

Properties of the Algorithm

The position update $x_{k|k}$ combines the one-step prediction $x_{k|k-1}$, on the basis of the ensemble spiking activity from $(0,(k-1)\Delta]$, with the weighted sum of $[n_k^c - \lambda_k^c \Delta]$, the innovation or error signal from neuron c for $c = 1,\ldots,C$ multiplied by the one-step prediction variance (Barbieri et al. 2004). The innovations are the new information from the ensemble spiking activity in $((k-1)\Delta, k\Delta]$. As is true for a Poisson process, $\lambda_k^c \Delta$ defines for a general point process model the probability of a spike in $((k-1)\Delta, k\Delta]$ (Brown et al. 2003; Daley and Vere-Jones 2003; Brown 2005). Each innovation compares the probability of a spike, $\lambda_k^c \Delta$, in $((k-1)\Delta, k\Delta]$ with n_k^c, which, for small Δ, is 1 if a spike is observed from neuron c in $((k-1)\Delta, k\Delta]$ and 0 otherwise. Thus, for small Δ, the innovation gives a weight in the interval $(-1,1)$. A large positive weight results if a spike occurs when a neuron has a low probability of spiking in $((k-1)\Delta, k\Delta]$, whereas a large negative weight arises if no spike occurs in $((k-1)\Delta, k\Delta]$ when the probability of a spike is high. In this way, the algorithm makes use of the times when a neuron either does or does not fire. The algorithm in Eqs. (7) and (8) is nonlinear and must be solved using Newton's method because the right- and left-hand sides of Eq. (7) depend on $x_{k|k}$.

DYNAMIC ANALYSIS OF POSITION REPRESENTATION IN ENSEMBLE HIPPOCAMPAL SPIKING ACTIVITY

Experimental Protocol

We applied the recursive filter paradigm to place cell spike train and position data recorded from a Long-Evans rat freely foraging in a circular environment 70 cm in diameter with 30-cm-high walls and a visual cue maintained in a fixed location. A multiunit electrode array was implanted into the CA1 region of the animal's hippocampus. The simultaneous activity of 34 place cells was recorded from the electrode array while the animal foraged in the environment for 25 minutes. Simultaneously with the recording of the place cell activity, the position of the animal was measured at 30 Hz by a camera tracking the location of two infrared diodes mounted on the animal's headstage (Brown et al. 1998; Barbieri et al. 2004).

Encoding Analysis

We modeled the spatial receptive field of each neuron by representing the conditional intensity function as a linear combination of Zernike polynomials defined as

$$\lambda_k^c = \exp\left\{\sum_{\ell=0}^{L}\sum_{m=-\ell}^{\ell}\theta_{\ell,m}^c z_\ell^m(\rho(k\Delta),\phi(k\Delta))\right\} \tag{9}$$

where $z_{\ell,m}$ is the mth component of the ℓth-order Zernike polynomial, $\theta_{\ell,m}^c$ is the associated coefficient, $\rho(k\Delta) = r^{-1}\left[(x_1(k\Delta) - \eta_1)^2 + (x_2(k\Delta) - \eta_2)^2\right]^{1/2}$, $\phi(k\Delta) = \tan^{-1}\left[(x_2(k\Delta) - \eta_2)(x_1(k\Delta) - \eta_1)^{-1}\right]$, (η_1,η_2) are the coordinates of the center of the circular environment, r is the radius of the circular environment, $\eta_1 = \eta_2 = r = 35$ cm, and $L = 3$. The Zernike polynomials are orthogonal polynomials on the disk and provide a mathematically efficient way to represent the range of shapes taken on by the spatial receptive fields. The order-3 model has ten nonzero coefficients. We estimated the spatial receptive field of each neuron by fitting the Zernike model to the spike train data of each neuron from the first 15 minutes of the experiment. That is, we used maximum likelihood to estimate the parameters $\theta_{\ell,m}^c$ for each neuron. The Zernike place field estimates, shown in Fig. 4, covered nearly the entire environment (Barbieri et al. 2004).

Decoding Analysis

We estimated the F and W_ε matrices (Eq. [1]) by maximum likelihood from the first 15 minutes of the path data in the encoding analysis. We decoded position from the ensemble spiking activity of the 34 neurons from the last 10 minutes of foraging with updates computed at $\Delta = 3.3$ msec. A representative example of the algorithm's performance is shown in Figure 5. In each of the four continuous 15-second segments, the estimated trajectories (Eq. [7]) (red) closely resembled the true (black) trajectories. The instantaneous 95% confidence ellipses were computed from Eq. (7) and the covariance matrix in Eq.

Place Receptive Fields

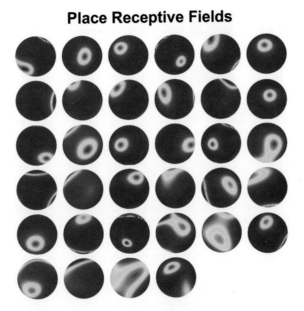

FIGURE 4. Pseudocolor maps of the spatial receptive field estimates from the Zernike models for the 34 place cells. (Adapted, with permission, from Barbieri et al. 2004 [©Massachusetts Institute of Technology].)

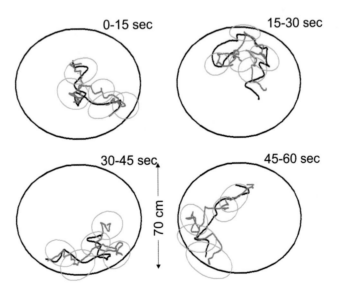

FIGURE 5. Continuous 60-sec segment of the true path (*black*) displayed in four 15-sec intervals with the estimated path (*red*) using the decoding algorithm in Eqs. (5)–(8) with the Zernike place field model (Eq. [9]) and the $AR(1)$ path model in Eq. (1).

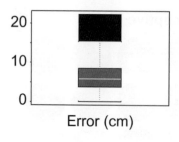

FIGURE 6. Box plot summary of decoding error for the Zernike model distribution (error = true position – estimated position). The lower border of the box is the 25th percentile of the distribution and the upper border is the 75th percentile. The white bar within the box is the median of distribution. The distance between the 25th and 75th percentiles is the interquartile range (IQR). The lower (upper) whisker is at 1.5 x the IQR below (above) the 25th (75th) percentile. All of the black bars above the whiskers are far outliers. (Adapted, with permission, from Barbieri et al. 2004 [©Massachusetts Institute of Technology].)

(8). The performance of the algorithm in decoding the spike train data can be best appreciated in the videos of this analysis on our Web site http://neurostat.mgh.harvard.edu/.

To quantify further the performance of the decoding algorithm, we computed the decoding error as the difference between the true and estimated path at each 33-msec time step for the 10 minutes of decoding. The box plot summaries of this error distribution are shown in Figure 6. The median decoding error for the Zernike model was 5.9 cm, with a 25th percentile of 3.9 cm, a 75th percentile of 8.6 cm, a minimum of 0.8 cm, and a maximum of 22.3 cm. The Zernike model was highly accurate near the borders of the environment (Fig. 6) because, by Eq. (5), the support of the Zernike model is restricted to the circle. The median error of 5.9 cm is an improvement of 2 cm from the 7.9 cm using spatial Gaussian functions to model the place receptive fields (Brown et al. 1998). Of this improvement, 1.2 cm was attributable to optimizing the learning rate, whereas 0.7 cm was because of the switch from the spatial Gaussian to the Zernike model. These results show that the CA1 region of the hippocampus maintains a dynamic representation of the animal's position in its environment. They also illustrate how the state-space modeling paradigm may be used to analyze how well a rat's internal representation of its position in its environment may be recovered from ensemble spiking activity of approximately 30 hippocampal neurons.

POINT PROCESS FILTER ALGORITHM FOR TRACKING HIPPOCAMPAL PLACE RECEPTIVE FIELD DYNAMICS

Steepest-descent Point Process Filter Algorithm

We can use the state-space paradigm to devise an algorithm to track the dynamics of a neuron's receptive field. The general formulation of this state-space algorithm for tracking neural plasticity is given in Eden et al. (2004). Here, we illustrate the idea with the special case of the steepest-descent point process filter (Brown et al. 2001). To derive the steepest-descent algorithm from the point process filter algorithm in Eqs. (5)–(8), we take the $C = 1$ neuron and set $F = I$, $W_\varepsilon = 0$, and $W_{k|k-1} = D(\varepsilon)$, where $D(\varepsilon)$ is a diagonal matrix with possibly distinct constant value ε_i on each ith diagonal. In this case, the point process filter algorithm simplifies to a single formula in which Eq. (7) becomes

$$\theta_{k|k} = \theta_{k-1|k-1} + D(\varepsilon)\nabla \log \lambda_{k-1}[n_k - \lambda_{k-1} \Delta] \tag{10}$$

where θ is the parameter of the conditional intensity function that is evolving in time. In the previous example, θ was the vector of coefficients in the Zernike polynomials. Because in our paradigm the conditional intensity function defines the receptive field of the neuron, and because θ parametrizes the conditional intensity function, tracking the evolution of θ through time allows us to track the temporal evolution of the receptive field. We term Eq. (10) the steepest-descent point process adaptive filter because it can also be derived using a steepest-descent argument applied to an appropriate local criterion function as is true for continuous-valued measurements (Brown et al. 2001). Unlike Eq. (7), in Eq. (10), the conditional intensity function is evaluated at time $k - 1$ instead of time k, making Eq. (10) a linear algorithm.

To illustrate the steepest-descent point process filter algorithm, we consider a spike train from a pyramidal cell in the CA1 region of the rat hippocampus recorded while the animal ran back and forth on a U-shaped track (Fig. 2). As stated above, the spatial receptive fields of hippocampal neurons have well-documented dynamic properties when an animal executes either a familiar or novel behavioral task (Mehta et al. 2000). In our analysis, we consider only the dynamic properties of place field position and shape, not other factors such as θ rhythm oscillations, and the interaction between the θ rhythm and position called phase precession (O'Keefe and Reece 1993). If $x(k\Delta)$ is the animal's position at time k, then we define the conditional intensity function for the place field model as

$$\lambda(k\Delta|\theta) = \exp\{\alpha - (2\sigma^2)^{-1} (x(k\Delta) - \mu)^2\} \tag{11}$$

where μ is the center of the spatial receptive field, σ is a scale factor, and $\exp\{\alpha\}$ is the neuron's maximum firing rate that occurs at the center of the receptive field. For this model, $\theta = (\alpha,\sigma,\mu)'$ is the three-dimensional parameter vector. From Eqs. (10) and (11), the steepest-descent point process filter algorithm at time k is

$$\theta_{k|k} = \theta_{k-1|k-1} - \begin{bmatrix} \varepsilon_\alpha \\ \varepsilon_\sigma(\sigma^{-3}_{k-1|k-1})(x(k\Delta) - \mu_{k-1|k-1})^2 \\ \varepsilon_\mu \sigma^{-2}_{k-1|k-1}(x(k\Delta) - \mu_{k-1|k-1}) \end{bmatrix} (n_k - \lambda((k - 1)\Delta|\theta_{k-1|k-1})) \tag{12}$$

where ε_α, ε_σ, and ε_μ are, respectively, the diagonal elements of $D(\varepsilon)$ for α, σ, and μ, and $k = 1,\ldots,K$. The elements of $D(\varepsilon)$ are the learning rate parameters and control how much the innovations, i.e., the new spiking information, are weighted.

Experimental Protocol

We applied the steepest-descent point process filter algorithm to an actual place cell spike train recorded from a Long-Evans rat running back and forth for 1200 sec on a 300-cm U-shaped track (Fig. 2) (Brown et al. 2001). Multiple single units as well as the position of the animal on the track were recorded as described in the previous example. The actual trajectory was at times irregular because the animal started and stopped several times, and in two instances (50 and 650 sec), turned around shortly after it initiated its run. On several of the upward passes, particularly in the latter part of the experiment, the animal slowed as it approached the curve in the U-shaped track at approximately 150 cm. The strong place-specific firing of the neuron is easily visible because the spiking activity occurred almost exclusively between 50 and 100 cm. The spiking activity of the neuron was entirely unidirectional because the cell discharged only as the animal ran up and not down the track (Fig. 2, inset).

Tracking Hippocampal Spatial Receptive Field Dynamics

We applied the steepest-descent point process filter algorithm (Eq. [12]) to these spike train data, updating the estimates every 1 msec. We used the learning rate parameters, $D(\varepsilon)$, chosen in Brown et al. (2001). The starting estimates for the parameter θ were the maximum likelihood estimates computed from the first 50 spikes (~200 sec). The time course of $\exp(\alpha)$, the maximum spike rate, showed a steady increase from 3 to almost 30 spikes/sec during the 1200 sec of the experiment (Fig. 7A). The increase was apparent in the spike train data in Figure 2. The scale parameter σ showed the greatest fluctuations during the experiment; it rose during the first 500 sec from 10 to 16 cm, and fluctuated between 15 and 16 cm for the balance of the experiment (Fig. 7B). This fluctuation in scale was also easily visible in the spike train data in Fig. 2. The place cell center migrated during the first 700 sec from 85 to 65 cm and stayed near 65 cm for the remainder of the experiment (Fig. 7C).

We illustrated the evolution of the entire field in Figure 8 by plotting the instantaneous place field estimates at 300, 550, 800, and 1150 sec. The sequence of place field estimates showed the temporal evolution of the cell's spatial receptive field. In contrast, the

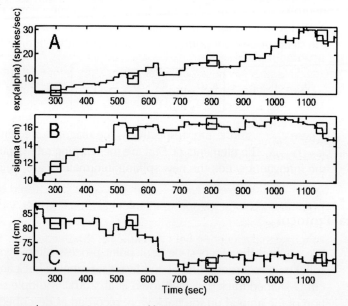

FIGURE 7. Steepest-descent point process filter estimates of the trajectories: (A) Maximum spike rate, $\exp(\alpha)$; (B) place field scale, σ; and (C) place field center, μ. The estimates were updated at 1-msec intervals. The squares in the panels at 300, 550, 800, and 1150 sec are the times at which the place fields are displayed in Fig. 8. The growth of the maximum spike rate (A), the variability of the place field scale (B), and the migration of the place field center (C) are all readily visible. (Reprinted, with permission, from Brown et al. 2001 [©National Academy of Sciences, U.S.A.].)

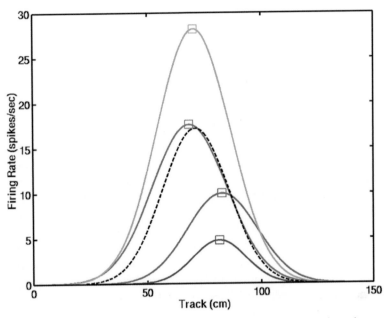

FIGURE 8. Estimated place fields at times 300 (*blue*), 550 (*green*), 800 (*red*), and 1150 (*aqua*) sec from the steepest-descent point process algorithm. The black dashed line is the maximum likelihood estimate of the place field obtained by using all of the spikes in the experiment. The maximum likelihood estimate ignores the temporal evolution of the place field. (Reprinted, with permission, from Brown et al. 2001 [©National Academy of Sciences, U.S.A.].)

maximum likelihood estimate based on the entire 1200 sec of data obscured the temporal dynamics by overestimating (underestimating) the place field's spatial extent, and underestimating (overestimating) its maximum firing rate at the experiment's beginning (end). This example illustrates that the dynamic of actual place cell receptive fields can be tracked instantaneously from recorded spiking activity. The migration of the center as well as the growth in scale and in maximum firing rate were consistent with previous reports (Mehta et al. 2000). The current algorithm could not track the skewing of the place fields because the Gaussian model is symmetric. This feature as well as more complex spatial and temporal dynamics have been well captured by the spline extensions of this point process filter algorithm (Frank et al. 2002, 2004). Video presentations of our analyses can be found in the supplemental information at www.pnas.org along with analyses of the spatial receptive field dynamics of three other CA1 hippocampal neurons.

SUMMARY

We have presented a state-space modeling approach for analyzing neural ensemble representations of biological signals and for analyzing neural receptive field plasticity. These algorithms are part of the paradigm that we have been developing to conduct state estimation from point process measurements (Smith and Brown 2003; Barbieri et al. 2004; Eden et al. 2004). We previously compared our state-space paradigm with several other approaches including the widely used reverse correlation algorithm (Brown et al. 1998; Barbieri et al. 2004). The reverse correlation algorithm, although simple to implement, performed significantly worse than our algorithm, because it made no use of the spatial and temporal structure in this decoding problem. This finding suggests that application of our algorithm to other decoding and neural prosthetic control problems where the reverse correlation methods have performed successfully (Bialek et al. 1991; Stanley et al. 1999; Wessberg et al. 2000; Serruya et al. 2002) could yield important improvements. Furthermore, our state-space approach to estimating receptive field dynamics offers a way of tracking their temporal evolution on a millisecond timescale (Brown et al. 2001; Frank et al. 2002, 2004; Eden et al. 2004).

GAPS AND OPPORTUNITIES

Further technical improvements in our algorithm can be readily incorporated into the current framework. These include combining the Zernike spatial model with models of known hippocampal temporal dynamics such as bursting (Quirk and Wilson 1999), θ phase modulation (Skaggs et al. 1996; Brown et al. 1998), and phase precession (O'Keefe and Reece 1993). Next, we can consider both higher-order linear state-space models (Shoham 2001; Smith and Brown 2003; Wu et al. 2004) and nonlinear state-space models to describe path dynamics more accurately (Kitagawa and Gersh 1996). Finally, we have developed Gaussian approximations to evaluate the Bayes' rule and Chapman-Kolmogorov equations (Eqs. [3] and [4]). We can improve our algorithm and evaluate the accuracy of this approximation by applying non-Gaussian approximations (Pawitan 2001) and exact numerical integration methods (Kitagawa and Gersh 1996) and/or Monte Carlo techniques (Doucet et al. 2001; Shoham 2001; Brockwell et al. 2004) to evaluate these equations.

We foresee several future applications of the state-space modeling paradigm. First, we will continue to use this modeling approach to study ensemble neural representations in the hippocampus. More detailed models that take account of bursting, θ phase precession, and the spiking history of the ensemble should allow us to gain more insight into how information in these ensembles of neurons is being represented (Okatan et al. 2005). Second, as mentioned above, the state-space approach provides a natural framework for developing algorithms to control brain-machine interfaces and neural prosthetic devices. The appropriate application of these algorithms to these problems will depend critically on the nature of the encoding model used to describe the relationship between neural spiking activity and the signal being represented, as well as the kinematic model for the movement (Eden et al. 2004; Wu et al. 2004; Truccolo et al. 2005). Third, we plan to further develop our state-space models to study receptive field dynamics (Frank et al. 2002, 2004). In particular, our approach will allow us to study the specific dynamics of hippocampal place fields when they form for the first time as an animal learns about a novel environment. Finally, our state-space framework has led to a new approach for characterizing learning in behavioral experiments and for relating it to simultaneously recorded changes in neural activity (Wirth et al. 2003; Smith et al. 2004). This work has allowed us to identify specific neurons in the monkey hippocampus that change their spiking properties in close relation to the time course of the animal's learning of a new behavioral task (Wirth et al. 2003).

All of these applications have direct implications for the study of drug addiction, to the extent that drugs alter neural circuits, and these drug-induced changes may affect the representation of information in the neural circuit and/or the normal plasticity of the neurons comprising the circuit. Depending on the nature of the neurophysiological investigation, these drug-induced impairments may be manifest in the activity of individual neurons, groups of neurons, or in the behavior of the organism. For example, the state-space methods we have presented could be used to analyze differences in the responses of neural circuits with and without a pharmacologic agent to help characterize such drug-induced effects. As a second illustration, we recently showed how our state-space paradigm applied to learning experiments could be used to dynamically quantify the differences in learning between a control group of rats and a second group treated with an NMDA receptor antagonist, MK801 (Smith et al. 2005). This latter application illustrates how our paradigm may be used to characterize dynamics on the timescale of an organism's behavior. For these reasons, we believe that the state-space paradigm offers a rich framework for extracting dynamic information from neurophysiological experimental investigations of both normal and pathological conditions.

ACKNOWLEDGMENTS

This work was supported in part by National Institute on Drug Abuse Grant No. DA-015644 and National Institute of Mental Health Grant Nos. MH-59733 and MH-61637. We are grateful to Julie Scott for technical assistance in preparing this chapter.

REFERENCES

Barbieri R., Frank L.M., Nguyen D.P., Quirk M.C., Solo V., Wilson M.A., and Brown E.N. 2004. Dynamic analyses of information encoding by neural ensembles. *Neural Comput.* **16:** 277–308.

Bialek W., Rieke F., de Ruyter van Stevenick R.R., and Warland D. 1991. Reading a neural code. *Science* **252:** 1854–1857.

Brockwell A.E., Rojas A.L., and Kass R.E. 2004. Recursive Bayesian encoding of motor cortical signals by particle filtering. *J. Neurophysiol.* **91:** 1899–1907.

Brown E.N. 2005. Theory of point processes for neural systems. In *Methods and models in neurophysics* (ed. C.C. Chow et al.), Chap. 14, pp. 691–726. Elsevier, Paris.

Brown E.N., Kass R.E., and Mitra P.P. 2004. Multiple neural spike train data analysis: State-of-the-art and future challenges. *Nat. Neurosci.* **7:** 456–461.

Brown E.N., Barbieri R., Eden U.T., and Frank L.M. 2003. Likelihood methods for neural data analysis. In *Computational neuroscience: A comprehensive approach* (ed. J. Feng), pp. 253–286. CRC, London.

Brown E.N., Frank L.M., Tang D., Quirk M.C., and Wilson M.A. 1998. A statistical paradigm for neural spike train decoding applied to position prediction from ensemble firing patterns of rat hippocampal place cells. *J. Neurosci.* **18:** 7411–7425.

Brown E.N., Nguyen D.P., Frank L.M., Wilson M.A., and Solo V.

2001. An analysis of neural receptive field plasticity by point process adaptive filtering. *Proc. Natl. Acad. Sci.* **98:** 12261–12266.

Chapin J.K., Moxon K.A., Markowitz R.S., and Nicolelis M.A.L. 1999. Real-time control of a robot arm using simultaneously recorded neurons in the motor cortex. *Nat. Neurosci.* **7:** 664–670.

Cohen N.J. and Eichenbaum H. 1993. *Memory, amnesia, and the hippocampal system.* MIT Press, Cambridge.

Daley D. and Vere-Jones D. 2003. *An introduction to the theory of point process,* 2nd edition. Springer-Verlag, New York.

Donoghue J.P. 2002. Connecting cortex to machines: Recent advances in brain interfaces. *Nat. Neurosci.* (suppl.) **5:** 1085–1088.

Doucet A., DeFreitas N., Gordon N., and Smith A. 2001. *Sequential Monte Carlo methods in practice.* Springer, New York.

Eden U.T., Frank L.M., Barbieri R., Solo V., and Brown E.N. 2004. Dynamic analyses of neural encoding by point process adaptive filtering. *Neural Comput.* **16:** 971–998.

Fahrmeir L. and Tutz G. 2001. *Multivariate statistical modeling based on generalized linear models,* 2nd edition. Springer-Verlag, New York.

Frank L.M., Stanley G.B., and Brown E.N. 2004. Hippocampal

plasticity across multiple days of exposure to novel environments. *J. Neurosci.* **24:** 7681–7689.

Frank L.M., Eden U.T., Solo V., Wilson M.A., and Brown E.N. 2002. Contrasting patterns of receptive field plasticity in the hippocampus and the entorhinal cortex: An adaptive filtering approach. *J. Neurosci.* **22:** 3817–3830.

Gandolfo F., Li C., Benda B.J., Schioppa C.P., and Bizzi E. 2000. Cortical correlates of learning in monkeys adapting to a new dynamical environment. *Proc. Natl. Acad. Sci.* **97:** 2259–2263.

Georgopoulos A.P., Kettner R.E., and Schwartz A.B. 1986. Neuronal population coding of movement direction. *Science* **233:** 1416–1419.

Jog M.S., Kubota Y., Connolly C.I., Hillegaart V., and Graybiel A.M. 1999. Building neural representations of habits. *Science* **286:** 1745–1749.

Kaas J.H., Merzenich M.M., and Killackey H.P. 1983. The reorganization of somatosensory cortex following peripheral nerve damage in adult and developing mammals. *Annu. Rev. Neurosci.* **6:** 325–356.

Kitagawa G. and Gersh W. 1996. *Smoothness priors analysis of time series.* Springer-Verlag, New York.

Mehta M.R., Quirk M.C., and Wilson M.A. 2000. Experience-dependent asymmetric shape of hippocampal receptive fields. *Neuron* **25:** 707–715.

Merzenich M.M., Nelson R.J., Stryker M.P., Cyander M.S., Schoppmann A., and Zook J.M. 1984. Somatosensory cortical map changes following digit amputation in adult monkeys. *J. Comput. Neurol.* **224:** 591–605.

Mussallam S., Corneil B.D., Greger B., Scherberger H., and Andersen R.A. 2004. Cognitive control signals for neural prosthetics. *Science* **305:** 258–262.

Okatan M., Wilson M.A., and Brown E.N. 2005. Analyzing functional connectivity using a network likelihood model of ensemble neural spiking activity. *Neural Comput.* **17:** 1927–1961.

O'Keefe J. and Dostrovsky J. 1971. The hippocampus as a spatial map: Preliminary evidence from unit activity in the freely-moving rat. *Brain Res.* **34:** 171–175.

O'Keefe J. and Reece M.L. 1993. Phase relationship between hippocampal place units and the EEG theta rhythm. *Hippocampus* **3:** 317–330.

Pawitan Y. 2001. *All likelihood: Statistical modelling and inference using likelihood.* Oxford University Press, London.

Quirk M.C. and Wilson M.A. 1999. Interaction between spike waveform classification and temporal sequence detection. *J. Neurosci. Methods* **94:** 41–52.

Rudy J.W. and Sutherland R.J. 1995. Configural association theory and the hippocampal formation: An appraisal and reconfiguration. *Hippocampus* **5:** 375–389.

Scoville W.B. and Milner B. 1957. Loss of recent memory after bilateral hippocampal lesions. *J. Neurol. Neurosurg. Psychiatry* **20:** 11–21.

Serruya M.D., Hatsopoulos N.G., Paninski L., Fellows M.R., and Donoghue J.P. 2002. Instant neural control of a movement signal. *Nature* **416:** 141–142.

Shoham S. 2001. "Advances towards an implantable motor cortical interface." Ph.D. thesis, University of Utah, Salt Lake City.

Skaggs W.E., McNaughton B.L., Wilson M.A., and Barnes C.A. 1996. Theta phase precession in hippocampal neuronal populations and the compression of temporal sequences. *Hippocampus* **6:** 149–172.

Smith A.C. and Brown E.N. 2003. Estimating a state-space model from point process observations. *Neural Comput.* **15:** 965–991.

Smith A.C., Stefani M.R., Moghaddam B., and Brown E.N. 2005. Analysis and design of behavioral experiments to characterize population learning. *J. Neurophysiol.* **93:** 1776–1792.

Smith A.C., Frank L.M., Wirth S., Yanike M., Hu D., Kubota Y., Graybiel A.M., Suzuki W.A., and Brown E.N. 2004. Dynamic analysis of learning in behavioral experiments. *J. Neurosci.* **24:** 447–461.

Squire L.R. 1982. The neuropsychology of human memory. *Annu. Rev. Neurosci.* **5:** 241–273.

Stanley G.B., Li F.F., and Dan Y. 1999. Reconstruction of natural scenes from ensemble responses in the lateral geniculate nucleus. *J. Neurosci.* **19:** 8036–8042.

Taylor D.M., Tillery S.I.H., and Schwartz A.B. 2002. Direct cortical control of 3D neuroprosthetic devices. *Science* **296:** 1829–1832.

Truccolo W., Eden U.T., Fellow M., Donoghue J.P., and Brown E.N. 2005. A point process framework for relating neural spiking activity to spiking history, neural ensemble and covariate effects. *J. Neurophysiol.* **93:** 1074–1089.

Wessberg J., Stambaugh C.R., Kralik J.D., Beck P.D., Laubach M., Chapin J.K., Kim J., Biggs S.J., Srinivasan M.A., and Nicolelis M.A. 2000. Real-time prediction of hand trajectory by ensembles of cortical neurons in primates. *Nature* **408:** 361–365.

Wilson M.A. and McNaughton B.L. 1993. Dynamics of the hippocampal ensemble code for space. *Science* **261:** 1055–1058.

Wirth S., Yanike M., Frank L.M., Smith A.C., Brown E.N., and Suzuki W.A. 2003. Single neurons in the monkey hippocampus and learning of new associations. *Science* **300:** 1578–1581.

Wu W., Black M.J., Mumford D., Gao Y., Bienenstock E., and Donoghue J.P. 2004. Modeling and decoding motor cortical activity using a switching Kalman filter. *IEEE Trans. Biomed. Eng.* **51:** 933–942.

Quantitative Functional Genomics and Proteomics of Drug Abuse

Willard M. Freeman and Kent E. Vrana

Department of Pharmacology, Penn State College of Medicine H078,
Milton S. Hershey Medical Center, Hershey, Pennsylvania 17033-0850

ABSTRACT

Addiction research has a long history of investigation into the molecular-biological and biochemical bases of behavior. For many years, specific genes or proteins have been examined for their quantitative relationship to drug abuse behaviors such as abuse susceptibility, locomotor sensitization, and relapse liability. This basic approach remains unchanged. The change that warrants new terms such as functional genomics and proteomics is the scale of the research. New technologies allow investigators to examine the quantitative levels of tens of thousands of genes and thousands of proteins instead of just one or two. This is an exciting development and promises to give investigators a panoramic view into the workings of the cell to replace the old tunnel vision. This review examines the technologies used in quantitative functional genomic and proteomic studies, and the analysis of data from these large studies, and gives examples of their application to drug abuse research. Quantitative technologies are the focus because they are often most applicable to examining the changes that occur in the brain with drug administration or behavior. Lastly, future developments and technical hurdles in these fields are discussed.

INTRODUCTION

Functional genomics and proteomics have emerged as important new tools for characterizing the biology of substance abuse. Indeed, it is clear that genes and their functional products (mRNA [messenger RNA] and protein) will have a central role in (1) the predisposition to abuse liability and (2) the long-term changes that underlie phenomena such as dependence, tolerance, craving, and withdrawal. Predisposition to drug abuse could take the form of genetic polymorphisms that alter functional properties of genes or change their expression in a manner that contributes to behaviors that place an individual at risk for addictive behaviors. Second, the repetitive and chronic use (and abuse) of drugs and alcohol can alter the homeostatic gene and protein expression of an individual—a process called allostasis—in a way that creates a dependent or "addicted" state. A better understanding of the role of patterns of gene expression that predispose an individual to drug abuse or changes that occur with chronic drug use will ultimately lead to a broader, and more effective, panoply of treatment options.

With the completion of the human genome sequencing project (as well as other ongoing sequencing projects), we are faced with determining the respective roles of more than 30,000 genes and well over 1,000,000 protein products[1] in a complex behavioral and psychosocial problem. Indeed, it is abundantly clear that a constellation of genes will each contribute, in some small way, to the addiction process. Drug abuse will not be a single-gene disorder such as sickle-cell anemia. Moreover, different genes will be undoubtedly involved with different abused substances (although there is likely to be a unified cadre of generic "abuse" genes). How then are we to make sense of this dizzying set of biological effectors and behavioral phenotypes?

GOALS OF THE CHAPTER

The goal of this chapter is to present the approaches by which this question can be answered and provide examples of attempts to answer it. Specifically, we outline the methods by which transcriptomes and proteomes can be quantitatively examined, the challenges of analyzing these large data sets, and what to do with them. Lastly, examples of published studies using these approaches are given and future directions of the field are discussed.

FUNCTIONAL GENOMICS VERSUS PROTEOMICS

During the past ten years, new technologies have been developed that permit simultaneous analysis of expression levels of thousands to tens of thousands of gene products. This analysis can occur at the level of RNA—variously referred to as functional genomics or transcriptomics—or at the level of protein, a specialty designated proteomics (Fig. 1). The former uses the tools of DNA microarrays, whereas the latter uses two-dimensional electrophoresis with differential protein labeling and/or liquid chromatography coupled with sophisticated mass spectrometry. Before embarking on a detailed consideration of functional genomics and proteomics, it is important to examine the relative strengths and weaknesses of the two approaches. The central message of such a consideration, however, is that no technique will universally prove to be the best. Rather, DNA microarrays and proteomics techniques are important and complementary approaches to better understand the biology of addiction (Nestler 2001).

Figure 1 defines some of the terminology involved in functional genomics and proteomics and, more importantly, provides an assessment of the merits and disadvantages of the two approaches. Although every cell in the body contains essentially the same genetic heritage (the genome), different cells express different subsets of genes. Therefore, the liver expresses a set of genes and the brain expresses another set of brain genes

---

[1]This number of different proteins is impossible to validate, but consider the following discussion. First, for a large number of genes, there are alternative splicing events with RNA and multiple transcriptional start sites. Then, proteins can have phosphorylation sites, glycosylation sites, and lipid addition sites, or be proteolytically modified. By way of example, tyrosine hydroxylase is the rate-limiting enzyme in catecholamine biosynthesis (dopamine, norepinephrine, and epinephrine). In humans, a single gene has four different splice variants. In addition, four different phosphorylation sites exist. In their various combinations, these could theoretically give rise to 60 distinct molecular forms. In practice, this undoubtedly does not occur, but it provides evidence for how a single gene could give rise to a tremendously complex set of protein products.

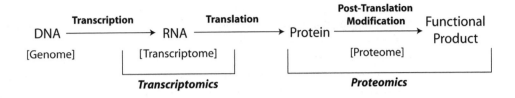

Scientific Issues:
(1) Analysis of complete transcriptome versus 1% to 10% of the proteome of a cell/tissue
(2) Changes in RNA do not necessarily reflect changes in protein
(3) Importance of post-translational processing
(4) RNA is not a functional product
(5) Overall sensitivty (how much of transcriptome or proteome do we 'miss')

FIGURE 1. Definition of transcriptomics and proteomics within the flow of genetic information.

through a process called transcription. This results in a tissue-specific transcriptome comprised of all the genes required to create a cell or tissue. Note, at this point, that although there are genes that are common to all tissues (e.g., intermediary metabolism, structural genes, ribosomal components), clearly cell- and tissue-specific genes exist as well (e.g., neurotransmitter receptor and biosynthetic genes). It is the study of gene expression (and alterations in expression that lead to addictive phenotypes and behaviors) that comprises the field of transcriptomics or functional genomics. We use these terms interchangeably. Two fundamental questions addressed by transcriptomics include the following: What genes are changed in their expression by chronic drug abuse and what inherited patterns of gene expression lead to aberrant drug-seeking behavior?

mRNAs are, generally, only an intermediate step and must be converted to a "functional copy"—the protein. Translation is the biological process by which the ribosome converts mRNA sequence information to protein sequence information. In an analogous fashion to the transcriptome, therefore, cell- and tissue-specific proteomes exist. A complication to this process is the fact that, once synthesized, many proteins are posttranslationally modified (glycosylated, phosphorylated, lipidated, proteolytically processed, etc.) to achieve functionality. This accounts for much of the observation that 30,000 genes can produce perhaps millions of different protein products. The field of study that looks at levels of proteins and posttranslational modification of protein products is referred to as proteomics. Sometimes the term functional proteomics is used to delineate quantitative studies from nonquantitative studies, but we use the term proteomics here. It is this final culmination of processed protein products that accounts for the structure and function of the cell and that contributes, ultimately, to behaviors that underlie addiction.

So, if the proteome is the functional end point of genetic information, why would we bother to examine the levels of mRNA? As we discuss below, DNA microarray technologies have developed to a point where—in a single experiment—the expression of essentially every gene in the human genome can be examined. In other words, we can examine the mRNA expression of essentially all genes in an RNA sample. Unfortunately, this is not the case for proteins or their posttranslation products. As noted in Figure 1, however, this does not mean that transcriptomics is without its share of concerns.

The first concern with functional genomics studies is that just because the entire genome is on a slide or chip does not mean that the investigator will have the requisite sensitivity to "see" it all. Proteomics is currently the more limited technology in this area, with its ability to detect a mere 1–10% of the proteome examined in most studies (we delve into this issue in more detail later in the chapter). Second, the major complaint against functional genomics is the concern that levels of mRNA, or more specifically, changes in mRNA, need not reflect the status of the protein. Although it is mistakenly assumed that the lack of absolute functional correlation between mRNA and protein is a common occurrence, it remains an important consideration—especially given the role of posttranslational regulation in enzyme activity. Indeed, this is the third major consideration in comparing transcriptomics and proteomics. For instance, a membrane-bound protein may acquire extensive modification as it traverses the Golgi apparatus and is transported to the cell membrane. None of this processing would be reflected in the mRNA. At some levels of analysis, in fact, this would not be reflected by traditional proteomics techniques either. Imagine, therefore, an experiment that examined the levels of a specific receptor protein in a cell. Simply disrupting the cell and performing a quantitative immunoblot would provide no insight into the levels of functional cell-surface receptor versus that which had been internalized or was in the process of being transported to the cell surface. On balance, then, it is important to bear in mind that transcriptomics and proteomics actually address different aspects of the flow of genetic information; notably, the RNA is not a functional product, but is, for the most part, merely an informational intermediate.

Finally, and most importantly, there is the issue of sensitivity. Given the availability of amplification technologies for nucleic acids (e.g., reverse transcriptase–polymerase chain reaction [RT-PCR] or linear, T7-based aRNA amplification) and the lack of such methods for proteins, transcriptomics is much more sensitive than proteomics. With amplification technologies, the transcriptome of just a few cells (hundreds to thousands) can be analyzed. In addition, when analyzing individual genes through RT-PCR, sensitivity can be "pushed" to the point of examining 1–10 mRNA copies per cell. Unfortunately, this level of sensitivity is not yet possible in proteomics, and improvements will most likely come from enhancing signal amplification because protein is not amenable to sample amplification techniques such as PCR.

Taken together, it is important to realize that transcriptomics and proteomics serve complementary roles based on technological development and the underlying biology that is to be probed. All other things being equal, it would clearly be most advantageous to examine the quantitative status of all proteins and their posttranslational modifications in a cell or tissue. This cannot currently be accomplished, but through the judicious use of DNA microarray or proteomic tools, valuable aspects of the process can be illuminated. As we further explore below, one of the key areas for future development is how to best integrate these two powerful and important tools.

HIGH-THROUGHPUT TRANSCRIPTOMIC TECHNOLOGIES

Transcriptomics has been an area of active research since the development of the northern blot (Alwine et al. 1977) and RNA dot blot (Kafatos et al. 1979; Sim et al. 1979). In the intervening years, powerful and sensitive tools of RNase protection (RPA), in situ hybridization, and RT-PCR have been developed. The major limiting factor in all of these techniques, however, is the fact that they assess only a single gene at a time. Given our preceding discussion regarding the fact that 30,000 different genes exist, these approaches are clearly of limited value for broad scale "discovery" science (i.e., the ability to screen, in an unbiased manner, all of the genes that might be associated with a disease state or treatment condition). Along the same lines, the ability to analyze many different samples at one time, in parallel, was developed with the creation of the dot (or slot) blot. In this case, independent samples are applied directly to nitrocellulose in an application manifold. The technique then evolved to the "reverse dot blot" in which different known genetic sequences were applied to the nitrocellulose and then all of the gene sequences were "probed" with an unknown (and complex) biological sample.

The field made a dramatic advance when technologies were developed to create massively parallel arrays of gene sequences (numbering in the hundreds to thousands) on a miniaturized scale. The first of these technologies was developed by Pat Brown and his colleagues at Stanford University (Brown and Botstein 1999). Essentially, cDNA clones of many individual genes were isolated and robotically spotted onto glass microscope slides (DNA microarrays) at predetermined and addressable locations. Messenger RNA populations were then labeled with fluorescent dyes and used to query the cDNAs on the slide. The principle remains the same today, in that labeled copies of RNA are exposed to an array of gene sequences and the intensity of the resulting hybridization at each "spot" is indicative of the amount of specific mRNA for that corresponding gene that was present in the original sample. Interestingly, there have been a number of different approaches to "visualizing" signal, including radioactive labeling/phosphorimaging, fluorescence labeling, dual fluorescence labeling, and chemiluminescence. Over time, a variety of different technologies and physical platforms have also been developed and/or marketed. Currently, the field has evolved to the point where the predominant technology is the oligonucleotide array (Fig. 2). However, companies and laboratories have had success with spotted cDNA fragments or small PCR amplicons.

In Situ Synthesized Oligonucleotides

The current market leader in the field of DNA microarrays is clearly Affymetrix. One of the original array companies, it started with a powerful and innovative technology that currently enjoys the widest use. The Affymetrix platform is based on an innovative and patented photolithography technology (Lipshutz et al. 1999). In brief, laser-induced chemical activation procedures (photolithography) are used to build short oligonucleotides directly on the surface of a chip. The resulting GeneChip has more than 1 million features, with each gene represented by 11–20 distinct sequences (taken from diverse locations in the linear sequence of the gene), as well as a set of single nucleotide mismatch probes (Fig. 2). These multiple probes per gene, therefore, sample each gene with an 11-fold redundancy, while simultaneously checking the hybridization fidelity by comparing probe signals with mismatch control signals. This provides a great deal of confidence in the resulting experimental output. In this particular technology, hybridization is detected through fluorescence, and individual samples are analyzed on different arrays.

What then are the limitations of this technology? First is the expense—this remains one of the most expensive approaches to undertake within a laboratory (costs include hardware, dedicated technical support, and the high cost of consumables). Many institutions have gone to centralized array facilities because of the cost of infrastructure and, more importantly, the need for dedicated (and talented) technical personnel. Indeed, the primary problem with successful array experiments is that a high level of technical expert-

Applied Biosystems - 60mer oligos spotted onto a nylon coated slide, detection by chemiluminescence

Affymetrix - 25mer *in situ* synthesized oligos with mismatch probes; detection by a biotin/avidin label binding

Mismatch probe

Agilent - 25mer or 60mer *in situ* synthesized oligos bound to slide by linker segment; 2 color detection (Cy3/Cy5)

25mer   60mer

| Biotin tag

• Streptavidin fluorescent label

➖ Digoxigenin label

⋺ Alkalin phosphatase conjugated antibody

Cy3 or Cy5 direct labled cRNA or cDNA

FIGURE 2. Commercially available oligonucleotide microarray formats. These different technologies use alternate physical substrates, oligonucleotide lengths, and synthesis methods, as well as different hybridization and detection methods.

ise is required for high-quality, reproducible data generation. Another limitation of this technology is the growing realization that the 25-residue oligonucleotide that is directly affixed to a solid surface provides a less than optimal hybridization target. Specifically, the tethering of the 25mer directly to the surface provides a small hybridization target that is constrained in its ability to bind to probes in solution. As a result, although there is a great deal of confidence in the signal that is detected, it is clear that the Affymetrix GeneChip is less sensitive overall.

On balance, then, the Affymetrix GeneChip is a very powerful microarray platform with distinct advantages and disadvantages. It is one of several formats that offers complete genome arrays for rat, mouse, and human (as well as other model organisms such as zebrafish, *Drosophila*, and pigs). In addition, as the market leader, it has an extended history, a variety of "turnkey" data analysis suites, and widespread acceptance in the substance abuse field.

An alternative method of in situ probe synthesis has been developed by Agilent Technologies. Using modified ink jet printing technology (Hughes et al. 2001), bases are sequentially added to oligonucleotides to make 60 base probes on the surface of the array. Linker sequences are used to limit steric hindrance and provide improved hybridization kinetics (Fig. 2).

Spotted Oligonucleotides

The second major approach to DNA microarrays is the deposition of prefabricated oligonucleotides onto a surface. In this case, the oligonucleotides are synthesized before manufacture of the array and are chemically attached to slides in a variety of different formats. The major advantage is that, because they are built on high-efficiency synthesizers, these oligos can be much longer than the short oligos constructed on the Affymetrix GeneChip through photolithography. This, then, provides an increase in hybridization strength affinity with a concomitant increase in sensitivity. This technology is available through several suppliers, and these approaches are being used in substance abuse research.

CodeLink (GE Healthcare/Amersham)

GE Healthcare/Amersham, a relatively new presence in the DNA microarray field, has developed CodeLink DNA microarrays. These oligo arrays are based on a novel slide technology in which 30mer oligonucleotides are affixed to a hydrophilic polyacrylamide three-dimensional slide surface (Fig. 2) (Ramakrishnan et al. 2002). This has benefits in several respects. First, the longer and more exposed oligonucleotides provide a larger target for hybridization, thus increasing binding strength and sensitivity. Second, being affixed in a manner that separates the oligonucleotides from the surface makes them more accessible for hybridization. Finally, the three-dimensional organization provides for a higher target density that also enhances sensitivity. The CodeLink system has also joined the ranks of companies that provide complete genome arrays for mouse, rat, and human. An added advantage of the system is that it is compatible with preexisting fluorescent array scanners and so requires minimal investment in equipment infrastructure for facilities that are already scanning arrays.

On the negative side, the CodeLink slide does not enjoy the technical redundancy (multiple probe sequences per gene and mismatch controls for hybridization fidelity) that is a hallmark of the Affymetrix GeneChip. They are, however, more sensitive because of longer oligos and a unique surface geometry. Several reports have suggested that this platform is more sensitive and has less inherent technical variation than the Affymetrix platform (Tan et al. 2003; Shippy et al. 2004).

Applied Biosystems

Perhaps the most recent addition to the major commercially offered microarrays is the Expression Array System from Applied Biosystems. This platform takes the novel approach of depositing presynthesized 60-base oligos onto a nitrocellulose array surface (affixed to a microscope slide) (Fig. 2). The detection technology uses chemiluminescent detection, and human, mouse, and rat whole genome arrays are available. This platform offers the potential for very high sensitivity, but currently little data are publicly available from the system.

Custom Synthesized (In-house Production)

In the 1990s, many universities and large companies had initiatives to make their own microarrays in-house using either oligo or cDNA probes and contact printing technologies like those initially popularized by Brown and colleagues at Stanford (Schena et al. 1995). The advantage of this approach is the practice of using dual labeling (e.g., Cy3 and Cy5) and competitive hybridization to a single slide. The disadvantage is the potential introduction of dye bias and the need for quality-control dye switching experiments. Printing microarrays has always been an expensive and difficult operation, and it seems that for many researchers, the era of producing one's own microarrays is passing. The economies of scale and reliability offered by commercial suppliers far exceed that of most core facilities. It is likely, in the future, that in-house produced arrays will be limited to a few well-established facilities and/or they will focus on special projects such as unusual model organisms for which commercial arrays are not available.

Concerns with DNA Microarray Analysis (Consistency and Reproducibility)

A potential source of concern regarding microarray studies is raised when comparing results from different array platforms using the same or similar samples (Marshall 2004). A number of studies have attempted to address this technical issue with mixed results (Barczak et al. 2003; Tan et al. 2003; Mah et al. 2004; Shippy et al. 2004; Woo et al. 2004; Yauk et al. 2004). Some cases produced very little correlation between platforms and other cases showed a high concordance. One of the problems with these platform comparisons is that they are often comparing apples to oranges, because correctly matching the different gene identifiers on the various platforms is complicated. In addition, some array platforms must use their individualized analysis techniques, which means that when comparing lists of "changed" genes, you are also comparing statistical analysis methods. Lastly, many of these platform comparison studies are underpowered statistically and it is not certain how much of the lack of agreement between platforms is just variance and noise. We do not dispute that this issue of different results from alternate platforms/laboratories is of critical importance and that strides must be made to allow the comparison of different microarray databases. However, these comparison studies do not condemn microarrays to irrelevance because the data in these platform comparison studies are far outweighed by the many studies that have confirmed array data by post hoc methods such as RT-PCR and immunoblotting, as we discuss later.

QUANTITATIVE PROTEOMIC APPROACHES

As the field of proteomics moves from solely technology development to routine scientific application, drug abuse researchers will have to determine the most appropriate uses for this technology. Addiction research has a great need for large-scale analysis of quantitative differences in protein expression and modification; this is the arena of quantitative proteomics. Many factors, similar to those in functional genomic studies, must be considered. These include limited sample amounts and samples with heterogeneous cellular populations. We mention here particular technologies for performing quantitative assessments of protein expression on the scale of hundreds to thousands of proteins, namely, two-dimensional in-gel electrophoresis (2DIGE) and isotopic labeling reagents (isotope tagging for relative and absolute protein quantication [iTRAQ] and isotope-coded affinity tags [ICAT]).

The term "proteome" was first used less than a decade ago (Wilkins et al. 1996), and proteomic research refers to the examination of protein expression, modification, or interaction on a large, parallel scale—hundreds to thousands of proteins simultaneously. Examination of protein expression, modification, and interaction is hardly novel, for this is basic biochemistry. It is the scale on which these studies can now be performed (hundreds to thousands of proteins) that warrants the new term "proteomics." Many possible applications are available for using proteomics in addiction research: determination of the neuroproteome, protein expression profiling, posttranslational protein modification profiling, and mapping of protein–protein interactions. We focus here on quantitative protein expression profiling approaches, which allow for the simultaneous examination of multiple samples. This is a critical point, as has been seen with functional genomics, for generating data sets with enough statistical power. These are the most developed approaches and can currently be applied to drug abuse research. Several books that discuss these methodologies in depth are important resources (Siuzdak 1996; Liebler 2002; Westermeier and Naven 2002). In addition, a detailed discussion of posttranslation assessment approaches can be found in a recent review by Nairn and colleagues (Williams et al. 2004).

All proteomic expression profiling methods require that proteins be separated to single species, thus permitting the species to be quantified and identified. Currently, two primary approaches to performing quantitative proteomic comparisons on brain tissue exist: 2DIGE and isotopic labeling such as ICAT and iTRAQ of liquid chromatography separated peptides. Broadly speaking, these two methodologies can be differentiated by their protein separation and quantitation approaches. Each has difficulties and strengths, and are discussed below in reference to separating, quantifying, and identifying proteins.

2DE

Two-dimensional electrophoresis (2DE) is not a new technology. Since its introduction as the O'Farrell and Klose techniques (Klose 1975; O'Farrell 1975), 2DE has enjoyed varying popularity. At its most basic, 2DE separates proteins in two ways: (1) by isoelectric point (pI) through isoelectric focusing (IEF) and (2) by relative size through sodium dodecylsulfate–polyacrylamide gel electrophoresis (SDS-PAGE). The pI of a protein is primarily a function of its amino-acid sequence, although posttranslational modifications also contribute to the pI. Isoelectric focusing takes advantage of this property by placing proteins in a pH gradient and applying a voltage potential. The proteins will migrate toward the anode or cathode, depending on their net charge. Eventually, the protein will reach a point in the pH gradient where the net charge of the protein is zero (the pI) and it will stop migrating. IEF has historically been technically difficult, tedious, and challenging to reproduce, but technological advancements have made this a much more robust and simple technique (Bjellqvist et al. 1982; Corbett et al. 1994; Gorg et al. 1995). Most current applications of IEF use immobilized gradients in a dedicated instrument that controls both gel current and temperature (Gorg et al. 2000). After IEF, the pH strip is placed atop a second-dimension SDS-PAGE gel for separation by molecular weight. These two degrees of separation are required to resolve proteins to individual species.

Stains

After electrophoresis, the separated proteins in the gel must be quantified. Traditionally, visible stains such as silver staining or coomassie have been used, whereas newer approaches use fluorescent stains (Rabilloud 2000). Silver staining is the "gold standard" for sensitivity; however, silver staining is difficult to use quantitatively because of complications with background, nonlinear signal, reproducibility, and mass spectrometry compatibility (Nishihara and Champion 2002). Other visible stains have been developed such as zinc imidazol (Fernandez-Patron et al. 1998) and Coomassie Blue (Neuhoff et al. 1988). Although both are more compatible with mass spectrometry, zinc imidazol has limited quantitation abilities and coomassie staining is much less sensitive than silver staining. All of these visual stains can be converted to digital images by either scanning or camera densitometry. Because of the aforementioned limitations, fluorescent stains and dyes have been developed that attempt to combine sensitivity, quantitation, and mass spectrometry compatibility. Fluorescent stains such as ruthenium bathophenanthroline disulfonate (Sypro Ruby) and Deep Purple (Rabilloud et al. 2001; Smejkal et al. 2004) are simple to use, quantitative, and compatible with mass spectrometry. Notably, the sensitivity of Sypro Ruby approaches that of silver staining. Some of these fluorescent stains have been developed such that they are selective for posttranslational modifications of proteins (phosphorylation and glycosylation) (Steinberg et al. 2001, 2003). With these developments, another possibility emerged: labeling of proteins before electrophoresis with fluorescent dyes (see next section below). In total, staining methods continue to improve, but no single stain has been found to be the most sensitive, quantitative, and compatible with mass spectrometry. The problems that have bedeviled the use of 2DE gel stains for applications such as protein expression profiling in addiction research have been those of reproducibility and quantitation. It is difficult to use staining to compare the large number of gels needed for quantitation of multigroup studies. In our experience, staining was too difficult to use as the only method of detection, and this lead us to use protein labeling as an alternative strategy.

Dyes (2DIGE)

To address some of these problems mentioned above, a new method, two-dimensional in-gel difference electrophoresis (2DIGE; Fig. 3), has been developed that uses direct labeling of proteins with cyanine dyes before IEF (Tonge et al. 2001; Alban et al. 2003; see also Westermeier and Naven 2002 for a detailed methodological discussion). This method relies on cyanine dyes that react with lysine groups on proteins (Unlu et al. 1997). Dye concentrations are kept low, such that approximately one dye molecule is added per

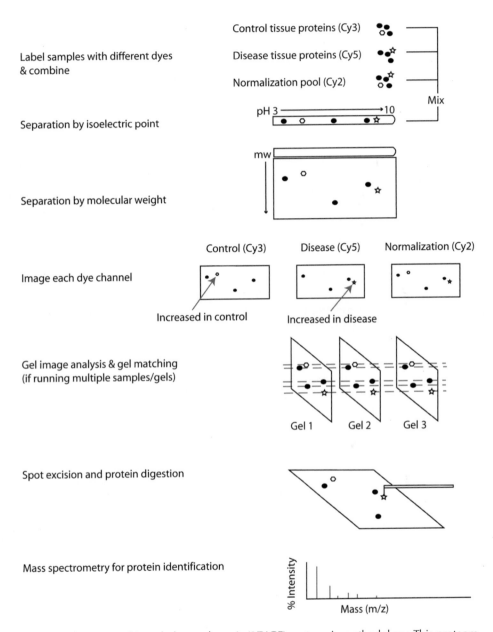

FIGURE 3. Two-dimensional in-gel electrophoresis (2DIGE) proteomic methodology. This proteomic approach uses two-dimensional electrophoresis for protein separation and fluorescent dyes for quantitation. Proteins are identified by mass spectrometric methods.

protein. The critical aspect of the use of 2DIGE technology is the ability to label two or more samples with different dyes and then resolve them on the same IEF and SDS-PAGE gels. This makes spot matching and quantitation much more simple and accurate. As it has been developed, 2DIGE uses Cy3, Cy5, and Cy2 dyes, with Cy2 being used for a normalization pool created from a mix of all samples in the experiment. This Cy2 pool is applied to all gels in a multiple gel experiment, facilitating spot matching and normalization of signal among different gels. The 2DIGE approach offers great promise to researchers using 2DE. The dyes are comparable in sensitivity to silver staining methods, compatible with mass spectrometry, and offer the best quantitation of any 2DE method. The major drawback of this technique is that it requires expensive, dedicated equipment, and software. A modification of 2DIGE, in which cyanine dyes label all of the cysteine residues of proteins, has recently been introduced to increase sensitivity and reduce sample amounts needed, but it is limited to only two dyes (Cy3 and Cy5), reducing its potential application (Shaw et al. 2003). Many more technical points deal with the quantitation of 2D gels (Freeman and Hemby 2004), but let us assume that the quantification is complete, and differentially regulated protein spots have been found. One key piece of information is obviously missing: the identity of the proteins in these spots.

Protein Identification

Mass spectrometry uses ionization to enable very accurate measurement of the mass of proteins or peptides. The scientific basis of mass spectrometry is beyond this discussion and the technique has been reviewed in reference to neuroscience elsewhere (Freeman and Hemby 2004; Williams et al. 2004). Moreover, for in-depth descriptions, several excellent books are available, written with the biologist in mind (Siuzdak 1996; Liebler 2002; Westermeier and Naven 2002). Briefly, peptides or proteins are ionized and then mass analyzers are used to separate ions according to their mass-to-charge (m/z) ratio. The end point of this analysis is very accurate measurement of the mass of a protein or peptide (to +/– 0.3 Da). Tandem mass spectrometry (MS/MS) (in which peptides are further fragmented) is then used to generate protein sequence information. All of these data are combined and submitted to an algorithm that compares the data to genomic databases to identify the protein (Perkins et al. 1999; Jonsson 2001). For gel-based separations, such as 2DIGE, mass spectrometry is usually conducted on in-gel digested proteins. Simply, spots to be identified are excised from gels and then the proteins within the spots are digested with trypsin. These tryptic peptides are what is analyzed by mass spectrometry. With protein identification complete, we would now have a list of quantities along with their respective identities. What to do with this data and subsequent steps are described later in the chapter.

Liquid Chromatography Separations

Because 2DE has a number of limitations, alternative approaches to quantitative proteomic profiling have been developed. In particular, 2DE is subject to variability and sensitivity problems. Therefore, tools have been created that rely on a form of liquid chromatography for separation. The two main methods are ICAT (Gygi et al. 1999) and iTRAQ (Fig. 4; Ross et al. 2004). ICAT, at its most basic, consists of two different tagging reagents that react with cysteine residues. The two different tags are exactly the same in that they have a reactive group and a biotin tag with a linker group in between. The difference between the two tags is that the linker segment of the heavy tag contains deuterium and the light tag con-

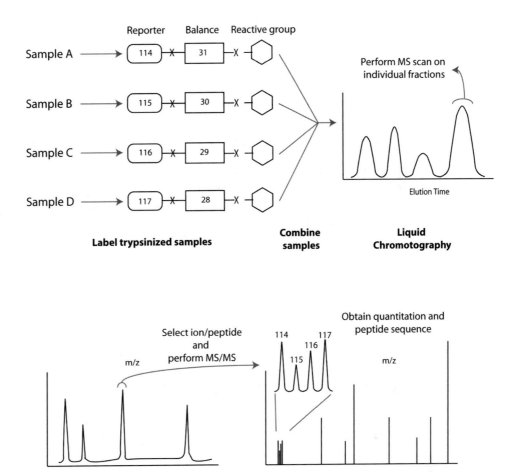

FIGURE 4. Isotope tagging for relative and absolute protein quantitation (iTRAQ) methodology. As opposed to gel-based methods, iTRAQ uses liquid chromatography to separate labeled tryptic peptides, and tandem mass spectrometry to quantify and identify proteins.

tains hydrogen. Two different samples are compared by labeling one sample with the heavy tag and one with the light tag. The two samples are mixed and digested with trypsin to produce peptides. Avidin is then used to capture only those peptides labeled with the ICAT reagent. These peptides are usually subjected to two-dimensional liquid chromatography (LC) such as strong cation exchange (SCX) followed by reverse phase hydrophobic chromotography. The resulting fractions are then analyzed by mass spectrometry. In the initial MS scan, the heavy and light labeled peptides will be exactly 8 Da apart. The peak size of these MS doublets provides relative quantitation. These MS peaks are then selected for MS/MS analysis to obtain protein sequence information.

As we have stated before, one of the major obstacles to simultaneously analyzing hundreds or thousands of RNAs or protein is obtaining sufficient statistical power. In addition, with complex behavioral studies such as those in drug abuse research, many different groups and controls are often present. While ICAT was a major advance, it did not provide adequate ability to simultaneously run multiple samples from multiple groups. The same is true of other isotopic quantitation methodologies (Oda et al. 1999; Wu et al. 2004). To help overcome issues such as these, iTRAQ was developed by Applied Biosystems (Ross et al. 2004). As described in Figure 4, these reagents label all amine residues and consist of a tag, a balancing element, and a reactive group. The tags add the same weight to the samples to avoid a shift in LC mobility. Each of the four tags is reacted with a different trypsinized sample. The labeled samples are then mixed together and separated by SCX and reverse phase LC. Standard MS is performed on the LC fractions and individual peptides are selected for MS/MS. In MS/MS, the tag breaks apart (at the positions marked with an "X" in Fig. 4). Standard protein sequence data on the peptide are collected, but quantitation is achieved through analysis of the 114, 115, 116, and 117 peaks. Unlike iCAT, this method allows four samples to be examined simultaneously, is more sensitive because quantitation occurs in MS/MS, and the protein identification is cumulative of all the samples. It is possible to envision the use of iTRAQ with one channel being used for normalization, in a manner similar to 2DIGE.

BIOINFORMATIC TOOLS FOR ANALYZING LARGE DATA SETS

So far, we have reviewed mechanisms by which large-scale analysis of RNA and protein expression can be performed. These studies produce large databases of quantitative information. A major question then is what to do with all of the data. The four areas of particular importance in the analysis of these large data sets are normalization, differential expression, pattern recognition, and biological relevance (Fig. 5). Each is a complex and involved issue, but we describe the bare basics for each.

Whether using many microarrays in a functional genomic experiment or multiple gels or liquid chromatographs in a proteomic experiment, data normalization is usually required. Direct comparisons among microarrays are often not possible because of differences arising from technical variability in labeling efficiency, hybridization, or microarray fabrication. A number of normalization strategies (e.g., median, mean, Gaussian, and Lowess) can correct the distribution in the various data samples, enabling comparison. There is currently no single standard method and generally, the normalization strategy chosen is dependent on the type and degree of technical variance.

Perhaps the most obvious bioinformatics challenge is determining the difference between samples and when this difference exists between samples. As with normalization, no single, most accepted method exists. In early microarray experiments, one method for "calling" a gene differentially expressed between samples was based on the magnitude of change. Generally, a twofold (i.e., a 100% increase or 50% decrease) change was commonly accepted as a difference. This method has fallen from favor because it lacks statistical rigor and eliminates the possibility of detecting smaller magnitude changes (a common occurrence when examining brain tissue, in particular). As use of microarray technology has evolved, standard parametric tests (such as the t-test and ANOVA [analysis of variance]) were applied to data analysis. However, with the hundreds to tens of thousands of dependent measures in a microarray experiment, the standard methods for multiple testing correction (e.g., Bonferonni correction) are often too conservative. This led to the development of

Methods for clustering samples or genes

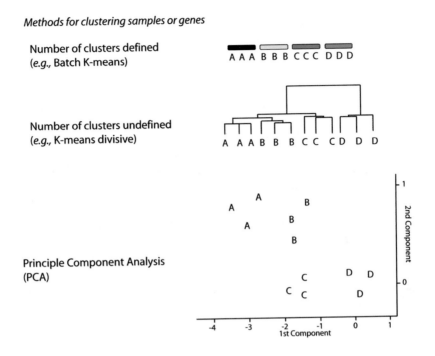

Principle Component Analysis (PCA)

Methods for determining functional properties of changes

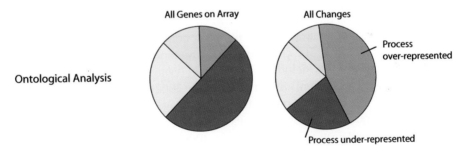

FIGURE 5. Bioinformatic tools for analysis of quantitative functional genomic and proteomic data. A wide variety of bioinformatics tools are available for analyzing data from functional genomic and proteomic studies. Examples are given that allow for separation of similarly regulated genes or samples and for determination of processes effected.

tools such as z-score analysis (Cheadle et al. 2003), statistical analysis of microarrays (SAM) (Tusher et al. 2001), and false-discovery rates (Shedden et al. 2005), to cite just a few examples. These tools offer a better approach to determining significant changes because they are designed for the specific requirements of examining thousands of dependent measures. For proteomic analysis, approaches for determining differential expression or modification resemble those of early microarray experiments with either fold-change thresholds or standard parametric analysis, but this is quickly evolving (Karp et al. 2004; Kreil et al. 2004).

A further approach to analyzing functional genomic and proteomic data is to determine not just individual differences between two groups, but the identification of patterns of gene expression. For example, in a time course experiment, it is of interest to find sets of genes that behave in a similar manner across the time points. Some standard methods for determining patterns of gene expression include clustering, principle component analysis (PCA), and self-organizing maps (SOMs) (Eisen et al. 1998; Toronen et al. 1999; Sherlock 2001). There have been many applications of these algorithms, but in the end their output is a smaller set of genes or proteins. We still have not answered the question of what this means for the biology of the system.

Whether examining a list of gene expression changes or a group of coregulated genes, one problem that has plagued researchers is understanding what these lists mean biologically for the system being studied. A methodology gaining acceptance is the use of gene ontologies (the function, process, or pathway that contains a particular gene) to look for specific biological processes that are affected. A number of tools have recently been developed to aid in this work (Jenssen et al. 2001; Doniger et al. 2003; Thomas et al. 2003a,b; Zeeberg et al. 2003). This is an encouraging development because it allows investigators to take lists of genes, find a biological story, and follow that story in further experiments. This is a natural approach to the issue of bioinformatic analysis because it goes to the goal of creating greater biological insight.

FOLLOW-UP/CONFIRMATION STUDIES

When examining thousands of mRNAs or proteins, it is almost a certainty that even with statistical methods, false positives will occur (i.e., changes will be observed by microarrays or proteomic screens that in fact do not exist). One of the best methods for dealing with this problem is to confirm changes by another method.

For microarray results, the most common method for confirmation is to use quantitative RT-PCR. The current preferred method is to use real-time techniques. A range of techniques exists including Syber Green (Wittwer et al. 1997), fluorescent primers (LUX [light upon extension]) (Nazarenko et al. 2002), and TaqMan fluorescent probes (Heid et al. 1996). Real-time methods take advantage of the kinetics of the PCR reaction to accurately quantify transcript levels. Several good resources are available for the specifics on quantitation using PCR (Bustin 2004). With microarrays now capable of examining the whole genome, high-throughput confirmation by quantitative PCR is needed. New instruments and protocols are making PCR confirmation of a large number of genes possible (Pinhasov et al. 2004). Other traditional methods such as in situ hybridization can be used for microarray confirmation (Mirnics et al. 2000), with the additional benefit that the transcript is anatomically localized. However, this method is more time-consuming and requires more samples than does quantitative PCR.

For confirmation of proteomic data, the preferred method is immunoblots. Immunoblots or "Westerns" can be used to confirm changes in protein expression as well as modification in the case of phosphorylation. The major limitation is that an adequate antibody is required for the specific protein of interest, and, in the case of phosphorylation, a phospho-specific antibody must be available. Currently no widely used alternatives to antibody-based methods exist. Immunoblotting can also be used to confirm (indirectly) microarray data. The advantage is that protein confirmation of mRNA data takes the analysis to the functional end point, protein. Although immunocytochemical localization is also an attractive tool, to date, quantitation of immunocytochemical slides remains problematic. However, it does provide anatomical localization not possible with immunoblots.

CASE STUDIES

Functional genomic and proteomic methods are beginning to be applied to drug abuse research. Unlike some other fields such as cancer research, the drug abuse field has been slower to adopt the new tools of functional genomics and proteomics, owing, in part, to the modest nature of such changes. However, use of these tools is growing dramatically. Functional genomic studies have been conducted on a number of different drugs and although all references cannot be given here, key examples are provided in Table 1. Functional genomic studies have examined human, nonhuman primate, and rodent samples. The accumulation of these data has produced reviews and meta-analyses of the results (Freeman et al. 2002; Yuferov et al. 2005).

Currently, very few discovery proteomic studies in the drug abuse field use high-throughput screening methodologies. To our knowledge, initial reports of proteomic profiling with nicotine (Yeom et al. 2005), amphetamine (Freeman et al. 2005), and morphine (Kim et al. 2005) are available (Table 2). It is anticipated that the trend will continue and

TABLE 1. Major Characteristics of Common Microarray Methodologies

Format/ supplier	Substrate	Probe type	Probe length	Detection	Available arrays
Applied Biosystems Microarray	Nitrocellulose pad mounted on glass slide	Oligonucleotide	60mer	Chemiluminescence	Human, mouse, and rat whole genome
Affymetrix GeneChip	Glass slide	Oligonucleotide generated in situ by photolithography	25mer	Fluorescent streptavidin/ biotin	Human, mouse, and rat whole genome, and other model organisms
Agilent	Glass slide	Oligonucleotide generated in situ by ink jet	60mer	Two-color fluorescence	Human, mouse, and rat whole genome, and other model organisms
GE Healthcare CodeLink	Polyacrylamide gel pad mounted on glass slide	Oligonucleotide spotted by ink jet	30mer	Fluorescent streptavidin/ biotin	Human, mouse, and rat whole genome
Core-facility produced	Typically, glass slide	Typically, oligo-nucleotides spotted by contact printing	25–60mer	Fluorescent, often two colors	Human, mouse, rat, and other model organisms

TABLE 2. Selected Examples of Functional Genomic and Proteomic Studies of Drug Abuse

Cocaine	Morphine	Amphetamine and methamphetamine	Nicotine	MDMA (ecstasy)	THC	Psychedelics
Freeman et al. 2001b	Loguinov et al. 2001	Jayanthi et al. 2001	Konu et al. 2001	Thiriet et al. 2002	Kittler et al. 2000	Nichols and Sanders-Bush 2002
Bibb et al. 2001	Hemby 2004	Xie et al. 2002	Li et al. 2004	Xie et al. 2004	Parmentier-Batteur et al. 2002	Toyooka et al. 2002
McClung and Nestler 2003	Ammon-Treiber et al. 2004	Gonzalez-Nicolini and McGinty 2002	Yeom et al. 2005			Kaiser et al. 2004
Yuferov et al. 2003	Rodriguez Parkitna et al. 2004	Sokolov et al. 2003				
Lehrmann et al. 2003	Kim et al. 2005	Thomas et al. 2004				
Albertson et al. 2004	Prokai et al. 2005	Freeman et al. 2005				
Yao et al. 2004						

This is not a comprehensive listing and is intended to provide examples with different technological platforms and experimental models.

that examples of proteomic studies will accumulate rapidly.

The first published microarray experiments are little more than 5 years old and it is no surprise that no new therapeutics have yet been developed as a result of this research. One area of success from microarray studies has been to identify new targets for further research. We describe two examples here in which a gene discovered through functional genomic experiments has lead to further research and characterization in other labs.

ΔFosB is an addiction-related protein extensively studied by Dr. Eric Nestler and colleagues (for a review, see McClung et al. 2004). In a study that used microarrays in conjunction with a inducible, striatal-specific, ΔFosB transgenic mouse, cyclin-dependent kinase 5 (CDK5) was found to be induced by ΔFosB overexpression (Bibb et al. 2001). This increase was also found to occur in response to cocaine administration (Bibb et al. 2001; Lu et al. 2003) and inhibition of CDK5 potentiated locomotor activity in response to cocaine (Bibb et al. 2001). Cocaine and amphetamine are known to alter spine density in the nucleus accumbens (NAc) (Robinson and Kolb 1999), and it has subsequently been found that CDK5 inhibition attenuates cocaine-induced outgrowths (Norrholm et al. 2003). The potential implications of CDK5 for addiction research have been further characterized by the reports that CDK5 phosphorylates tyrosine hydroxylase (Kansy et al. 2004) and mediates dopamine signaling (Takahashi et al. 2005). Furthermore, CDK5 alterations have been discovered in human opioid addicts (Ferrer-Alcon et al. 2003) and methamphetamine-treated rats (Chen and Chen 2005), suggesting a potential role for this protein across abused drugs.

We originally found that the *wnt* pathway gene β-catenin was cocaine responsive in the NAc of chronically cocaine-treated nonhuman primates using microarrays and we confirmed this induction at the protein level (Freeman et al. 2001b). Later, we also found that β-catenin was induced in other brain regions of rats (Freeman et al. 2001a). This finding was extended and found to be dopamine receptor D1-dependent using knockout mice (Zhang et al. 2002). This induction was then found in a rabbit fetal exposure model (Gil et al. 2003). In a mouse fetal exposure model, a microarray study also found an induction of β-catenin and confirmed this at the level of protein (Novikova et al. 2005). The examples of CDK5 and β-catenin show how genes originally found in a functional genomics discovery experiment can be characterized further in different models, brain regions, and with different drugs for functional relevance to improve our understanding of addiction biology.

CONCLUSION

Recent advances in functional genomics (transcriptomics) and proteomics open new windows into understanding substance abuse. On the surface, they are perfectly suited to characterizing such multigenic disorders. Given the complexity in executing and analyzing these experiments (including the extensive monetary and time commitments), it is critical that the studies be carefully designed. There is also a growing realization that, when possible, complementary proteomic and functional genomic studies are optimal. Only in this way will we gain the fullest insights into the roles of mRNA and protein in the predisposition to drug abuse and the allostatic response to chronic drug use.

GAPS AND FUTURE DIRECTIONS

Functional genomic studies of drug abuse are still quite young and the proteomic era of addiction research is just dawning—many changes are yet to come. Primary challenges include making studies more anatomically specific, improving data analysis, performing larger studies, and bringing together the expertise needed to undertake these complex projects.

Because the brain is the most heterogeneous organ in the body, studies examining dissected tissue contain many different cell types and processes. In this regard, the complexity with regard to neuroscience applications of functional genomics and proteomics may not be seen in other fields. Specifically, the systems biology of cellular interactions must also be considered—clearly, the gene and protein expression of a cell must be understood in the context of the anatomical localization of the cell and the behavior of the animal as a whole (Choudhary and Grant 2004). To help examine specific cell types and improve biological understanding at the cellular level, laser dissection techniques (laser capture microdissection [LCM] or laser pressure catapulting [LPC]) have been developed to microdissect specific regions. More specifically, by using cell-specific markers, such as antibodies to a particular protein, defined cell types can be collected. These applications are currently possible and examples exist of transcriptional profiling from laser-captured cells (Backes and Hemby 2003; Grimm et al. 2004; Hemby 2004; Meguro et al. 2004). The biggest challenge with these samples is that very limited amounts of samples are obtained and mRNA amplification is needed to obtain enough material for microarrays. On the other hand, LCM samples can often be directly applied to quantitative PCR approaches (Meguro et al. 2004). Proteomic experiments will pose more obstacles, e.g., no protein amplification methods are available to aid in generating sufficient samples for proteomic profiling. Initial steps are being taken to overcome this hurdle (Mouledous et al. 2003a,b).

A common complaint from those performing functional genomic and proteomic research is that the data analysis methods have not kept pace with ability to generate increasing data. Although a large number of different normalization, differential expression, and clustering algorithms exist, no commonly accepted standards for the correct types of data analysis or effective approaches to determine why one algorithm is better than another are available. Aid from the statistical field is greatly needed. In addition, although an accepted data sharing format is available (minimum information about a microarray experiment [MIAME]; Brazma et al. 2001), this has not eliminated all of the barriers to data sharing (Pollock 2002). Similar questions will have to be solved for the proteomics field (Orchard et al. 2003a,b).

Initial functional genomic and proteomic studies have often suffered from being underpowered statistically and examining too few conditions or brain regions. This is natural in early studies that are particularly difficult and costly. But, as technology becomes more standardized and easier to access, the addiction field will need to undertake studies such as those seen in the cancer field, for example, in which very large numbers of samples are examined. This is the step needed to transform drug abuse research from looking for a few interesting genes to true systems biology research.

Lastly, a legitimate complaint of researchers who are just becoming adept at functional genomic studies is that they now have to learn yet another field, proteomics. Although it is true that more training will be needed, the benefits make the effort worthwhile. A current discussion is whether proteomics should be outsourced to core facilities or if individual laboratories should specialize in addiction proteomics. Most likely, an intermediate approach will be needed, with some drug abuse researchers becoming well versed in the technology so that the most gain can be achieved from these studies.

ACKNOWLEDGMENTS

This work was supported by Department of Health and Human Services Grant No. DA-013770 (K.E.V.).

REFERENCES

Alban A., David S.O., Bjorkesten L., Andersson C., Sloge E., Lewis S., and Currie I. 2003. A novel experimental design for comparative two-dimensional gel analysis: Two-dimensional difference gel electrophoresis incorporating a pooled internal standard. *Proteomics* 3: 36–44.

Albertson D.N., Pruetz B., Schmidt C.J., Kuhn D.M., Kapatos G., and Bannon M.J. 2004. Gene expression profile of the nucleus accumbens of human cocaine abusers: Evidence for dys-

regulation of myelin. *J. Neurochem.* 88: 1211–1219.

Alwine J.C., Kemp D.J., and Stark G.R. 1977. Method for detection of specific RNAs in agarose gels by transfer to diazobenzyloxymethyl-paper and hybridization with DNA probes. *Proc. Natl. Acad. Sci.* 74: 5350–5354.

Ammon-Treiber S., Tischmeyer H., Riechert U., and Höllt V. 2004. Gene expression of transcription factors in the rat brain after morphine withdrawal. *Neurochem. Res.* 29: 1267–1273.

Backes E. and Hemby S.E. 2003. Discrete cell gene profiling of ventral tegmental dopamine neurons after acute and chronic cocaine self-administration. *J. Pharmacol. Exp. Ther.* **307:** 450–459.

Barczak A., Rodriguez M.W., Hanspers K., Koth L.L., Tai Y.C., Bolstad B.M., Speed T.P., and Erle D.J. 2003. Spotted long oligonucleotide arrays for human gene expression analysis. *Genome Res.* **13:** 1775–1785.

Bibb J.A., Chen J., Taylor J.R., Svenningsson P., Nishi A., Snyder G.L., Yan Z., Sagawa Z.K., Ouimet C.C., Nairn A.C., Nestler E.J., and Greengard P. 2001. Effects of chronic exposure to cocaine are regulated by the neuronal protein Cdk5. *Nature* **410:** 376–380.

Bjellqvist B., Ek K., Righetti P.G., Gianazza E., Gorg A., Westermeier R., and Postel W. 1982. Isoelectric focusing in immobilized pH gradients: Principle, methodology and some applications. *J. Biochem. Biophys. Methods* **6:** 317–339.

Brazma A., Hingamp P., Quackenbush J., Sherlock G., Spellman P., Stoeckert C., et al. 2001. Minimum information about a microarray experiment (MIAME)-toward standards for microarray data. *Nat. Genet.* **29:** 365–371.

Brown P.O. and Botstein D. 1999. Exploring the new world of the genome with DNA microarrays. *Nat. Genet.* **21:** 33–37.

Bustin S.A. 2004. *A–Z of quantitative PCR.* International University Line, La Jolla, California.

Cheadle C., Vawter M.P., Freed W.J., and Becker K.G. 2003. Analysis of microarray data using Z score transformation. *J. Mol. Diagn.* **5:** 73–81.

Chen P.C. and Chen J.C. 2005. Enhanced Cdk5 activity and p35 translocation in the ventral striatum of acute and chronic methamphetamine-treated rats. *Neuropsychopharmacology* **30:** 538–549.

Choudhary J. and Grant S.G. 2004. Proteomics in postgenomic neuroscience: The end of the beginning. *Nat. Neurosci.* **7:** 440–445.

Corbett J.M., Dunn M.J., Posch A., and Gorg A. 1994. Positional reproducibility of protein spots in two-dimensional polyacrylamide gel electrophoresis using immobilised pH gradient isoelectric focusing in the first dimension: An interlaboratory comparison. *Electrophoresis* **15:** 1205–1211.

Doniger S.W., Salomonis N., Dahlquist K.D., Vranizan K., Lawlor S.C., and Conklin B.R. 2003. MAPPFinder: Using Gene Ontology and GenMAPP to create a global gene-expression profile from microarray data. *Genome Biol.* **4:** R7.

Eisen M.B., Spellman P.T., Brown P.O., and Botstein D. 1998. Cluster analysis and display of genome-wide expression patterns. *Proc. Natl. Acad. Sci.* **95:** 14863–14868.

Fernandez-Patron C., Castellanos-Serra L., Hardy E., Guerra M., Estevez E., Mehl E., and Frank R.W. 1998. Understanding the mechanism of the zinc-ion stains of biomacromolecules in electrophoresis gels: Generalization of the reverse-staining technique. *Electrophoresis* **19:** 2398–2406.

Ferrer-Alcon M., La H.R., Guimon J., and Garcia-Sevilla J.A. 2003. Downregulation of neuronal cdk5/p35 in opioid addicts and opiate-treated rats: Relation to neurofilament phosphorylation. *Neuropsychopharmacology* **28:** 947–955.

Freeman W.M. and Hemby S.E. 2004. Proteomics for protein expression profiling in neuroscience. *Neurochem. Res.* **29:** 1065–1081.

Freeman W.M., Dougherty K.E., Vacca S.E., and Vrana K.E. 2002. An interactive database of cocaine-responsive gene expression. *Sci. World J.* **2:** 701–706.

Freeman W.M., Brebner K., Amara S.G., Reed M.S., Pohl J., and Phillips A.G. 2005. Distinct proteomic profiles of amphetamine self-administration transitional states. *Pharmacogenomics J.* **5:** 203–214.

Freeman W.M., Brebner K., Lynch W.J., Robertson D.J., Roberts D.C., and Vrana K.E. 2001a. Cocaine-responsive gene expression changes in rat hippocampus. *Neuroscience* **108:** 371–380.

Freeman W.M., Nader M.A., Nader S.H., Robertson D.J., Gioia L., Mitchell S.M., Daunais J.B., Porrino L.J., Friedman D.P., and Vrana K.E. 2001b. Chronic cocaine-mediated changes in non-human primate nucleus accumbens gene expression. *J. Neurochem.* **77:** 542–549.

Gil M., Zhen X., and Friedman E. 2003. Prenatal cocaine exposure alters glycogen synthase kinase-3beta (GSK3beta) pathway in select rabbit brain areas. *Neurosci. Lett.* **349:** 143–146.

Gonzalez-Nicolini V. and McGinty J.F. 2002. Gene expression profile from the striatum of amphetamine-treated rats: A cDNA array and in situ hybridization histochemical study. *Brain Res. Gene Expr. Patterns* **1:** 193–198.

Gorg A., Boguth G., Obermaier C., Posch A., and Weiss W. 1995. Two-dimensional polyacrylamide gel electrophoresis with immobilized pH gradients in the first dimension (IPG-Dalt): The state of the art and the controversy of vertical versus horizontal systems. *Electrophoresis* **16:** 1079–1086.

Gorg A., Obermaier C., Boguth G., Harder A., Scheibe B., Wildgruber R., and Weiss W. 2000. The current state of two-dimensional electrophoresis with immobilized pH gradients. *Electrophoresis* **21:** 1037–1053.

Grimm J., Mueller A., Hefti F., and Rosenthal A. 2004. Molecular basis for catecholaminergic neuron diversity. *Proc. Natl. Acad. Sci.* **101:** 13891–13896.

Gygi S.P., Rist B., Gerber S.A., Turecek F., Gelb M.H., and Aebersold R. 1999. Quantitative analysis of complex protein mixtures using isotope-coded affinity tags. *Nat. Biotechnol.* **17:** 994–999.

Heid C.A., Stevens J., Livak K.J., and Williams P.M. 1996. Real time quantitative PCR. *Genome Res.* **6:** 986–994.

Hemby S.E. 2004. Morphine-induced alterations in gene expression of calbindin immunopositive neurons in nucleus accumbens shell and core. *Neuroscience* **126:** 689–703.

Hughes T.R., Mao M., Jones A.R., Burchard J., Marton M.J., Shannon K.W., et al. 2001. Expression profiling using microarrays fabricated by an ink-jet oligonucleotide synthesizer. *Nat. Biotechnol.* **19:** 342–347.

Jayanthi S., Deng X., Bordelon M., McCoy M.T., and Cadet J.L. 2001. Methamphetamine causes differential regulation of prodeath and anti-death Bcl-2 genes in the mouse neocortex. *FASEB J.* **15:** 1745–1752.

Jenssen T.K., Laegreid A., Komorowski J., and Hovig E. 2001. A literature network of human genes for high-throughput analysis of gene expression. *Nat. Genet.* **28:** 21–28.

Jonsson A.P. 2001. Mass spectrometry for protein and peptide characterisation. *Cell Mol. Life Sci.* **58:** 868–884.

Kafatos F.C., Jones C.W., and Efstratiadis A. 1979. Determination of nucleic acid sequence homologies and relative concentrations by a dot hybridization procedure. *Nucleic Acids Res.* **7:** 1541–1552.

Kaiser S., Foltz L.A., George C.A., Kirkwood S.C., Bemis K.G., Lin X., Gelbert L.M., and Nisenbaum L.K. 2004. Phencyclidine-induced changes in rat cortical gene expression identified by microarray analysis: Implications for schizophrenia. *Neurobiol. Dis.* **16:** 220–235.

Kansy J.W., Daubner S.C., Nishi A., Sotogaku N., Lloyd M.D., Nguyen C., Lu L., Haycock J.W., Hope B.T., Fitzpatrick P.F., and Bibb J.A. 2004. Identification of tyrosine hydroxylase as a physiological substrate for Cdk5. *J. Neurochem.* **91:** 374–384.

Karp N.A., Kreil D.P., and Lilley K.S. 2004. Determining a significant change in protein expression with DeCyder during a pairwise comparison using two-dimensional difference gel electrophoresis. *Proteomics* **4:** 1421–1432.

Kim S.Y., Chudapongse N., Lee S.M., Levin M.C., Oh J.T., Park H.J., and Ho I.K. 2005. Proteomic analysis of phosphotyrosyl proteins in morphine-dependent rat brains. *Brain Res. Mol. Brain Res.* **133:** 58–70.

Kittler J.T., Grigorenko E.V., Clayton C., Zhuang S.Y., Bundey S.C., Trower M.M., Wallace D., Hampson R., and Deadwyler S. 2000. Large-scale analysis of gene expression changes during acute and chronic exposure to Δ^9-THC in rats. *Physiol. Genomics* **3:** 175–185.

Klose J. 1975. Protein mapping by combined isoelectric focusing and electrophoresis of mouse tissues. A novel approach to testing for induced point mutations in mammals. *Humangenetik* **26:** 231–243.

Konu O., Kane J.K., Barrett T., Vawter M.P., Chang R., Ma J.Z., Donovan D.M., Sharp B., Becker K.G., and Li M.D. 2001. Region-specific transcriptional response to chronic nicotine in rat brain. *Brain Res.* **909:** 194–203.

Kreil D.P., Karp N.A., and Lilley K.S. 2004. DNA microarray normalization methods can remove bias from differential protein expression analysis of 2D difference gel electrophoresis results. *Bioinformatics* **20:** 2026–2034.

Lehrmann E., Oyler J., Vawter M.P., Hyde T.M., Kolachana B., Kleinman J.E., Huestis M.A., Becker K.G., and Freed W.J. 2003. Transcriptional profiling in the human prefrontal cortex: Evidence for two activational states associated with cocaine abuse. *Pharmacogenomics J.* **3:** 27–40.

Li M.D., Kane J.K., Wang J., and Ma J.Z. 2004. Time-dependent changes in transcriptional profiles within five rat brain regions in response to nicotine treatment. *Brain Res. Mol. Brain Res.* **132:** 168–180.

Liebler D.C., ed. 2002. *Introduction to proteomics.* Humana Press, Totowa, New Jersey.

Lipshutz R.J., Fodor S.P., Gingeras T.R., and Lockhart D.J. 1999. High density synthetic oligonucleotide arrays. *Nat. Genet.* **21:** 20–24.

Loguinov A.V., Anderson L.M., Crosby G.J,. and Yukhananov R.Y. 2001. Gene expression following acute morphine administration. *Physiol. Genomics* **6:** 169–181.

Lu L., Grimm J.W., Shaham Y., and Hope B.T. 2003. Molecular neuroadaptations in the accumbens and ventral tegmental area during the first 90 days of forced abstinence from cocaine self-administration in rats. *J. Neurochem.* **85:** 1604–1613.

Mah N., Thelin A., Lu T., Nikolaus S., Kuhbacher T., Gurbuz Y., Eickhoff H., Kloppel G., Lehrach H., Mellgard B., Costello C.M., and Schreiber S. 2004. A comparison of oligonucleotide and cDNA-based microarray systems. *Physiol. Genomics* **16:** 361–370.

Marshall E. 2004. Getting the noise out of gene arrays. *Science* **306:** 630–631.

McClung C.A. and Nestler E.J. 2003. Regulation of gene expression and cocaine reward by CREB and ΔFosB. *Nat. Neurosci.* **6:** 1208–1215.

McClung C.A., Ulery P.G., Perrotti L.I., Zachariou V., Berton O., and Nestler E.J. 2004. ΔFosB: A molecular switch for long-term adaptation in the brain. *Brain Res. Mol. Brain Res.* **132:** 146–154.

Meguro R., Lu J., Gavrilovici C., and Poulter M.O. 2004. Static, transient and permanent organization of GABA receptor expression in calbindin-positive interneurons in response to amygdala kindled seizures. *J. Neurochem.* **91:** 144–154.

Mirnics K., Middleton F.A., Marquez A., Lewis D.A., and Levitt P. 2000. Molecular characterization of schizophrenia viewed by microarray analysis of gene expression in prefrontal cortex. *Neuron* **28:** 53–67.

Mouledous L., Hunt S., Harcourt R., Harry J., Williams K.L., and Gutstein H.B. 2003a. Navigated laser capture microdissection as an alternative to direct histological staining for proteomic analysis of brain samples. *Proteomics* **3:** 610–615.

———. 2003b. Proteomic analysis of immunostained, laser-capture microdissected brain samples. *Electrophoresis* **24:** 296–302.

Nazarenko I., Lowe B., Darfler M., Ikonomi P., Schuster D., and Rashtchian A. 2002. Multiplex quantitative PCR using self-quenched primers labeled with a single fluorophore. *Nucleic Acids Res.* **30:** e37.

Nestler E.J. 2001. Psychogenomics: Opportunities for understanding addiction. *J. Neurosci.* **21:** 8324–8327.

Neuhoff V., Arold N., Taube D., and Ehrhardt W. 1988. Improved staining of proteins in polyacrylamide gels including isoelectric focusing gels with clear background at nanogram sensitivity using Coomassie Brilliant Blue G-250 and R-250. *Electrophoresis* **9:** 255–262.

Nichols C.D. and Sanders-Bush E. 2002. A single dose of lysergic acid diethylamide influences gene expression patterns within the mammalian brain. *Neuropsychopharmacology* **26:** 634–642.

Nishihara J.C. and Champion K.M. 2002. Quantitative evaluation of proteins in one- and two-dimensional polyacrylamide gels using a fluorescent stain. *Electrophoresis* **23:** 2203–2215.

Norrholm S.D., Bibb J.A., Nestler E.J., Ouimet C.C., Taylor J.R., and Greengard P. 2003. Cocaine-induced proliferation of dendritic spines in nucleus accumbens is dependent on the activity of cyclin-dependent kinase-5. *Neuroscience* **116:** 19–22.

Novikova S.I., He F., Bai J., and Lidow M.S. 2005. Neuropathology of the cerebral cortex observed in a range of animal models of prenatal cocaine exposure may reflect alterations in genes involved in the Wnt and cadherin systems. *Synapse* **56:** 105–116.

O'Farrell P.H. 1975. High resolution two-dimensional electrophoresis of proteins. *J. Biol. Chem.* **250:** 4007–4021.

Oda Y., Huang K., Cross F.R., Cowburn D., and Chait B.T. 1999. Accurate quantitation of protein expression and site-specific phosphorylation. *Proc. Natl. Acad. Sci.* **96:** 6591–6596.

Orchard S., Hermjakob H., and Apweiler R. 2003a. The proteomics standards initiative. *Proteomics* **3:** 1374–1376.

Orchard S., Zhu W., Julian R.K., Jr., Hermjakob H., and Apweiler R. 2003b. Further advances in the development of a data interchange standard for proteomics data. *Proteomics* **3:** 2065–2066.

Parmentier-Batteur S., Jin K., Xie L., Mao X.O., and Greenberg D.A. 2002. DNA microarray analysis of cannabinoid signaling in mouse brain in vivo. *Mol. Pharmacol.* **62:** 828–835.

Perkins D.N., Pappin D.J., Creasy D.M., and Cottrell J.S. 1999. Probability-based protein identification by searching sequence databases using mass spectrometry data. *Electrophoresis* **20:** 3551–3567.

Pinhasov A., Mei J., Amaratunga D., Amato F.A., Lu H., Kauffman J., Xin H., Brenneman D.E., Johnson D.L., Andrade-Gordon P., and Ilyin S.E. 2004. Gene expression analysis for high throughput screening applications. *Comb. Chem. High Throughput Screen* **7:** 133–140.

Pollock J.D. 2002. Gene expression profiling: Methodological challenges, results, and prospects for addiction research. *Chem. Phys. Lipids* **121:** 241–256.

Prokai L., Zharikova A.D., and Stevens S.M., Jr. 2005. Effect of chronic morphine exposure on the synaptic plasma-membrane subproteome of rats: A quantitative protein profiling study based on isotope-coded affinity tags and liquid chromatography/mass spectrometry. *J. Mass. Spectrom.* **40:** 169–175.

Rabilloud T. 2000. Detecting proteins separated by 2-D gel electrophoresis. *Anal. Chem.* **72:** 48A–55A.

Rabilloud T., Strub J.M., Luche S., van Dorsselaer A., and Lunardi J. 2001. A comparison between Sypro Ruby and ruthenium II tris (bathophenanthroline disulfonate) as fluorescent stains for protein detection in gels. *Proteomics* **1:** 699–704.

Ramakrishnan R., Dorris D., Lublinsky A., Nguyen A., Domanus M., Prokhorova A., Gieser L., Touma E., Lockner R., Tata M., Zhu X., Patterson M., Shippy R., Sendera T.J., and Mazumder A. 2002. An assessment of Motorola CodeLink microarray performance for gene expression profiling applications. *Nucleic Acids Res.* **30:** e30.

Robinson T.E. and Kolb B. 1999. Alterations in the morphology of dendrites and dendritic spines in the nucleus accumbens and prefrontal cortex following repeated treatment with ampheta-mine or cocaine. *Eur. J. Neurosci.* **11:** 1598–1604.

Rodriguez Parkitna J.M., Bilecki W., Mierzejewski P., Stefanski R., Ligeza A., Bargiela A., Ziolkowska B., Kostowski W., and Przewlocki R. 2004. Effects of morphine on gene expression in the rat amygdala. *J. Neurochem.* **91:** 38–48.

Ross P.L., Huang Y.N., Marchese J.N., Williamson B., Parker K., Hattan S., Khainovski N., Pillai S., Dey S., Daniels S., Purkayastha S., Juhasz P., Martin S., Bartlet-Jones M., He F., Jacobson A., and Pappin D.J. 2004. Multiplexed protein quantitation in *Saccharomyces cerevisiae* using amine-reactive isobaric tagging reagents. *Mol. Cell. Proteomics* **3:** 1154–1169.

Schena M., Shalon D., Davis R.W., and Brown P.O. 1995. Quantitative monitoring of gene expression patterns with a complementary DNA microarray. *Science* **270:** 467–470.

Shaw J., Rowlinson R., Nickson J., Stone T., Sweet A., Williams K., and Tonge R. 2003. Evaluation of saturation labelling two-dimensional difference gel electrophoresis fluorescent dyes. *Proteomics* **3:** 1181–1195.

Shedden K., Chen W., Kuick R., Ghosh D., Macdonald J., Cho K.R., Giordano T.J., Gruber S.B., Fearon E.R., Taylor J.M., and Hanash S. 2005. Comparison of seven methods for producing Affymetrix expression scores based on False Discovery Rates in disease profiling data. *BMC Bioinformatics* **6:** 26.

Sherlock G. 2001. Analysis of large-scale gene expression data. *Brief Bioinform.* **2:** 350–362.

Shippy R., Sendera T.J., Lockner R., Palaniappan C., Kaysser-Kranich T., Watts G., and Alsobrook J. 2004. Performance evaluation of commercial short-oligonucleotide microarrays and the impact of noise in making cross-platform correlations. *BMC Genomics* **5:** 61.

Sim G.K., Kafatos F.C., Jones C.W., Koehler M.D., Efstratiadis A., and Maniatis T. 1979. Use of a cDNA library for studies on evolution and developmental expression of the chorion multigene families. *Cell* **18:** 1303–1316.

Siuzdak G. 1996. *Mass spectrometry for biotechnology.* Academic Press, San Diego.

Smejkal G.B., Robinson M.H., and Lazarev A. 2004. Comparison of fluorescent stains: Relative photostability and differential staining of proteins in two-dimensional gels. *Electrophoresis* **25:** 2511–2519.

Sokolov B.P., Polesskaya O.O., and Uhl G.R. 2003. Mouse brain gene expression changes after acute and chronic amphetamine. *J. Neurochem.* **84:** 244–252.

Steinberg T.H., Top K.P.O., Berggren K.N., Kemper C., Jones L., Diwu Z.J., Haugland R.P., and Patton W.F. 2001. Rapid and simple single nanogram detection of glycoproteins in polyacrylamide gels and on electroblots. *Proteomics* **1:** 841–855.

Steinberg T.H., Agnew B.J., Gee K.R., Leung W.Y., Goodman T., Schulenberg B., Hendrickson J., Beechem J.M., Haugland R.P., and Patton W.F. 2003. Global quantitative phosphoprotein analysis using Multiplexed Proteomics technology. *Proteomics* **3:** 1128–1144.

Takahashi S., Ohshima T., Cho A., Sreenath T., Iadarola M.J., Pant H.C., Kim Y., Nairn A.C., Brady R.O., Greengard P., and Kulkarni A.B. 2005. Increased activity of cyclin-dependent kinase 5 leads to attenuation of cocaine-mediated dopamine signaling. *Proc. Natl. Acad. Sci.* **102:** 1737–1742.

Tan P.K., Downey T.J., Spitznagel E.L., Jr., Xu P., Fu D., Dimitrov D.S., Lempicki R.A., Raaka B.M., and Cam M.C. 2003. Evaluation of gene expression measurements from commercial microarray platforms. *Nucleic Acids Res.* **31:** 5676–5684.

Thiriet N., Ladenheim B., McCoy M.T., and Cadet J.L. 2002. Analysis of ecstasy (MDMA)-induced transcriptional responses in the rat cortex. *FASEB J.* **16:** 1887–1894.

Thomas D.M., Francescutti-Verbeem D.M., Liu X., and Kuhn D.M. 2004. Identification of differentially regulated transcripts in mouse striatum following methamphetamine treatment—An oligonucleotide microarray approach. *J. Neurochem.* **88:** 380–393.

Thomas P.D., Campbell M.J., Kejariwal A., Mi H., Karlak B., Daverman R., Diemer K., Muruganujan A., and Narechania A. 2003a. PANTHER: A library of protein families and subfamilies indexed by function. *Genome Res.* **13:** 2129–2141.

Thomas P.D., Kejariwal A., Campbell M.J., Mi H., Diemer K., Guo N., Ladunga I., Ulitsky-Lazareva B., Muruganujan A., Rabkin S., Vandergriff J.A., and Doremieux O. 2003b. PANTHER: A browsable database of gene products organized by biological function, using curated protein family and subfamily classification. *Nucleic Acids Res.* **31:** 334–341.

Tonge R., Shaw J., Middleton B., Rowlinson R., Rayner S., Young J., Pognan F., Hawkins E., Currie I., and Davison M. 2001. Validation and development of fluorescence two-dimensional differential gel electrophoresis proteomics technology. *Proteomics* **1:** 377–396.

Toronen P., Kolehmainen M., Wong G., and Castren E. 1999. Analysis of gene expression data using self-organizing maps. *FEBS Lett.* **451:** 142–146.

Toyooka K., Usui M., Washiyama K., Kumanishi T., and Takahashi Y. 2002. Gene expression profiles in the brain from phencyclidine-treated mouse by using DNA microarray. *Ann. N.Y. Acad. Sci.* **965:** 10–20.

Tusher V.G., Tibshirani R., and Chu G. 2001. Significance analysis of microarrays applied to the ionizing radiation response. *Proc. Natl. Acad. Sci.* **98:** 5116–5121.

Unlu M., Morgan M.E., and Minden J.S. 1997. Difference gel electrophoresis: A single gel method for detecting changes in protein extracts. *Electrophoresis* **18:** 2071–2077.

Westermeier R. and Naven T. 2002. *Proteomics in practice.* Wiley-VCH, Weinheim.

Wilkins M.R., Sanchez J.C., Gooley A.A., Appel R.D., Humphery-Smith I., Hochstrasser D.F., and Williams K.L. 1996. Progress with proteome projects: Why all proteins expressed by a genome should be identified and how to do it. *Biotechnol. Genet. Eng. Rev.* **13:** 19–50.

Williams K., Wu T., Colangelo C., and Nairn A.C. 2004. Recent advances in neuroproteomics and potential application to studies of drug addiction. *Neuropharmacology* (suppl. 1) **47:** 148–166.

Wittwer C.T., Herrmann M.G., Moss A.A., and Rasmussen R.P. 1997. Continuous fluorescence monitoring of rapid cycle DNA amplification. *BioTechniques* **22:** 130–138.

Woo Y., Affourtit J., Daigle S., Viale A., Johnson K., Naggert J., and Churchill G. 2004. A comparison of cDNA, oligonucleotide, and Affymetrix GeneChip gene expression microarray platforms. *J. Biomol. Tech.* **15:** 276–284.

Wu C.C., MacCoss M.J., Howell K.E., Matthews D.E., and Yates J.R., III. 2004. Metabolic labeling of mammalian organisms with stable isotopes for quantitative proteomic analysis. *Anal. Chem.* **76:** 4951–4959.

Xie T., Tong L., Barrett T., Yuan J., Hatzidimitriou G., McCann U.D., Becker K.G., Donovan D.M., and Ricaurte G.A. 2002. Changes in gene expression linked to methamphetamine-induced dopaminergic neurotoxicity. *J. Neurosci.* **22:** 274–283.

Xie T., Tong L., McCann U.D., Yuan J., Becker K.G., Mechan A.O., Cheadle C., Donovan D.M., and Ricaurte G.A. 2004. Identification and characterization of metallothionein-1 and -2 gene expression in the context of (+/–)3,4-methylene-dioxymethamphetamine-induced toxicity to brain dopaminergic neurons. *J. Neurosci.* **24:** 7043–7050.

Yao W.D., Gainetdinov R.R., Arbuckle M.I., Sotnikova T.D., Cyr M., Beaulieu J.M., Torres G.E., Grant S.G., and Caron M.G. 2004. Identification of PSD-95 as a regulator of dopamine-mediated synaptic and behavioral plasticity. *Neuron* **41:** 625–638.

Yauk C.L., Berndt M.L., Williams A., and Douglas G.R. 2004. Comprehensive comparison of six microarray technologies. *Nucleic Acids Res.* **32:** e124.

Yeom M., Shim I., Lee H.J., and Hahm D.H. 2005. Proteomic analysis of nicotine-associated protein expression in the striatum of repeated nicotine-treated rats. *Biochem. Biophys. Res. Commun.* **326:** 321–328.

Yuferov V., Nielsen D.A., Butelman E.R., and Kreek M.J. 2005. Microarray studies of psychostimulant-induced changes in gene expression. *Addict. Biol.* **10:** 101–118.

Yuferov V., Kroslak T., LaForge K.S., Zhou Y., Ho A., and Kreek M.J. 2003. Differential gene expression in the rat caudate putamen after "binge" cocaine administration: Advantage of triplicate microarray analysis. *Synapse* **48:** 157–169.

Zeeberg B.R., Feng W., Wang G., Wang M.D., Fojo A.T., Sunshine M., Narasimhan S., Kane D.W., Reinhold W.C., Lababidi S., Bussey K.J., Riss J., Barrett J.C., and Weinstein J.N. 2003. GoMiner: A resource for biological interpretation of genomic and proteomic data. *Genome Biol.* **4:** R28.

Zhang D., Zhang L., Lou D.W., Nakabeppu Y., Zhang J., and Xu M. 2002. The dopamine D1 receptor is a critical mediator for cocaine-induced gene expression. *J. Neurochem.* **82:** 1453–1464.

Index